T0328614

Food Safety in the 21st Century

Food Safety in the 21st Century
Public Health Perspective

Editors

Rajul Kumar Gupta

Puja Dudeja

Amarjeet Singh Minhas

AMSTERDAM • BOSTON • HEIDELBERG • LONDON
NEW YORK • OXFORD • PARIS • SAN DIEGO
SAN FRANCISCO • SINGAPORE • SYDNEY • TOKYO

Academic Press is an imprint of Elsevier

Academic Press is an imprint of Elsevier
125 London Wall, London EC2Y 5AS, United Kingdom
525 B Street, Suite 1800, San Diego, CA 92101-4495, United States
50 Hampshire Street, 5th Floor, Cambridge, MA 02139, United States
The Boulevard, Langford Lane, Kidlington, Oxford OX5 1GB, United Kingdom

Notices

Knowledge and best practice in this field are constantly changing. As new research and experience broaden our understanding, changes in research methods, professional practices, or medical treatment may become necessary.

Practitioners and researchers must always rely on their own experience and knowledge in evaluating and using any information, methods, compounds, or experiments described herein. In using such information or methods they should be mindful of their own safety and the safety of others, including parties for whom they have a professional responsibility.

To the fullest extent of the law, neither the Publisher nor the authors, contributors, or editors, assume any liability for any injury and/or damage to persons or property as a matter of products liability, negligence or otherwise, or from any use or operation of any methods, products, instructions, or ideas contained in the material herein.

Library of Congress Cataloging-in-Publication Data
A catalog record for this book is available from the Library of Congress

British Library Cataloguing-in-Publication Data
A catalogue record for this book is available from the British Library

ISBN: 978-0-12-801773-9

For information on all Academic Press publications
visit our website at https://www.elsevier.com/

Working together
to grow libraries in
developing countries

www.elsevier.com • www.bookaid.org

Publisher: Nikki Levy
Acquisition Editor: Patricia Osborn
Editorial Project Manager: Jaclyn Truesdell
Production Project Manager: Caroline Johnson
Designer: Mark Rogers

Typeset by Thomson Digital

Contents

SECTION 1 EPIDEMIOLOGICAL ASPECTS OF FOODBORNE DISEASES

CHAPTER 9 Toxicological Profile of Indian Foods—Ensuring Food Safety in India........................... 111

S.P. Singh, S. Kaur, D. Singh

CHAPTER 10 Detection of Food Adulterants/Contaminants............ 129

D.P. Attrey

CHAPTER 17 Safe Storage and Cooking Practices for Foods of Animal Origin in Home Kitchen Before Consumption 229

D.P. Attrey

SECTION 5 FOOD SAFETY IS A SHARED RESPONSIBILITY: ROLE OF VARIOUS STAKEHOLDERS IN IMPLEMENTING FOOD SAFETY

CHAPTER 18 Role of Government Authorities in Food Safety 243

P. Dudeja, A. Singh

SECTION 6 FOOD SAFETY IN LARGE EATING ESTABLISHMENTS

SECTION 7 FOOD SAFETY IN SMALL EATING ESTABLISHMENTS AND IN SPECIAL SITUATIONS

CHAPTER 30 Food Safety in Schools, Canteens, Hostel Messes, Mid-Day Meal Scheme, ICDS 387
R.K. Gupta

CHAPTER 31 Food Safety Issues Related to Street Vendors.......... 395
S. Malhotra

SECTION 8 DOMESTIC REGULATORY SCENARIO OF FOOD SAFETY AND INTERFACE OF FOOD SAFETY LAWS, STANDARDS, REGULATIONS, AND POLICIES AT THE INTERNATIONAL LEVEL

CHAPTER 35 Relevant Food Safety Regulations and Policies....... 437

D.P. Attrey

CHAPTER 36 Food Safety Policies in Agriculture and Food Security with Traceability 449

D.P. Attrey

CHAPTER 45 Ready to Eat Meals 541

R.K. Gupta, P. Dudeja

List of Contributors

P.K. Ahuja
Postgraduate Institute of Medical Education and Research, Chandigarh, India

D.P. Attrey
Central Military Veterinary Laboratory, Meerut, Uttar Pradesh; High Altitude Research, Defence Research and Development Organisation, Leh, Jammu and Kashmir; Amity Institute of Pharmacy; Innovation and Research Food Technology; Amity Institute of Seabuckthorn Research, Amity University, Noida, Uttar Pradesh; Lala Lajpat Rai University of Veterinary and Animal Sciences, Hisar, Haryana, India

S. Bajaj
Department of Community Medicine, Armed Forces Medical College, Pune, Maharashtra, India

D. Banerjee
Experimental Medicine and Biotechnology Department, Postgraduate Institute of Medical Education and Research, Chandigarh, India

M. Bansal
School of Public Health, Post Graduate Institute of Medical Education and Research, Chandigarh, India

S.V. Bhaskar
Armed Forces Medical Services, New Delhi, India

R. Bhattacharyya
Biotechnology Department, Maharishi Markandeshwar University, Mullana, Haryana, India

S. Chakraborty
Experimental Medicine and Biotechnology Department, Postgraduate Institute of Medical Education and Research, Chandigarh, India

S. Chowdhary
Experimental Medicine and Biotechnology Department, Postgraduate Institute of Medical Education and Research, Chandigarh, India

P. Dudeja
Department of Community Medicine, Armed Forces Medical College, Pune, Maharashtra, India

J. Dutta
Human Resource Development Centre, Panjab University, Chandigarh, India

G. Ghose
Department of Community Medicine, IQ City Medical College, Durgapur, West Bengal, India

R.K. Gupta
Department of Community Medicine, Army College of Medical Sciences, New Delhi, India

S. Kathirvel
Department of Community Medicine, School of Public Health, Post Graduate Institute of Medical Education and Research, Chandigarh, India

I. Kaur
Post Graduate Institute of Medical Education and Research (PGIMER), Community Medicine (School of Public Health), Chandigarh, India

S. Kaur
Department of Biochemistry, Government Medical College and Hospital, Chandigarh, India

A. Khera
Department of Community Medicine, Armed Forces Medical College, Pune, Maharashtra, India

J. Kumar
Centre for Public Health, Panjab University, Chandigarh, India

S. Malhotra
Department of Dietetics, Postgraduate Institute of Medical Education and Research, Chandigarh, India

I. Saha
Department of Community Medicine, IQ City Medical College, Durgapur, West Bengal, India

A. Singh
School of Public Health, Post Graduate Institute of Medical Education and Research, Chandigarh, India

D. Singh
Department of Forensic Medicine, Post Graduate Institute of Medical Education and Research, Chandigarh, India

S.P. Singh
Department of Forensic Medicine and Toxicology, Government Medical College and Hospital, Chandigarh, India

About the Editors

Rajul K Gupta, Professor, Community Medicine, Army College of Medical Sciences, New Delhi, India

Dr. Rajul K Gupta is alumnus of Armed Forces Medical College, Pune and PhD from National Institute of Nutrition, Hyderabad. A professor of Public Health & Nutrition, he is also Director (Health) for Indian Armed Forces. He was member of "Prime Minister's National Council on Prevention of Malnutrition" at Planning Commission. He is advisor to Government of India on "Nutrition in Disasters" and "Nutrition in High Altitude." He is a standing member of UN "Standing Committee on Nutrition," UN expert groups on "Nutrition & Non-Communicable Diseases" and "Nutrition and Global Climatic Change." He has edited and co-authored 5 books, including WHO "Standard Treatment Guidelines" and WHO "Text book of Public Health & Community Medicine," authoring 60 chapters.

Puja Dudeja, Research Scholar, School of Public Health, Postgraduate Institute of Medical Education and Research, Chandigarh, India

Dr. Puja Dudeja is Associate Professor, Community Medicine, AFMC, Pune, India. She has pursued her PhD in Community Medicine on the topic of food safety from PGIMER, Chandigarh. She has authored 90 chapters in various public health related books. She has developed two short films titled "Gravy Extra" and "Food Safety: Farm to Fork" for training food handlers on food safety.

Amarjeet Singh Minhas, School of Public Health, Postgraduate Institute of Medical Education and Research, Chandigarh, India

Dr. Amarjeet Singh is Professor, Community Medicine, PGIMER, Chandigarh, India. He has published 170 research articles, written 85 chapters in books, and coauthored 20 books. He has completed 55 research projects. He has published books on the following topics "Demedicalizing Women's Health," "Health Promotion," "Care of Dependent and Elderly," and "Salutogenesis." His work on mobile-phone-based health care (with Dr. Surya Bali) was listed as an "Innovative approaches" in HSPROD database. He has initiated a new concept of establishing multipurpose behavior therapy (MPBT) room in OPD to enhance satisfaction of patients and doctors. He has also produced eight short films on various topics of public health along with their script writing.

Acknowledgments

The contents of this book emerge from the collective wisdom of multitude of experts from the fields of Nutrition, Public Health, Law, Veterinary Science, Agriculture and many others. I wish to acknowledge them all, who have been my co-editors, co-authors, reviewers or well wishers. I wish to express my deep sense of personal gratitude to my teacher, mentor, guide and *guru*, Dr K Vijayaraghavan, Scientist Emeritus and former Dy Director (Senior Grade) from the prestigious National Institute of Nutrition, Hyderabad, India, for having always inspired and encouraged me to take on new ventures. I sincerely acknowledge Prof Rajvir Bhalvar, Director Medical Services, Apollo Hospitals and former HOD, Community Medicine, AFMC, Pune who taught me and Dr Puja, the art and craft of writing a book. It's owing to the blessings of my grandfather, Late Prof RD Gupta, former HOD, Chemistry, Birla Institute of Technology & Sciences, Pilani, India, that I have been able to accomplish major academic projects.

I acknowledge the support and profound academic environment rendered by Armed Forces Medical College (AFMC), Pune and Post Graduate Institute of Medical Education & Research (PGIMER), Chandigarh, without which this book would have merely remained a dream.

I wish to acknowledge the Office of the Director General Armed Forces Medical Services, India, to have granted permission to author and edit this esteemed book.

I praise the God Almighty to have granted us the strength to have put in innumerable hours and days of concentration to have completed this first edition of the book of a kind.

Rajul Kumar Gupta

Introduction

Like most tribal people who live self-sufficient on their ancestral lands, they hunt pig and turtle and fish with bows and arrows in the coral-fringed reefs for crabs and fish, including striped catfish-eel and the toothed pony fish. They collect wild honey from lofty trees. They also gather fruits, wild roots, tubers and honey. The bows are made from the chooi wood... (Survival International, in press)

Imagine a closed family/community not dependent on anyone else, even for food. Food is grown or produced at home (own fields), milk obtained from own cattle, water from a nearby stream, game is hunted from the jungle...and life goes on the hunter–gatherer way. Well, this was the story a few centuries back in many parts of the world and is still true for some closed tribal communities, like the Jarawas of Andaman islands and some San tribes of Africa. With this lifestyle, the threat to food safety would be minimal. It's the constant evolution of civilization and advancing food practices that have systematically compromised food safety. Man's quest to explore new arenas at work, profession, adventure, and war took him out of the comfort zone and ushered him into the unknown. This exposed him to uncertainty with respect to food availability and food safety. Whether one is on a voyage or simply out of the house in today's urbanized world to the office (both spouses working); he eats out compulsorily or for fun in a restaurant, food safety is at stake. When there is pressure to increase food production for the rising population (using pesticides, fertilizers, and GM seeds), food safety might be overlooked. When there is a craving to enjoy distant exotic fruits and sea food, thousands of miles away at home, food safety may be under question.

Different connotations are applied to food safety by different people. For a housemaid having washed vegetables and fruits is food safety; for a housewife having served fresh hot food is food safety; for a husband, having picked up fresh, good-looking apples and grapes is food safety; for a farmer somehow pushing his tomatoes into the market before they get spoiled is food safety. Even for the medical professionals, the idea of food safety might just be a food product that doesn't cause disease. For a microbiologist, it stops at culturing microbes and for the Public Health specialist, it may not be more than investigating an outbreak of food poisoning and for a scientist, developing a particular product, a reagent, a new packaging or an app is food safety! Thus for a common man, safety means merely clean and hygienic food; however in technical terms, the umbrella of food safety encompasses chemical, physical, and biological safety of the food consumed.

A HISTORICAL PERSPECTIVE

To offset threats to food safety, rules have been in position since time immemorial. About 2000 BC, the book of Leviticus records that Moses made laws to protect people against infectious disease. There were rules on slaughtering animals and hand washing. Swine and small creatures (like insects, lizards, mice, rats, snakes) were prohibited. Of water creatures, only those with scales and fins could be eaten, which ruled out shellfish and crustaceans. Eating scavenging birds (crow, vulture, eagles, sea gulls, herons) was prohibited. Food was not supposed to be stored for more than a day's supply, and when the Israelites disobeyed this rule, they were struck with "deadly plague." The Bible also speaks of the Hebrews receiving *manna* (a flat wafer) from heaven every morning. It got spoiled ("bred worms and stank") after 24 to 48 h and was no longer edible. Confucius, in 500 BC, warned against eating "sour rice." Ancient Egyptians were possibly the first to develop the silo, a storage tank designed to hold grain. Storing grain in a silo kept it cool, dry and able to last into the nonharvest months or longer. Ancient Rome focussed on freshness of fruit. Romans also practised "salting" to preserve foods (Hobbs and Roberts, 1989).

Classical Indian medical literature of *Ayurveda* is very rich and refined, dating back to more than 4000 years. *Charak Samhita (400 BC)*, a treatise on *Ayurveda*, not only contains ample knowledge on gross concepts of food safety (like poisonings, etc.), but it also elaborates in detail, the fine elements of food, that are probably not encountered in any other system of medicine. Subtle concepts like certain "category" of foods not to be consumed by people of a particular "body-constitution" (*prakriti*) or age; certain foods not be used in a given season or region; certain foods to be avoided at noon or night; incompatibility, cross-reactions or antagonism of various foods; the order of eating various foods; and quantity of food to be eaten are integral to *Ayurvedic* principles. *Parhej* or restriction of eating certain foods in a particular disease, is also an elementary concept of *Ayurvedic* medical treatment. Texts have elaborated (food related) poisons and safety concerns in great detail. The diseases produced, clinical features, testing techniques, and protection from poisonous foods have been described.

Traditionally, food has always been worshipped in Indian culture. On any religious occasion, a small part of cooked food is first offered to gods. In traditional Indian society, the kitchen was always given the place of a temple. However, in the same society incidents of food adulteration can also be traced back to the times of Kautilya (Chanakya) in 4th century BC. Elaborate stringent laws, regulation, and procedures were evolved by Kautilya to ensure protection of the king from any poisoning attempts through the Royal Kitchen. He even described how to suspect a cook who might poison a king's meals (shifty eyes; not making eye contact). The laws related to adulteration are as old as the crime itself. In the past, there were rules in *Arthshastra,* a classic text by Kautilya. Kagle has translated *Arthshastra* into English and he quotes *"As to difference in weight or measure or difference in price or quality, for the weigher and measurer who by the trick of the hand brings about (difference to the extent of) one-eight part in (an article) priced at one panna, the fine is two hundred (pannas)... For mixing things of similar kind with objects such as grains, fats, sugar, salt, perfumes and medicines the fine is twelve pannas."* Adulteration was the gravest of socio-economic crime.

People have long recognized that keeping some foods cold could make them last longer. Meat or fish was kept in the creek or waterfall to keep it fresh. Others recognized that snow and ice were natural refrigerants. Napoleon offered a reward to anyone who could keep the food from spoiling. Appert put the food into jars with lids on them, and boiled it until cooked, the first version of canning.

It was only in 1675 that germs were discovered by Anton von Leeuwenhoek. This was the beginning of granting a "scientific basis" to food-safety procedures. The great French chemist Louis Pasteur revolutionized the thought process with demonstration of the role of bacteria in fermentation; that caused food spoilage but also produced wine! His subsequent work on pasteurization made an enormous impact on food safety. In 1888, when 57 people who had eaten beef from a cow slaughtered while it was ill, became sick, it was Gartner who studied the symptoms of all the patients, and concluded that some bacteria from the diarrhea of the cow must be responsible. Gartner diagnosed a foodborne illness bacteria, *Bacillus enteritidis*. This highlighted the importance of handling meat safely. Dr E Salmon isolated the hog cholera bacillus in 1885, and many bacteria causing food poisoning were grouped under *Salmonella* from 1909 to 1923. Around the same time, relevant bacteria like *Clostridium botulinum (by E van Ermengem, 1896), Staphylococcus* (1914), *Clostridium perfringens* (1945), and later *Bacillus cereus, E. coli,* and *Campylobacter* were also shown to be causes for food poisoning.

The sanitary conditions in Britain being poor, the era of "Great Sanitary Awakening" had begun in the 1840s with the *"Report on the Sanitary Conditions of the Labouring Population of Great Britain"* by Edwin Chadwick. Relationship between dirty conditions and disease was established and measures were taken to dispose of sewage and purify water. In 1854, John Snow recognized the role of drinking water in cholera and William Budd concluded that typhoid was spread by milk/water contaminated with sewage. Pasteurization eliminated bovine TB and with elimination of Brucellosis in cattle, undulant fever also ebbed (Hobbs and Roberts, 1989).

With the discovery of more bacteria and new disease, and the potential to prevent them, emerged new laws and labs. A Public Health bacteriological laboratory service was instituted in England in 1939, conducting analysis, keeping records of food poisoning and issuing notifications; thus an era of surveillance began. Teaching food hygiene became popular, which was equally useful to all—the health/medicine cadre, local government, food industry, catering services, groups of public, and private organizations.

Besides germs, the ancient historical accounts of food poisoning and "killings" attributable to chemicals in foods, spanning over centuries have also been real. It was Socrates in ancient Greece (399 BC) who was killed with Hemlock poison and Napoleon with Arsenic in 1821. Various heavy metals like lead, mercury, arsenic, cadmium could reach food and cause illness or death. The popular phrase *"Mad as a hatter,"* comes from a condition (of insanity) suffered by hat makers as a result of the long-term use of mercury products in the hat-making trade.

On April 21, 1956, a five-year-old girl was examined at the Chisso Corporation's factory hospital in Kyushu, Japan. The physicians were puzzled by her symptoms: convulsions, difficulty in walking, seeing, hearing, and numbness of hands/feet.

Later, her younger sister also began to exhibit the same symptoms, followed by hospitalization of eight more patients and soon, an "epidemic" of an unknown disease of the nervous system was declared (Harada, 1972).

The "*Strange Disease Counter-measures Committee*" investigated this epidemic and at the end of May 1956 a "contagion" was suspected and as a precaution patients were isolated and their homes disinfected. The committee uncovered surprising anecdotal evidence of strange behavior of local cats having convulsions, going "mad," and dying. Locals called it the "*cat dancing disease.*" Crows had fallen from the sky, seaweed no longer grew on the sea bed, and fish floated dead on the surface of the sea (Nicol, 2012). Victims, clustered in fishing hamlets along the shore of Minamata Bay, the staple food of whom was fish. It was declared that "Minamata disease was due to poisoning by a heavy metal (an industrial effluent), entering human body through eating fish and shellfish," which was later confirmed to be (organic) methyl mercury. As of March 2001, 2265 victims had been officially recognized (1784 of whom had died) and over 10,000 had received financial compensation from Chisso to the tune of $86 million. Many chemicals have since compromised food safety.

THE CURRENT SITUATION

Until a few decades back, food safety was not of as much concern, as most of us would eat home cooked food. But food safety is more likely to be compromised now, than ever before as new threats are constantly emerging in the 21st century. Food safety is vulnerable even before seeds are sown (genetically modified seeds) and continues so, till consumption. Produce can be rendered unsafe during cultivation (pesticides/insecticides), harvesting (contamination), storage (fungal infections), distribution (time delay, repackaging), transport (temperature fluctuation), packaging (mis-branding, contamination), and sale (hygiene, temperature). Food distribution through a long trail makes it vulnerable to temperature changes, physical, chemical and bacteriological contaminations and decomposition. Changed consumption patterns in the form of fastfoods, semiprocessed foods, and ready-to-eat meals also affect safety. Environmental changes like air and water pollution, climate change, falling water table, indirectly affect food safety. There is a threat to the crops from fertilizers, toxic industrial chemicals leached into soil, pesticides, and weedicides. New and emerging pathogens and antimicrobial resistance pose a challenge to food safety. Mass production, frequent travel, community feasting, industrialization, urbanization, SEZs, take away food parcels and home deliveries also expose food to threats. With advances in food storage, transport and processing technologies there has been a lengthening in the food chain from farm-to-fork. Food can be deliberately made unsafe through widespread practice of food adulteration, for financial gains.

In the past two decades, there has been an exponential rise in the growth of the food industry leading to an increase in international trade. The variety of food items available in the market has also increased tremendously. Food grown in one country and/or packaged in a different country is available across the globe for consumption. Shelves of supermarkets are flooded with a variety of such food products. Most of the fruits and vegetables are now available throughout the year. This has resulted in greater demand for more exotic foods leading to new opportunities for the food trade, but sometimes at the cost of food safety.

Other food processing/handling units like slaughterhouses, restaurants, flight catering installations, street vendors ... may also compromise food safety. Specific high-risk situations for food safety are the mass congregations, as in religious gatherings (*Haj, Kumbh Mela*), fairs, festivals, marriage parties, school feeding programmes, disasters, migrations, refugee crisis, famines, and impact of military conflicts.

Contrary to our beliefs, even our kitchens may not be safe! There are chances that contaminants may creep in literally from the walls, floors, or drains. There could be cross-contamination between vegetarian and nonvegetarian raw food, cooked and uncooked food, fresh and stale food. Food handlers too, always remain a threat to food safety.

With these changes, the so far "one off" localized safety issue could become a global concern, e.g., incidents of Bovine Spongiform Encephalopathy and adultera-tion of meat with horse meat, to name a few.

As per the WHO, unsafe food is linked to the death of an estimated 2 million people annually. Food containing harmful bacteria, viruses, parasites, or chemical substances is responsible for more than 200 diseases, ranging from diarrhea to cancers. Of these, foodborne infections are the most predominant. The global burden of infectious diarrhea involves 3–5 billion cases and nearly 1.8 million deaths annually, mainly in young children, caused by contaminated food and water, mostly from developing countries.

These given situations, emerging threats and the need to bridge the gap on food safety, right from farm-to-fork, prompted us to write this book. In fact one of the author–editors of the book, Dr Puja Dudeja, was so intrigued by the subject that she had embarked on her PhD on food safety, coincidentally when the WHO World Health Day theme too was on Food Safety! (Dudeja, 2015) This underlines the importance of this topic and the commitment of authors for the cause.

CHANGING FACE OF FOOD SAFETY IN THE 21ST CENTURY

The increased disease burden has necessitated newer and innovative ways of risk analysis, surveillance, detection, and preventive and control modalities. Now there is a rising stake of hitherto seemingly unrelated sciences like information technology, apps development, and nanotechnology. The ever-important need to strengthen, empower, and apply laboratory-based research in the fields of food adulteration detection, nutraceuticals, GM foods, food traceability, organic farming, and even food policy and laws continue. All this doesn't mean that the traditional measures like training of food handlers, managers, and sanitary inspections would go out of vogue.

To the aforementioned we may add another tool to fight this battle namely, formulation and implementation of legislative measures. At the highest level Codex, a joint body of WHO and FAO, has laid down certain principles of food safety. These cover the entire gamut of activities from farm-to-fork, namely harvesting, slaughter, milking, fishing, storage, transportation, etc. This also encompasses prevention of food safety hazards like chemical contaminants, pests, veterinary drugs, and other plant diseases. It covers the principles of HACCP (hazard analysis and critical control points) and emphasizes the development of standard procedures for different processes during manufacture of food. Under the category of Food Safety Certification Systems; International Standards (ISO 22000) deals with Food Safety Management System (FSMS).

A good food safety law consists of documented procedures for preventing problems before they occur rather than relying on a reactive approach once problems have occurred. A comprehensive law instills confidence.

In India, a new era in food safety has been initiated by formulation of the Food Safety and Standards Act (FSSA) 2006. While eating our favorite *panipuri* at the street, we tend to overlook hygiene but, the government doesn't. It has included all (even the small vendors) under the purview of this act. The act brings out a single statute to provide for comprehensive, scientific, based on international legislations, and Codex Alimentarius commission providing a policy framework through a single window to regulate those engaged in the food industry. It regulates and monitors the manufacturing, processing, storage, distribution, sale, and import of food. The standards are at par with those of the developed countries like FDA, EFSA, and Australia and New Zealand Food Laws, etc.

While it is the overall responsibility of the government to ensure food safety, there are multiple stakeholders in its management, namely farmers, transporters, processors, food handlers, food business operators (FBOs), consumers, government, and policy makers. A misplaced perception among FBOs is that good practices will increase the cost. However, the fact is that safe food is a result of correct management of farm-to-fork chain. The success of ensuring food safety depends on implementation of food safety laws.

Food safety takes into account all those hazards which make food unsafe. Unsafe food creates a vicious cycle of disease and malnutrition. Although microbiological contamination and chemical hazards have received most attention, it is recognized that food adulteration can affect health and could deprive essential nutrients for growth and development. Food adulteration is a global economic problem, which is comparable with organized crime. Baby formula feed adulteration with melamine in China, microbial contamination of vegetables throughout Europe and adulteration of milk and milk products with synthetic milk in India, etc. are some common examples. In fact no food is spared: milk and milk products, flour, edible oil, cereals, condiments, pulses, coffee, tea, confectionary, baking powder, nonalcoholic beverages, vinegar, *besan*, curry powder, vegetables, fish, etc. all can be adulterated.

There is an inescapable requirement to have food testing laboratories. The endeavor of research is to develop sensitive and reliable methods for food testing besides simple tests capable of detecting adulteration at domestic level; since even famous food brands too have not escaped the adulteration controversy (recent detection of MSG in noodles and potassium bromate in bread in 2016).

At the macro level food safety seems to be merely a public health issue, but disregard to good agricultural practices (GAP), good hygiene practices (GHP), good distribution practices (GDP), and good manufacturing practices (GMP) generates and perpetuates the problem.

There is also emphasis on product information through nutrition labeling. A label on the packaged product is the means of communication between manufacturer and consumer. It elaborates information like date of manufacture, best before date, nutrient contents, etc. which have implications on food safety. These traits cannot be detected by sight, smell, or taste. The future labels in India would also include GM labeling and labeling for allergens.

The future of food safety lies in advancement in science, innovations, and changes in society. In times to come we may have to respond to novel microorganisms, chemical, physical, and radiological hazards. The prospect of genetically modified foods is under debate. The information technology sector has a promising role to play in this field. Newer advances in technology like the implications of nanotechnology, biotechnology, RFID (radio-frequency identification) technology, to name a few, might change the canvas of food safety.

Since the consumer is the ultimate sufferer, he has to be most aware of the risks and how to offset them. Risk assessment (with components of risk analysis, management, and communication) is vital to food safety. And, the fact remains, no matter who we are, we are all consumers and all of us are at the "end of the food chain" of food safety ... most vulnerable to someone else's deeds. This calls for a joint effort from all of us, citizens, housewives, farmers, animal handlers, food industry, food business operators, food handlers, consumers, public health experts, veterinarians, civil societies, government, law/policy makers, local health authorities, national and international organizations, laboratory workers, medical facilities and researchers, to be vigilant and sincerely contribute to food safety, which is as vital to our lives as food itself.

After all, in the midst of this flourishing 21st century, to maintain food safety, we can't afford to be overwhelmingly "self-sufficient" on our ancestral lands to be hunting pig and turtle like them! But we could be sure enough to enjoy the best possible food from the world over with confidence and without fear of food poisoning if the basic tenets of food safety are ensured and the implementation of the rapidly advancing technology keeps pace with emerging and reemerging threats.

REFERENCES

Dudeja, P., 2015. Impact of an intervention package on status of conformance to food safety and standard regulations 2011 and quality of services in eating establishments of PGIMER Chandigarh: an intervention vs control comparison study. PhD Thesis, Post Graduate Institute of Medical Education & Research, Chandigarh, India.

Harada, M., 1972. Minamata Disease. Kumamoto Nichinichi Shinbun Centre & Information Center/Iwanami Shoten Publishers, Tokyo, Japan.

Hobbs, B.C., Roberts, D., 1989. Food Poisoning and Food Hygiene, fifth ed. Edward Arnold Publishers, London, 3-9.

Nicol, C.W., 2012. Minamata: a saga of suffering and hope. Japan Times. 7 October 2012.

Survival International, in press. Available from: http://www.survivalinternational.org/tribes/jarawa

Epidemiological aspects of foodborne diseases

1

Foodborne diseases— disease burden

S.V. Bhaskar

Armed Forces Medical Services, New Delhi, India

1.1 INTRODUCTION

Every year, thousands of people suffer from foodborne diseases (FBD). These diseases are globally important because of their high incidence and the costs that they impose on the society. There looms a potential threat of large outbreaks of FBD in both developing and developed countries.

More than 250 different FBD have been described. Most of these diseases are infections, caused by a variety of bacteria, viruses, and parasites that can be foodborne. *Salmonella*, *Campylobacter*, and enterohemorrhagic *Escherichia coli* are among the most common foodborne pathogens that affect millions of people annually— sometimes with severe and fatal outcomes. Other diseases are poisonings, caused by harmful toxins or chemicals that have contaminated the food, for example, poisonous mushrooms. Health departments commonly conduct surveillance for diseases potentially transmitted by food to monitor trends and outbreaks to aid prevention efforts. However, only a small proportion of cases are reported to the health departments because infections due to some pathogens are not notifiable and patients do not seek medical attention or receive a specific diagnosis for their illness.

The achievement of various development goals, including the overarching goal of poverty reduction, in part depends on a successful reduction of the burden of FBD, particularly among vulnerable groups. Without reliable information on disease burden, policy-makers cannot assess the effectiveness of their investment in FBD prevention and control nor reduce their burden.

1.2 FACTORS AFFECTING BURDEN OF FBD

The factors responsible for emergence of FBD are almost similar to those, which affect an emergence of other infectious diseases. Few of the factors contributing to the increase in the FBD burden are: (1) increase in trade and travel, (2) newer food technologies, (3) changes in life style, (4) changes in animal husbandry, and (5) increase in the susceptible population. Changes in human demographics and behavior, industrial growth, international travel and commerce, microbial adaptation, economic development, and probable lacunae in the public health setup has played a major role in the emergence of these diseases.

The rise in cases due to *Salmonella* has been attributed to centralized food production and large-scale distribution in many countries. *E. coli* O571:H7 was one of the first recognized human pathogen and was found during an outbreak associated with consumption of undercooked hamburgers from a fastfood restaurant chain. Increased susceptibility of population to FBD is associated with comorbities, namely, immunodeficiency states; chronic illnesses; changes in eating pattern among individuals; emergence of food eateries, such as, fast food restaurants; poor awareness of food safety; and inadequate public health infrastructure.

The food we eat is sourced from several places around the globe and distributed over large distances. This global trade provides opportunities for the exporting countries to earn foreign exchange and drives the increase in the standard-of-living in the developing countries. Not only have supply chains become longer, but the global trade in food has become more specialized also. International travel has also contributed to the global burden of FBD. Changes in the globalization of food trade have important implications for food safety. The increase in imported foods and food ingredients suggests that we depend on the food-safety systems of other countries. The reliance on the centralized production of foods has a disadvantage; when a problem occurs, it can lead to a widespread outbreak. In this setting, contaminated food can rapidly cause a geographically widespread or "dispersed scenario" type of FBD outbreak.

Pathogenic (disease-causing) microorganisms can be introduced at any point in the food chain. Some pathogen contamination is the result of production/agriculture conditions. The farm environment provides many opportunities for food contamination and its complete control is impossible. At other times, contamination can occur from environmental sources during processing. Some pathogens are introduced during handling and food preparation, either through inadequate human sanitation or through cross-contamination by contact with other foods. Antimicrobial resistance and the re-emergence of pathogens have also contributed to the rise in number of cases.

Although the processing methods may inactivate many pathogens, proper food handling after processing remains an important part of microbial control. The consumer demand for foods that seem "fresher" has prompted development of a number of alternative processing technologies that use mechanisms other than heat to control pathogens. Although some of these technologies have tremendous potential, it is unlikely that any single technology can be expected to control all pathogens in all food products. The processing technologies used to inactivate pathogens vary by food product. Foods may be washed or rinsed with organic acids, sanitizers, or other antimicrobial agents. Thermal processing heats a particular food item to a specific temperature and holds it at that temperature for a specified time. Examples of thermal processing include cooking and pasteurization. Nonthermal processing technologies use other means to inactivate pathogens, such as, pulsed electric fields, high pressure, or ultraviolet radiation. Other processing methods may change the characteristics of the food in a way to control pathogens. For example, fermentation, drying, and salting all change parameters of food in a controlled manner to inhibit pathogen growth. As with any technology, alternative processing methods have their limitations. Some methods work extremely well for some foods and not so well for others, depending on the characteristics of the food. Similarly, the mechanism used to inactivate pathogens can be more effective for some pathogens than others.

1.3 FOODBORNE ZOONOTIC DISEASES

Foodborne zoonotic diseases are caused by consuming food or drinking water contaminated by pathogenic microorganisms, such as, bacteria and their toxins, viruses, and parasites. Many of these microorganisms are commonly found in the intestines of healthy food-producing animals. The risks of contamination are present from farm-to-fork and require prevention and control throughout the food chain. Approximately 500 species of commensal bacteria colonize the human gastrointestinal tract, producing disease only when normal anatomic or immunologic defenses are abrogated. The principal invasive intestinal bacterial pathogens of food–animal origin are *Campylobacter*, *Salmonella*, *Listeria*, *E. coli* O157 (and other Shiga toxin and enterotoxin-producing strains of *E. coli*), *Yersinia*, and *Vibrio*. Nearly all are common commensals in cattle, swine, and poultry sometimes cause invasive infection in animals and humans (except for *E. coli* O157, a colonizer of cattle). *Vibrio*, an exception, is found in seawater and shellfish. Other microorganisms of food–animal origin, such as, *Enterococcus* species and *E. coli* strains that produce neither Shiga toxin nor enterotoxin, also may enter and mix with commensal bacteria in the human gastrointestinal tract. Foodborne zoonotic diseases are a significant and widespread global public health threat.

Animal feed is at the beginning of the food-safety chain. The emergence of variant Creutzfeldt–Jakob disease has raised awareness of the importance of contaminated animal feed; but less attention has been paid to the role of bacterial contamination of animal feed in human foodborne illness. Although tracing contamination to its ultimate source is difficult, several large outbreaks have been traced back to contaminated animal feed. Food-producing animals (e.g., cattle, chickens, pigs, and turkeys) are the major reservoirs for many of these organisms, which include *Campylobacter* species and non-Typhi serotypes of *Salmonella enterica*, Shiga toxin-producing strains of *E. coli*, and *Yersinia enterocolitica*. Food-producing animals acquire these pathogens by ingestion. Contamination of animal feed before arrival and while on the farm contributes to infection and colonization of food-producing animals with these pathogens. Pathogens are then transmitted through the food chain to humans and cause human FBD.

Several incidents have been reported; in which human illness was traced back to contaminated animal feed. In 1958, an outbreak of infection with foodborne *S. enterica* serotype Hadar in Israel was linked to the consumption of chicken liver. An investigation of the chicken farm found that bonemeal fed to the chickens was contaminated with the same serotype of *Salmonella* (Hirsch and Sapiro-Hirsch, 1958). A milkborne outbreak of infection due to *S. enterica* serotype Heidelberg in England in 1963 resulted in 77 human illnesses and was traced to a cow with bovine mastitis due to the same organism. Investigation revealed that meat and bonemeal fed to the cow was contaminated with the same organism (Knox et al., 1963). During 1968, frozen chickens from a packing station in Cheshire, England, were implicated in a large outbreak of infection with *S. enterica* serotype Virchow (Semple et al., 1968). Investigation showed that the hatchery and the majority of rearing farms that supplied the packing station contained chickens colonized with *S. enterica* serotype Virchow, and the organism was isolated from feed fed to the chickens (Pennington et al., 1968). What happens on farms, in feedlots, during transport, and lairage before slaughter, as well as, during slaughter and further processing can have a major effect on human health. Domesticated food animals can also serve as a source of contamination of nearby produce-growing fields.

1.4 NONINFECTION-BASED FBD

Contamination of food due to chemicals at various levels is an important reason for the incidence of FBD. The contamination may be unintentional or due to intentional adulteration of the foodstuffs. Naturally occurring toxins, such as, mycotoxins, marine biotoxins, cyanogenic glycosides, and toxins found in poisonous mushrooms are few of the common causes for FBD. Staple foods, such as, corn or cereals may contain high levels of mycotoxins, such as aflatoxin. A long-term exposure can affect the immune system and normal development or cause cancer. Persistent organic pollutants, such as, dioxins and polychlorinated biphenyls that accumulate in the environment and human body are by-products of industrial processes and waste incineration. They are found worldwide in the environment and accumulate in animal food chains. Dioxins are highly toxic and can cause reproductive and developmental problems, damage the immune system, interfere with hormones, and cause cancer. Role of heavy metals in FBD is well-documented. Metals, such as, lead, cadmium, and mercury cause neurologic and kidney damage. Contamination by heavy metal in food occurs mainly through pollution of air, water, and soil.

1.5 SEASONAL VARIATION OF FBD

In temperate climates, waterborne or foodborne enteric infections typically alternate between periods of low endemic levels and periods with outbreaks, forming a typical seasonal pattern. For example, illness caused by *Salmonella* spp. or *Campylobacter jejuni* rises in the summer and declines in the winter (Amin, 2002) Enteric infections caused by the protozoans *Giardia* and *Cryptosporidium* also exhibit a seasonal variation, although they shift toward autumn (Birkhead and Vogt, 1989). In contrast, the seasonality of hepatitis A and shigellosis is not marked (Bowman et al., 2003).

Consistent temporal fluctuations for diseases with similar sources for exposure or similar routes of transmission, suggest the presence of environmental factors that synchronize this seasonal variation (Hald and Andersen, 2001). Deviations from an established seasonal pattern may provide important clues to the factors that influence the disease occurrence. These factors may include changes in the sources of exposure and spread, changes in the affected population, or differences in the pathogen itself. Ecological disturbances, perhaps from climate change, may influence the emergence and proliferation of parasitic diseases, including cryptosporidiosis and giardiasis. Ambient temperature has been associated with short-term temporal variations (week-to-week and month-to-month) in reported cases of food poisoning in the UK, often caused by *Salmonella*. Increased temperatures and extreme precipitation events have also been shown to have a short-term effect on health outcomes.

1.6 CHALLENGES IN ASSESSMENT OF BURDEN OF FBD

The challenge inherent in measuring the burden of FBD is that only a fraction of the people who become sick from contaminated food, seek medical care. Only a fraction of those cases are recognized as having been caused by a hazard in food. Few are treated accordingly; and even fewer are reported to the public health authorities and recorded in official disease statistics. Certain chronic diseases, such as, cancer, kidney or liver failures that result from contaminated food appear long after the ingestion of food and the causal link is never discovered for each case. The assessment of the total economics of the disease burden would involve the measurement of all costs related to a given disease in the population. These include the costs related to the resources used within the healthcare sector, to the resources used by patients and their families, to productivity losses due to work absence of patients and care givers, and other nonhealth care costs indirectly related to illness (Flint et al., 2005).

1.7 ASSESSMENT OF DISEASE BURDEN

Burden of disease is the impact of a health problem as measured by financial cost, mortality, morbidity, or other indicators. It is often quantified in terms of quality-adjusted life years or disability-adjusted life years (DALYs), both of which quantify the number of years lost due to disease. One DALY can be thought of as one year of healthy life lost; and the overall disease burden can be thought of as a measure of the gap between current health status and the ideal health status (where the individual lives to old age free from disease and disability).

Estimations of the burden of FBD are complicated by the fact that very few illnesses can be definitively linked to food. Often these links are only made during outbreak situations. Studies determining the burden of acute gastroenteritis provide the basis for estimating the burdens due to food and specific pathogens commonly transmitted by food. Analysis of foodborne outbreak data is one approach to estimate the proportion of human cases of specific enteric diseases attributable to a specific food item (food attribution). Although for a variety of reasons reported outbreaks represent only a small portion of all actual outbreaks, using outbreak data for food attribution is the only methodological approach, wherein theoretically, there is an actual direct link between the pathogen, its source, and each infected person.

The burden of FBD is not well-defined in many countries or regions or on a global level. The WHO, in conjunction with other national public health agencies, has been coordinating a number of international activities designed to assist countries in the strengthening of disease surveillance and in determining the burden of acute gastroenteritis. Although a number of countries have conducted studies to determine the burden of foodborne disease, global estimates are lacking. The enormity of the problem is evident, however, from estimates of the incidence of acute gastroenteritis during childhood, for which an important proportion of cases are caused by foodborne pathogens. The globalization of the food supply has presented new challenges for food safety and has contributed to the international public health problem of FBD. To initiate and sustain efforts aimed at preventing foodborne disease at national and international levels, the magnitude of the problem needs to be determined (WHO, 2007). Foodborne outbreaks have led to claims that the number of FBD outbreaks and concomitant illnesses has increased in recent years.

The WHO has been involved in several initiatives designed to enhance laboratory-based surveillance and to determine the burden of disease in countries and regions lacking such estimates. WHO Global Salm–Surv Strategic Plan was launched in Jan. 2000, as an international capacity-building program that strengthens national laboratory-based surveillance, outbreak detection, and response to diseases commonly transmitted by food (WHO, 2006). WHO Sentinel Sites Project of Mar. 2002 was convened to discuss the feasibility of establishing sentinel sites to determine the burden of FBD in regions lacking estimates. In response to a worldwide interest, an international collaboration on enteric diseases: the "Burden of Illness Studies" estimated the burden of acute gastroenteritis and FBD. WHO, together with its partners, launched the "Initiative to Estimate the Global Burden of FBD" in 2006. Foodborne Disease Burden Epidemiology Reference Group (FERG) was established to lead the initiative. The objective of the initiative was to provide estimates on the global burden of FBD for a defined list of causative agents of microbial, parasitic, and chemical origin; to strengthen the capacity of countries to conduct assessments of the burden of FBD; and to encourage them to use the burden of FBD estimates for cost-effectiveness analyses of prevention, intervention, and control measures including implementation of food-safety standards in an effort to improve national food-safety systems (WHO, 2006).

1.8 WHO ESTIMATES OF THE GLOBAL BURDEN OF FBD

WHO published a comprehensive report on the impact of contaminated food on health and well-being named "Estimates of the Global Burden of FBD" in 2015. This report, resulting from the WHO initiative and prepared by the WHO FERG, provides the first estimates of global FBD incidence, mortality, and disease burden in terms of DALYs. As per the projections, almost one-third (30%) of all deaths from FBD occur in children under the age of 5 years, despite the fact that they make up only 9% of the global population. Unsafe food containing harmful bacteria, viruses, parasites, or chemical substances, causes more than 200 diseases—ranging from diarrhea to cancer. An estimated 600 million, almost 1 in 10 people in the world, fall ill after eating contaminated food and 420,000 die every year, resulting in the loss of 33 million healthy life years (DALYs). It is estimated that children under 5 years of age carry 40% of the foodborne disease burden, with 125,000 deaths every year. Diarrheal diseases are the most common illnesses resulting from the consumption of contaminated food, causing 550 million people to fall ill and 230,000 deaths every year.

1.9 WHO REGION-WISE DISEASE DISTRIBUTION

The report has brought out the region-wise disease burden of FBD. The WHO African Region was estimated to have the highest burden of FBD per population. More than 91 million people are estimated to fall ill and 137,000 die each year. Diarrheal diseases were responsible for 70% of FBD in the African Region. Nontyphoidal *Salmonella*, which can be caused by contaminated eggs and poultry, causes the most deaths, killing 32,000 a year in the region—more than half of the global deaths from the disease. Ten percent of the overall FBD burden in this region is caused by *Taenia solium* (the pork tapeworm).

The WHO Southeast Asia Region had the second highest burden of FBD per population, after the African Region. However, in terms of absolute numbers, more people living in the WHO Southeast Asia Region fall ill and die from FBD every year than in any other WHO Region, with more than 150 million cases and 175,000 deaths a year. Around 60 million children under the age of 5 years fall ill and 50,000 die from FBD in the Southeast Asia Region every year.

The WHO Region of the Americas is estimated to have the second lowest burden of FBD globally. Nevertheless, 77 million people fall ill every year from contaminated food, with an estimated 9000 deaths annually in the region. Of those who fall ill, 31 million are under the age of 5 years, resulting in more than 2000 of these children dying each year.

The WHO European Region showed the lowest estimated burden of FBD globally, more than 23 million people in the region fall ill from unsafe food every year, resulting in 5000 deaths. Diarrheal diseases account for the majority of FBD in the WHO European Region, with the most common being due to norovirus infections, causing an estimated 15 million cases, followed by campylobacteriosis, causing close to 5 million cases. Nontyphoid salmonellosis causes the highest number of deaths—almost 2000 annually. Foodborne toxoplasmosis, a severe parasitic disease spread through undercooked or raw meat and fresh produce, may cause up to 20% of the total FBD burden and affects more than 1 million people in the region each year. *Listeria* infection has a severe impact also on the health of people who contract it and causes an estimated 400 deaths in the European Region annually. *Listeria* can result in septicemia and meningitis and is usually spread by consuming contaminated raw vegetables, ready-to-eat meals, processed meats, smoked fish, or soft cheeses.

The global burden of FBD caused by the 31 hazards in 2010 was 33 million DALYs; children under 5 years old bore 40% of this burden. The report emphasizes the global threat posed by FBD and reinforces the need for governments, the food industry, and individuals to put in more efforts to make food safe and prevent FBD. There remains a significant need for education and training on the prevention of FBD among food producers, suppliers, handlers, and the general public (WHO, 2016).

1.10 **FBD IN INDIA**

In India, food poisoning affecting 78 personnel was reported in 1998 at high altitude, wherein *Salmonella enteritidis* was identified as the etiological agent and frozen fowl was the implicated food source for the outbreak. A food poisoning outbreak due to *Salmonella paratyphi A* that affected 33 people, due to vegetarian food was reported from Yavatmal (Maharashtra) in 1995. Two separate food poisoning outbreaks due to S*almonella weltevreden* and *Salmonella wein* affecting 34 and 10 people, respectively, due to nonvegetarian food (chicken and fish) were reported from Mangalore in 2008–09. A food poisoning outbreak due to *Y. enterocolitica* was reported in 1997 from Tamil Nadu affecting 25 people; in which buttermilk was incriminated as the food source. An outbreak of foodborne botulism due to *Clostridium butyricum* affecting 34 students from a residential school in Gujarat was reported in 1996; the food sample found to be contaminated was "*sevu*" (crisp made from gram flour). An outbreak of *Staphylococcal aureus* food poisoning due to contaminated "*bhalla*" (a snack made up of soaked and fermented black lentil balls fried in vegetable oil) affected more than 100 children and adults in Madhya Pradesh in 2007.

There have been reports of food contamination due to chemicals and pesticides. The first report of pesticide poisoning in India was from Kerala in 1958, where over 100 people died after consuming food made from wheat flour contaminated with parathion. In 1997, a foodborne outbreak of organophosphate (Malathion) poisoning affected 60 men (and was fatal for one) who ate a communal lunch prepared from food stored in open jute bags, which was contaminated with the pesticide sprayed in the kitchen that morning. It is estimated that 51% of food commodities are contaminated with pesticide residues in India. An outbreak of food poisoning due to epidemic dropsy (mustard oil contaminated with Argemone oil) was reported in Delhi in 1998; in which 60 persons lost their lives and more than 3000 cases were hospitalized. Under the Integrated Disease Surveillance Project in India, food poisoning outbreaks reported from all over India in 2009 increased by 140% as compared to the previous year (120 outbreaks in 2009, as compared to 50 in the year 2008) (National Centre for Disease Control, 2009).

1.11 PUBLIC HEALTH IMPACT OF THE BURDEN

After estimating the public health impact of different FBD, regulators and industries need to implement effective interventions to improve food safety. To identify these and allocate resources, they need to know the major food sources of illness. By attributing the estimated number of foodborne illnesses to particular sources (e.g., animals and foods), we can introduce measures to prevent food contamination and set goals for improvement. Knowledge on the burden of FBD is essential to set public health goals, allocate resources, and to measure the public health and economic impact of disease. FBD can be prevented as long as we know their causes and methods of intervening in the food chain to change the population's current exposure to these causes.

The assessment of the impact of the global burden would be an active process with an ever-changing disease profile. The epidemiology of FBD has changed in recent decades as new pathogens have emerged, the food supply has changed, and the number of people with heightened susceptibility to FBD has increased. Emergence is often the consequence of changes in some aspects of the social environment. The global economy, for example, has facilitated the rapid transport of perishable foods, increasing the potential for exposure to foodborne pathogens from other parts of the world. Other factors altering FBD patterns are the types of food that people eat, the sources of those foods, and the possible decline in public awareness of safe food-preparation practices. There would be a need to update the countries with the latest research in the field (Altekruse et al., 1996).

There is a need to foster communication between researchers and to create a forum for sharing information about the design, implementation, and analysis of studies on the burden of illness. This would help in providing necessary assistance to countries intending to conduct burden-of-illness studies and enable countries to contribute to the global FBD burden estimates. It is evident that the overall challenge is the generation and maintenance of constructive dialogue and collaboration between public health and veterinary and food-safety experts, bringing together multidisciplinary skills and multipathogen expertise. Such collaboration is essential in monitoring changing trends of the well-recognized diseases and in detecting emerging pathogens (Newella et al., 2010). While researchers continue to improve estimates of the burden of FBD, numerous studies are also attempting to attribute disease to specific food–animal sources. With more accurate information about the relative contribution of different foods to the total disease burden, and with more precise estimates of the burden of FBD, the studies would support the overall goal of reducing the socioeconomic burden of essentially preventable diseases.

REFERENCES

Altekruse, S.F., Swerdlow, D.L., 1996. The changing epidemiology of foodborne diseases. Am. J. Med. Sci. 311 (1), 23–29.

Amin, O.M., 2002. Seasonal prevalence of intestinal parasites in the United States during 2000. Am. J. Trop. Med. Hyg. 66, 799–803.

Birkhead, G., Vogt, R.L., 1989. Epidemiologic surveillance for endemic *Giardia lamblia* infection in Vermont. The roles of waterborne and person-to-person transmission. Am. J. Epidemiol. 129, 762–768.

Bowman, C., Flint, J., Pollari, F., 2003. Canadian integrated surveillance report: *Salmonella*, *Campylobacter*, pathogenic *E. coli* and *Shigella*, from 1996 to 1999. Can. Commun. Dis. Rep. Wkly. 29 (Suppl. 1), 1–6.

Flint, J.A., van Duynhoven, Y.T., Angulo, F.J., DeLong, S.M., Braun, P., Kirk, M., et al., 2005. Estimating the burden of acute gastroenteritis, foodborne disease, and pathogens commonly transmitted by food: an international review. Clin. Infect. Dis. 41 (5), 698–704.

Hald, T., Andersen, J.S., 2001. Trends and seasonal variations in the occurrence of *Salmonella* in pigs, pork and humans in Denmark, 1995-2000. Berl. Munch. Tierarztl. Wochenschr. 114, 346–349.

Hirsch, W., Sapiro-Hirsch, R., 1958. The role of certain animal feeding stuffs, especially bone meal, in the epidemiology of salmonellosis. Bull. Hyg. 33, 647.

Knox, W.A., Galbraith, N.C., Lewis, M.J., Hickie, G.C., Johnston, H.H., 1963. A milkborne outbreak of food poisoning due to *Salmonella Heidelberg*. J. Hyg. 61, 175–185.

National Centre for Disease Control, 2009. Foodborne diseases. Quarterly Newsletter from the National Centre for Disease Control 13 (4), 1–12.

Newella, D.G., Koopmansb, M., Verhoefb, L., Duizerb, E., et al., 2010. Foodborne diseases—the challenges of 20 years ago still persist while new ones continue to emerge. Int. J. Food Microbiol. 139 (Suppl. 1), S3–S15.

Pennington, J.H., Brooksbank, N.H., Pool, P.M., Seymour, F., 1968. *Salmonella Virchow* in a chicken-packing station and associated rearing units. Br. Med. J. 4, 804–806.

Semple, A.B., Turner, G.C., Lowry, D.M., 1968. Outbreak of food poisoning caused by *Salmonella Virchow* in spit-roasted chicken. Br. Med. J. 4, 801–803.

World Health Organization, 2006. The global burden of foodborne diseases: taking stock and charting the way forward: WHO consultation to develop a strategy to Estimate the Global Burden of Foodborne Diseases. Geneva, 25–27 September, 2006.

World Health Organization, 2007. WHO Initiative to Estimate the Global Burden of Foodborne Diseases. Geneva, 26–28 November, 2007.

World Health Organization, 2016. WHO estimates of the global burden of foodborne diseases. Foodborne diseases burden epidemiology reference group 2007–2015.

Foodborne infectious diseases

2

R.K. Gupta

Department of Community Medicine, Army College of Medical Sciences, New Delhi, India

Prevention depends on "clean food handled by clean people in clean premises with clean equipment and protection from flies and vermin"

Food nourishes us and is vital to the survival of species. But food is an important source of pathogens and toxins as well. Good hygiene; cooking and preservation practices; and adequate processing and eating discipline keep these pathogens and toxins at bay. But there are times when the food may be contaminated with infectious or other agents; the toxin might be an inherent part of the food (mushroom poisoning) or extraneous toxins/bacteria might gain access to food leading to disease, which might even cause death. Foodborne diseases occur frequently in the community, but their disease burden is hardly reflected in the morbidity statistics, as many of these are self-limiting, thus, conforming to the classical iceberg phenomenon.

A foodborne disease is one where the disease agent (which might be toxic or infectious; inherent to food or extraneous) is transmitted to the human body through food.

A broad classification of foodborne hazards is outlined here. The chapter focuses primarily on foodborne infections/infestations.

2.1 CLASSIFICATION

The foodborne hazards can be broadly classified as natural or contaminants. Since foodborne infectious diseases generally fall under the "natural" category, the same are elaborated in Fig. 2.1. The second group, that is, "contaminants" could be added intentionally (food adulteration) or there could be an unintentional contamination. Details of these conditions are discussed in a different chapter.

It is clear that, foodborne infectious diseases are a subset of foodborne diseases. The dangers of infective pathogenic organisms in food far exceed those of toxic agents (natural toxins or man-made chemicals). Therefore, this chapter deals with the infectious foodborne diseases before those caused by toxic agents.

Food Safety in the 21st Century. http://dx.doi.org/10.1016/B978-0-12-801773-9.00002-9

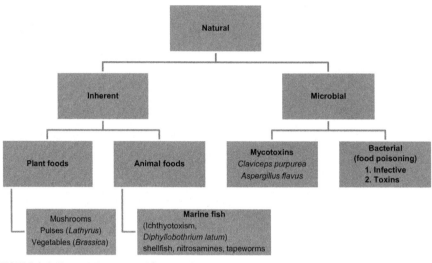

FIGURE 2.1 Natural Foodborne Diseases

2.2 INFECTIOUS DISEASES SPREAD BY FOOD

The main foodborne infectious diseases are listed in Table 2.1 (Gandy et al., 2007).

Multitude of infections, etiologically as varied as worms, protozoa, bacteria, and viruses are well known to gain access to food and enter the body to cause illness.

The infective diseases can also be classified on their etiological basis:

Bacterial: *More than 1600 bacterial species are known. Bacteria are the commonest causes of food poisoning, and can also cause tuberculosis, brucellosis, etc.*
Helminthic: *Tapeworm (*Taenia solium and Taenia saginata*), roundworms (*Ascaris*),* Trichinella spiralis.
Viral: *Hepatitis A and E, polio.*

2.3 INFECTIONS VERSUS INTOXICATION

There is a subtle difference between infection and intoxication. A foodborne infection is the disease caused when food containing microorganisms is consumed, and the latter proliferate further in the human intestinal tract and cause illness. Some bacteria, all viruses, and all parasites cause illness via infection, for example, *Salmonella, Listeria monocytogenes, Campylobacter jejuni, Vibrio parahaemolyticus,* and *Yersinia enterocolitica*, etc. Common viruses are: Hepatitis A, norovirus, and rotavirus and the most common foodborne parasites are: *T. spiralis, Giardia, Toxoplasma gondii, Cryptosporidium parvum,* and *Cyclospora cayetanensis*.

Table 2.1 Infectious Diseases Spread by Food

Disease	Organism	Mode of Spread
Food Poisoning		
Salmonellosis	*Salmonella typhimurium*	Farm animals, poultry, rats, mice
Staphylococcal	*S. aureus*	Human carrier
Perfringens enteritis	*C. perfringens (welchii)*	Dust, feces, flies
Botulism	*C. botulinum*	Dust, soil
Viral Gastroenteritis	*Norovirus, rotavirus*	Feco-oral, person to person, aerosolization
From Milk		
Brucellosis (undulant fever)	*Brucella abortus, Brucella melitensis*	Infected cows or goats
Bovine tuberculosis	*Mycobacterium bovis*	Infected cows or goats
From Infected Meat		
Trichinosis	*T. spiralis*	Pork
Tapeworms	*T. saginata, T. solium*	Beef
Cysticercosis	*T. solium*	Pork
Clonorchiasis	*Clonorchis sinensis*	Fish in Far East
Diphyllobothriasis	*Diphyllobothrium latum*	Fish
Balantidiasis	*Balantidium coli*	Pigs
Paragonimiasis	*Paragonimus westermani*	Cray fish, crabs
From Vegetables		
Fascioliasis	*Fasciolopsis buski*	Pigs and ruminants
From Infested Grain/Nuts		
Aflatoxicosis	*Aspergillus* species	Groundnuts
Ergotism	*Claviceps* species	Bajra, sorghum, rice, rye

On the other hand, an intoxication results when toxins that cause illness are ingested. Toxins are produced by harmful microorganisms, the result of a chemical contamination, or are naturally part of a plant or seafood. Some bacteria cause an intoxication, for example, *Clostridium botulinum, Staphylococcus aureus, C. perfringens*, and *B. cereus*. Viruses and parasites, though do cause foodborne illnesses, but do not cause foodborne "intoxication."

Certain noninfectious toxins would include seafood toxins (ciguatera toxin, scombroid toxin, shellfish toxins, and systemic fish toxins). Plants and mushrooms can also cause an intoxication. Certain nonnatural substances, for example, chemicals can also cause intoxication such as cleaning products, sanitizers, pesticides, and metals (lead, copper, brass, zinc, antimony, and cadmium) (APHA, 2000). However, this chapter deals primarily with natural foodborne diseases.

2.4 **TRANSMISSION**

Illness can be transmitted through various mechanisms. These could be following:

1. *Consumption of inherently toxic food*: Consumption of food that inherently contains the toxin, such as poisonous mushrooms, fish, Lathyrism, etc. can cause illness.
2. *Feco-oral transmission*: Many pathogenic organisms are excreted in human feces. Infection can then spread through contaminated hands, food handlers, flies, or fomites. Such infections are the feco-oral infections. They include acute gastroenteritis, diarrhea, dysentery, and some types of food poisoning (Table 2.2).
3. *Consumption of contaminated food*: Consumption of contaminated food can also cause illness, for example, adulterated food (epidemic dropsy), food/water contaminated by chemicals (Minamata disease, arsenic poisoning, etc.).

2.5 **FOOD POISONING**

Food poisoning may be defined as an acute gastroenteritis caused by ingestion of food or drink contaminated with bacteria/their toxins/chemicals or poisons derived from plants and animals.

2.5.1 **TYPICAL CHARACTERISTICS**

An episode of food poisoning is generally characterized by the following features:

1. A group of people present with a history of ingestion of a common food.
2. Many persons are affected at the same time.
3. Most of the cases involved, present with similar clinical features.

Table 2.2 Feco-oral Infections Passing From Man to Man by Indirect Spread

Disease	Organism
Gastroenteritis	Enteroviruses, bacteria, e.g., *E. coli*
Bacillary dysentery	*Shigella*
Typhoid	*Salmonella typhi*
Cholera	*Vibrio cholerae*
Amoebic dysentery	*Entamoeba histolytica*
Infectious hepatitis	*Hepatitis virus (A, E)*
Acute gastroenteritis	*Norovirus, Rotavirus*
Poliomyelitis	*Poliovirus*
Ascariasis (round worm)	*Ascaris lumbricoides*
Enterobiasis (threadworm)	*Enterobius vermicularis*
Trichuriasis (whip worm)	*Trichuris trichiura*
Giardiasis	*Giardia intestinalis*
Hymenolepiasis (dwarf tape worm)	*Hymenolepis nana*

Even though, by convention the term "food poisoning" is applied only to the most commonly occurring food poisoning—the bacterial food poisoning; there are other illnesses that fall under this broad term.

2.5.2 CLASSIFICATION OF FOOD POISONING

Food poisoning can be classified as bacterial, viral, inherently poisonous foods, allergic/sensitivity reactions to certain foods, from chemical contamination, and caused by parasites (in animals/meat), for example, *Trichinella*. Details are given in Table 2.3.

The common food poisonings are discussed in detail in subsequent sections.

2.6 SALMONELLOSIS

Salmonella is one of the commonest causes of food poisoning.

2.6.1 SOURCE

Salmonella is common to human and animal gut and is excreted in stools. One may get infected through consumption of raw foods of animal origin, for example, meat,

Table 2.3 Classification of Food Poisoning

Type of Poisoning	Subtype	Selected Specific Agents
Bacterial	Salmonellosis	*S. typhimurium*
		S. enteritidis
		Salmonella choleraesuis
	Staphylococcal	*S. aureus*
	Botulism	*C. botulinum*
	Perfringens enteritis	*C. perfringens* (*welchii*)
	B. cereus	*B. cereus*
	E. coli	*E. coli*
Viral		*Norovirus, Rotavirus*
Inherently poisonous foods	Mushrooms	
	Shellfish	
	Some plants	*Dhatura*
Allergic foods		Prawns
Chemicals	From utensils	Zinc, Copper
	Industrial metals	Mercury, Lead, Arsenic
	Heavy metals	Mercury, Lead,
	Pesticides, insecticides	DDT, BHC, Malathion
	Alkaloids	
Parasites	From animals/meat	*Trichinella, Taenia*

poultry, sausages, egg products, or through human/animal excreta. Insects, birds, vermin and domestic pets may also cause transmission.

2.6.2 CAUSE OF ILLNESS

Once food is contaminated, the bacteria multiply to cause illness. The live bacteria cause this illness.

2.6.3 CLINICAL FEATURES

The illness begins after about 6–36 h of consuming contaminated food. Symptoms include fever, headache, abdominal pain, diarrhea, and vomiting. The illness may last for 1–7 days. Most cases recover well but illness may rarely be fatal as in the elderly, young children, or sick people.

2.7 STAPHYLOCOCCAL FOOD POISONING

S. aureus is another common cause of food poisoning.

2.7.1 SOURCE

The commonest source is the skin of human carriers. It is also found in the nose, on the hands, in the throat, in boils, carbuncles, and other septic lesions. The raw milk could also be a source. A cow suffering from mastitis is also capable of transferring the organisms and toxin to raw milk, cream, and products made from them.

2.7.2 CAUSE OF ILLNESS

The bacteria produce a toxin which causes the illness.

2.7.3 CLINICAL FEATURES

The symptoms begin 2–6 h after eating contaminated food. The patient presents with acute vomiting, pain in abdomen, cramps, and diarrhea. There is no fever. The illness lasts for not more than 24 h.

2.8 *C. PERFRINGENS* FOOD POISONING

C. perfringens is also common cause of food poisoning as the bacteria is found routinely in the feces of humans.

2.8.1 SOURCE

The bacterium is an anaerobic organism found in the human and animal intestine, soil, dust, flies, etc. It is known to frequently contaminate raw meat, poultry, and some dried food products. The spores of *C. pelfringens* can survive normal cooking temperatures.

2.8.2 CAUSE OF ILLNESS

The food gets contaminated by the bacterial spores. Even if the food is cooked, the spores continue to be active and grow readily, if food is left for few hours in a warm place. Large numbers of bacilli grow in this food. A toxin is produced and released in the intestine.

2.8.3 CLINICAL FEATURES

The symptoms start 8–24 h after eating contaminated food. Symptoms include abdominal pain, cramps, and diarrhea, but rarely vomiting. Fever is also not usual. The illness lasts for 1–2 days.

2.9 BOTULISM FOOD POISONING

Botulism is an uncommon but the most serious cause of food poisoning.

2.9.1 SOURCE

Clostridium botulinum is an anaerobic spore bearing bacilli. The bacteria are commonly found in soil, meat, and fish in some areas. The spores survive normal cooking processes.

2.9.2 CAUSE OF ILLNESS

The toxin is produced by the bacilli as they grow in food. Such food, when ingested, causes the illness.

2.9.3 CLINICAL FEATURES

The symptoms begin 12–96 h after eating contaminated food. Its clinical symptoms are different from any other type of food poisoning as the nervous symptoms are predominant rather than the gastrointestinal symptoms. There is headache and dizziness. There may be dysphagia (difficulty in swallowing), diplopia (double vision), ptosis (drooping of eyelids), dysarthria, and weakness. In severe cases quadriplegia (paralysis) may result, eventually leading to cardiorespiratory failure and death. The gastrointestinal symptoms are very slight; may be nausea and diarrhea in the beginning followed by the neurological symptoms particularly the disturbance of vision and speech. Death occurs within 8 days, unless antitoxin is given soon after onset.

2.10 *B. CEREUS* FOOD POISONING

B. cereus may be a common cause of food poisoning.

2.10.1 **SOURCE**

B. cereus is a large spore-bearing aerobic bacilli, found in soil, dust, cereals, spices, vegetables, dairy products, and many other foods. Some spores may survive cooking.

2.10.2 **CAUSE OF ILLNESS**

The spores can survive cooking. On getting favorable conditions the spores germinate, bacilli multiply and produce a toxin. Two distinct types of toxins are produced causing either of the types of illness discussed subsequently.

2.10.3 **CLINICAL FEATURES**

1. *Emetic form or vomiting type*: It begins early (1–6 h) after eating contaminated food. A cooked rice dish is often incriminated. Symptoms are predominantly upper gastrointestinal type, including nausea and vomiting. Sometimes there may be diarrhea little later. The illness lasts not more than 24 h.
2. *Diarrheal type:* It has a longer incubation period, beginning 8–16 h after eating contaminated food. Symptoms are predominantly of the lower gastrointestinal type, that is, diarrhea and abdominal pain, but rarely vomiting. The illness lasts for not more than 24 h. Fever is generally absent.

2.11 *E. COLI* FOOD POISONING

E. coli is often thought to be associated with Traveler's diarrhea. It may also be a common cause of food poisoning.

2.11.1 **SOURCE**

E. coli is commonly present in human and animal gut and excreted in stools. The kitchen is likely to get infected through human and animal excreta. The raw foods of animal origin such as meat, poultry, and eggs could also be the source. The insects, birds, vermin, and domestic pets could also be playing some part in its transmission.

2.11.2 **CAUSE OF ILLNESS**

The living bacteria having access to food in the aforementioned ways is the cause of illness.

2.11.3 **CLINICAL FEATURES**

The symptoms may start after 18–48 h of consuming contaminated food or water. Symptoms include abdominal pain, cramps, and diarrhea. Fever and vomiting may also be there. The illness lasts for up to 5 days.

In approximately 15% of the children in North America, who are infected with *E. coli* O157, the hemolytic–uremic syndrome develops soon after onset of diarrhea. This syndrome is characterized by thrombocytopenia, hemolytic anemia, and nephropathy and is believed to be caused by Shiga toxins elaborated by *E. coli* O157:H7 or other infecting *E. coli* that have been absorbed into the systemic circulation (Wong et al., 2000).

2.12 NOROVIRUS GASTROENTERITIS

Norovirus, (earlier called as Norwalk virus) sometimes also known as winter vomiting bug in the United Kingdom, is one of the most common causes of viral gastroenteritis in humans, affecting people of all ages. It causes approximately 90% of epidemic nonbacterial outbreaks of gastroenteritis around the world (Said et al., 2008).

2.12.1 SOURCE

Infected person, contaminated food or water, or contaminated surfaces could be common sources. Infection may be commonly encountered in closed eating facilities such as hostels, messes, canteens, camps, prisons, cruise ships, etc.

2.12.2 CAUSE

Infection may often occur by touching ready-to-eat foods (raw fruits and vegetables) with contaminated bare hands. The virus is transmitted through feco-oral route via contaminated food or water, or by person-to-person contact. Aerosolization of the virus and subsequent contamination of surfaces could also lead to infection.

2.12.3 CLINICAL FEATURES

Norovirus infection is characterized by gastroenteritis (nausea, vomiting, and diarrhea) along with abdominal pain and loss of appetite. Lassitude, lethargy, weakness, muscle aches, headaches, and low-grade fevers may also follow. In severe cases dehydration may also occur. The disease is usually self-limiting with full recovery within 2–3 days. Some cases (especially children and elderly) may develop serious illness with occasional fatalities.

2.12.4 PREVENTIVE MEASURES FOR BACTERIAL FOOD POISONING AT HOUSEHOLD LEVEL

1. *Kitchen hygiene*
   a. Raw foods should not be allowed to mix with cooked foods and they must be deliberately separated. This is particularly true for the nonvegetarian foods.

Different surfaces and equipment (boards, cutting mincing machines, cloths, and kitchen tools) must be used to prevent cross-contamination.

b. All surfaces, equipment, and tools should be thoroughly scrubbed and cleaned.

2. *Care of personal hygiene* by washing hands before starting cooking and after visiting the toilet. It is a good practice to wash hands before and after handling food—especially raw meat and poultry.

3. *Storage of food*: Avoid storing food as far as possible. If it has to be stored, resort to cold or hot storage, to prevent multiplication of bacteria. Warm food should be rapidly cooled to prevent multiplication of bacteria.

4. *Avoid home preserved meat*: It is important for prevention of the dangerous botulism food poisoning that home preservation of meat, poultry, and fish must be avoided except by freezing. Careful inspection of cans and their contents is also vital.

Further details of kitchen hygiene are elaborated in Chapters 15 and 17.

2.13 SANITATION OF SOME SPECIFIC FOODS

Some foods require special attention. These are mainly the nonvegetarian foods and fresh salads.

Poultry: Poultry may harbor food-poisoning organisms on the skin, offal, and inside the carcass. Thus, care should be taken where and how birds are dressed. Surfaces and utensils should be well cleaned after use, and hands should be washed well after handling the raw materials. Frozen meat and poultry should be thawed properly before cooking

Sausages: Sausages, raw scraps, and minced meat may be contaminated with *Salmonellae*. Great care should be taken when preparing sausages for cooking. Sausages should be well cooked.

Meat: Dishes should be prepared fresh from raw meat. If there is likely to be any delay in using cooked meat, steaming under pressure is the best way to ensure the destruction heat-resistant organisms; it is a safe method of cooking. If leftovers are used after warming, they must be cooked thoroughly to boiling. It must not be allowed to cool slowly and stored at atmospheric temperature, as that promotes rapid multiplication of bacteria.

Salads and fruits: Salad vegetables including spinach, cabbage, and lettuce could have been grown in sewage farming. Salads and cut fruits (which cannot be peeled), should be washed well, preferably with water containing hypochlorite.

Sandwiches: Sandwiches should be made as near the time of consumption as possible. If prepared early they should be refrigerated in a bread box.

Cooked rice: Cooked rice should not be stored overnight without refrigeration.

Milk/eggs: Milk should be consumed after pasteurization or boiling. The most hygienic way to consume eggs is to boil them and eat.

2.14 **FOODBORNE DISEASES OF FUNGAL ORIGIN**

Certain fungal infestations in grain (cereals/millets) or legumes and nuts (groundnut) are also known to cause foodborne illness. These, though not very common today, do occur sporadically and baffle the medical community. Some of the commoner conditions are discussed here (Gupta, 2010).

2.15 **AFLATOXICOSIS**

Aflatoxicosis was first reported in England in 1960, among young turkeys fed on infested groundnut meal which had hepatitis and enteritis. The groundnuts concerned were harvested, stored, and processed in high-humidity conditions. The toxic effects were produced by a fungus *Aspergillus flavus,* contaminating the nuts. Human cases have occurred off and on since then.

A. flavus or *Aspergillus parasiticus* are storage fungi that affect foods in poor storage conditions with high temperature (30–37°C) and humidity, as is seen in rainy season and during floods and cyclones. The fungus infests improperly stored foods such as maize, groundnut, soya, sorghum, rice, wheat, sunflower, tree nuts, spices, and even milk and cheese.

Toxic furanocoumarin compounds known as the "aflatoxins" are responsible for the condition. Aflatoxin B1 and G1 are potent natural hepatocarcinogens. They cause hepatitis (jaundice), ascites, portal hypertension, liver cirrhosis, and hepatocellular carcinoma.

2.15.1 **ERGOTISM**

Unlike Aspergillus (a storage fungus), *Claviceps fusiformis and Claviceps purpurea* are field fungi. Crops get infested in flowering or seeding stages. Bajra, rice, sorghum, wheat, and rye get commonly affected. Ergotamine is the toxin responsible for clinical symptoms of nausea, vomiting, abdominal cramps, muscular cramps, giddiness, burning, itching, and gangrene of digits and limbs.

Epidemics of ergotism were known as *St. Anthony's fire* in France in the 11th century. The disease was referred to as "fire" because of the intolerable burning pain in the limbs, which became black and shriveled (gangrenous) and eventually dropped off. The legend also says that the condition used to improve when the patients visited St Anthony's shrine located a distance away. The patients probably improved because of the discontinuation of consumption of ergot-affected cereals, as they shifted to the new location (of the shrine).

Epidemics occurred in Germany, Poland, England, and Russia till the late 18th century, when it was related to the consumption of fungus (*C. purpura*) infested rye. Besides the symptoms enumerated previously, convulsions, palsies, and discordant movements were also known, indicating the affliction of the nervous system.

The *C. fusiformis* infestation of bajra in India is a milder clinical entity, which leads to nausea, vomiting, abdominal cramps, and drowsiness. The recovery is usually complete.

2.15.2 FUSARIUM

Fusarium incarnatum is another field fungus affecting crops such as sorghum, rice, and maize. It is seen in the subtropical and temperate regions. The fungus produces toxins such as deoxynivalenol and fumonisin which are responsible for certain clinical symptoms such as vomiting and diarrhea. The episodes of mouldy *ragi* poisoning in India (1929) and alimentary toxic aleukia (hemorrhagic rash, bleeding nose, leucopoenia) seen in Russia during the Second World War were due to fusarium.

Detection of mycotoxins: Many sophisticated methods are available for the detection of mycotoxins. Thin layer chromatography (TLC), radioimmuno assay (RIA), and ELISA tests are available. Several rapid kits are also available for detection of aflatoxicosis, etc.

Prevention of foodborne diseases of fungal: For prevention of these mycotoxins, four broad steps should be taken (Bamji et al., 2003):

1. *Plant breeding*: Cultivating fungus-resistant varieties of rye, bajra, millets, and wheat can substantially minimize the problem.
2. *Good agricultural practices during pre- and postharvest period*: Good preharvest agricultural practices such as avoiding water stress, minimizing insect infestation, are effective in reducing aflatoxin contamination in groundnuts and maize. Good postharvest and storage conditions for grains and nuts are also of paramount importance. These foods must be stored under ideal humidity and temperature conditions. Appropriate drying, storage, and reducing the chances of moisture entry in the stores also limit the probability of contamination with storage fungi. If contamination does occur, the infested grains can be removed using the floatation method in which the grain is allowed to float in 20% salt water. The infested grains floats and can be easily removed. Air floatation and hand picking techniques can also be used.
3. *Detoxification*: Ammonia process is being used to detoxify aflatoxin-affected groundnuts and remove the mycotoxin. The detoxified product is available only for animal feeds and is not suitable for human consumption.
4. *Health education*: The community must be educated about the ill effects of the conditions and the importance of the preventive measures described previously.

2.16 FOODBORNE DISEASES CAUSED BY TOXIC AGENTS

Many a times toxins are inherent to food, or they may be acquired by the food from outside (as a contaminant or adulterant). One common example of each is discussed here in detail.

Table 2.4 Some Possible Toxic Effects of Common Foods

Food Stuff	Active Toxic Ingredient	Effects on
Some bananas	5-Hydroxytryptamine, adrenaline, noradrenaline	Central or and peripheral nervous systems
Some types of cheese	Tyramine	Blood pressure
Almond, cassava	Cyanide	Tissue respiration
Some fish/meat	Nitrosamines	Cancers
Khesri dal (Lathyrus)	Beta-N-oxalyl-amino-l-alanine and others	Neurolathyrism
Brassica species (seeds)	Glucosinolates, thiocyanate	Goiter
Green potato	Solanine	Gastrointestinal upset
Mushrooms (Amanita muscaria, Amanita phalloides)	Various toxins	Central Nervous System effects

Toxins inherent to food: However surprising it might appear, but commonly used fruits, vegetables, nuts, legumes, and tubers may contain inherent toxins that may become manifest and cause disease, under certain conditions. A brief on these is given in Table 2.4, and *Lathyrus* is discussed in detail in subsequent paragraphs.

2.17 LATHYRUS TOXIN

Lathyrus sativus (*Khesari dal*) looks like *Arhar dal* (*toor dal*) red gram or *Bengal* gram. However, its seeds are triangular in shape and grayish in color. It is cheaper and a rich source of protein. It has been described in old historical Hindu texts, Bible, and was also elaborated by Hippocrates. *Cantani* coined the term Lathyrism in Italy.

It is deliberately sown with wheat in the dry districts. If the rains are good, wheat overgrows the *Lathyrus* (and it is not harvested), but if rain fails and there is poor crop of wheat, a reasonable crop of *Lathyrus* is reaped. Lathyrus is a tasty and high-protein pulse. If eaten in small quantities it does not cause toxicity. If consumed in larger amounts (providing more than 30% of energy) a severe disease of the nervous system may result, leading to spastic paralysis. For the poor, *Khesri dal* might become the staple diet and cause neurolathyrism.

Besides India, the disease is seen in some European, African, and Asian countries. Cases are reported from Spain, Algeria, Ethiopia, Mexico, Afghanistan, and India. In India, it is seen in some districts of Madhya Pradesh, Maharashtra, Uttar Pradesh, Bihar, Rajasthan, Assam, and Gujarat. Old literature reports thousands of cases of neurolathyrism occurring in epidemic proportions in some given regions, but now it occurs only sporadically.

In 1962 a neurotoxin, β-N-oxayl amino-l-alanine (BOAA) was isolated from the common vetch (vicia sativa), which frequently grows as a weed in *L. sativus*. *In 1963,* another toxin β-N-oxayl-l-α,β di-aminopropionic acid was isolated from the seeds *of L. sativus.* Both of these can cause neurological lesions in primates. These toxins are neuroexitants and can be removed by soaking in hot water and rejecting it.

2.17.1 CLINICAL FEATURES

If more than 30% energy is obtained from *Khesri dal* for more than 6 months, the signs and symptoms may appear in the form of spastic paralysis. The condition is known as *Lathyrism.* It is most commonly seen in men of the age group 15–45 years. The onset of *Lathyrism* is sudden, and is often preceded by exertion or exposure to cold. A patient may find himself paralyzed on getting up in the morning. Sometimes backache and stiffness of legs precede the paralysis of legs (Passmore and Eastwood, 1986).

The patient may pass through progressive stages of severity. In the latent stage, the patient may be apparently healthy. In the mild stage, there is stiffness and weakness of legs. As the disease progresses, the gait may be affected and the patient walks with bent knees on tiptoe. The legs may become crossed and patient may develop scissor gait. The patient may be able to walk only with one stick and later with two sticks. Later on when paraplegia develops, walking may become impossible. The patient has to support his body on his hands, buttocks, and heels for moving about (crawling stage). In the most severe stage patient can only move on "all fours," supported by his hands.

Detection of toxin: The toxin can be detected through laboratory methods using the ninhydrin reaction, which gives a purple color. Electrophoresis and biological methods (bioassay) can also be used.

Prevention: The condition is preventable if pulse is removed from diet, at the earliest. Use of the legislation to limit consumption of crop must be encouraged. In case the crop has to be consumed, the toxin can be removed by steeping. At the household level, steeping can be done by soaking the pulse in hot water for 2 h. Water is then drained and pulse is dried in sun. The disadvantage of using this method is that the taste and nutrients are lost to an extent. At a large scale, parboiling can be done. The process is the same as used for parboiling rice. Soaking the pulse overnight in limewater and subsequently boiling or cooking it also helps in removing the toxin.

High-dose vitamin C (1000 mg/day) prophylaxis for few weeks is also found to be useful. Nutritional education in the form of abstaining from the use of crop or using it in the manner prescribed previously would be useful in preventing the consumption of toxic crops.

Some toxins can be acquired by the food from outside, as a contaminant or adulterant. One such example of epidemic dropsy is discussed subsequently in detail.

2.18 EPIDEMIC DROPSY

Several cases of epidemic dropsy were reported from many states of India in the year 1996. It was discovered by Indian scientists in the early 20th century that the condition is attributable to contamination of mustard oil with argemone oil. Subsequently, the toxic alkaloid *sanguinarine* was isolated from argemone oil and was chemically analyzed. It was also determined that sanguinarine interferes with the oxidation of pyruvic acid, which is responsible for dropsy.

Argemone mexicana or prickly poppy plant grows indiscriminately and wild in India, which has large prickly leaves and bright yellow flowers. The argemone seeds look like mustard seeds. They mature with mustard crop and may be harvested along with mustard. Argemone seeds (or oil) can be mixed with mustard (oil) to adulterate it. The contamination may sometimes be accidental.

Clinical features: The patient gets generalized swelling, seen as bilateral pedal edema of sudden onset. Patient may get diarrhea. In advanced stage, dyspnea and signs of congestive heart failure are seen. If not treated death may ensue. A mortality rate of 5–50% has been reported.

Detection of toxin: The toxin can be detected by nitric acid test or using paper chromatography

Prevention: The growth of argemone plant must be discouraged and must be weeded out. Unscrupulous traders deliberately adulterating argemone oil to mustard oil must be tried under the law. Early detection and institution of control measures must be encouraged to limit severity and further spread of morbidity. Educating and making the public aware of the problem and likely solutions will also help.

2.19 CONCLUSIONS

Food borne diseases are caused by a multitude of agents such as bacteria, viruses, parasites, and fungus which enter the body and cause illness. Transmission of these diseases can take place by consumption of inherently toxic or contaminated food or feco oral route. Clinical presentation of food poisoning varies with the causative agent. Preventive measures for food borne diseases include maintenance of kitchen hygiene, specific sanitation measures for some foods. Illness of fungal origin (Ergotism, Afltoxicosis) can be prevented by good agricultural and storage practices.

REFERENCES

American Public Health Association (APHA), 2000. Control of Communicable Diseases Manual. In: Clin, J. (ed.). American Public Health Association, Washington, DC, p. 624.

Bamji, M.S., Rao, N.P., Reddy, V., 2003. Textbook of Human Nutrition, second ed. Oxford & IBH Publishing Co. Pvt. Ltd, New Delhi.

Gandy, J.W., Madden, A., Holdsworth, M., 2007. Oxford Handbook of Nutrition and Dietetics. Oxford University Press, New Delhi.

Gupta, R.K., 2010. A Textbook of Community Medicine. Food Processing, Food Adulteration, Food Additives, Preservatives, Food Toxicants and Food Fortification. 1209–1219. WHO, Pune.

Passmore, R., Eastwood, M.A., 1986. Human Nutrition and Dietetics, eighth ed. Churchill Livingstone, ELBS, London.

Said, M.A., Perl, T.M., Sears, C.L., 2008. Healthcare epidemiology: gastrointestinal flu: norovirus in health care and long-term care facilities. Clin. Infect. Dis. 47 (9), 1202–1208.

Wong, C.S., Jelacic, S., Habeeb, R.L., Watkins, S.L., Tarr, P.I., 2000. The risk of the hemolytic–uremic syndrome after antibiotic treatment of *Escherichia coli* O157:H7 INFECTIONS. New Engl. J. Med. 342 (26), 1930–1936.

Outbreak investigation of foodborne illnesses

A. Khera

Department of Community Medicine, Armed Forces Medical College, Pune, Maharashtra, India

3.1 INTRODUCTION

An outbreak is defined as unusual occurrence of disease or health related events clearly in excess of the usual occurrence in a defined community, region, or season (Park, 2011; World Health Organisation, 2015b). This unusual occurrence is defined as a number of cases two standard deviations in excess of the endemic state of the disease. For diseases which are under elimination or eradication or absent from a population or caused by agents not previously identified in that community or area, for example, polio and measles, even a single case is considered as an outbreak (World Health Organisation, 2015b).

Outbreak investigation is one of the most challenging tasks faced by a public health specialist. There is an urgency among the health care administrators of controlling the outbreak at the earliest. Frequently, the cause of the outbreak is unknown and there is a concern among the susceptible population about the disease in terms of whether they or their children are at risk of the disease or not.

Outbreaks cannot always be predicted or prevented. But recognition of early warning signs, timely and prompt investigations to identify the cause, and the use of specific preventive and control measures can reduce morbidity/mortality and duration of the outbreak. The primary aim of an outbreak investigation is to control/limit the spread, and identify the strategies for prevention or elimination of the risk of such outbreaks in the future.

3.2 TRIGGER EVENTS

Based on the definition, which provides a broad framework of what should be labelled as an outbreak, there are certain trigger events/warning signs which point toward an impending outbreak (Table 3.1). Once the trigger event has been identified or an outbreak has been reported, there is a need to investigate the same to identify the cause and source of the outbreak and institute preventive measures. The various steps followed during outbreak investigation are (Table 3.2):

Table 3.1 Trigger Events for a Foodborne/Waterborne Disease Outbreak

- Time clustering of cases of diarrhea with or without fever
- Unusual increase in cases or deaths of food borne diseases, that is, above the upper control limit for the area/season/population
- Occurrence of two or more epidemiologically linked cases of viral hepatitis
- Natural disasters
- A single case of cholera

Table 3.2 Steps of an Outbreak Investigation

Verify and confirm the existence of the outbreak and prepare for field work
Case-finding through active surveillance and community surveys
Line listing of cases
Descriptive epidemiology
Determining who is at risk of disease
Formulation of hypothesis
Testing of hypothesis
Refining the hypothesis and conducting additional studies
Deciding when the outbreak is over and documentation
Prevention and control

1. *Verify and confirm the existence of the outbreak and prepare for field work*
   Initial reports of an outbreak or trigger events of an impending outbreak are generally reported by laypersons, media or primary health care workers. The first step of outbreak investigation is to verify the number of cases and confirm the diagnosis of as many reported cases as possible. The reported cases should be investigated by a medical officer to confirm the diagnosis. The majority of the cases are expected to fall within the broad standard case definitions. The reported number of cases is then subjected to the case definition of an outbreak, identified as part of trigger events. Second, since, the first report about an outbreak is generally based on clinical syndrome, the diagnosis is presumptive (Centers for Disease Control and Prevention, 2015; Division of Scientific Education and Professional Development CDC, 2012).
   Laboratory confirmation of clinically diagnosed cases and the identification of causative agent may sometimes be necessary. In an outbreak, only a few samples should be collected carefully from selected cases. It is not necessary to collect samples from all cases as it unnecessarily places a heavy load on the laboratory. The common foodborne disease outbreaks present as syndromes; which may include:
   a. acute diarrhoeal diseases
      - acute watery diarrhoea in young children
      - acute watery diarrhoea in patients > 5 years with dehydration
      - dysentery (bloody diarrhoea)

    b. jaundice (mostly waterborne)
- hepatitis A
- hepatitis E

Before embarking on field work the investigators needs to be armed with all the relevant items/equipment for the outbreak investigation (Cuomo, 2016). Preparations include the following (the list is not exclusive):

a. sample questionnaires

b. consult laboratory staff for culture media, transport media and specific temperature requirement for transport media

c. laptop

d. tape recorder, camera

e. stationary including calculator

f. torch

g. reference material, for example, journal articles/books/old case histories

h. permission from administrative authorities

2. *Case-finding through active surveillance and community surveys* (Centers for Disease Control and Prevention, 2015; Cuomo, 2016)

   After establishing the existence of an outbreak and verifying the diagnosis, it becomes important to accurately define and count the cases. During the period of the outbreak all the cases of the foodborne disease under consideration occurring in that area/region should be identified and listed. Active surveillance, that is, active search for cases may include visits to the susceptible population or telephone calls to the hospitals that might be expected to admit or attend the cases of the disease. Active surveillance must be maintained until the outbreak is over. It is generally calculated as double the incubation period of the disease identified as the cause from the reported last case. Initial case definitions may be based on "suspect cases" so that the sensitive definition should not miss out any case. Later, based on the grouping of clinical and laboratory features the definition may be made more specific as "probable" or "confirmed" case. The case definition should include the clinical as well as the epidemiological features.

   Suspect case: These have fewer or atypical clinical features in a defined area and time period.

   For example, any diarrhoea of acute onset in place X, with onset period being between dates A and B.

   Probable case: They have the typical clinical features of the illness but without laboratory confirmation, for example, acute onset of diarrhea with rice water stools with temperature greater than 99.0°F or 37.2°C and pain abdomen, with onset in place X, with onset period being between dates A and B

   Confirmed case: Positive laboratory result (isolation of the causative agent or positive serological test), for example, a case who has laboratory evidence of *Vibrio* in culture and confirmed serology with onset in place X, with onset period being between dates A and B.

Search for more cases:

Visits to the community and health institutions are important especially during the declining phase of the outbreak to not to miss last cases and evaluate control strategy applied as general principles during the initial reporting and assessment of the outbreak.

3. *Line listing of the cases* (Table 3.3)

   It is important that clinical and epidemiological information be recorded in a standardized manner in an epidemiological case sheet as line listing. The epidemiological case sheet includes persons who are ill and meet the case definition for the outbreak. It also includes the details of the intake of the various food items likely to cause the outbreak, outcome (dead/alive/hospitalized/OPD treatment), and date of (admission/discharge/death) etc.

4. *Descriptive epidemiology* (Park, 2011; Cuomo, 2016; World Health Organisation, 2015b)

   In investigating an outbreak, it is necessary to provide a detailed description of outbreaks in terms of time, place, and person. This provides clues to the etiology and in generation of hypotheses.

   a. *Distribution by time (Epidemic Curve)* (Fig. 3.1)

      The onset of illness of the cases is graphed using a histogram by hours, days, weeks or months depending on the type of outbreak. This graph is known as epidemic curve. It provides information on the magnitude of outbreak and its mode of spread. In the single exposure point source epidemics, the epidemic curve aids in identification of possible incubation period, in cases of unknown outbreaks and provides clues for the likely agent and the source. In outbreaks, in which the agent is known, the epidemic curve helps in identifying the serving time of the culprit food item. The method to calculate the incubation period in a point source outbreak is by using the maximum (last case of the cluster), minimum (first case of the cluster) and median (midpoint of the cluster). The data are then compared with the the incubation period of known agents to implicate the time of serving or identify the possible agent, if the food timing is known.

   b. *Distribution by person*

      Cases should be described in terms of age, sex, occupation, socio-economic parameters, food history, travel history, and other relevant characteristics. It is usually better to enter completed years rather than to group cases by age groups by 0–11 months, 1–4 years, 5–14 years, 15–44 years, and >45 years as it can be done while analysis.

   c. *Distribution by place*

      A map of the area or even a rough sketch can be drawn showing where each reported case resides to indicate geographical distribution of cases and to identify high-risk pockets. It enables the investigator to identify possible area of contamination (majority cases from a single canteen or dining hall). The map which describes the distribution of cases as per place distribution is known as a spot map (Fig. 3.2).

Table 3.3 Line Listing of Cases (Epidemiological Case Sheet)

Case #	Name	Age	Sex	Address	Date of Adm	Date of Onset	Diagnosis	Nausea	Vomiting	Fever	Lab Investigations	Food Item A	Food Item B
1													
2													
.													
.													

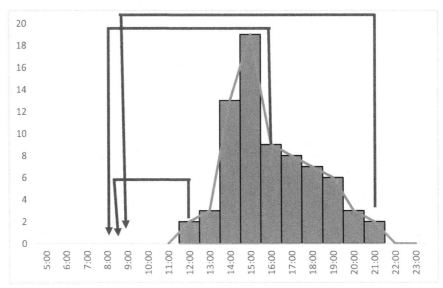

FIGURE 3.1 Finding the Possible Meal When the Agent is Known Using the Epidemic Curve

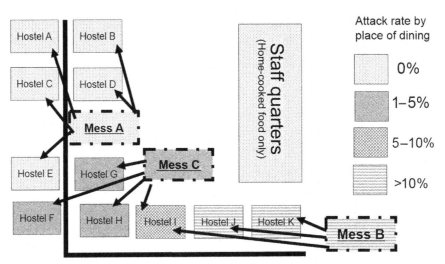

FIGURE 3.2 Spot Map of an Outbreak of Foodborne Disease in a Hostel With Multiple Dining Areas

d. *Description of environmental conditions*
The study of environmental conditions and the dynamics of its interaction with the population and causative agents will help in formulating the hypothesis, which will make the basis for the control measures to be taken. Information of ambient temperatures/food

storage facility/cutting and chopping equipment and washing facility as available may be utilized for the description of the epidemic. Different colors or shades can be used to present high and low attack rate areas on the spot map.

  e. *Laboratory investigation*

The results of laboratory investigations on the identification of suspected agents from food items, cooks, vomitus, etc. should be included in the outbreak investigation report. The measures to prevent or limit the spread of the disease and treatment of patients should not be delayed pending laboratory confirmation of diagnosis. Action should be initiated based on clinical, epidemiological, and environmental findings. However, special emphasis must be put on proper collection using the correct culture and transportation media, labeling, transportation, and storage of food and clinical samples.

5. *Determining who is at risk of disease*

The descriptive epidemiology will help to define the population groups at high-risk of disease in terms of age, place, gender, etc. Identification of the high-risk population, place, or gender is more appropriately done by measuring attack rates (in cohort analysis) or odds ratio (in case–control analysis) because these measures of association take into account the variations in the population size of different groups.

6. *Formulation of hypothesis*: (Fig. 3.3)

On the basis of distribution of the outbreak in terms of time, place, and person or the host–agent–environment model, a hypothesis is formulated to explain the epidemic in terms of possible source, causative agent, and mode of spread.

7. *Testing of hypothesis* (Division of Scientific Education and Professional Development CDC, 2012; Gordis, 2014)

The next important step in outbreak investigation is to test all the hypotheses generated, based on the preliminary evaluation of the descriptive studies with

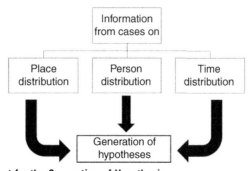

FIGURE 3.3 Flowchart for the Generation of Hypothesis

Table 3.4 Odds Ratio in Case–Control Studies

Food	Cases			Controls			
	Ate (A)	Did Not Eat (B)	Odds of Eating (C) = (A/B)	Ate (A)	Did Not Eat (B)	Odds of Eating (C) = (A/B)	Odds Ratio
Noodles	20	5	4.00	10	40	0.25	16.00
Chicken chilly	17	8	2.13	15	35	0.43	4.96
Chocolate fudge	20	5	4.00	20	30	0.67	6.00

analytical studies. The analytical studies which can be used for investigating an outbreak include:

a. *Case–control study*: The classical case–control study is a retrospective design, in which information regarding possible exposures (food items) is taken from both the cases (those who have the disease) and controls (those who do not have the disease). The ratio of cases to controls may be maximum up to 1:4, especially when the cases are less in order to improve the power of the study. Any ratio of cases to controls over this, offers no advantage on improving the power of the study. The analysis is done to first calculate the odds of disease in cases and the odds of disease in controls. The ratio of these is the measure of association in case control studies, that is, odds ratio or "OR". This design is commonly used when the population denominator is not known or difficult to obtain, for example, hepatitis E outbreak in a city/municipality (Table 3.4).

b. *Cohort studies*: The other study design which is commonly used to analyze the outbreak data is the "retrospective cohort study" in which the investigator hypothetically travels in time to the "onset" of the outbreak and then classifies the population into exposed and unexposed groups. Further, the "follow up" is done to the development of the outcome in both the groups and compared. The risk of disease is identified in both the exposed and the unexposed group based on exposure (food item). The measure of association used in retrospective cohort studies is relative risk or "RR," which is the ratio of the risk of disease in exposed and the unexposed group. The other measures which can be calculated include the following:

$$\text{Attributable risk} = \text{Risk in exposed} - \text{Risk in unexposed}$$

$$\text{Attributable risk}\% = \frac{\text{Risk in exposed} - \text{Risk in unexposed}}{\text{Risk in exposed}} \times 100$$

Retrospective cohort studies are ideal for outbreaks in which the population details are available, that is, outbreak of food poisoning in a hostel (Table 3.5). In situations, where the odds ratio can be assumed to be the same as relative

Table 3.5 Analysis Using Retrospective Cohort Study Using Same Data as Used in Case–Control study

| Food | Number of Persons Who Ate Specified Food | | | | Number of Persons Who Did Not Eat Specified Food | | | | Attributable Risk (C − Z) | Attributable Risk % (C − Z)/C*100 | Risk Ratio (C/Z) |
	III (A)	Well	Total (B)	Risk (A/B)=C	III (X)	Well	Total (Y)	Risk (X/Y)=Z			
Noodles	20	10	30	0.67	5	40	45	0.11	0.56	83.33%	6.00
Chicken chilly	17	15	32	0.53	8	35	43	0.19	0.35	64.98%	2.86
Chocolate fudge	20	20	40	0.50	5	30	35	0.14	0.36	71.43%	3.50

risk (rare disease and both cases and controls representing the source population), both attributable risk and attributable risk % can be calculated by replacing RR with OR.

In situations where more than one food item is being implicated (situations where both the items are eaten together by most of the participants), cross tabulation or stratification based on one food item being eaten or not may be used to identify and implicate the causative food item.

8. *Refining the hypothesis and conducting additional studies*

Many a time there are situations where the hypotheses planned do not bring out the causative food item. One should reconsider the presence of confounders or meet the cases again to elicit the responses with other possible hypotheses (Table 3.6).

9. *Deciding when the outbreak is over and Documentation* (World Health Organisation, 2015a)

An outbreak is presumed to be over once the number of incident cases of the disease reported each the endemic levels in that area and time. The epidemic curve aids the investigators for the same. The time beyond which, the surveillance needs to carried out after the last case has been identified as related to the outbreak is generally twice the time of the maximum incubation period.

Documentation of the outbreak is an important step in outbreak investigation. The findings of the outbreak should be discussed and shared with all the shareholders. The document should summarize the circumstances of the outbreak, the causative agent, the chain of events which led to the outbreaks, the impact of the control measures and measures for prevention of similar episodes in future.

10. *Prevention and control*

The prevention and control of the outbreak should not wait for the result of the investigation, which may take time to reach a conclusion. The general measures based on the initial findings on the basis of strong epidemiological evidence on the possible source of infection, causative agent may be instituted as early as possible. An outbreak may be controlled by eliminating or reducing the source of infection, interrupting the transmission, and protecting the susceptible persons. Once the cause is confirmed, specific measures can be undertaken. The control measure may be aimed at any of the following:

a. *Controlling the source of infection*: If the food item is identified attempt must be made to stop its consumption. For example, discarding the suspected food, clean and disinfect food facilities, recalling food items, etc. (Centers for Disease Control and Prevention, 2015)

b. *Breaking the chain of transmission*: This may involve prompt diagnosis and treatment of cases (e.g., cholera) health education (hand washing, how to make the food safer, protect the food from flies and cockroaches, or to avoid it completely), improvements in environmental (chlorinated water) and personal hygiene. The common agents involved in foodborne outbreaks are listed in Table 3.5.

Table 3.6 Incubation Period and Clinical Features of Commonly Encountered bacteria Responsible for Food Borne Outbreaks

Organism	Incubation Period (in Hours)	Diarrhoea	Vomiting	Fever	Cramps	Any Other Symptom	Food Item
Staphylococcus aureus	1–4	+	+++	+	+	–	Unrefrigerated meat, potato, creams, and milk products
Bacillus cereus (preformed toxin)	1–6	+	+++	–	–	–	Fried rice, meat
Bacillus cereus (diarroheal toxin)	6–24	+	+	–	++	–	Meat and gravies
Clostridium perfringens (toxin)	6–24	++	+/–	–	+	–	Time and temperature abused gravies and meat
Shigella spp	24–108	+	–	–	–	Bloody diarrhea	Faeco–oral transmission
Clostridium botulinum	12–48	+	+	–	–	Blurred vision, diplopia, dysphagia	Canned and tinned food
Escherichia coli enterotoxigenic (ETEC)	6–48	++	+/–	–	+	–	Food and water contamination
Salmonella typhimurium	6–48	+++	+	+	–	–	Meat, milk products

c. *Prevention of disease in susceptible population*: The first and an important step in an outbreak is to prevent the disease from spreading to other members of the community. High-risk groups are offered vaccination or chemoprophylaxis in many diseases.

REFERENCES

Centers for Disease Control and Prevention, 2015. Investigating Outbreaks. National Center for Emerging and Zoonotic Infectious Diseases (NCEZID): Division of Foodborne, Waterborne, and Environmental Diseases (DFWED), Atlanta.

Cuomo M.J. 2016. Steps of an outbreak investigation. Available from: http://www.phsource. us/PH/FBI/Steps%20of%20an%20Outbreak%20Investigation.htm

Division of Scientific Education and Professional Development CDC, 2012. Epidemic Disease Occurrence, Principles of Epidemiology in Public Health Practice: An Introduction to Applied Epidemiology and Biostatistics, third ed. Centre for Disease Control, Atlanta.

Gordis, L., 2014. The dynamics of disease transmission. In: Gordis, L. (Ed.), Epidemiology. fifth ed. Elsevier, Canada, pp. 34–36.

Park, K., 2011. Principles of epidemiology and epidemiologic measures. In: Park, K. (Ed.), Park's Textbook of Preventive and Social Medicine. twenty first ed. M/S Banarsidas Bhanot, Jabalpur, India, pp. 118–121.

World Health Organisation, 2015a. Criteria for declaring the end of the Ebola outbreak in Guinea, Liberia or Sierra Leone. World Health Organisation, Geneva.

World Health Organisation, 2015b. Guiding Principles for International Outbreak Alert and Response, Emergencies Preparedness, Response. World Health Organisation, Geneva.

Surveillance of foodborne illnesses

4

A. Khera

Department of Community Medicine, Armed Forces Medical College, Pune, Maharashtra, India

4.1 DEFINITION

The Merriam–Webster Dictionary defines surveillance as "the act of carefully watching someone or something especially in order to prevent or detect a crime." In simplest words, surveillance is "information for action" (Merriam–Webster, 2015; WHO, 1998). Alexander Langmuir "the father of shoe leather epidemiology" was one of the first people to define public health surveillance as the ongoing, systematic collection, consolidation, and analysis of data and the dissemination of information (Langmuir, 1963). This definition was further expanded by Thacker and Stroup (1994) after almost 30 years to include "the application of these data to prevention and control." The WHO consolidated the above definitions to bring out the currently used definition in 2005, *"the systematic ongoing collection, collation, and analysis of data for public health purposes; and the timely dissemination of public health information for assessment and public health response as necessary"* (WHO, 2005).

The vital component of surveillance is the use of information for action (WHO, 1998). Food-based surveillance systems use these data to:

- mainly identify foodborne disease outbreaks;
- identify high-risk groups, place (areas), and time; and
- monitor and evaluate the public health program.

The overall goal of surveillance is to reduce the incidence and prevalence of disease by providing relevant, timely, and complete data and information to decision makers, professionals, and healthcare workers. The primary role of surveillance, in general, is to provide information on changing trends in types of foods associated with outbreaks and to offer insight into the effectiveness of prevention and control measures. It also provides data on identification of specific "agent–food" pairs repeatedly linked to outbreaks and illnesses. Timely information aids in implementation of prompt preventive measures to limit the outbreak. Over a period of time, human (food preparation and handling practices) and food factors can be identified and policy makers can direct implementation of either health education or legislative measures against them. Surveillance data provide inputs for measuring progress toward food-safety goals.

Food Safety in the 21st Century. http://dx.doi.org/10.1016/B978-0-12-801773-9.00004-2

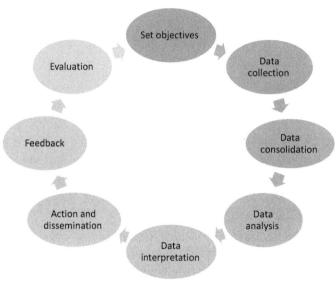

FIGURE 4.1 Surveillance Cycle

Under the foodborne disease surveillance, the first step is for health providers or public health professionals to identify the diseases to be put under surveillance; those having high burden, high morbidity, and mortality. Many a times, these identified diseases are mandated as notifiable diseases through a legislative process or by law. This ensures completeness of reporting by the healthcare providers. Many a times, additional diseases, which need to be put under surveillance, are identified by the states and added on to the diseases identified at the national level (Amato-Gauci and Ammon, 2008).

Graphically, the surveillance cycle can be depicted as shown in Fig. 4.1.

4.2 **SET OBJECTIVES**

The ultimate objective of any surveillance system is to provide information to guide policies and interventions. The objectives and actions needed to make successful policies and interventions determine the design and type of surveillance systems. For example, if the objective is to prevent the spread of epidemics of foodborne diseases, the surveillance-system managers need to identify common food preparation practices, types of foods related to a specific agent in a region, and intervene quickly to prevent or stop the spread of disease. Therefore, a surveillance system that provides rapid early warning information from basic surveillance units and laboratories is ideal. Another major objective of the surveillance system is to capture data on success or failure of the measures implemented for the prevention or control of the disease under surveillance (Nsubuga et al., 2006).

The objectives of foodborne illness surveillance can include any or many of the following points (Centers for Disease Control and Prevention, 2015a; WHO, 2012).

1. To determine the magnitude of the public health problem caused by foodborne diseases.
2. To monitor trends of foodborne diseases over time.
3. To identify outbreaks of foodborne disease early for timely remedial action.
4. To determine the extent to which food acts as a route of transmission for specific pathogens.
5. To identify high-risk foods, improper food production, processing, and handling practices.
6. To determine the risk factors and behaviors promoting illness in food handlers and cooks.
7. To determine the risk factors and behaviors promoting illness in vulnerable populations.
8. To provide information to enable the formulation of health policies regarding foodborne diseases.
9. To assess and evaluate the effectiveness of programs to improve food safety.
10. To generate hypotheses for research.

4.3 DATA COLLECTION AND CONSOLIDATION

Once the objective is identified for the foodborne disease-surveillance system, the next step is to plan data collection. The key elements under the data collection are:

1. case definition,
2. population under surveillance,
3. data providers and data sources,
4. data elements and data-collecting tools,
5. flow charts and data transmission,
6. human and financial resources, and
7. data security and confidentiality.

The data for any surveillance system can be obtained in two ways: passive and active. In the passive surveillance system, the data are collected as a routine by the designated nodes and forwarded to the next-level surveillance setup, hence the analogy *"the data comes to you."* The problem with the passive surveillance system is that data collected are simple and do not unnecessarily burden the data provider and hence are likely to be incomplete. Examples include laboratory reporting of foodborne diseases and healthcare-provider reporting of clustering of cases under the Integrated Disease Surveillance Project (IDSP) (IDSP, 2009).

In the active surveillance system, the data-collection setup is created to collect the data, hence the analogy *"you go toward the data."* The data collected for active surveillance require inputs of manpower, money, and material. With these inputs, the data quality is superior to the passive surveillance. An example of active surveillance for foodborne diseases is the FoodNet (Centers for Disease Control and Prevention, 2015b), which is an "active" surveillance system, in which the public health officials regularly contact laboratories to find new cases of foodborne diseases. In addition, it is also intended to monitor events that occur along the foodborne diseases pyramid. FoodNet's active surveillance focuses on persons who have a diarrheal illness since, most foodborne infections cause diarrheal illnesses.

Another aspect for the surveillance system is to provide case definitions for the diseases under surveillance. The case definitions are the "heart" of the surveillance system. They should include clinical, biologic, laboratory, and epidemiologic (time–place–person) factors. The combination of the above factors classifies the case definition into possible/suspect, probable, or confirmed cases.

Various methods of data transmission exist, which range from postal services to email and even uploading the data directly on a web server. The advent of faster modes of data transmission has improved the data compilation for surveillance. The frequency of the data transmission should depend on the rapidity of the response required for the surveillance system. Frequency used under IDSP is a weekly reporting by district- and state-surveillance units (DSU and SSU, respectively) to the central surveillance unit (CSU) (IDSP, 2009). A daily reporting may lead to unnecessary burden on the data providers and a monthly report may be too late to respond and break the chain of transmission in a common source continuous exposure or an intermittent exposure outbreak of foodborne illnesses. The reporting units should also ensure completeness and timeliness of the data. In the "zero reporting" surveillance all units reporting data to the upper level surveillance unit have to report even if no case occurs in their zone of responsibility. This ensures no loss of data occurs in transit.

The data-collection tools and the elements in it should be long enough to completely collect the relevant data and brief enough to be filled up and entered quickly. Ideally, the data-collecting tool should be pilot tested and, only then, should be put in the field. Another important aspect is to identify elements as part of the initial feedback from the periphery, which might be unique for the zone, that is, choices provided for the source of water may be different in different regions and may not appear in the initial elements.

The requirement of trained manpower is essential for quality data collection. Regular training on the data-collecting forms and mechanisms ensures quality reporting by the surveillance system. Many surveillance systems encrypt the data during transmission to avoid loss of patient confidentiality. The reports should also ensure that they do not identify individuals as it can have a negative effect on the data collected in the future.

4.4 DATA ANALYSIS AND INTERPRETATION

The data at the CSU are entered in computer-based surveillance programs; collated, validated, and obvious data entry errors are identified. The data are validated for the entries and any outliers are identified and revalidated. Ongoing analysis of surveillance data is important for detecting changes in patterns of diseases, monitoring the trends, and evaluating the effectiveness of control programs. The data analyzed are also used to determine the most appropriate and efficient allocation of health resources and personnel. It is also important to find the denominators for the data if they are planned to be presented as rates or risks. Usually the data in such situations are also presented as person and place distributions for identifying any geographical or age clustering.

The presentation of the report by the CSU depends upon the objectives set for the surveillance system. It is important that the data are disseminated timely and effectively. Communicating data for action is vital for an effective surveillance system. The data need to be translated into a form that the intended audience, which includes stakeholders, other professionals, public, media, and decision makers, can easily understand. The IDSP, for example, presents the data as descriptive analyses as clustering of cases in a district (IDSP, 2009).

4.5 EVALUATION (SILVERMAN ET AL., 2009; GERMAN ET AL., 2001; THACKER AND STROUP, 1994; WHO, 2013)

The evaluation of the surveillance system (Fig. 4.2) includes routine assessment of the individual components (in form of an internal assessment) with the goal to manage the system, to ensure the quality of the information produced, and to help in midcourse corrections during the surveillance. Evaluation of surveillance systems, per se, is a more formal process, which aims to ensure and improve the quality and efficiency of the system. It also helps in identifying useful variables in the surveillance data without overburdening the data collection. The quality, efficiency and effectiveness factors, optimal use of resources, and design/protocol adjustments should be an integral part of the evaluation.

Logical framework approach (LFA) can be used as an indicator for evaluation of surveillance systems. There are five types of indicators under the LFA (Fig. 4.3).

1. *Input indicators* are the resources, which are needed to implement the surveillance system. These include trained personnel, financial inputs, guidelines, identified communication facilities, data-collection forms for surveillance, surveillance aids (such as, computers), and any other logistics.

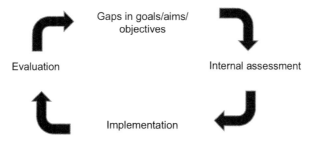

FIGURE 4.2 Evaluation is a Continuous Process

FIGURE 4.3 Evaluation Using LFA

2. *Process indicators* include core surveillance functions, such as, training, supervision, development of guidelines and tools, etc.
3. *Output indicators* measure the immediate results of the activities. These include reports from surveillance data, the number of times feedback given to the data providers, number of trained health staff, etc.
4. *Outcome indicators* measure the extent to which the surveillance objectives are achieved, usefulness of the system, completeness of reporting, use of surveillance data for policy and program decisions, and appropriateness of outbreak response.
5. *Impact indicators* measure the extent to which the overall objectives of the system are being achieved, which include changes in morbidity patterns, behavior changes in the health staff in implementing the system, and changes in health-related behaviors of the target population.

4.6 DISSEMINATION (ORENSTEIN AND BERNIER, 1990; WHO, 1998, 2012)

The most important task for a surveillance system is to timely disseminate the conclusions derived based on the data received. This may hold true even for an acute onset foodborne illness, even though by the time the data are even collected the outbreak might be over. Tracing the implicated food may help prevent future outbreaks. Similarly, an outbreak related to packaged food, the dissemination of data is of utmost importance. The contaminated food item may be recalled and destroyed.

4.7 INTEGRATED DISEASE SURVEILLANCE PROJECT (IDSP, 2009)

The IDSP, a multidisease passive surveillance network, was launched in 2004 with the assistance of the World Bank in India, to quickly detect and respond to disease outbreaks. Since, Mar. 2012, IDSP has been continuing under the National Rural

Table 4.1 Reported Outbreaks by the IDSP, India (2012–2014)

Year	Number of Outbreaks	Number of Foodborne Outbreaks	Percentage
2012	1584	600 (approx)	37.8
2013	1964	750 (approx)	38.2
2014	1562	525 (approx.)	33.6

Health Mission. Since May 2013, the IDSP has been brought under the National Health Mission. Over the past 3 years, approximately 35% of the total outbreaks were related to foodborne diseases (Table 4.1).

The diseases identified for surveillance include malaria, cholera, typhoid, tuberculosis, polio, plague, meningoencephalitis, and unusual clinical syndromes causing death and hospitalization and state-specific diseases. The CSU is established and integrated in the National Centre for Disease Control (NCDC), Delhi. The surveillance units in India under the IDSP are the SSU at the state level and DSU at the district level. The states/districts have their own State/District Surveillance Teams and Rapid Response Teams (RRT), which are responsible for the outbreak investigations. For data entry, training, video conferencing, and outbreak discussion, all the 776 sites in the States/District headquarters and premier institutes have been connected with an IT network with the help of the National Informatics Centre (NIC) and EDUSAT connectivity by the Indian Space Research Organization (ISRO).

Weekly disease-surveillance data on epidemic prone diseases are being collected under the IDSP from the reporting units, such as, subcenters, primary health centers, community health centers, government and private sector hospitals, and medical colleges. The data on "S" syndromic; "P" probable; and "L" laboratory formats are collected using standard case definitions. The data, which are collected from the subcenters, are analyzed weekly by the SSU/DSU for disease trends. The role of the RRT is to investigate and control the outbreaks. In order to improve the surveillance and fill in the gaps in reporting from remote locations, a 24×7 call center was established in Feb. 2008 to receive disease alerts on a toll-free telephone number (1075). This information on alerts is shared with the SSU/DSU for investigation and response.

Laboratories from medical colleges and district have been identified and strengthened for diagnosis of epidemic prone diseases. In addition, a network of 12 laboratories has been developed for influenza surveillance in the country. The "L" forms are filled by these laboratories and submitted to the SSU/DSU for collation and confirmation of diagnosis. Manpower to improve the quality of data collection, analysis, and response include health professionals in the field of epidemiology, microbiology, and entomology at district and state levels. The Ministry of Health and Family Welfare (MOHFW) under the National Health Mission (NHM) has approved the recruitment and training of these health professionals. Under the IDSP, one epidemiologist, each at state/district headquarters, one microbiologist and entomologist each at the state headquarters are being placed.

4.8 SURVEILLANCE SYSTEMS IN THE UNITED STATES OF AMERICA (CENTERS FOR DISEASE CONTROL AND PREVENTION, 2015A,B)

4.8.1 FOODNET

FoodNet, an active surveillance system, was established in 1996 as a multipronged foodborne surveillance program that includes 10 participating state-health departments in the United States; working in collaboration with the Center for Disease Control (CDC) and Food and Drug Administration (FDA) (Fig. 4.4). The FoodNet surveillance area covers 15% of the US population. The investigators under the FoodNet regularly conduct population-based surveillance and contact laboratories to enhance reporting of the selected seven bacterial and two parasitic infections transmitted commonly through food. FoodNet provides data for the incidence and trends in foodborne and diarrheal diseases.

Over the years, the data generated by the FoodNet surveillance have identified trends and evaluated the effectiveness of control strategies. The components of FoodNet include the following:

1. *Active Surveillance:* Under active surveillance, the data are collected from each identified site of the actual number of laboratory-confirmed cases of illness caused by the seven-targeted bacteria (*Salmonella, Shigella, Campylobacter, Escherichia coli O157, Yersinia, Listeria*, and *Vibrio*) and two targeted parasites (*Cryptosporidium* and *Cyclospora*).
2. *Laboratory Survey:* The role of the laboratory surveillance is to identify the proportion of the samples received, which have been subjected to culture, including those that did and did not yield a pathogen. The collected data are analyzed by the CDC.
3. *Physician Survey:* The physician survey is conducted to determine whether or not physicians who see patients for symptoms of diarrheal diseases are referred for laboratory analysis.
4. *Population Survey:* The population (both adults and children) in the area under FoodNet survey forms the sampling frame for the survey. The primary aim for the population survey is to determine population behavior regarding food consumption and preparation practices.
   On identification of a case, the FoodNet personnel at each site collect information about core variables, enter the information into a database, and transmit it to the CDC. The data, which are collected, include the following:
   a. hospitalizations occurring within 7 days of the specimen collection date;
   b. the patient's status (alive or dead) at hospital discharge (or at 7 days after the specimen collection date if the patient is not hospitalized);
   c. travel history of the patient abroad in the 7 days before onset of illness; and
   d. selected food and environmental exposures for select pathogens.
5. *Special Surveillance projects:* FoodNet also conducts special surveillance projects for new emerging and reemerging diseases, for example, FoodNet

conducted population-based surveillance for reactive arthritis associated with *Campylobacter, Salmonella, Shigella, Yersinia,* and Shiga toxin-producing *E. coli* (STEC) infections in 2002. In 2009, it conducted a pilot surveillance program for community-acquired *Clostridium difficile.*

4.8.2 OTHER SURVEILLANCE NETWORKS SPECIFIC TO FOODBORNE ILLNESSES IN THE UNITED STATES (FIG. 4.4)

National Antimicrobial Resistance Monitoring System—enteric bacteria (NARMS): The NARMS, which was established in 1996, is responsible for conducting surveillance for antimicrobial resistance among foodborne bacteria in humans, retail meat, and animals. The role of NARMS is to detect, respond, and prevent antimicrobial resistance among foodborne bacteria.

National Electronic Norovirus Outbreak Network (CaliciNet): Noroviruses, which belong to Caliciviridae, are responsible for the large number of foodborne outbreaks in the United States. In order to increase the quality of national norovirus surveillance, the CDC developed an electronic norovirus outbreak-surveillance network in 2009 with the help of public health laboratories of the states. Norovirus sequences are used to compare and rapidly link norovirus outbreaks with food items and to identify emerging norovirus strains.

National Molecular Subtyping Network for Foodborne Disease Surveillance (PulseNet): PulseNet is a national network of laboratories, which are coordinated by the CDC and the Association of Public Health Laboratories (APHL). PulseNet uses methods to perform pulsed-field gel electrophoresis (PFGE) on foodborne bacterial agents. Laboratories upload PFGE patterns to an electronic database and compare

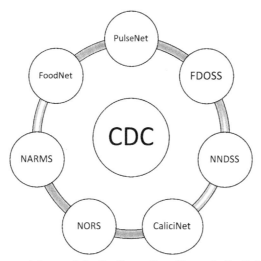

FIGURE 4.4 Multipronged Approach for Foodborne Surveillance in the United States

them with patterns of pathogens isolated from humans, animals, and foods with an aim to identify possible linkages between pathogens. PulseNet has revolutionized the detection and investigation of foodborne disease outbreaks, especially those occurring at multiple sites across the country.

National Notifiable Diseases Surveillance System (NNDSS): Healthcare providers and microbiologists/pathologists are required by law to report cases of selected diseases or clinical syndromes identified during clinical examination or patient specimens. This information is then transmitted to the NNDSS. This notifiable disease surveillance is "passive." Foodborne diseases reported include botulism, hemolytic uremic syndrome, listeriosis, salmonellosis, STEC infections, etc.

National Outbreak Reporting System (NORS): CDC collects reports of foodborne outbreaks from the NORS on enteric bacterial, viral, parasitic, and chemical agents. The NORS-surveillance team conducts analyses of these data to improve the understanding of the impact of foodborne outbreaks on human health and the pathogens, foods, settings, and contributing factors involved in these outbreaks.

Foodborne Disease Outbreak Surveillance System (FDOSS): Data on outbreaks caused by bacterial, viral, parasitic, and chemical agents are captured by the FDOSS since 1973. It also conducts analyses of these data to improve understanding of the health impact of foodborne outbreaks, the pathogens, foods, settings, and contributing factors associated with an outbreak.

4.9 CONCLUSIONS

Foodborne diseases are the most common diseases reported by any surveillance system. A multipronged approach with an intersectorial coordination involving microbiologists, public health specialists, food and agricultural organizations, geographical information system, epidemiologists, and the policy makers will go a long way in preventing and controlling foodborne illnesses.

REFERENCES

Amato-Gauci, A., Ammon, A.P.A.O., 2008. Surveillance of communicable diseases in the European Union: a long-term strategy (2008–2013). Euro. Surveill. 13 (26), 717–727.

Centers for Disease Control and Prevention, 2015a. Foodborne illness surveillance, response, and data systems. National Center for Emerging and Zoonotic Infectious Diseases (NCE-ZID), Division of Foodborne, Waterborne, and Environmental Diseases (DFWED), USA.

Centers for Disease Control and Prevention, 2015b. FoodNet surveillance. National Center for Emerging and Zoonotic Infectious Diseases (NCEZID), Division of Foodborne, Waterborne, and Environmental Diseases (DFWED), USA.

German, R.R., Lee, L.M., Horan, J.M., Milstein, R., Pertowski, C., Waller, M., 2001. Updated guidelines for evaluating public health surveillance systems. MMWR Recomm. Rep. 50 (1–35).

Integrated Disease Surveillance Project (IDSP), 2009. Restructuring Mission Government of India. New Delhi, India.

Langmuir, A.D., 1963. The surveillance of communicable diseases of national importance. N. Engl. J. Med. 268, 182–192.

Merriam–Webster, 2015. Surveillance, An Encyclopedia Britannica Company.

Nsubuga, P., White, M.E., Thacker, S.B., Anderson, M.A., Blount, S.B., Broome, C.V., Chiller, T.M., Espitia, V., Imtiaz, R., Sosin, D., Stroup, D.F., 2006. Public health surveillance: a tool for targeting and monitoring interventions. Disease Control Priorities in Developing Countries, second ed. World Bank, Washington (DC), USA.

Orenstein, W.A., Bernier, R.H., 1990. Surveillance. Information for action. Pediatr. Clin. North Am. 37, 709–734.

Silverman, B., Mai, C., Boulet, S., O'Leary, L., 2009. Logic models for planning and evaluation, Centers for Disease Control and Prevention (CDC), United States.

Thacker, S.B., Stroup, D.F., 1994. Future directions for comprehensive public health surveillance and health information systems in the United States. Am. J. Epidemiol. 140, 383–397.

World Health Organization, 1998. Information for action: developing a computer-based information system for the surveillance of EPI and other diseases. Geneva.

World Health Organization, 2005. World Health Assembly Revision of the International Health Regulations, WHA58.3. Geneva, 23 May, 2005.

World Health Organization, 2012. Manual for Integrated Foodborne Disease Surveillance in the WHO African Region. WHO Regional Office for Africa, Brazzaville, Republic of Congo.

World Health Organization, 2013. Evaluating a national surveillance system. Geneva, November, 2013.

Role of risk analysis and risk communication in food safety management

D.P. Attrey

Central Military Veterinary Laboratory, Meerut, Uttar Pradesh, India; High Altitude Research, Defence Research and Development Organisation, Leh, Jammu and Kashmir, India; Amity Institute of Pharmacy, Amity University, Noida, Uttar Pradesh, India; Innovation and Research Food Technology, Amity University, Noida, Uttar Pradesh, India; Amity Institute of Seabuckthorn Research, Amity University, Noida, Uttar Pradesh, India; Lala Lajpat Rai University of Veterinary and Animal Sciences, Hisar, Haryana, India

5.1 INTRODUCTION

Getting safe, nutritious, and balanced food is essential for a healthy life. The 1996 declaration of World Food Summit at Rome had reaffirmed the right of everyone to have access to safe and nutritious food, consistent with the right to adequate food, and fundamental right of everyone to be free from hunger (FAO, 1996).

After independence in 1947, India created the basic document for effective governance in 1950, that is, Constitution of India, which provided every Indian citizen a fundamental "right to food." "Food Security Act" was implemented in September 2013, which entitled almost two-third of 1.2 billion Indian citizens, the right to sufficient, nutritious and quality food at affordable price (highly subsidized through public distribution system). In addition, Supreme Court of India in 2013, also directed that food given to Indian citizens should be safe. Food Safety and Standards Authority of India (FSSAI) has recently started working on how to implement Food Law in India.

WHO gives high priority to food safety, quality and consumer protection programs and food safety is an essential public health function. WHO aims to develop sustainable, integrated food-safety systems for reduction of health risks from the entire food chain. Food safety, in fact, is an essential activity and an integral part of any public health programme (WHO, 2000). Food-borne diseases (FBD) result in substantial burden on healthcare systems, trade and tourism, market access; and reduce economic productivity and threaten livelihood. But FBDs generally go undetected, mostly due to lack of communication among the human, veterinary, agriculture, and food sectors.

As per WHO (2015) global human movements are increasing as also the international food trade, resulting in new risks to food safety since more and more people want to eat out to try new varieties of foods. Such new trends are also modifying the patterns of food production, distribution and consumption, integrating agricultural and food industries. Although international food trade offers many benefits to consumers, by bringing wider variety of foods to the market which are accessible, affordable, and meet consumer demands and provide opportunities for food-exporting countries to earn foreign exchange. The global value of food trade now exceeds US$500 billion. However, globalization of food trade has also brought new challenges and risks to food control and regulations and even a single source of contamination may become widespread, with global consequences. Antimicrobial resistance to food-borne pathogens; emergence of new pathogens, chemical contaminants, new technologies in food production and processing, like genetic engineering and nanotechnology, changing animal food production and animal husbandry practices, are impacting consumer safety (FAO/WHO-Provisional Edition, 2005).

On the World Health Day (Apr. 7, 2015), WHO gave a slogan "From farm to plate, make food safe." According to Director of WHO's Department of Food Safety and Zoonoses, it often takes a crisis for the collective consciousness on food safety to be stirred and any serious response to be taken. Impacts on public health and economies can be great. Almost one third (30%) of all deaths from FBDs, caused by 31 agents—bacteria, viruses, parasites, toxins, and chemicals, are in children under 5 years, despite the fact that they make up only 9% of the global population. The report, states that each year as many as 600 million, or almost 1 in 10 people in the world, fall ill after consuming contaminated food. Of these, 420,000 people die, including 125,000 children under the age of 5 years. Until now, such estimates of FBDs were vague and imprecise (WHO, 2015). As mentioned by Director-General of WHO "Knowing which food-borne pathogens are causing biggest problems in which parts of world can generate targeted public health action by the governments, food industry, and public."

WHO had established Global Food-borne Infections Network (GFN) to strengthen national and regional integrated surveillance, investigation, prevention, and control of food-borne and other enteric infections. The network promotes integrated, laboratory-based surveillance and fosters intersectoral collaboration and communication among microbiologists and epidemiologists in human health, veterinary and food-related disciplines (WHO, 2013). Early warning system, based on robust risk assessment to inform action, is very useful in addressing health threats and their timely communication. In response to threats such as the avian influenza virus H5N1 and severe acute respiratory syndrome (SARS), the Global Early Warning System (GLEWS) for Transboundary Animal Diseases, including Zoonoses, was jointly established by WHO, FAO, and OIE. GLEWS builds on existing internal systems of three participating organizations to confidentially track and verify events in order to improve harmonization and decrease duplication. It embodies a unique cross-sectoral and multidisciplinary partnership for early identification and assessment of health risks at the human–animal–ecosystem interface. Joint WHO-FAO International Food Safety Authorities Network (INFOSAN), is a voluntary global network of food safety authorities and provides an important platform for rapid exchange of information in case of food safety crises and for sharing data on both routine and emerging food safety issues. It tries to insure effective and rapid communication during food safety emergencies. To insure a comprehensive approach, GLEWS links with INFOSAN to insure and promote seamless action throughout food value chain within human–animal–ecosystem interface (WHO, 2013).

Enforcing food safety standards requires a food chain that is highly controlled and supplied with appropriate data on contaminants, hazards, and risk management strategies. Experts at FAO and WHO have observed that when incidence of FBD and number of disruptions to international trade in foodstuffs are increasing, it has never been more important for countries to implement an effective food safety system, guided by modern concept of risk analysis, to respond to current challenges (FAO/WHO-Provisional Edition, 2005), namely:

1. Traditional food safety systems are inadequate to cope with complex, persistent, pervasive, and evolving array of food safety issues existing today.
2. Modern food safety systems need to be science based to effectively cope with, and respond to wide range of food safety challenges presently confronting countries.
3. Science-based approaches are an essential part of risk analysis framework and crucial to creating a modern and effective food safety system.

Concept of "risk analysis" in food safety management needs to be understood properly to assess, manage, and communicate risks for proper implementation of food safety policies, food laws, and standards in a country. As such it is essential to understand the definitions of certain related terms. Some of the definitions related to risk analysis are enclosed as Appendix A.

The performance of an organization improves through the use of Quality Management Principles and adoption of a "process approach" besides emphasis on the role of concerned controlling authority. As per Attrey (2008), a Quality Management System (QMS) is directly linked to organizational processes. Process approach is a method of obtaining desired results, by managing activities, related resources as a process and is a key element of the QMS. It takes in to consideration the statutory and regulatory requirements and pays adequate attention to resource availability. Before discussing Food Quality, it is essential to understand certain important definitions about QMS, process, etc., which are also enclosed as Appendix A.

Food quality relates to quality characteristics of food that is acceptable to consumers. This includes external factors as appearance (size, shape, color, gloss, and consistency), texture, and flavor; factors such as federal grade standards (e.g., of eggs) and internal (chemical, physical, and microbial). Food safety is a scientific discipline describing handling, preparation, and storage of food in ways that prevent food-borne illness. This includes a number of routines that should be followed to avoid potentially severe health hazards (Wikipedia, 2015).

Food Quality Analysis is conducted for ascertaining content of nutritional, mineral, volatile and semivolatile compounds, and additives' components, etc. Food Safety Analysis is conducted to analyze potential hazards like pesticide residues, heavy metals, veterinary drug residues, fertilizers, growing aids, nonpermitted food additives, mycotoxins and other naturally occurring food toxicants, residues of persistent organic compounds—dioxins, PAH, PCB, etc., microbial contaminants, allergens, mycotoxins including aflatoxins, enzymes, and hormones, genetically modified content, pollutants, defective packaging and labeling, adulteration and tampering, extraneous matter (physical hazards), animal feed additives, acrylamide, etc.

New international trade agreements developed under World Trade Organization (WTO) have emphasized need for regulations governing international trade in foods to be based on scientific principles. Sanitary and Phytosanitary Agreement (SPS) permits countries to take legitimate measures to protect the life and health of consumers, animals, and plants, provided such measures can be justified scientifically and do not unnecessarily impede trade. SPS directs all countries to insure that their sanitary and phytosanitary measures are based on an assessment of the risk to human, animal, or plant life or health, taking into account risk assessment techniques developed by relevant international organizations and defines obligation of developed countries to assist less developed countries to improve their food safety systems (FSSAI, 2010).

Science-based food safety controls were controlling the food safety policies initially till early 1990s. As the awareness among the food scientists and concerned stakeholders grew, newer systems of achieving food safety were innovated during mid 1990s. Hazard Analysis Critical Control Point (HACCP), developed during mid 1990s, soon became the most popular method of achieving food safety. It is still being implemented by most organizations involved in food production across the world. However, during late 1990s, the Food Safety Approach for hygienic production of food was shifted from the routine "Food Safety Controls" to the "Risk-Based Controls." Although this was accepted by most developed countries as the future method of achieving food safety, it has still not been adopted by developed countries (McKenzie and Hathaway, 2006).

In this context, European Union and United States have taken important steps and enacted relevant legislations for adopting risk-based preventive controls, for example, FDA under its new Act, the "Food Safety Modernization Act" (FSMA), proposed a new rule, that is, "Current Good Manufacturing Practice & Hazard Analysis and Risk-Based Preventive Controls (HARPC)" (FDA, 2013). While Food Safety Controls are based on Current Good Manufacturing Practice (cGMP) and Hazard Analysis, "risk-based preventive controls" are based on "risk assessment."

According to Buchanan (2011), last two decades have seen emergence of risk analysis as foundation for developing food safety systems and policies. This period has witnessed a gradual shift from a "hazards-based approach" to food safety (i.e., mere presence of a hazard in a food was deemed unsafe) to a "risk-based approach" (determination whether exposure to a hazard has a meaningful impact on public health). Standard methods of analysis help in confirming that a food system is controlling a specific hazard. Through the selection of analytical method, sampling plan, and frequency of testing, a risk-based decision is introduced. Analysis of collected information is another challenge. Traditionally, food safety systems look at the individual steps along food chain, treating each step in isolation from others. However, as the systems become more complex, interactions and synergies between components become increasingly important and decrease utility of simple analyses of individual steps. When a high level of complexity is reached, impact of individual steps cannot be analyzed without considering the entire system.

5.2 **RISK ANALYSIS**

As per CAC (2013), risk analysis is a process which follows a structured approach comprising three distinct but closely linked components, namely, risk assessment, risk management, and risk communication, each component being integral to overall risk analysis and playing an essential and complementary role in risk analysis process. The overall objective of risk analysis applied to food safety is to insure human health protection and it should be based on all available scientific evidence, information on perceptions, costs, environmental, cultural factors, etc., which is gathered and analyzed according to scientific principles.

As such, it should always be open and transparent and must be documented according to principles of QMS (to preserve confidentiality and accessibility to all interested parties) in a transparent manner and should be applied within the framework for management of food related risks to human health. Effective communication and consultation with all interested parties should be insured throughout the risk analysis. Precaution is an inherent part of risk analysis. Many sources of uncertainty exist in process of risk assessment and risk management of food related hazards to human health. Degree of uncertainty and variability in available scientific information should be explicitly considered in risk analysis. Needs and situations of developing countries should be specifically identified and taken into account by the responsible bodies in different stages of risk analysis.

According to FSSAI (2010) the risk analysis principles apply equally to issues of national food control and food trade, should be applied in a nondiscriminatory manner and should be made an integral part of a national food safety system. Implementation of risk management decisions at the national level should be supported by an adequately functioning food control system/program. Interaction between risk managers and risk assessors is essential for practical application of risk analysis.

National government should also use guidance and information available with Codex Alimentarius Commission (CAC), Food and Agriculture Organization (FAO), World Health Organization (WHO), and other relevant international intergovernmental organizations, including World Organization for Animal Health [formerly Office International des Epizooties (OIE)] and International Plant Protection Convention (IPPC). Accordingly appropriate programs for training, information, and capacity building should be designed. In order to perform successful risk analysis, country needs to have a well functioning food safety system, the support and participation of key stakeholders (government, industry, academia, consumers), and basic knowledge about the three main components of risk analysis.

5.3 **RISK ANALYSIS PROCESS**

Risk Assessment is central scientific component of risk analysis (preferably performed by Veterinary Public Health) but risk management, defines the problem, articulates the goals of the risk analysis, and identifies the questions to be answered by the risk assessment (preferably performed by Public Health). Science-based tasks of measuring and describing nature of risk being analyzed (i.e., risk characterization) are performed during risk assessment. Risk management and risk assessment are performed within an open and transparent environment based on communication and dialog. Risk communication encompasses an interactive exchange of information and opinions among risk managers, risk assessors, the risk analysis team, consumers, and other stakeholders. Process often culminates with implementation and continuous monitoring of a course of action by risk managers (FAO/WHO-Provisional Edition, 2005).

Risk analysis is just one part of an effective food safety system. It will also be essential to develop and improve components of food safety systems including food safety policies, food legislation (encompassing food law, regulations and standards), food inspection, laboratory analysis, epidemiological surveillance of FBDs, monitoring systems for chemical and microbiological contamination in foods, and information, education, and communication. Thus use of a science-based approach shall enable governments to develop and implement a range of general improvements and interventions tailored to specific high-risk areas, which will ultimately improve food safety and reduce the burden of FBD.

As a concept, a science-based approach to food safety is not completely new. It is related to processes such as good agricultural practices, good hygienic practices, good manufacturing practices, and hazard analysis and critical control point (HACCP) system, as part of food safety management system (FSMS), which are already used in many countries. Scientific assessment of chemicals in general has also a rather long "tradition." What is new is use of risk analysis as a framework to view and respond to food safety problems in a systematic, structured and scientific way in order to enhance the quality of decision-making throughout food chain (FAO/WHO-Provisional Edition, 2005).

An effective FSMS helps in developing an effective "risk analysis framework" to collect and analyze the best available scientific information on a hazard that presents a risk to people, animals, or plants and consider the information along with other important nonscientific information, about a chemical, biological or physical hazard, possibly associated with food in order to select the best option to manage that risk based on various alternatives identified (FAO/WHO-Provisional Edition, 2005).

Risk analysis process begins with risk management defining problem, articulating goals of risk analysis and defining questions to be answered by risk assessment. Next step is to develop a risk profile, that is, science-based tasks of "measuring" and "describing" nature of risk being analyzed (i.e., risk characterization) are performed during risk assessment. Risk management and assessment are performed in an open and transparent environment based on communication and dialog. The process often culminates with the decision for implementation and continuous monitoring of a course of action by risk managers (FAO/WHO-Provisional Edition, 2005). But the process does not end with the decision itself. Members of the risk analysis team regularly monitor the success and impact of their decision. Modifications are made as required—on the basis of new data or information or changes in the context of the problem—to achieve further reductions in adverse human health effects (FSSAI, 2010).

5.4 RISK ASSESSMENT

Risk assessment is a process consisting of (1) hazard identification, (2) hazard characterization, (3) exposure assessment, and (4) risk characterization. Risk Assessment Policy should be laid down as per the FSMS and documented as per the requirements, maintaining the scientific integrity of the process. A risk profile describes food safety problem, which identifies, characterizes and assesses exposure to hazard and its effect as potential risk.

Hazard identification is the identification of biological, chemical, and physical agents capable of causing adverse health effects and which may be present in a particular food or group of foods. Specific identification of the hazard(s) of concern is a key step in risk assessment and to begin the process of estimation of risks specifically due to that hazard(s). Hazard identification may have already been carried out to a sufficient level during risk profiling; this generally is the case for risks due to chemical hazards. For microbial hazards, the risk profile may have identified specific risk factors associated with different strains of pathogens, and subsequent risk assessment may focus on particular subtypes. Risk managers are the primary arbiters of such decisions (CAC, 2013).

Hazard characterization is qualitative and/or quantitative evaluation of nature of adverse health effects associated with biological, chemical, and physical agents, which may be present in food. For chemical agents, a dose–response assessment should be performed. For biological or physical agents, a dose–response assessment should be performed if the data are obtainable. During hazard characterization, risk assessors describe the nature and extent of the adverse health effects known to be associated with specific hazard. A dose–response relationship is established between different levels of exposure to hazard in food at the point of consumption and likelihood of different adverse health effects. Dose–response assessment is the determination of relationship between magnitude of exposure (dose) to a chemical, biological, or physical agent and severity and/or frequency of associated adverse health effects (response). Types of data that can be used to establish dose–response relationships include animal toxicity studies, clinical human exposure studies, and epidemiological data from investigations of illness (CAC, 2013).

Exposure assessment is qualitative and/or quantitative evaluation of likely intake of biological, chemical, and physical agents via food as well as exposures from other sources if relevant. It characterizes amount of hazard that is consumed by various members of the exposed population(s). Analysis makes use of levels of hazard in raw materials, in food ingredients added to primary food and in general food environment to track changes in levels throughout food production chain. These data are combined with food consumption patterns of the target consumer population to assess exposure to hazard over a particular period of time in foods as actually consumed (CAC, 2013).

Risk characterization is where outputs from the previous three steps are integrated to generate an estimate of risk. Estimates can take a number of forms and uncertainty and variability must also be described if possible. A risk characterization often includes narrative on other aspects of the risk assessment, such as comparative rankings with risks from other foods, impacts on risk of various—what if scenarios, and further scientific work needed to reduce gaps. Risk characterization for chronic exposure to chemical hazards does not typically include estimates of the likelihood and severity of adverse health effects associated with different levels of exposure. A notional zero risk approach is generally taken and where possible the goal is to limit exposure to levels judged unlikely to have any adverse effects at all (CAC, 2013).

5.5 RISK ASSESSMENT POLICY

According to CAC (2013), determination of risk assessment policy should be included as a specific component of risk management. Risk assessment policy should be established as per the principles of FSMS by risk managers in advance of risk assessment, in consultation with risk assessors and all other interested parties. This procedure aims at ensuring that risk assessment is systematic, complete, unbiased, and transparent. The mandate given by risk managers to risk assessors should be as clear as possible. Where necessary, risk managers should ask risk assessors to evaluate the potential changes in risk resulting from different risk management options in accordance with risk assessment policy.

5.5.1 SAFETY ASSESSMENT AS RISK ASSESSMENT

The well-known no-observed-adverse-effect-level/safety factor-uncertainty factor (NOAEL/SF-UF) process has long been used to understand and regulate exposure to any potentially toxic substance. In controlled exposures, substance has no apparent or observable adverse health effect. Application of SF-UF(s), which typically composed of multiples of 10, produces a level of exposure that may lead to development of a regulatory standard such as an Acceptable Daily Intake (ADI), a Provisional Tolerable Weekly Intake (PTWI), Reference Dose (RfD), or Minimal Risk Level (MRL). Although it was first introduced by US Food and Drug Administration for the purpose of regulating food additives, NOAEL/SF-UF procedure is now widely used for other potentially toxic substances also. A key feature of NOAEL/SF- UF procedure is that at no point does it yield a quantitative prediction of harm. NOAEL/ SF-UF procedure is intended to establish safety. In a legal sense, procedure often defines what the word "safe" means for the potentially toxic substance (Carrington and Bolger, 2000).

Uncertainty remains an important part of safety assessment. It is usually understood that magnitude of uncertainty increases with degree of uncertainty, since NAOEL/SF-UF procedure is designed to establish "certainty," that a substance is safe (e.g., a food additive). However, in a safety assessment there is no attempt to state either how great the uncertainty is or precisely what the impact of the uncertainty on risk management decisions.

Carrington and Bolger (2000) have observed that ADI concept is flawed because in practice, the ADI is viewed as an "acceptable" level of exposure, and, by inference, any exposure greater than ADI is seen as "unacceptable." ADI was the basis for a regulation on food additives. It was used to calculate how much of the additive could be added to food, with acceptance of the agency as a matter of policy. In order to deal with this "problem," the ADI was renamed as "RfD."

5.6 **RISK MANAGEMENT**

Risk management is the process, distinct from risk assessment, of weighing policy alternatives, in consultation with all interested parties, considering risk assessment and other factors relevant for health protection of consumers and for promotion of fair trade practices and for selecting appropriate prevention and control options, that is, monitoring measures, etc. The management of food-related risks therefore involves balancing of the recommendations formulated by experts and resources of all types that social and commercial groups and manufacturers can set aside for dealing with these risks (CAC, 2013).

Risk management should follow a structured approach including preliminary risk management activities, evaluation of risk management options, monitoring, and review of decision taken. Decisions should be based on risk assessment, and taking into account, where appropriate, other legitimate factors relevant for health protection of consumers and for promotion of fair practices in food trade. Risk assessment should be presented before making final proposals or decisions on available risk management options. Relevant production, storage, and practices used throughout the food chain (including traditional practices, methods of analysis, sampling and inspection, feasibility of enforcement and compliance, and prevalence of specific adverse health effects) must be a part of the risk assessment process.

Risk management process should be transparent, consistent, and fully documented. Preliminary risk management activities include: identification of a food safety problem; establishment of a risk profile; ranking of the hazard for risk assessment and risk management priority; establishment of risk assessment policy for conduct of risk assessment; commissioning of risk assessment; and consideration of result of risk assessment.

There should be a functional separation of risk assessment and risk management, in order to insure the scientific integrity of the risk assessment, to avoid confusion over the functions to be performed by risk assessors and risk managers and to reduce any conflict of interest. However, it is recognized that risk analysis is an iterative process, and interaction between risk managers and risk assessors is essential for practical application.

5.7 GENERAL PRINCIPLES OF FOOD SAFETY RISK MANAGEMENT

1. Protection of human health should be primary objective in risk management decisions.
2. Risk management should follow a structured approach.
3. Risk management decisions and practices should be transparent, consistent, and fully documented.
4. Risk management should take into account whole food chain.
5. Risk management should insure scientific integrity of risk assessment process by maintaining the functional separation of risk management and risk assessment.
6. Risk managers should take account of risks resulting from regional differences in hazards in the food chain and regional differences in available risk management options.
7. Risk management should include clear, interactive communication with consumers and other interested parties in all aspects of the process.
8. Risk management should be a continuing process that takes into account all newly generated data in the evaluation and review of risk management decisions (CAC, 2013).

5.8 RISK MANAGEMENT FRAMEWORK

A generic risk management framework for food safety risk management must be functional in both strategic, long-term situations (e.g., development of international and national standards when sufficient time is available) and in the short-term work of national food safety authorities (e.g., responding rapidly to a disease outbreak). In all cases, it is necessary to strive to obtain the best scientific information available. There are four components of risk management framework:

1. Preliminary risk management activities which comprise initial process, including establishment of a risk profile to facilitate consideration of issue

within a particular context, and provide as much information as possible to guide further action. As a result of this process, risk manager may commission a risk assessment as an independent scientific process to inform decision-making.

2. Evaluation of risk management options which involves weighing of available options for managing a food safety issue in the light of scientific information on risks and other factors, and may include reaching a decision on an appropriate level of consumer protection. Optimization of food control measures in terms of their efficiency, effectiveness, technological feasibility and practicality at selected points throughout the food chain is an important goal. A cost-benefit analysis could be performed at this stage.

3. Implementation of risk management decision will usually involve regulatory food safety measures like using HACCP. Flexibility in choice of individual measures applied by industry is a desirable element, as long as overall program can be objectively shown to achieve stated goals. Ongoing verification of application of food safety measures is essential.

4. Monitoring and review is gathering and analyzing of data so as to give an overview of food safety and consumer health. Monitoring of contaminants in food and FBD surveillance should identify new food safety problems as they emerge. Where there is evidence that required public health goals are not being achieved, redesigning of food safety measures will be needed (CAC, 2013).

5.8.1 PRELIMINARY RISK MANAGEMENT ACTIVITIES

- identify food safety issues
- develop risk profile
- establish goals of risk management
- decide on need for risk assessment
- establish risk assessment policy
- commission risk assessment, if necessary
- consider results of risk assessment
- rank risks, if necessary
- identification and selection of risk management options
- identify possible options
- evaluate options
- select preferred option (s)
- rank risks, if necessary

5.8.2 IMPLEMENTATION OF RISK MANAGEMENT DECISIONS

- validate control (s) where necessary
- implement selected control (s)
- verify implementation
- rank risks, if necessary

Monitoring and Review

- monitor outcome of control(s)
- review control (s) where indicated (FAO/WHO-Provisional Edition, 2005)

5.9 RISK COMMUNICATION

Risk communication is interactive exchange of information and opinions throughout risk analysis process concerning risk, risk-related factors and risk perceptions, among risk assessors, risk managers, consumers, industry, the academic community and other interested parties, including explanation of risk assessment findings and basis of risk management decisions.

Risk communication should:

1. promote awareness and understanding of specific issues under consideration during the risk analysis;
2. promote consistency and transparency in formulating risk management options/recommendations;
3. provide a sound basis for understanding risk management decisions proposed;
4. improve overall effectiveness and efficiency of risk analysis;
5. strengthen the working relationships among participants;
6. foster public understanding of the process, so as to enhance trust and confidence in safety of food supply;
7. promote appropriate involvement of all interested parties; and
8. exchange information in relation to concerns of interested parties about risks associated with food.

Risk analysis should include clear, interactive, and documented communication, among risk assessors and risk managers. In fact risk communication should be more than dissemination of information. Its major function should be to insure that all information and opinion required for effective risk management is incorporated into decision-making process.

Risk communication involving interested parties should include a transparent explanation of the risk assessment policy and of assessment of risk, including uncertainty. Need for specific standards or related texts and procedures followed to determine them, including how uncertainty was dealt with, should also be clearly explained. It should indicate any constraints, uncertainties, assumptions and their impact on the risk analysis, and minority opinions that had been expressed in the course of the risk assessment (FAO/WHO-Provisional Edition, 2005).

5.10 CONCLUSIONS

Role of risk analysis and risk communication in food safety management has been discussed and all components of risk analysis need to be adopted for effective implementation of Food Control Policies. Role of various stakeholders in performing their respective tasks like those of Public Health and Veterinary Public Health, etc. must be clearly identified and assigned.

APPENDIX A

Some relevant definitions according to Indian FSS Act 2006 are

- "Food" means any substance, whether processed, partially processed or unprocessed, which is intended for human consumption and includes primary food as defined in the Act.
- "Food safety" means assurance that food is acceptable for human consumption according to its intended use.
- "Food Safety Management System" means the adoption of Good Manufacturing Practices, Good Hygienic Practices, Hazard Analysis and Critical Control Point and such other practices as may be specified by regulation, for the food business.
- "Primary food" means an article of food, being a produce of agriculture or horticulture or animal husbandry and dairying or aquaculture in its natural form, resulting from the growing, raising, cultivation, picking, harvesting, collection, or catching in the hands of a person other than a farmer or fisherman.
- "Risk" in relation to any article of food, means the probability of an adverse effect on the health of consumers of such food and the severity of that effect, consequential to a food hazard.
- "Risk analysis" in relation to any article of food, means a process consisting of three components, that is, risk assessment, risk management, and risk communication.
- "Risk assessment" means a scientifically based process consisting of the following steps: (1) hazard identification, (2) hazard characterization, (3) exposure assessment, and (4) risk characterization.
- "Risk communication" means the interactive exchange of information and opinions throughout the risk analysis process concerning risks, risk-related factors and risk perceptions, among risk assessors, risk managers, consumers, industry, the academic community and other interested parties, including the explanation of risk assessment findings and the basis of risk management decisions.
- "Risk management" means the process, distinct from risk assessment, of evaluating policy alternatives, in consultation with all interested parties considering risk assessment and other factors relevant for the protection of health of consumers and for the promotion of fair trade practices, and, if needed, selecting appropriate prevention and control options.
- "Unsafe food" means an article of food whose nature, substance, or quality is so affected as to render it injurious to health.

CAC (2013) has provided the authentic definitions related to Food Safety as under:

Hazard: a biological, chemical, or physical agent in, or condition of, food with the potential to cause an adverse health effect.

Risk: a function of the probability of an adverse health effect and the severity of that effect, consequential to a hazard(s) in food.

Risk analysis: a process consisting of three components: risk assessment, risk management, and risk communication.

Risk assessment: a scientifically based process consisting of the following steps: (1) hazard identification, (2) hazard characterization, (3) exposure assessment, and (4) risk characterization.

Risk management: the process, distinct from risk assessment, of weighing policy alternatives, in consultation with all interested parties, considering risk assessment and other factors relevant for the health protection of consumers and for the promotion of fair trade practices, and, if needed, selecting appropriate prevention and control options.

Interested parties: are risk assessors, risk managers, consumers, industry, the academic community and, as appropriate, other relevant parties and their representative organizations.

Risk communication: the interactive exchange of information and opinions throughout the risk analysis process concerning risk, risk-related factors, and risk perceptions, among risk assessors, risk managers, consumers, industry, the academic community and other interested parties, including the explanation of risk assessment findings and the basis of risk management decisions.

Risk assessment policy: documented guidelines on the choice of options and associated judgments for their application at appropriate decision points in the risk assessment such that the scientific integrity of the process is maintained.

Risk profile: the description of the food safety problem and its context.

Risk characterization: the qualitative and/or quantitative estimation, including attendant uncertainties, of the probability of occurrence and severity of known or potential adverse health effects in a given population based on hazard identification, hazard characterization, and exposure assessment.

Risk estimate: the quantitative estimation of risk resulting from risk characterization.

Hazard identification: the identification of biological, chemical, and physical agents capable of causing adverse health effects and which may be present in a particular food or group of foods.

Hazard characterization: the qualitative and/or quantitative evaluation of the nature of the adverse health effects associated with biological, chemical, and physical agents, which may be present in food. For chemical agents, a dose–response assessment should be performed. For biological or physical agents, a dose–response assessment should be performed if the data are obtainable.

Dose–response assessment: the determination of the relationship between the magnitude of exposure (dose) to a chemical, biological, or physical agent and the severity and/or frequency of associated adverse health effects (response).

Exposure assessment: the qualitative and/or quantitative evaluation of the likely intake of biological, chemical, and physical agents via food as well as exposures from other sources if relevant.

Food safety objective (FSO): the maximum frequency and/or concentration of a hazard in a food at the time of consumption that provides or contributes to the appropriate level of protection (ALOP).

Performance criterion (PC): the effect in frequency and/or concentration of a hazard in a food that must be achieved by the application of one or more control measures to provide or contribute to a PO or an FSO.

Performance objective (PO): the maximum frequency and/or concentration of a hazard in a food at a specified step in the food chain before the time of consumption that provides or contributes to an FSO or ALOP, as applicable.

Attrey (2008) has identified the definitions from QMS resources as under:

- *Quality*: is the totality of features and characteristics of a product showing its ability to meet stated or implied needs or degree to which a set of inherent characteristics fulfils customer requirements.
- *Quality management*: are the coordinated activities to direct and control an organization with regard to quality.
- *System*: is a set of interrelated and interfacing elements. Quality System is the organizational structure, responsibilities, procedures, processes, and resources for implementing quality management.
- *Management system*: management system refers to what the organization does to manage its process or activities. It is a system to establish a Quality Policy and achieve quality objectives in an organization.
- *Quality management system (QMS)*: it is a management system, which directs and controls an organization with regard to quality. QMS has a set of requirements that deal with each aspect of the activities of the organization that affect quality.
- *Quality requirements*: ISO family of standards represents an international consensus on good management practices with the aim of ensuring that the organization can time and again

deliver the product or services that meet the client's quality requirements. These good practices are combined in a set of standardized requirements for QMS (e.g., ISO 22000, the Food Safety Management System (FSMS), which has two parts, namely, "System Requirements" and "Product Quality Requirements").

- *Quality assurance (QA)*: all activities needed to provide adequate confidence that a product or service will fulfill the requirements for quality. All activities associated with the attainment of quality, constitute Quality Assurance. It is that part of quality management which focuses on providing confidence that quality requirements will be fulfilled.
- *Quality control (QC)*: actions and systems to measure and regulate the "quality," constitute Quality Control.
- *Implementation*: means putting all the intentions (Quality Policy and Objectives) into practice.
- *Process*: consists of a series of actions, which produce a change or a set of interrelated or interacting activities which transforms inputs into outputs. A series of operations or steps that results in a product or service or a set of causes and conditions that work together to transform inputs into an output.
- *Product*: in relation to the QMS, a product is either a tangible product obtained from raw materials through a series of standardized processes or it can be an intangible process/technology, a service or an information or scientific data, etc.

REFERENCES

Attrey, D.P., 2008. A Mini Text for the beginners. ISO 9001: 2000, Quality Management System. Attrey's Mini Text Series on Quality Management Systems and Good Quality Practices under a project on "Design, development and establishment of suitable quality management system for INMAS for subsequent certification in ISO 9001:2000 & NABL Accreditation by a certifying body"; Submitted to Institute of Nuclear Medicine and Allied Sciences (INMAS), (DRDO) as Faculty, Institute of Defence Scientist and Technologists; CEFEES, Defence Research and Development Organization (DRDO), Ministry of Defense, Government of India, Brig. S. K. Mazumdar Road, Timarpur, Delhi.

Buchanan, R.L., 2011. Understanding and Managing Food Safety Risks. In: Food Safety Magazine; December 2010/January 2011. Available from: http://www.foodsafetymagazine.com/magazine-archive1/december-2010january-2011/understanding-and-managing-food-safety-risks/

CAC, 2013. Procedural Manual—Codex Alimentarius Commission; ISSN 1020-8070. Codex Alimentarius; Joint FAO/WHO Food Standards Programme; twenty first ed. Issued by the Secretariat of the Joint FAO/WHO Food Standards Programme, FAO, Rome. World Health Organization; Food and Agriculture Organization of the United Nations; Rome.

Carrington, C., Bolger, M., 2000. In Sometimes More Is Less. United States Department of Agriculture Office of Risk Assessment and Cost-Benefit Analysis Safety Assessment and Risk Assessment. ORACBA News vol. 4 (4), 5 (2) Spring 2000 Fall 1999. Available from: http://permanent.access.gpo.gov/lps38806/2000/news52.pdf

FAO, 1996. Rome Declaration on World Food Security and World Food Summit Plan of Action World Food Summit—13 to 17 Nov' 1996, Rome, Italy. FAO Corporate Document Repository; Produced by: Deputy Director general (Operations). Available from: http://www.fao.org/docrep/003/w3613e/w3613e00.HTM

FAO/WHO-Provisional Edition, 2005. Food Safety Risk Analysis—Part I—An Overview and Framework Manual—Provisional Edition. http://www.fao.org/es/ESN/index_en.stm. Available from: https://www.fsc.go.jp/sonota/foodsafety_riskanalysis.pdf

FDA, 2013. FDA Food Safety Modernization Act". Under the provision of the Federal Food, Drug, and Cosmetic Act (21 U.S.C. 301 et seq.). U.S. Department of Health and Human Services, U.S. Food and Drug Administration (FDA). Available from: http://www.fda.gov/Food/GuidanceRegulation/FSMA/ucm247548.htm and http://www.fda.gov/Food/GuidanceRegulation/FSMA/ucm359436.htm

FSSAI, 2010. Training Manual for Food Safety Regulators—vol. 5. Key Aspects to ensure Food safety; 2010. The Training Manual for Food Safety Regulators who are involved in implementing Food Safety and Standards Act 2006 across the Country. Available from: http://www.fssai.gov.in/Portals/0/Training_Manual/Volume%20V-%20Key%20Aspects%20to%20ensure%20Food%20Safety.pdf

McKenzie, A.I., Hathaway, S.C., 2006. The role and functionality of Veterinary Services in food safety throughout the food chain, New Zealand Food Safety Authority, P.O. Box 2835, Wellington, New Zealand. Rev. Sci. Tech. Off. Int. Epiz., 2006, 25 (2), 837–848. Available from: http://citeseerx.ist.psu.edu/viewdoc/download?doi=10.1.1.118.746&rep=rep1&type=pdf

WHO, 2000. WHO Global Strategy for Food Security. Safer food for better health: Food Safety Programme—2002, World Health Organization. http://www.who.int/fsf; Eighth plenary meeting, 20 May 2000 – Committee A, second report. ISBN 92 4 154574 7 (NLM Classification: WA 695). Available from: http://apps.who.int/iris/bitstream/10665/42559/1/9241545747.pdf

WHO, 2013. Strategic Plan for Food Safety, Including Food-borne Zoonoses 2013–2022; Advancing Food Safety Initiatives. ISBN 978 92 4 150628 1 (NLM classification: WA 701). Available from: http://www.searo.who.int/entity/foodsafety/global-strategies.pdf

WHO, 2015. WHO estimates of the Global Burden of Food-borne Diseases (FBDs) and Risk Assessment in Food Safety; Available from: http://www.who.int/foodsafety/risk-analysis/riskassessment/en/; and http://apps.who.int/iris/bitstream/10665/200046/1/WHO_FOS_15.02_eng.pdf?ua=1

Wikipedia, 2015. Available from: https://en.wikipedia.org/wiki/Food_quality and https://en.wikipedia.org/wiki/Food_safety

Food safety issues in contemporary society

An ayurvedic perspective on food safety

J. Kumar

Centre for Public Health, Panjab University, Chandigarh, India

Ayurveda is one of the great gifts of the sages of ancient India to mankind. Ayurveda indicates the science by which life understood in its totality. Thousands of years ago, ayurveda pointed out the importance of preventive over curative approach. Ayurveda, regarded as a holistic manual of life and age, describes a lifestyle that is in harmony with nature. The ayurvedic description of health is:

> *"Samadosha, Samadhatu Samagnischa, malkriyah,*
> **(समदोषः समधातु समाग्निश्चा मलक्रियः)**
> *Prasannatmendriyamanah, Swastha Ityabhidhiyate"*
> **(प्रसन्नात्मेंद्रियामनः स्वस्थ इत्याभिधियते)"**

that is, only he, whose *dosas* (*vata, pitta, kapha*) *dhatus* [physical components— *rasa (plasma), rakta* (blood), *mansa* (flesh), *meda* (fat), *asthi* (bones), *majja* (bone marrow), *shukra* (semen), and *agni* (digestive fire) is balanced, appetite is good, all tissues of the body and all natural urges are functioning properly, and whose mind, body, and spirit (self) are cheerful or full of bliss, is a perfectly healthy person.

Ayurveda considers the individual as whole and seeks to re-establish harmony between all the constituents of the body and a perfect balance of the tripod—mind, body, and spirit. Basically ayurveda is health promotive, preventive, curative, and nutritive—all self-contained.

The two principle objectives of ayurveda are

1. *Swasthasya Swasthya Rakshanam* (स्वस्थस्य स्वास्थ्यरक्षणं)': To prolong life and promote perfect health (add years to life and life to years).
2. *Aturasya Vikar Prashamanamcha* (अतुरस्य विकार प्रशमनाम्चा)': To completely eradicate the disease and dysfunction of the body.

Ayurveda, the age-old science of life, has always emphasized to maintain health and prevent diseases by following proper diet and lifestyle regimen rather than treatment and cure of the diseases.

Ayurveda is grounded in a metaphysics of the *panchmahabhut* ("five great elements"; earth, water, fire, air, and ether)—all of which forms the universe, including the human body. Ayurveda stresses a balance of three humors or energies: *vata* (wind/air), *pitta* (bile), and *kapha* (phlegm). According to ayurveda, these three regulatory principles—*dosas* (*tridosa*)—are important for health, because when they are in balanced state, the body is healthy, and when imbalanced, the body has diseases.

Ayurveda gives elaborate guidelines for achieving perfect health and remaining healthy through *dinacharya* (daily routine) and *ritucharya* (seasonal regimens). Comprehensive instructions are given on specific food/dietary schedules (for different times of the day, different seasons, according to one's age and most importantly, to suit one's individual constitution (*prakriti*).

Ayurveda has views on how individuals should be involved in their own health and health care. The essentials of health care, namely, health education, personal hygiene and habits, exercise, dietary practices, food, sanitation, environmental sanitation, code of conduct and self-discipline, civic and spiritual values, treatment of minor ailments and injuries, etc. are emphasized and advocated in ayurveda.

In the treatment of diseases, observance of a diet regime is a very essential part of treatment. Ayurveda has given stress on *ahara* (food) for good health. Diet has been scientifically and extensively linked to disease.

According to ayurveda food is medicine and medicine is food. Eating correctly is the most important aspect of ayurvedic lifestyle in both the short term and the long term. What is so-called "correct" or "compatible" depends on the individual and as the saying goes *"one man's meat is another man's poison."* Different foods suit different people. Thus, the health of a community depends upon the adequate availability of the food and intelligent consumption of food.

Charaka Samhita, Sushruta Samhita, and *Ashtanga Hridaya* are considered as the major triad of ayurveda. In all these classical texts of ayurveda, concepts of food along with its safety issues are described in detailed. In this chapter we have elaborated the food safety concept as discussed in *Charaka Samhita.*

According to *Charaka Samhita,* among three *upastambhas* (supporting factors) of life, the *ahara* (food), *nidra* (sleep) and *brahmacharya,* the diet is an essential factor for maintenance of healthy life. Being supported by these three well-regulated factors of life, the body is endowed with strength, complexion and growth and continues for life.

Charaka had mentioned that, *ahara* (food) is the best sustainer of life. *Charaka* says that it is the *Ahara* (food) which maintains the equilibrium of bodily *dhatus* and helps in promotion of health and prevention of diseases. He classified the food articles in different ways.

Charaka Samhita explained intensely about health, hygiene, diet, lifestyle, and medicine. According to this *Samhita,* the objects of the Ayurvedic science are twofold, namely, the treatment of patients suffering from the disease and maintenance of positive health. Of all the factors for the maintenance of positive health, food taken in proper quantity occupies the most important position.

Collection of plants products: According to *Charaka,* drugs and foods related with plants are required to be collected keeping in view the appropriate habitat, appropriate season, and their effective attributes. He mentioned that these should be collected in the appropriate season when they attained maturity in respect of their size, taste, potency, and smell.

Proper storage: The collected plant products should be kept in appropriate containers well covered with a lid, and hung on a swing. The store room should have doors facing toward the east or the north and be free from hazards of fire, water, moisture, smoke, dust, mice, and quadrupeds.

Quantity of food: The quantity of food to be taken again depends upon the individual power of digestion and metabolism. The amount of food which, without disturbing the equilibrium of *dhatus* and *dosas* of the body, gets digested and metabolized is considered adequate.

Food safety according to taste of food: The concept of taste is very important in context to food safety according to ayurvedic perception. If food ingredients with different tastes are not in prescribed ratio, the equilibrium of *dhatus* and *dosas* gets disturbed. When employed properly, *rasas* maintained the body and their incorrect utilization results in the vitiation of *dosas.*

Charaka has mentioned the six *rasas:* sweet, astringent, bitter, sour, salty and pungent. Just like the *dosas,* each taste is composed of two elements. For instance, the sweet taste is composed of earth and water, the sour taste of fire and water, the salty of earth and fire, pungent of fire and air. The qualities are then expanded. Each taste then has qualities and also actions it has on substance. Each taste has a specific effect on the human body and has its specific characteristics as mentioned in Table 6.1.

Utility of food: Charaka has given the description of the eight factors which determine the utility of various types of food. These include: *prakriti* (nature of food articles); *karana* (methods of their processing); *samyoga* (combination); *rasi* (quantity); *desa* (habitat); *kala* (time, i.e., stage of the disease or the stages of the individual); *upayogasamstha* (rules governing the intake of food), and *upayokirn* (wholesomeness to the individuals who takes it). These eight factors are associated specifically with useful and harmful effects and they are conditioned by one another.

Table 6.1 Composition of Tastes, their Qualities and Effects on *Dosas*

Taste	Element Composition	Qualities	Alleviates	Vitiates
Sweet	Earth + water	Heavy, slow, cold, and oily	*Vata, pitta*	*Kapha*
Sour	Fire + earth	Hot, liquid, light, oily	*Vata*	*Pitta, kapha*
Salty	Fire + water	Light, sharp, subtle, oily, and hot	*Vata*	*Pitta, kapha*
Pungent	Fire + air	Light, sharp, rough, hot, and subtle	*Kapha*	*Pitta, vata*
Bitter	Ether + air	Light, rough, cold	*Kapha, pitta*	*Vata*
Astringent	Earth + air	Heavy, rough, cold, penetrating	*Kapha, pitta*	*Vata*

Suitable and unsuitable foods according to Prakriti (body constitution): Ayurveda gives emphasis to *prakriti* or individual body constitution in consideration of food intake. Three basic *dosas—vata, pitta,* and *kapha* form seven types of *prakriti.* In this world each person is a unique being of unique heredity, environment, biochemical structure, and mental status. For this reason all the natural and good food items cannot be effective with all the individuals to the same extent. Every individual should take a diet suitable to his predominant constitutional *dosas,* to balance them in different seasons (Tables 6.2–6.4).

Charaka has described various food articles as most wholesome and unwholesome *ahara* by nature. These are given in Table 6.5.

Table 6.2 Diet for *Vata Prakriti*

	Foods to Favor (*Vata* Alleviating)	Foods to Avoid (*Vata* Vitiating)
Food	Foods of sweet, sour and salty tastes, warm and unctuous foods	Foods of pungent, bitter, and astringent tastes, light, dry, rough, cold foods
Fruits	Sweet and sour fruits, apricots, banana, cherries, figs, grapes, grape fruit, lemon, mango, orange, pineapple, papaya, pomegranate, peach, plum, raspberries, strawberries	Apple, watermelon
Vegetables	Cooked vegetables, brinjal, beets, cooked onion, garlic, green beans, lady finger, potato, radish, sweet-potato, tomato	Raw vegetables, cabbage, cauliflower, peas, carrot, cucumber, leafy greens, lettuce, mushrooms, raw onion, spinach
Cereals	Rice, wheat, oats (cooked)	Barley, corn, millet, rye, oats (dry)
Pulses	Mudga	Chick peas, kidney beans, lentils, peanuts, soybeans, split peas
Nuts and seed	Almond, cashew, coconut, walnut, pumpkin seeds, sunflower seeds, sesame seeds, mustard seeds	Large quantity of nuts and seeds
Dairy products	Milk, ghee, butter, cream, cheese, butter milk, yogurt	Ice cream
Animal foods	Chicken egg, fish, shellfish, turkey	Beef, lamb, pork
Oils	Sesame, mustard, coconut, peanut, olive, almond	Corn, soya, safflower
Condiments	Asafetida, black-pepper, basil, cumin, coriander, cinnamon, cardamom, cloves, fennel, fenugreek, garlic, ginger, mint, mustard, nutmeg, turmeric	None
Sweeteners	Raw sugar, honey, jaggery, molasses, fruit sugar	White sugar

Table 6.3 Diet for *Pitta Prakriti*

	Foods to Favor (*Pitta* Alleviating)	Foods to Avoid (*Pitta* Vitiating)
Food	Foods of sweet, bitter and astringent tastes, cool, slightly dry, and little heavy foods	Foods of sour salty and pungent tastes, hot, sharp, and light foods
Fruits	Sweet fruits, apple, dates, figs, grapes, mango, melons, orange, plums, pear, pineapple, pomegranate, raspberries	Sour fruits, apricot, banana, cherries, grape-fruit, lemon, papaya, peach, strawberries
Vegetables	Cabbage, cauliflower, potato cucumber, fresh peas, green beans, lady finger, lettuce, mushrooms,	Beets, brinjal, chilies, garlic, onion, radish, spinach, turnip, tomato
Cereals	Rice, wheat, barley, oat	Corn, millet, rye
Pulses	All beans, namely, kidney, soya, chick peas, split peas, mudga	Peanuts, lentils
Nuts and Seeds	Coconut, sunflower	Almond, cashew, peanut, pumpkin seeds, sesame, walnut
Dairy products	Milk ghee, cream, cheese (unsalted)	Cheese (salted), butter milk, yogurt, ice-cream
Animal foods	Egg- white, chicken, turkey	Egg- yolk, beef, fish, shell fish, lamb, pork
Oils	Coconut, soya, sunflower	Corn, mustard, olive, peanut, sesame, almond, safflower
Condiments	Coriander, cumin, cardamom, fennel, mint, turmeric	Asafetida, basil, black pepper, cinnamon, cloves, fenugreek, garlic, ginger, mustard, nutmeg.
Sweeteners	Raw sugar, fresh honey, jaggery, fruit sugar	White sugar, old honey, molasses

Apart from elemental constitution of food various dietary rules and other factors such as *matra* (quantity), *kala* (time or season), *kriya* (mode of preparation), *bhumi* (habitat or climate), *deha* (constitution of person), *desa* (body humor and environment), etc. also play a significant role in the acceptability of wholesome diet. *Charaka* had defined this concept in the form of antagonism foods. According to him, the substances which are contrary to "*deha-dhatus*" (the body tissues) behave with "*virodha*" (antagonism) to them (the tissues) and are defined as antagonistic/incompatible food.

Charaka has described eighteen factors responsible for food incompatibility: (1) *desa* (climate), (2) *kala* (season), (3) *agni* (digestive power), (4) *matra* (quantity), (5) *satmya* (accustom), (6) *dosas* (tridosha), (6) *samskara* (mode of processing), (7) *aharavirya* (potency of food), (8) *kostha* (bowel habits), (9) *avastha* (state of health), (10) *krama* (order of food intake), (11) *parihara* (restriction), (12) *parihara* (proscription), (13) *upachara* (prescription), (14) *paka* (cooking), (15) *sanyoga*

Table 6.4 Diet for *Kapha Prakriti*

	Foods to Favor (*Kapha* Alleviating)	Foods to Avoid (*Kapha* Vitiating)
Food	Foods of pungent, bitter and astringent tastes, warm, light, dry foods	Foods of sweet, sour and salty tastes, cold, heavy, unctuous foods
Fruits	Apple, pomegranate	Sweet and sour fruits, banana, berries, cherries, dates, grapes, grape fruit, lemon, mango, melons, orange, pineapple papaya, plum, pears
Vegetables	Cabbage, cauliflower chilies, fresh peas, green beans, garlic, lettuce, mushroom, onion, radish, turnip, spinach	Brinjal, cucumber, carrot, lady finger, sweet potato, tomato
Cereals	Barley, corn, millet, rye, dry oats	Rice, wheat, khus, cooked oats
Pulses	All beans, namely, kidney, soya, lentils, mudga, peanut split peas	Chick peas
Nuts and seeds	Sunflower, pumpkin	Almond, coconut, cashew, peanut, sesame, walnut
Dairy products	Goat milk, butter milk, soya milk	Milk, ghee, butter, cheese, cream, yogurt, ice-cream
Animal foods	Chicken, turkey	Egg, fish, beef, lamb, pork, shell fish
Oils	Corn, mustard, sunflower, safflower	Almond, olive, peanut, soya, sesame
Condiments	Asafetida, black pepper, basil, cloves, cumin, cardamom, cinnamon, coriander, fennel, fenugreek, garlic, ginger, mint, mustard nutmeg, turmeric	None
Sweeteners	Honey	White sugar, brown sugar, jaggery, molasses, fruit sugar

(combination), (16) *hridya* (palatability), (17) *sampad* (richness of quality), and (18) *vidhi* (rules of eating). Some of them are illustrated with examples in Table 6.6.

Antagonistic foods and associated diseases: According to *Charaka*, intake of antagonistic food is responsible for causation of sterility, blindness, *visarpa* (an obstinate skin disease), *kilasa* type of skin disease, leprosy, ascitis, eruptions, insanity, fistula, fainting, intoxication, tympanitis, spasmodic obstruction in throat, anemia, poisoning due to *ama*, sprue, edema, acid dyspepsia, fever, rhinitis, fetal diseases, and even death. According to him:

- One must not take milk along with fish especially with *cilicima* type of fish. If this fish is taken with milk, it produces *amavisa* (toxin due to improper digestion/metabolism), causing constipation.

Table 6.5 *Hitatama* (Wholesome) and *Ahitatama* (Unwholesome) *Ahara* (Food)

Group	Hitatama Ahara	Ahitatama Ahara
Cereals	*Red shali rice (Oryza sativa)*	*Yavaka* (a variety of *Hordeum vulgare*)
Pulses	*Mudga* (green gram)	*Masha* (black gram)
Green vegetables	*Jivanti* (Leptadenia reticulata)	*Sarshapa* (mustard)
Rhizomes	*Sringavera* (ginger)	*Aluka* (Potato)
Fruits	*Mridvika* (dry grapes)	*Nikucha (Artocarpus nikucha)*
Salts	*Saindhava* (rock salt)	*Usara*
Sugarcane products	*Sarkara*	*Phanita*
Drinking water	*Antriksha jala*	River water in rainy season
Milk	*Gavya* (of cow)	*Avika* (of sheep)
Ghrita	*Gavya* (of cow)	*Avika* (of sheep)
Vegetable fats	*Tila oil* (sesamum)	*Kusumbha oil (Canthamus tinctorius)*

- Meat of domestic, marshy, and aquatic animals should not be taken together with honey, sesamum seeds, sugar candy, milk, *masa* (*Phaseolus radiates* Linn.) radish, lotus stalk, or germinated grains. By doing so, one gets afflicted with deafness, blindness, trembling, loss of intelligence, loss of voice, and nasal voice; it may even cause death.
- One should not take vegetable of *puskara* (*Nelumbo nucifera* Gaertn.) and meat of *kapota* (dove) fried in mustard oil together with honey and milk. This obstructs channels of circulation and causes dilatation of blood vessels, epilepsy, *sankhaka* (a disease characterized by acute pain in temporal region), *galaganda* (scrofula), *rohini* (diphtheria) or even death.
- Milk should not be taken after the intake of radish, garlic, *krsnagandha* (*Moringa oleifera* Lam.), *arjaka* (*Ocimum gratissimum* Lirm.), *surasa* (*Ocimum sanctum* Linn.), etc.; this may cause obstinate skin diseases including leprosy.
- Diseases caused by the intake of unwholesome diets and drugs can be cured by emesis, purgation or administration of antidotes and by taking prophylactic measures.

Quality of food creating *ama* (toxin) are described by *Charaka* as: heavy, rough, cold, dry, disliked, distending, burning, unclean, antagonistic, taken untimely, and also while afflicted by emotions (passion, anger, greed, confusion, envy, bashfulness, grief, conceit, excitement, and fear).

Indicated and contraindicated foods: Descriptions of indicated and contraindicated foods are also given in *Charaka Samhita*. It is mentioned that one should not regularly take heavy articles such as *vallura* (dried meat), dry vegetables. One should never take meat of a diseased animal. Moreover, one should not regularly take *kurchika* (boiled buttermilk), pork, beef, meat of buffalo, fish, curd, etc. On the other

Table 6.6 Some Examples of Antagonistic Foods

Place	Intake of dry and sharp substance in deserts; unctuous and cold substance in marshy land.
Time	Intake of cold and dry substance in winter; pungent and hot substance in the summer.
Power of digestion	Intake of heavy food when the power of digestion is mild (*mandagni*); intake of light food when the power of digestion is sharp (*tiksagni*).
Dosage	Intake of honey and ghee in equal quantity.
Habit	Intake of sweet and cold substance by persons accustomed to pungent and hot substance.
Dosa	Utilization of drugs, diets and regimen having similar qualities with clops but at variance with the habit of the individual.
Mode of preparation	Diets which when prepared in a particular way produce poisonous effects, for example, meat of peacock roasted on a castor spit.
Potency	Substances having cold potency in combination with those of hot potency.
Bowel	Administration of a mild purgative in a small dose for a person of costive bowel and administration of strong purgatives in strong doses for a person having taxed bowel.
State of health	Intake of *vata* aggravating food by a person after exhaustion, sexual act and physical exercise or intake of *kappa* aggravating food by a person after sleep or drowsiness.
Order	If a person takes food before his bowel and urinary bladder are clear (empty) or when he does not have appetite or after his hunger has been aggravated.
Proscriptions and prescriptions	Intake of hot things after taking pork etc., and cold things after taking ghee.
Cooking	Preparation of food, etc., with bad or rotten fuel and undercooking, overcooking, or burning during the process of preparation.
Combination	Intake of sour substance with milk.
Palatability	Any substance which is not pleasant in taste.
Richness of quality	Intake of substances that are not matured, over matured or putrefied.
Rules for eating	Taking meals in public.

hand, regularly intake of *swastika* (a kind of rice harvested in 60 days), *sali, mudga,* rock salts, rain water, *amalaka,* ghee, meat of animal dwelling in arid climate and honey is recommended.

Quality of food with reference to their heaviness and lightness is elaborated by *Charaka*. If the food article is heavy, only three-fourth or half of the stomach capacity is to be filled. After taking food, one should never take such heavy articles like pastries, rice, *prthuka* (boiled and flattened rice).

Quality of food: Charaka had described the aspects of unwholesomeness of diets:

Vegetable: Vegetable infested with insects, exposed to wind, exposed to the sun for long, dried up, old and unseasonal are unwholesome. When they are cooked without adding fat and residual water after boiling is not filtered out, vegetable become unwholesome for use.

Fruits: Those which are old, unripe, afflicted by insects and serpents, putrefied, exposed to snow and sun for long time, growing in the land and season other than the normal habitat and time are unwholesome.

Corn and grains: Corn and grains, 1 year after their harvesting, are wholesome. Old corn and grains are mostly light for digestion whereas fresh corn and grains are heavy for digestion. Corn and grains which take a shorter time for cultivation are lighter than those taking longer time.

Animal food: Meats of animals which have died a natural death, emaciated or dried up after death, fat, cold, too young, killed by poisonous arrow, graze in a land not commensurate with their natural habitat or are bitten by snakes and tigers, etc. are unwholesome.

Water: Water is devoid of merits when it is excessively deranged in respect of smell, color, taste and touch, is too slimy, deserted by aquatic birds, and aquatic animals.

Concept of poisoned food: According to *Charaka*, the king is exposed to danger of being poisoned through food and regimens by the attendants secretly employed in his palace by enemy or his own wives.

Features of poison giver: Charaka had mentioned that poison giver is likely to be a person who might behave in an extremely suspicious manner, who is garrulous or who speaks very little, who has lost luster of his face or who exhibits changes in his characteristic features.

Examination of poisoned food: According to *Charaka*, suspected food should not be taken immediately, but a part of it should be thrown over fire. If the food is poisoned, then the flame exhibits abnormal characteristics like variegated color of peacock feather and makes a cracking noise; it moves spirally or it gets extinguished. The smoke is sharp, intolerable and smells of a dead body.

The poisoned food when kept in a pot gets discolored, and flies sitting on it succumb to death. When this poisoned food is seen by crows, their voice becomes feeble, and when the *chakor* bird sees it, its eyes become discolored.

Symptoms of poisoned person: According to *Charaka*, if the poison is added to drinks such as alcohol, then blue lines appear over its surface or it becomes discolored. A person's own shadow is not reflected through such drinks or the shadow is reflected in a distorted manner.

The smell of poisoned food and drinks causes headache, pain in the cardiac region, and fainting. It may cause edema and numbness in the hands, burning sensation and pinching pain in the fingers, and cracking of the nails.

It may cause tingling in the lips, swelling, numbness and discoloration of the tongue, tingling sensation in the teeth/gums, stiffness of the jaw bones, burning sensation in face, salivation and morbidity in the throat.

Further, the patient may suffer from discoloration, sweating, asthenia, nausea, impairment of the vision, arrest of cardiac functions and appearance of drop-like pimples all over the limbs. If the poisoned food and drinks enter the colon, then the patient suffers from fainting, intoxication, unconsciousness, burning sensation, weakness, drowsiness and emaciation. The patient suffers from anemia when the poisoned food and drinks get localized in the abdomen.

If the water of wells and ponds are poisoned, then the water becomes foul smelling, dirty, and discolored. Intake of this poisoned water causes edema, urticaria and pimples, and even death.

If the tooth brushing twig is poisoned, when the brush-like tip of it gets withered, and the patient suffers from edema of the teeth, lips and muscles of the mouth. If the oil for application over the head is poisoned, then the patient suffers from hair-fall, headache and tumors in the head.

Intake of poisoned food vitiates the *kostha* (gastrointestinal tract) and external application of poisoned material afflicts the skin in the beginning. If the poison has reached the stomach, then the physician in the beginning should administer emetic therapy. If the poisonous material is located in the skin, then ointments and fomentation therapy, etc., should be administered. These therapeutic measures should be administered, keeping in view the nature of *dosas* and the strength of the patient.

Seasonal diets: The sixth chapter of *Sutrasthana* of *Charaka Samhita* explains the seasonal regimen in detail. It is called as *Tasyashiteeya Adhyaya*. It literally means qualitative dietetics explained on the basis of seasons. Suitable diets for every season are described in *Charaka Samhita*.

Rules for taking food: Rules for taking food are pointed out in *Charaka Samhita*. According to this, healthy individuals as well as some of the patients should eat only that food in proper quantity which is hot, unctuous and not contradictory in potency and that too, after the digestion of previous food. Food should be taken in proper place equipped with all the accessories, without talking and laughing, with concentration of mind and paying due regards to oneself. Thus, the food taken in prescribed manners helps in bringing about the strength, complexion, happiness, and longevity.

Due to its great significance, the food has attained the highest position in this universe. Charaka has given much detail of the religious practices; those increase devotion towards diet and have good impact on the mind and body. Hence, before taking food, for the betterment of life, it should be worshipped.

Charaka had not only described the general food safety concerns, but he also illustrated it with individualized focus according to the nature and body type of person. The concept of taste and analogism of food in correlation to foods safety are extensively elaborated. Attention is required to these unexplored issues, so as to accommodate them in current food safety instruction both at community as well as individual level.

REFERENCES

Datta, N., Chowdhury, K., Badal, J., 2014. Diet in elderly: an ayurvedic perspective. Scholars J. Appl. Med. Sci., 660–663, Retrieved from: http://saspublisher.com/wp-content/uploads/2014/03/SJAMS-22B660-663.pdf.

Nathani, N., 2013. An appraisal of the concept of diet and dietetics in ayurveda. Asian Journal of Modern and Ayurvedic Medical Science 2 (1), Retrieved from: http://ajmams.com/viewpaper.aspx?pcode=ef5cc006-6583-4806-b293-7d170f36d956.

Sharma, R.K., Dash, B., 2011. Charaka Samhita, vols. 1–5, Chowkhamba Sanskrit Series Office, Varanasi, Uttar Pradesh, India.

Shukla, V., Tripathi, R.D. (Eds.), 2003. Carak Samhita of Agnivesa, vol. II, Chaukhamba Sanskrit Pratishthan, Delhi, India.

Thakkar, J., Chaudhari, S., Sarkar, P.K., 2011. Ritucharya: answer to the lifestyle disorders. Ayu 32 (4), 466–471, Retrieved from: http://www.ncbi.nlm.nih.gov/pmc/articles/PMC3361919/.

Stayavati G.V. (n.d.). Ayurvedic concepts of nutrition and dietary guidelines for promoting/preserving health and longevity. Retrieved from: http://nutritionfoundationofindia.res.in; http://nutritionfoundationofindia.res.in/FetchScriptpdf/festschrift%20%20for%20%20Dr%20Gopalan/Section%201-scientific%20papers/Satyavati%204.11.pdf

Food safety in modern society—changing trends of food production and consumption

7

P. Dudeja*, A. Singh**

*Department of Community Medicine, Armed Forces Medical College, Pune, Maharashtra, India; **School of Public Health, Post Graduate Institute of Medical Education and Research, Chandigarh, India

7.1 INTRODUCTION

Food is an integral part of every human culture. In the preindustrialization era majority of the Indians lived in small villages, working either in agriculture or as skilled craftsmen. Farming was the predominant occupation. Each meal at home was cooked fresh and served to the entire family. However, the people were living at the mercy of Mother Nature as they faced famines and poverty. With industrialization, migration from villages to urban areas began at a fast pace. Inadequate income (from farm sources), complemented by the attraction of better jobs, employment opportunities, and better facilities (communication, roads, education, and health) resulted in the migration of a large number of people from villages to cities. Urban society had a contrasting lifestyle as compared to its rural past. Migration led to a shearing of the traditional joint family fabric, metamorphosing into nuclear family. Gradually, the society was weaned of the virtues of village life. The expenses of living in a city were high and the family resources were extremely stretched. Men made less money for working longer hours in factories and as laborers. This resulted in women also working to meet both ends. Families were forced to do this, since they desperately needed money. The factory owners, all this while, were happy to employ women and children at lower wages. This unprecedented growth and profit cycle was another social change that occurred at the cost of the worker. The cost was paid in terms of mental peace.

With passage of time, women started working outside the home for longer hours. These working women, majority of whom are from the middle class, prefer to spend quality time with children rather than in the kitchen with routine household work. They do not have enough time to shop for groceries or to eat with family at home.

This factor, along with easy availability of processed and ready-to-eat (RTE) products, brought a significant change in the type of food items in kitchens and shelves of refrigerators. These double income group families found it affordable to go out and eat. They are also forced to eat at canteens/street food at place of work, as there is not much time in the mornings to cook and pack food.

Eating at home was ingrained in our traditional society and culture. For a nation that used to be fastidious about its home cooked and fresh food, the use of food services and eating outside the home has increased. There has been widespread mushrooming of malls with food courts, highways with "drive and dine" and "drive through" locations. Opening of eating establishments (EE) in the cities continued unabated. Eating out in the present society is perceived as a common way to socialize outside home with friends, a pastime, or an outing with the family. McDonalds have specifically targeted the children in our society by providing toys along with the meal. Contemporary generations of permissive parents are willing targets of such aggressive markets. The range of products in restaurants attracts all ages, sex, classes, and income groups. People now enjoy eating out because it is affordable, convenient, tasty, and above all heavily promoted.

The fast pace of life coupled with liberalization of our economy has also brought the concept of fastfood culture in our society. Fastfood is the term given to food that can be prepared and served quickly. While any meal with low preparation time can be considered to be fastfood, the term generally refers to food sold in a restaurant or store and served to customer in form of take away. The menu for fastfoods in Indian context varies from pizza, burgers, and chips to *samosas, vadapav, panipuri*, and *dahibahllas*. McDonaldization has changed the eating culture of our society. The major players of fastfood in India are McDonalds, Dominos, Pizza Hut, Subway, KFC, Barista, and Café Coffee Day etc. These dynamic changes have also transpired due to the fast growing economy, a shift from traditional to modern technologies, increase in international travel, evolving tastes, and increased demands for "fast" and processed foods throughout our country. In the present Indian society, food is not only a means of life but also a meaningful investment for business. These changes in the eating habits of Indians are more likely referring to middle class and upward earning society of India (Fig. 7.1).

7.2 CHANGING TRENDS IN FOOD PRODUCTION

The landmark achievements in the history of food production in India have been the green, white, and yellow revolution, which saw a tremendous increase in grain, milk, and oil production, respectively. As per the Food and Agriculture Organization (FAO) 2010 world agriculture statistics, India is the world's largest producer of many fresh fruits and vegetables, milk, major spices, staples, such as, millets and castor oil seed (Murugasamy and Veerachamy, 2012). It is the second largest producer of the world's major food staples wheat and rice. As of 2011, it is one of the world's five largest producers of livestock and poultry meat (Hawkes et al., 2012).

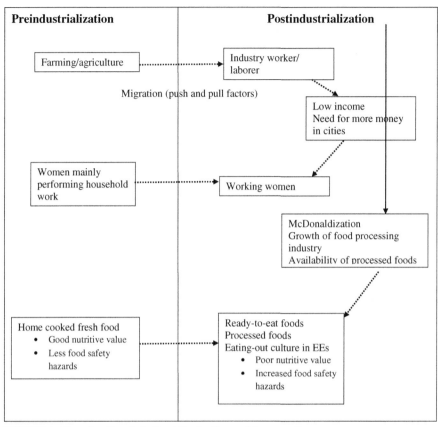

FIGURE 7.1 Changing Trends in Food Production and Consumption

In India, the first food-processing industry was established in 1942 with a large spectrum of industries, producing fruits and vegetables, bakery products, confectionary, marine products, and meat products. The food-processing industry formed a crucial link between farmers to final consumers in domestic, as well as, international markets. Food-processing combined with marketing could solve the basic problems of agricultural surplus, wastage, rural jobs, and better remuneration to the growers. India's food-processing technology has progressed much since postindependence and India is in a position to emerge as a world leader in food processing. The food industry in the current scenario is involved in mass production followed by marketing and distribution of food items for more profit, which was not present earlier.

Furthermore, the rising income of the middle class, with an increase in percentage of working women's demand for nonperishable and nonfood staples and processed foods, has risen. The profit-driven food retailers are ready to provide them with a year round availability of all fruits and vegetables through imports. The burgeoning processed-food industry introduces innovative new products to meet the demands of its consumers. Multinational companies that entered our economy balance the supply with the demand of processed foods. Data from the Organization of Economic Cooperation and Development, UN Food and Agriculture Organization (OECD-FAO) Agricultural Outlook 2014, show chicken consumption in India grew at an annual rate of 5.9% between 1992 and 2013. This makes India the fourth fastest-growing market for chicken while it is also the seventh for fish.

Another area, which is experiencing growth in food production, is the RTE market in India for food. RTE can be defined as food products that constitute complete meals; require minimal processing, if any, typically requiring reheating to desired temperature or addition of water. They are often termed as "convenience food" as they are positioned as "value for money" products that solve the issue of time constraints faced by consumers due to the pressures of urban life. RTEs are categorized into two product categories: "shelf-stable packaged food" and "frozen packaged."

7.3 CHANGING TRENDS IN FOOD CONSUMPTION

The contemporary Indian middle class has flouted the Indian food-eating practices by promoting eating out in bars and restaurants. Traditional Indian diets were of immense variety and the diversified preparations not only offered the whole range of nutrients but also maintained all physiologic functions. Until a few years ago, natural foods were preferred over refined foods, and light foods (less oily) over heavy foods (spicy, oily, and energy dense). Our traditional meals were mostly plant-based (as animal foods were expensive) with spices and were cooked and eaten fresh at home. They were a combination of cereals, millets, pulses, and spices, such as, pepper, cumin, asafetida, and coriander with curd and coconut, satisfying our energy and protein requirements. Those who could afford had milk, curd, eggs, and small amounts of animal meat added to the protein requirement. Vegetables and fruits contributed to the intake of vitamins, minerals, and antioxidants required for supportive functions. We used traditional oils from groundnut, sesame, mustard, and ghee, which are essential, in small quantities, to absorb fat-soluble vitamins and contribute to several hormonal functions. The foods were balanced, diversified, and freshly prepared but not stored.

However, the intake pattern has undergone a paradigm shift in the current society. The National Sample Survey Office report released in 2014 stated that the top 5% of urban India spends Rs 3000 per capita per month on groceries and eating out on average. This class consumes the least amount of whole grains and consumes and more of refined cereals in the form of noodles and bread (*pav, kulcha, burger, rolls,* buns, cakes, etc.). The consumption of milk, eggs, meat, and other processed foods has increased. On the other end of the spectrum, the bottom 5% of India spends just over Rs 400 per person per month on food and a quarter of this is on cereals. The National Sample Survey Office data over the years confirms some of these as long-standing trends. The share of cereals in the Indian household expenditure, for

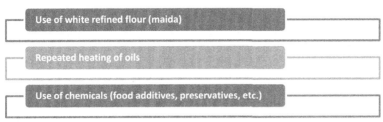

FIGURE 7.2 Hazards Due to Change in Nutritional Aspects of Food

instance, has decreased from 41% to 18% in rural India and from 23% to 10% in urban India between 1972–73 and 2004. Data from 2004 also shows that Indians were moving away from locally available seasonal vegetables toward high-value vegetable produce, such as, broccoli, cabbage, cauliflower, and capsicum. Needless to repeat, along with this that the consumption of processed foods has increased considerably, which lack nutritional quality. When food is processed, much of the nutritional value depreciates with cooking and also with the addition of sodium, sugar, and preservation with chemical additives. Most fastfoods have ingredients that contribute to obesity, heart disease, and overall poor health. The hazards of changes in the consumption pattern can be broadly classified into two groups:

1. hazards due to nutritional aspects of food (Fig. 7.2) and
2. hazards due to safety aspects of food.

Maida is actually a waste with low-nutritive value, which gets rejected while processing wheat. In EEs, oils are reused or kept hot for most of the day to supply freshly cooked food to the customers. However, polyunsaturated vegetable oils that have been reheated or used for long periods of time are hazardous as fatty acid-derived toxin 4-hydroxy-*trans*-2-nonenal accumulates in high amounts in them, which has been linked to the increased risk for a multitude of diseases, including strokes, Parkinson's, Alzheimer's, liver diseases, and cancers (Grootveld et al., 1998). Use of *food additives* prolong the shelf life of foods and make food more attractive, such as, colors and flavoring agents. The hidden use of mono sodium glutamate by the EEs has been on the rise as it enhances flavors in food. It virtually has no flavor of its own, but neurologically causes people to experience a more intense flavor from the foods that they eat containing the substance. This is used as a means to increase profit, a simple way to balance taste in a product line, mask unwanted tastes, and to make an otherwise unpalatable food acceptable. Trans fats are very common in RTE foods and can have devastating effects on the human body. High consumption of trans fats can lead to high cholesterol, which increases the risk of heart attack and other heart problems. Processed foods often have a high sodium content and can contribute to hypertension and heart disease (Miyoshi et al., 1997). The other class is artificial preservatives, which can cause allergies and cancer. Certain coloring agents impart colors to food or replace colors lost during food processing. Examples are, caramel, carotene, and food colorings (dyes), such as, red, orange, yellow, etc. Food dyes have been known to worsen attention deficit/hyperactive disorder (Boris and Mandel, 1994). High fructose corn syrup, used in place of sugar, is a common sweetener and preservative in prepared foods, such as, soda, juices, condiments, and breads. Excessive intake of high fructose corn syrup can lead to obesity, diabetes, high blood pressure, and heart disease (Ray et al., 2004; White, 2008).

7.4 HAZARDS DUE TO SAFETY ASPECTS OF FOOD

Changes in the consumption pattern of society have been identified as certain risk factors for foodborne illnesses, such as, improper holding temperatures, preparing food ahead of planned schedule, poor personal hygiene, inadequate cooking, inadequate cleaning and disinfecting of equipment, crosscontamination, use of left-over food, and contaminated raw material. There has been a lengthening of the food chain from farm-to-fork and contamination of food can occur at any stage of the food production. Poor standards of hygiene during food preparation and the lack of training in food safety by the food handlers have been the most important causes for the increase in foodborne illnesses. Mishandling of food and disregard of hygienic measures on the part of food handlers may enable pathogens to come into contact with food and, in some cases, to survive and multiply in sufficient numbers to cause illness in the consumer. Food safety has become an important element of consumer awareness these days. In case of food products, its quality depends not only on its nutritional value, but also on its safety for human consumption. With the increasing awareness and media coverage, the voice of the consumer has evolved in the country.

REFERENCES

Boris, M., Mandel, F.S., 1994. Foods and additives are common causes of the attention deficit hyperactive disorder in children. Ann. Allergy 72 (5), 462–467.

Grootveld, M., Atherton, M., Sheerin, A., Hawkes, J., Blake, D., Richens, T., et al., 1998. In vivo absorption, metabolism, and urinary excretion of alpha, beta-unsaturated aldehydes in experimental animals. Relevance to the development of cardiovascular diseases by the dietary ingestion of thermally stressed polyunsaturate-rich culinary oils. J. Clin. Invest. 101 (6), 1210.

Hawkes, C., Friel, S., Lobstein, T., Lang, T., 2012. Linking agricultural policies with obesity and noncommunicable diseases: a new perspective for a globalising world. Food Policy 37 (3), 343–353.

Miyoshi, A., Suzuki, H., Fujiwara, M., Masai, M., Iwasaki, T., 1997. Impairment of endothelial function in salt-sensitive hypertension in humans. Am. J. Hypertens. 10 (10), 1083–1090.

Murugasamy, M., Veerachamy, P., 2012. Resource use efficiency in agriculture—a critical survey of the literature. Language in India 12 (4).

Ray, G.A., Nielsen, S.J., Popkin, B.M., 2004. Consumption of high-fructose corn syrup in beverages may play a role in the epidemic of obesity. Am. J. Clin. Nutr. 79 (4), 537–543.

White, J.S., 2008. Straight talk about high-fructose corn syrup: what it is and what it ain't. Am. J. Clin. Nutr. 88 (6), 1716S–1721S.

Food contamination and adulteration

3

Food toxicology—past, present, and the future (the Indian perspective)

S.P. Singh*, S. Kaur**, D. Singh†

*Department of Forensic Medicine and Toxicology, Government Medical College and Hospital, Chandigarh, India; **Department of Biochemistry, Government Medical College and Hospital, Chandigarh, India; †Department of Forensic Medicine, Post Graduate Institute of Medical Education and Research, Chandigarh, India

8.1 INTRODUCTION

Food toxicology deals with the injurious effects of toxins present in the food on living beings. This subject is perhaps as old as the human civilization itself. The primitive man migrated to different places having diverse climatic conditions in search of food and shelter. In the pursuit of food, he must have eaten various foods of plant and animal origin. This experimentation with newer foods ultimately led to categorization of safe and unsafe foods. With passage of time, it became possible to shortlist a select group of plants and animals that were not only safe but also met most of his nutritional requirements as well. This experience gained over centuries led to a rich knowledge of the subject that still serves us in different aspects of life.

8.2 THE HISTORY OF FOOD TOXICOLOGY

8.2.1 GLOBAL PICTURE

The Vedas (c. 5000 BC) of India are among the most ancient medical records that contain valuable information on a number of poisons. *Shen Nung* (c. 2696 BC), who is often referred to as the Father of Chinese medicine was known for tasting 365 herbs to ascertain their qualities (Dikshith, 2008). He noted these findings in his treatise *On Herbal Medical Experiment Poisons*. He, however, died of a toxic overdose. A number of doctors over the next generations continued his work and formed basis of the vast Chinese knowledge of herbal remedies. In 994 BC, about 40,000 people died of gangrene that developed after consuming wheat/rye that was contaminated with ergot (*St. Anthony's Fire*). *Hippocrates* (c. 460–370 BC) referred to the relationship between emergence of diseases in humans and quality of air, water, food, the topography of land, and general habits. These references can still be found in his treatise *Air, Water and Places*. *Socrates* (469–399 BC), a famous Greek scholar, was accused of corrupting the minds of local youth with his philosophical ideas and therefore forced to drink poison hemlock, the active ingredient of which was coniine (Radenkova-Saeva, 2008). *Mithridates VI* (c. 131–63 BC) tested antidotes to various poisons (Dikshith, 2008). He used to test these on himself and prisoners. It is believed that he administered increasing sublethal doses of poisons to himself to develop immunity against poisons as he feared that poisons might be used to assassinate him. He produced a universal antidote known as *Antidotum Mithridaticum*. Ruler *Sulla* (81 BC) issued an edict known as *"Lex Cornelia"* which prohibited assassination by poisons. This edict became the first legislative enactment in history against poisons to be used as a tool of homicide (Goldflank and Flomenbaum, 2006). *Cleopatra* (c. 69–30 BC) experimented with various poisons including strychnine on prisoners and poor. She later committed suicide with the help of Asp bite (Radenkova-Saeva, 2008).

Pedanius Dioscorides wrote *De Materia Medica* (50–70 AD) in which he mentioned more than 600 plants and their medicinal uses. He also mentioned certain substances of animal and mineral origin (John O'Connell, 2015). *Commodus* (180–192 AD), a Roman emperor, is believed to have poisoned Motinenous, the Prefect of Praetorian Guard, with poisonous figs (Cilliers and Retief, 2000). *Moses Maimonides* (1135–1204 AD), a Jewish physician and philosopher talked about various poisons in his book, *Treatise on Poisons and Their Antidotes* (Rosner, 1968). In 1419 AD, an Italian group known as *Venetian "Council of Ten"* gained notoriety for killing people for a fee. The main ingredients of their poison recipe included mercuric chloride, arsenic trioxide, arsenic trisulfide, and arsenic trichloride with which they poisoned their targets. *Leonardo de Vinci* (1452–1519 AD) studied accumulation of poisons in animals. *Pope Clement VII* (1478–1534 AD) lost his life after consuming *Amanita phalloides* mushroom. *Paracelsus* (1493–1541 AD) stated "All substances are poisons; there is none, which is not a poison. The right dose differentiates a poison from a remedy." The queen of France, *Catherine Medici* (1519–89 AD) was a skilful assassin and used to test poisons on sick and the poor (Trestrail, 2007). In the 15th century, the members of Borgia clan (mainly Rodrigo and Cesare Borgia) of Italy were infamous for poisoning people. They used *"La Cantrella"* (a mixture of arsenic and phosphorus) to murder people for monetary and political gains. Between 15th and 17th century, special schools were run for students who wished to become professional poisoners. One such poisoner was a lady named Hieronyma Spara. In 1659 AD she formed a society of poisoners where she taught women how to kill their husbands by poisons. *Orfila* (1787–1853 AD) who is considered as the father of toxicology established toxicology as a distinct scientific discipline (Burcham, 2014). He attempted to build a systematic correlation between the biological and chemical knowledge of the poisons available at that time. He had gained this knowledge after observing the effects of poisons on many animals, especially dogs. Till 1800, most of poison-related homicides were reported among the wealthy or influential families of different countries of the world. After this era, as the life-insurance industry grew and human values changed over time, these incidences began to appear in circles of common man as well.

The modern times have also witnessed a lot of poison-related deaths. In most of these instances, people have been killed for property inheritance, insurance benefits, political assassinations, revenge, etc. A few cases have been cited as follows.

From 1892 to 1905, a serial killer named Johann Otto Hoch (USA) killed nearly 12 out of 24 women whom he married. He used to win the affection of new widows, marry them and eventually poison them with arsenic to inherit their properties. He then moved to another town to repeat his crime. However, the similarity of these deaths brought the attention of authorities and he was caught. Later he confessed to his crime and was awarded death penalty (Trestrail, 2007).

Janie Lou Gibbs was a serial killer who lived in Cordele, Georgia, USA. In 1966–67, she killed five members of her own family including her husband, three sons, and a grandson to receive US$ 31,000 in life insurance benefits (Vronsky, 2007). She fed rat poison to her family members and was later served five life imprisonments. She passed away in 2010.

In 2012, a teenager girl from Patna, Bihar, India was charged with killing of her father and younger brother who were against her love affair with a boy whom she intended to marry (Nelson, 2012). She later confessed that she deliberately mixed poison with vegetable curry and served it to her father and brother.

8.2.2 INDIAN MYTHOLOGY AND HISTORY

8.2.2.1 *The* Puranas

The *Puranas* are among the oldest books on religion in the world. Mainly focusing on the life of gods, these books contain many references of poisoning. According to Hindu mythology, the oldest incidence of poisoning was perhaps at a time when the gods and the demons collectively churned the oceans of the earth to extract the "nectar of immortality" (referred to as *Amrit*) which was to be shared among all (Kramrisch, 1992). The first "element" that came out of these waters as a result of this process was poison (known as *halahal*). This poison was very toxic and posed a great threat to all. Then Lord Shiva drank all of this poison to save the earth. His throat turned blue as a result of it and hence Lord Shiva is also known as *Neelkanth* (the one with a blue throat).

8.2.2.2 *Mahabharata*

Mahabharata is one of the finest epic poems of India. It focuses on the life of *Kauravas* and *Pandavas* (the two groups that shared a common ancestry but were against each other since childhood). *Duryodhan* was the eldest among *Kaurava* brothers. He was never able to defeat *Bhim* (one of *Pandava* brothers who was known for his physical strength and love for food). *Duryodhan* hated him so much that even as a child he hatched a conspiracy to kill *Bhim*. He fed *Bhim* a poison-laced *kheer* (a food article) and when he lost consciousness, threw his body into *Ganga* River. *Bhim*, however, managed to escape (Geraets, 2011).

8.2.2.3 Vish kanya *(vish = poison, kanya = damsel)*

The ancient Indian treatise *Arthashastra* mentions a unique kind of assassins known as *vish kanyas* or *poison damsels*. *Arthashastra* was written by *Chanakya* (c. 370–283 BC) who was a royal adviser to Indian emperor *Chandragupta Maurya*. These girls were given sublethal doses of poisons (Mithridatism) from the childhood onward in the belief that all their bodily secretions would become poisonous. The girls, however, remained immune to the effects of poisons. After these girls matured, the ancient kings/emperors used to send them to their enemies. They allured them to mate and kill them.

8.2.2.4 Juahar *(jiv = life, har = defeat)*

The women members of Rajput community in Rajasthan practiced *juahar* in case of an imminent defeat at the hands of invaders. All the male members of the community marched toward the invaders for their last fight while women practiced self-immolation to prevent capture, enslavement, sex slavery, and dishonor by the invaders. Though fire was the chief means to end the life, sometimes poisons were also employed for this purpose. Similar instances have been documented in other parts of the globe.

8.3 THE MODERN HISTORY

8.3.1 THE NAZI EXPERIMENTS

The Nazis conducted all kinds of experiments on their prisoners. Between 1943 and 1944, they administered various kinds of poisons in the foods to prisoners to study their effects. The Buchenwald concentration camp was a major German camp that was infamous for such experiments (Whitlock, 2013). As a result of these experiments, these "guinea pig" prisoners lost their lives. If they did not die by themselves due to poisoning, they were deliberately killed to facilitate postmortems. At the Dachau concentration camp, some prisoners were kept hungry and fed only seawater for experimentation purpose only.

8.3.2 THE DEATH OF INDIAN PRIME MINISTER LAL BAHADUR SHASTRI

Mr. Lal Bahadur Shastri was the second Prime Minister of India. In 1966, he went to Tashkent, USSR to sign an agreement with Pakistan after the 1965 war between India and Pakistan. The very next day of signing the agreement the news of his death from cardiac arrest came. It was not easy for any Indian to accept it as a natural death as he had been in good health. His wife later alleged that he was poisoned. It is believed that no postmortem was conducted on his dead body and the actual cause of death remains a mystery till date (Agrawal, 2012).

8.3.3 **MIDDAY MEAL DISASTERS**

Midday meal scheme is a very ambitious project of India. It was introduced to increase the enrollment and attendance in schools and to simultaneously raise their nutritional status of children as well. Today, it serves more than 120 million children with free and nutritious lunch in schools. It is the world's largest lunch program.

However, there have been many incidents that have marred the image of such an important health program. The most unfortunate of these instances occurred in July 2013, when 23 children died and many more fell ill after consuming lunch at the government school in Bihar. Later investigations showed that the cooking oil used in the preparation of meal was placed in a container that was previously used to store pesticides. The chemical examination report of food showed the presence of high levels of monocrotophos, a pesticide (Idrovo, 2014; Krause et al., 2013).

8.3.4 **PESTICIDE CONTAMINATION OF FOOD**

The problem of pesticides in foods in India is so widespread that many consider it extremely difficult, if not impossible to rectify. The pesticides have penetrated in almost all of Indian foods. There are many factors responsible for pesticide finding their way into foods. First, India is an agriculture-based economy with majority of its population employed in agriculture sector (Singh et al., 2013). This fact itself makes India vulnerable to pesticides (Abhilash and Singh, 2009). Second, most of the farmers are illiterate with little knowledge about the proper use of pesticides and time of harvesting after their use (Kesavachandran et al., 2009). This leads to premature harvesting and thus greater residues of pesticides in crops. Third, despite a ban, the toxic pesticides are freely available in the market, showing a blatant disregard for laws of public safety. Fourth, the public response to pesticides misuse is dismal as no one except a few wishes to come forward and build adequate pressure on the government to take remedial actions.

In the early years of last decade, a controversy surrounded Coca Cola and PepsiCo companies in India that their soft drinks contained high level of pesticides (lindane, malathion, DDT, chlorpyrifos). The tests were conducted by Center for Science and Environment (CSE), an independent nonprofit agency (Fernado, 2009). The soft drink samples from USA that were used as control did not show any pesticides. It was alleged that the water used in preparing soft drink contained pesticide residues. It was claimed that the Bureau of Indian Standards (BIS) did not have any standards for soft drinks and the standards set by Prevention of Food Adulteration (PFA) are far below than international standards. Similar incidences involving honey, milk, drinking water, and other foodstuffs are also reported now and then.

Many researchers and independent organizations have tested for pesticide residues in Indian diets and forwarded their findings to relevant authorities. Despite this fact, little has been done to take strict measures for its prevention. Therefore, India needs a very proactive approach to prevent the pesticides from further penetration into the foods.

8.3.5 **HEAVY METAL POISONING**

Heavy metal poisoning of Indian soils and water bodies is so widespread that if measures are not taken in time, we might cross the point of no return. The situation is particularly critical in and around the cities (Sharma et al., 2007, 1992). For example, Delhi is situated on the banks of Yamuna River which is heavily polluted. The factories located upstream (i.e., in and around Sonipat and Panipat regions of Haryana) are the main source of this pollution. According to some media reports, the drinking water in Delhi contains chromium, lead, and iron well above the permissible limits. A study conducted in 2013 found total coliform count of 1200 (against the permissible of 10) and biological oxygen demand (BOD) of 240 mg L^{-1} (against 30 mg L^{-1}) and chemical oxygen demand (COD) of 768 mg L^{-1} (against the permissible limits of 250 mg L^{-1}) in the water samples (Lalchandani, 2013). One of the main reasons for this contamination is that the effusion treatment plants (ETPs) at Wazirabad which is not equipped to detect and remove heavy metals, similar is the situation around most other cities in the country.

Some of the prominent cases of heavy metal poisoning in India are discussed as follows:

Arsenic toxicity in West Bengal: The high levels of arsenic in ground waters of the Bengal basin (mainly comprising of Bangladesh and the Indian state of West Bengal) have played havoc with the life of local inhabitants. The ground water with a depth of 35 m or less is particularly problematic. The arsenic level in groundwater in this region is more than 50 µg kg^{-1} (which is the safe limit set in India). The WHO sets a still lower value of 10 µg kg^{-1} as the safe limit for arsenic. A study reported that 9 districts of West Bengal and 50 districts of Bangladesh showed the features of chronic arsenic poisoning (Rahman et al., 2001). These features had developed over a period extending from 6 months to 2 years or more. The subjects reported features of sensory neuropathy and dermatopathy (leucomelanosis, melanosis, and keratosis). Features affecting other systems were also reported (Banerjee et al., 2013). The more serious cases presented with cancers of skin, lungs, liver, kidneys, and bladder. A majority of children below 11 years of age showed raised levels of arsenic in hair and nails.

No definite cause of arsenic poisoning of groundwater has been established till date. Previously it was believed that the artificial ponds of this region could be the possible source (Datta et al., 2011). However, these claims have been challenged by many. Now the most accepted theory is that the reductive dissolution of secondary iron oxide/oxyhydroxide along with oxidation of organic matter and consequent discharge of adsorbed or coprecipitated arsenic into the groundwater are the main processes in the mobilization of arsenic to groundwater. However, further studies are required to confirm it as more than 60 million people from Bangladesh and West Bengal depend on this water source.

Mercury poisoning: Before discussing the issues related to mercury poisoning in India, it is imperative to discuss in brief the Minimata incidence of Japan.

8.3.5.1 Minimata disease (Japan)

Minimata is a small town in the Kumamoto prefecture on Kyushu island of Southern Japan. This area first made headlines in 1956 when a young girl from region showed features of neurological damage (convulsions, difficulty in speaking and walking) that could not otherwise be explained. Following this incident many more cases started coming up with similar complaints. An inquiry was ordered that showed that the presence of what is now known as Minimata disease (Hachiya, 2006).

This disease is characterized by neurological manifestations as a result of consumption of fish and shellfish that are contaminated with methylmercury. This toxic by-product (methyl mercury) was getting generated during synthesis of acetaldehyde in a chemical factory owned by Chisso Company and drained into the sea. This methyl mercury entered the bodies of fish and shellfish and got accumulated. Consumption of significant quantities of contaminated fish on regular basis produced the disease. The sludge in Minimata Bay reported a very concentration (up to 2000 ppm) of methylmercury. In spite of this epidemic, no concrete preventive measures were taken and the disease surfaced again in 1965. This time it appeared along the Agano River basin in Niigata Prefecture. The culprit company in this second epidemic was Showa Denko's Kanose factory. Nearly 3000 people got the Minimata disease as a result of these two incidents. Huge compensations had to be paid to the victims of these two man-made disasters.

8.3.5.2 Mercury menace of India

All of mercury that is used in India is imported from foreign countries. The Chlor alkali industry, coalmines, thermometers, sphygmomanometers, dental amalgams, and cold candescent fluorescent lights (CFLs) are the main sources by which mercury is added to the environment. The chlor alkali industry, over the past few years, has now adopted different technologies and the mercury requirement has decreased. But there is no concrete proof that the sites where most of these industrial units were earlier located are totally decontaminated.

Coal is the most important fuel for electricity generation in India. Burning of coal besides other pollutants add the naturally occurring elements such as mercury into the environment (Li et al., 2009). Because of these concerns and rapidly depleting reserves, the government is exploring different ways to reduce the dependence on coal for electricity generation. One important step in this direction is the introduction of CFLs into the market as they consume less electricity and consequently reduce the burden on coal based thermal plants. This leads to decreased pollution and emission of mercury vapors into the environment. Today, India produces about 400 million units of CFLs every year and the demand is growing with each passing day. However, these CFLs require mercury for producing light. In spite of this fact, no limit of mercury usage in CFLs has been set. As a result, the mercury content in CFLs manufactured in India remains unregulated and far above than that found in CFLs manufactured in USA and other developed countries. For example, as per the government estimates, an average CFL in India contains about 3–12 mg of mercury. However, an independent study (Betne and Rajankar, 2011) found it to be 2.27–62.56 mg per CFL with an average of 21.21 mg per CFL which is far above of the levels set in USA and Europe. Improper disposal of CFLs also remains an issue of concern.

There have been many instances of mercury poisoning in India but they are largely getting ignored. The Kodaikanal region, a famous hill station in the Indian state of Tamil Nadu, is an excellent example. In 2001, the place got the attention of everyone when local residents found piles of broken glass thermometers in the nearby forest. They suspected that this waste was dumped by a nearby thermometer-manufacturing factory, owned by Hindustan Unilever. It led to widespread protests that forced the company to shut down the factory. Many local people were affected by mercury poisoning (Karunasagar et al., 2006). The protesters also forced the remaining mercury to be sent back to USA (widely hailed as "Reverse Dumping"). This incident is a fine example of unity of people to protect the environment from pollution.

Another case of mercury pollution came up in Uttar Pradesh (UP) state of India. The *Sonbhadra* region of UP has the one of the largest reservoirs of coal in the country. For quite some time, the residents of this region have been complaining of increased health problems. They blame that mercury from the nearby thermal plants is responsible as it has polluted not only the soil and air but also the groundwater. An independent study (Sahu et al., 2014) conducted by Center for Science and Environment (CSE) found that the human blood samples from this region showed a mercury level of 34.3 ppb (parts per billion) which is above the safe limits of 5.8 ppb as prescribed by United States Environment Protection Agency (USEPA). The fish from this region also showed the presence of methylmercury at a level of 0.447 ppm which is higher than the level set by Food Safety and Standards Authority of India (FSSAI), hence unfit for human consumption. The groundwater, especially that drawn by hand pumps had 26 times higher mercury content than set by Bureau of Indian Standards (BIS), that is, 0.001 ppm.

A previous study conducted by Indian Institute of Toxicological Research (IITR), a government agency, had also reported similar findings. However, no concrete remedial steps were taken in this direction. The issue of mercury poisoning needs an urgent intervention before it is too late.

8.4 RADIATION CONTAMINATION OF FOOD AND WATER
8.4.1 THE VANDERBILT NUTRITION STUDY (USA)

It was conducted at Vanderbilt University, USA from 1945 to 1947 with a purpose to study the effect of woman's diet and nutrition on pregnancy, delivery, and condition of the infant. More than 800 women were "enrolled" for this experiment and radioactive iron (Fe-59) was administered to them without their knowledge or consent (McCally et al., 1994). They were just told to drink a "cocktail" that would be beneficial for their babies. Contents of this cocktail were never revealed to them. These women, however, developed bruises, rashes, loss of hair and teeth. Four children who had been given prenatal radiation exposure developed fatal cancers. On the other hand, the children of mothers of unexposed group did not develop any cancers.

8.4.2 FUKUSHIMA NUCLEAR DISASTER (JAPAN)

The Fukushima nuclear disaster of Japan is one of the worst nuclear accidents ever. It occurred as a result of Tohoku earthquake that led to nuclear meltdown in the reactors of Fukushima Daiichi nuclear plant. The resulting widespread radioactive emissions contaminated the air, soil, food materials, and drinking water (United Nations Scientific Committee on the Effects of Atomic Radiation, 2014). Some experts believe that radioactive exposure through food is worse than that of air which dissipates over a few weeks. The International Atomic Energy Agency (IAEA) reported elevated levels of Iodine 131 and Caesium 137 in almost all vegetables and milk samples after 1 week of incident. The ingestion of such contaminated foods can cause cancer. Iodine 131, if ingested or inhaled, gets accumulated in the thyroid gland and starts emitting beta radiations during its decay and can cause cancer in higher doses. Though everyone is adversely affected by the effects of such foods, children are most susceptible because of greater body surface area, immature immune system, and higher metabolism. In the Fukushima region of Japan, the children showed excessive levels of Iodine 131 and Caesium 137 in their bodies and were found to be the worst affected.

The radioactive contamination of seawater during this episode was also higher than any nuclear accident. As the level of radioactivity builds up in the higher levels of food chains due to bioaccumulation, there is a serious threat to humans who feed on such fish. As a consequence of this incident, many countries banned the import of Japanese food.

8.4.3 URANIUM POISONING (INDIA)

Punjab is one of the leading agriculture states of India. There has been excessive use of pesticides over the past few decades that contaminated the environment. In addition, the draining of untreated industrial waste containing heavy metals into water bodies is also an issue (Thakur et al., 2010). In 2009, Punjab became first state in the country to report contamination of ground water with Uranium.

It was brought to the notice of certain researchers that the people of Malwa region of Punjab were reporting increased number of neurological defects and cancers. These researchers from South Africa, Germany, and New Zealand conducted a collaborative study and found that in 88% of children of 12 years of age or below and 85% of adults uranium levels exceeded the reference range (Blaurock-Busch et al., 2010). As the researchers examined the hair and urine samples of children of this region, it was almost presumed that heavy metals and pesticides would be detected in samples. But the detection of uranium in addition to pesticides and other metallic poisons rang the alarm bell.

After the aforementioned revelation, a search to detect the source of uranium contamination began. In 2010, Bhabha Atomic Research Center (BARC), Mumbai confirmed the presence of high concentration of uranium in drinking water samples from Malwa region of Punjab. This study, however, refuted the claims that the uranium levels were high enough to cause neurological problems and cancers. The BARC study also observed that the geological factors could be responsible for the water contamination. As there is no uranium mining in Punjab, the coalmines and fly ash (that contains high levels of uranium) from coal-fuelled power plants could be such possible sources. But no definite cause has been established till date.

Today, Punjab reports 90 cancer patients per 100,000 people against a national cancer average of 80. Some even call Punjab as the "Cancer Capital of India." In 2013, the Health minister of Punjab revealed that more than 33,000 people from Punjab alone had lost their lives to cancer in the last 5 years (The Times of India, 2013, Chandigarh). Out of these, 14682 were reported from Malwa region alone. A train that runs from Abohar, Punjab and ferries cancer patients to Bikaner, Rajasthan (where a famous government-owned cancer hospital is located) has been nicknamed as the "Cancer Train."

As a part of preventive measures, the government has declared the ground water unfit for human consumption and established many reverse osmosis (RO) plants in this region. But many doubt whether these measures are sufficient to bring any long-term respite to local people.

The mining of uranium in Jadugoda region of Bihar, India exposed the workers to the ill effects of radiation over a prolonged period of time (Mongillo and Warshaw, 2000).

8.5 TOXIC WASTE IMPORTS IN INDIA

India is fast emerging as a prominent dumping ground for toxic waste from many developed countries. Under the Basel convention, the export of such material from developed countries to less developed countries is banned. However, it is going on. There are many reasons for dumping toxic waste in India. First, it is cheaper to export the toxic waste products to India rather than to recycle it in the countries of origin. Second, the lax attitude of the Indian government in formulating strict protocols for such procedures and ensuring implementation of laws is a big problem. Third, it is a so-called "source of employment" for poorer sections of Indian society (Demaria, 2010). No one knows how much of toxic waste is imported into India every year. But many environmentalist claim it is enough to pose a significant health hazard to environment and health.

A typical example of dumping of toxic wastes is at Alang, Gujarat, the largest ship-breaking yard in the world. This industry has an annual turnover of around Rs 6000 crores. On an average about 200 ships are dismantled here each year. The old ships from all across the globe reach this site for their final disposal. These ships contain many hazardous substances including asbestos and heavy metals (Karpagam and Jaikumar, 2010). Because of poor regulatory machinery, there is frequent spillage of toxic wastes from this shipyard into the marine environment. Greenpeace organization has protested against the docking of foreign ships at the shores of Alang. Now efforts are being made to employ latest techniques so as to minimize the exposure of the surrounding ecosystems and protect the marine life from toxic wastes.

8.6 TOXICOLOGY AND CRIME

8.6.1 THE RAJNEESHEE BIOTERROR ATTACK

This is perhaps one of the most infamous incidents of bioterrorist attacks ever recorded. This attack happened during the election time in 1984, in Dallas, Oregon, USA. About 20 followers of a spiritual leader, Bhagwan Shri Rajneesh deliberately contaminated the salad bars at many restaurants. They used *Salmonella typhimurium* to incapacitate the voters in the hope that their own candidate would win the election (Crowe, 2007). The salmonella were obtained from a medical corporation owned by Rajneesh. As a result of contamination, 750 people fell sick and many had to be hospitalized. Later Rajneesh denied any involvement and blamed many of his disciples for the same. Many culprits were prosecuted in this case.

8.6.2 THE BISCUIT GANGS OF INDIA

For the past few years there have been increasing number of incidents where unsuspecting passengers have been duped of money and other valuables (Singh et al., 2015; The Hindu, 2006). Most of these stories revolve around a person who becomes too friendly with gullible passengers on buses or trains. He offers biscuits or other food articles that are laced with a sedative (benzodiazepines in most instances) to the unsuspecting passenger. Soon, the passenger becomes drowsy and loses his/her valuables to the miscreant. Many such incidents have been reported from trains bound for Chennai, Tamil Nadu and Mumbai, Maharashtra.

8.6.3 DATE RAPE DRUGS

These are used to incapacitate the person and facilitate drug-facilitated sexual assault (DFSA). Till the recent past, the use of such agents was restricted to developed world only but now this trend is evident in India as well. Most of these incidents are reported in discotheques and rave parties. These agents are frequently employed by the assailants to sedate the victim and to minimize her power to resist (Singh et al., 2015). Generally such drugs are mixed with food or drink to mask the taste and odor and served to the victim without her knowledge. The attacker then offers a ride to her home as she is not in her senses only to commit sexual assault/rape. The unsuspecting victim then falls prey to his misdeeds. A German study found that benzodiazepines are the most frequently used agents in such crimes. Most of the benzodiazepines produce anterograde amnesia and this might be one of the main reasons of choice of these agents (Curran, 1986). She might not be able to identify the assailant at a stage of criminal proceedings and he may escape free. Many such instances escape the official statistics and therefore complete data on these instances are not available.

Though a number of agents are employed for the purpose, the common ones in present days include:

1. benzodiazepines (flunitrazepam, lorazepam, etc.)
2. nonbenzodiazepines (zolpidem, zopiclone) and barbiturates
3. ketamine
4. alcohol

In the year 2010, an Agra-based Indian doctor along with his children and other accomplices was found to have sold in bulk, among other drugs, ketamine (which can be used as a "date rape drug") thorough an Internet pharmacy that he ran illegally. He had sold much of these drugs in USA which otherwise can only be obtained through a medical prescription.

In 2013, a Delhi court observed that India is not updated on the date rape drugs and hence urgent sensitization of the people is essential. The court also observed that testing for date rape drugs should be made a compulsory part of the rape kit while collecting exhibits in hospitals.

A strict governmental control over the sale of such agents is likely to produce rewarding results. The authors feel everyone in general and women in particular should avoid taking drinks/food items offered by strangers to prevent such incidents.

8.6.4 THE HOOCH TRAGEDIES

The production of illicit liquor is rampant in India (Divekar et al., 1974; Jarwani et al., 2013; Krishnamurthi et al., 1968; Ravichandran et al., 1984). In most such instances a substandard raw material is used for the purpose of production. Under these unregulated conditions, methanol is generated as a part of process of production which is responsible for many methanol-related causalities. Sometimes, the illegal brewers deliberately add methanol to ethanol under the impression that it will increase its potency. Adding methanol also serves to decrease the cost of liquor substantially. The reduced cost of the liquor attracts people from lower socioeconomic groups in whom the tragedies are most frequently reported. The problem is serious as it is difficult to differentiate the effects of methanol from ethanol in the initial phase.

8.6.5 THE PROBLEM OF ALCOHOL

The use of alcohol in India in the ancient and medieval times was limited as strict laws governed who is allowed to drink and under what circumstances. The Brahmin class of the Indian society was strictly forbidden to drink (Lochtefeld, 2002) whereas others were allowed to consume alcohol under certain situations such as festivals, religious events, and wars. However, under the British rule the consumption of alcohol began to rise slowly and steadily. Today, with increased purchasing power, changed value systems, greater independence, and easy availability more and more Indians are consuming alcohol. Nowadays, the consumption of alcohol is not limited to certain occasions. It usually begins as an "experiment" or curiosity among the peer groups (Bhullar et al., 2013) and then progresses to more frequent and chronic use. One of the other concerns regarding alcohol consumption is the falling mean age at which a person first starts alcohol intake. A study conducted in 2005 found that the mean age for starting alcohol consumption has come down from 28 years in 1920–30 to 20 years in 1980–90 (Benegal et al., 2003).

The official per capita consumption figures are not very accurate as they do not take into account the local and cheap beverages (whether legal or illegal) into account. Consequently, the actual adult per capita consumption (APC) is higher than stated. For example, the official figures for 2003 were 2 litres adult^{-1} per year. However, a study (Benegal, 2005) calculated the APC to be around 4 litres adult^{-1} per year for the same year after taking into account the unrecorded consumption (tax evaded products or illicit liquor).

Despite being aware of the ill effects of alcohol on health, the problem is far from control. One of the main reasons for this is the amount of revenue that is generated from the sale of alcohol. In some of the Indian states, the duties imposed on the sale of alcohol constitute a quarter of the total revenue of the state, making its sale a very lucrative business. The World Health Organization, South East Asia Regional Office (WHO SEARO) conducted a study and found that India generates a revenue of about Rs 216 billion a year from sale of alcohol but at the same time spends Rs 244 billion a year to manage the consequences of its use. These figures show that our country loses more than it gains from the sale of alcohol (Gururaj et al., 2006).

Taking all these factors into account, it can be stated that unless some serious steps are taken to curb the alcohol use, the problem will haunt India for a fairly long period of time.

8.7 POISON CONTROL/INFORMATION CENTERS IN INDIA

The 20th century saw the rise of pesticides that helped man to drastically increase the crop yield to feed the increasing population. However, as the time passed the ill effects of these chemical substances started coming to light. Improper use and a general lack of awareness about pesticides among the public were among the main reasons for acute and chronic toxicity/poisoning. Many newly invented medicinal drugs also caused toxicity in individuals. This led to an increased need to establish centers that provide quick information on the ways to deal with ever-rising poisoning cases. As a result, the poison information centers started coming into existence in the western countries in the early 1950s. However, India was quite late to establish such centers in spite of the fact that India reports a high number of the poisoning cases (Lall and Peshin, 1997). Many such centers have now been established in India and

Table 8.1 The Established Poison Information Centers in India

S. No.	Name of the Institute	Location	Established	24-Hours Service	Availability to Public
1	Poison Information Center, CEARCH	Ahmedabad, Gujarat	1993	Yes	No
2	National Poisons Information Center	AIIMS, New Delhi	1995	Yes	Yes
3	Poison Control Center	Cochin, Kerala	2003	Yes	Yes
4	Poison Control, Training and Research Center	Chennai, Tamil Nadu	2007	Yes	Yes

the number is increasing. Today, these centers provide quick information on prevention, early diagnosis, and management of poisoning cases and assist the doctors in treatment. These centers are also involved in activities to increase the public awareness on various poisons. Most of these centers operate round the clock to deal with poisoning cases. The centers mentioned subsequently are recognized by the WHO (Table 8.1).

8.8 FOOD SAFETY OF VVIPS

It is important that the food articles meant for the top leadership of the country passes through stringent quality tests before serving. Little information is disclosed about the actual protocols that are followed for the same. The general belief is that it is first tested for palatability and then for any poisons before it is served. Every step right from the preparation of food preparation till its serving is supervised. A certificate is issued by competent authority before it is served that the food is safe to eat.

8.9 FOOD SECURITY IN INDIA

Despite being a major producer of food grains, India battles with hunger and malnutrition. Almost half of world's hungry live in India. The Article 47 of Constitution of India provides the right to food. It states "The State shall regard the raising of the level of nutrition and the standard of living of its people and the improvement of public health as among its primary duties and, in particular, the State shall endeavor to bring about prohibition of the consumption except for medicinal purpose of intoxicating drinks and of drugs which are injurious to health."

8.10 INTERNATIONAL/NATIONAL ACTIVITIES FOR HEALTH PROTECTION

8.10.1 FREDERICK ACCUM (1769–1838)

He was probably the first to expose food adulteration. He was a German chemist who migrated to the United Kingdom in 1793. During his practice, he came to know about food adulteration that was rampant in the British society. His book titled "*A Treatise on Adulterations of Food and Culinary Poisons*" came to market in 1820 and sold more than a thousand copies in a month. He raised public interest by elaborating how different adulterant were mixed with pure food articles and sold in market (Hayes and Laudan, 2009). However, his endeavors against food adulteration brought him a number of enemies and ultimately he had to leave the United Kingdom and spend the rest of his life in Germany.

8.10.2 RACHEL CARSON (1907–64)

Through her literary work, Rachel led a crusade against pesticides (Sideris and Moore, 2008). In her book titled "*Silent Spring*" she explained the side effects of pesticides on human health. In most part of the book she not only focused on the detrimental effects of pesticides on environment but also discussed human pesticide poisoning, cancers, and other ill effects. She also accused the chemical industries of spreading misinformation about pesticides. She had particular interest in DDT that emerged as a revolutionary pesticide in the 1940s. The DDT at that time was awaiting tests for safety and effects on ecology. The American Chemical Society honored the book in 2012 as National Historic Chemical Landmark for giving impetus to environment protection movements.

8.10.3 GREENPEACE INDIA

Since 2001 this nonprofit organization has been actively involved in various environment-protection issues in India and raising public awareness for the same. Among its many other activities are movements against rampant and unscientific use of pesticides, heavy metal poisoning, and protection of forests. The organization also studied the health impact of pollution from coal-based thermal plants and manufacturing units associated with mercury.

8.10.4 TOXIC LINKS

Toxic Links is an NGO that is working toward the protection of environment from pollutants. Its main areas of focus include municipal, hazardous, biomedical waste management. The organization is based in Delhi and keeps a vigil over the environmental contamination from various pollutants and consequent ill effects on health. Toxic Links also actively participates in the bidirectional transfer of information across borders. Though pesticides, heavy metals, biomedical waste, pesticides, and persistent organic pollutants have been its traditional areas of work, electronic waste in India is the latest area of attention.

8.11 INDIAN ACTS/LAWS FOR FOOD SAFETY

There have been many laws in India since 1899 to regulate the quality of food. Till the framing of Prevention of Food adulteration Act (PFA) in 1954, many states followed their own laws to ensure adequate quality of food to consumers. But the differences in such rules and regulations hampered interprovincial trade. The Central Advisory Board (year 1937) and the Food Adulteration Committee (year 1943) recommended for a central legislation and the Prevention of Food adulteration Act (PFA) came into existence on June 15, 1955. This Act repealed all laws of different states regarding food adulteration that were in force at that time (FSSAI, 2015).

8.11.1 PREVENTION OF FOOD ADULTERATION ACT, 1954

This act came into existence with an objective of providing safe and wholesome food to the people and averts food frauds. This act was amended in 1964, 1976, and 1986 to make it more effective and punishments more stringent. The enforcement of the act lies with the government of respective states and union territories. The role of central government is mainly advisory in nature in addition to its statutory functions as per the Act.

8.11.2 INSECTICIDES ACT, 1968

It was enacted to regulate the import, production, sale, transport, distribution, and use of insecticides thereby minimize the safety threat to humans or animals. This act has its jurisdiction all over India.

8.11.3 THE FOOD SAFETY AND STANDARDS AUTHORITY OF INDIA (FSSAI)

It has been created under Food Safety and Standards Act, 2006 that merges a variety of acts and orders that have till now dealt with food related matters. The FSSAI puts in place standards for food articles and regulate their production, storage, distribution, and sale to ensure safe and nutritious food for consumption. There are many central acts that have been repealed after FSS Act, 2006. These include PFA, 1954; Fruit Products Order, 1955; Meat Food Products Order, 1973; Vegetable Oil Products (Control) Order, 1947; Edible Oils Packaging (Regulation) Order, 1988; Solvent Extracted Oil, De-Oiled Meal and Edible Flour (Control) Order, 1967; Milk and Milk Products Order, 1992, etc. The head quarters of Food Safety and Standards Authority of India are located at Delhi. The administrative control of FSSAI is with the Ministry of Health & Family Welfare, Government of India (FSSAI, 2015).

8.11.4 **INDIAN PENAL CODE**

The Indian Penal Code (IPC) prescribes the punishments for food related offences under the following sections:

272 IPC—Adulteration of food or drink intended for sale—"Whoever adulterates any article of food or drink, so as to make such article noxious as food or drink, intending to sell such article as food or drink, or knowing in to be likely that the same will be sold as food or drink, shall be punished with imprisonment of either description for a term which may extend to six months, or with fine which may extend to one thousand rupees, or with both" (Indian Penal Code, 1860).

273 IPC—Sale of noxious food or drink—"Whoever sells, or offers or exposes for sale, as food or drink, and article which has been rendered of has become noxious, or is in a state unfit for food or drink, knowing or having reason to believe that the same is noxious as food or drink, shall be punished with imprisonment of either description for a term which may extend to six months, of with fine which may extend to one thousand rupees, or with both" (Indian Penal Code, 1860).

8.11.5 **277 IPC—FOULING WATER OF PUBLIC SPRING OR RESERVOIR**

Whoever voluntarily corrupts or fouls the water of any public spring or reservoir, so as to render it less fit for the purpose for which it is ordinarily used, shall be punished with imprisonment of either description for a term which may extend to three months, or with fine which may extend to five hundred rupees, or with both (Indian Penal Code, 1860).

8.11.6 **278 IPC—MAKING ATMOSPHERE NOXIOUS TO HEALTH**

Whoever voluntarily vitiates the atmosphere in any place so as to make it noxious to the health of persons is general dwelling or carrying on business in the neighborhood or passing along a public way, shall be punished with fine which may extend to five hundred rupees (Indian Penal Code, 1860).

8.12 **REMEDIAL MEASURES**

The authors believe a safe and healthy environment is the responsibility of the both the state and its citizens. The government should implement all the existing laws in full spirit to bring down the levels of various toxins in the foods to a minimum and adequately punishes the offenders. Biodegradable pesticides should be encouraged. The government should also grant adequate financial and infrastructural support to research. This will provide impetus to development of newer and safer technologies for early detection and elimination of toxins from foods.

The people in general also need to be vigilant of any such offences occurring in open. For example, harvesting of crops on which pesticides have been sprayed before appropriate times should be brought to knowledge of the appropriate authority as these are likely to contain high levels of pesticide residues. Similarly, if banned pesticides are being sold in market, it should be reported. Drainage of battery water from old batteries directly into open drains, disposal of hazardous wastes by manufacturing units is extremely hazardous and such offenders should be punished.

The NGOs can increase the public awareness of about contamination of the foods with pesticides and heavy metals and can serve as an important link between the people and the government. These organizations can lead the masses in various environment protection activities.

It is hoped that with these remedial measures in place, the supply of safe food to the consumers will become better in future.

REFERENCES

Abhilash, P.C., Singh, N., 2009. Pesticide use and application: an Indian scenario. J. Hazard. Mater. 165, 1–12.

Agrawal, P., 2012. Silent Assassins Jan11, 1966. Rajesh M/s Agrawal Stationery House, Ramsagarpara, Raipur, Chhattisgarh, India.

Banerjee, M., Banerjee, N., Bhattacharjee, P., Mondal, D., Lythgoe, P.R., Martínez, M., Pan, J., Polya, D.A., Giri, A.K., 2013. High arsenic in rice is associated with elevated genotoxic effects in humans. Sci. Rep. 3, 2195.

Benegal, V., 2005. India: alcohol and public health. Addiction 100 (8), 1051–1056.

Benegal, V., Gururaj, G., Murthy, P., 2003. Report on a WHO collaborative project on unrecorded consumption of alcohol in Karnataka, India (monograph on the Internet). Available from: http://www.nimhans.kar.nic.in/Deaddiction/lit/UNDOC_Review.pdf

Betne, R., Rajankar, P., 2011. Toxics in that glow-mercury in compact fluorescent lamps (CFLs) in India. Toxics Link. Available from: http://toxicslink.org/docs/CFL-Booklet-Toxics-in-That-Glow.pdf

Bhullar, D.S., Singh, S.P., Thind, A.S., Aggarwal, K.K., Goyal, A., 2013. Alcohol drinking patterns: a sample study. J. Indian Acad. Forensic Med. 35, 37–39.

Blaurock-Busch, E., Friedle, A., Godfrey, M., Schulte-Uebbing, C.E.E., Smit, C., 2010. Metal exposure in the children of Punjab, India. Clin. Med. Insights Ther. 2, 655–661.

Burcham, P.C., 2014. An Introduction to Toxicology, first ed. Springer-Verlag, London.

Cilliers, L., Retief, F.P., 2000. Poisons, poisoning and the drug trade in ancient Rome. Akroterion 45, 88–100.

Crowe, K., 2007. Salad bar salmonella. Forensic Examiner 16 (2), 24–26.

Curran, H.V., 1986. Tranquillising memories: a review of the effects of benzodiazepines on human memory. Biol. Psychol. 23 (2), 179–213.

Datta, S., Neal, A.W., Mohajerin, T.J., Ocheltree, T., Rosenheim, B.E., White, C.D., Johannesson, K.H., 2011. Perennial ponds are not an important source of water or dissolved organic matter to groundwaters with high arsenic concentrations in West Bengal, India. Geophys. Res. Lett. 38, 1–5.

Demaria, F., 2010. Ship breaking at Alang–Sosiya (India): an ecological distribution conflict. Ecological Economics xxx (2010) xxx–xxx (article in press).

Dikshith, T.S.S., 2008. Safe Use of Chemicals: A Practical Guide. CRC Press, Boca Raton, FL.

Divekar, M.V., Mamnani, K.V., Tendolkar, U.R., Bilimoria, F.R., 1974. Acute methanol poisoning: report on a recent outbreak in Maharashtra. J. Assoc. Phys. India 22, 477–483.

Fernado, A.C., 2009. Business Ethics: An Indian Perspective, first ed. Dorling Kindersley (Ind) Pvt. Ltd, India.

Food Safety and Standards Authority of India (FSSAI), 2015. Available from: http://www.fssai.gov.in/AboutFssai/Introduction.aspx?RequestID=s8tMMSesiKu8seM4mi3_doAction=True

Geraets, W., 2011. The Wisdom Teachings of Harish Johari on the Mahabharata. Inner Traditions/Bear & Co, Vermont.

Goldflank, L.R., Flomenbaum, N., 2006. Goldfrank's Toxicologic Emergencies, eighth ed. McGraw-Hill Professional, New York.

Gururaj, G., Girish, N., Benegal, V., 2006. Burden and socio-economic impact of alcohol use: the Bangalore study. World Health Organization, Regional Office for South-East Asia, New Delhi.

Hachiya, N., 2006. The history and present of Minamata disease—entering the second half of century. Japan Med. Assoc. J. 49 (3), 112–118.

Hayes, D., Laudan, R., 2009. Food and Nutrition: Grains to legumes. Marshall Cavendish, New York.

Idrovo, A.J., 2014. Food poisoned with pesticide in Bihar, India: new disaster, same story. Occup. Env. Med. 71, 228.

Indian Penal Code, 1860. Chapter XIV. Of offences affecting the public health, safety, convenience, decency and morals. Available from: http://www.advocatekhoj.com/library/bareacts/indianpenalcode/index.php?Title=Indian%20Penal%20Code,%201860

Jarwani, B.S., Motiani, P.D., Sachdev, S., 2013. Study of various clinical and laboratory parameters among 178 patients affected by hooch tragedy in Ahmedabad, Gujarat (India): a single center experience. J. Emerg. Trauma Shock. 6 (2), 73–77.

John O'Connell, 2015. The Book of Spice: From Anise to Zedoary. Profile Books Ltd, London.

Karpagam, M., Jaikumar, G., 2010. Clemenceau Waste Disposal Controversy. Green management: Theory and Applications. Ane Books Pvt. Ltd., New Delhi.

Karunasagar, D., Balarama Krishna, M.V., Anjaneyulu, Y., Arunachalam, J., 2006. Studies of mercury pollution in a lake due to a thermometer factory situated in a tourist resort: Kodaikanal, India. Environ. Pollut. 143 (1), 153–158.

Kesavachandran, C.N., Fareed, M., Pathak, M.K., Bihari, V., Mathur, N., Srivastava, A.K., 2009. Adverse health effects of pesticides in agrarian populations of developing countries. Rev. Environ. Contam. Toxicol. 200, 33–52.

Kramrisch, P., 1992. The Presence of Siva. Princeton University Press, New Jersey.

Krause, K.H., Van, T.C., De Sousa, P.A., Leist, M., Hengstler, J.G., 2013. Monocrotophos in Gandaman village: India school lunch deaths and need for improved toxicity testing. Ach. Toxicol. 87, 1877–1881.

Krishnamurthi, M.V., Natarajan, A.R., Shanmugasundaram, K., Padmanabhan, K., Nityanandan, K., 1968. Acute methyl alcohol poisoning. (A review of an outbreak of 89 cases). J. Assoc. Phys. India 16, 801–805.

Lalchandani, N., 2013. Heavy metals in Delhi's drinking water. Times of India, Delhi, March 11.

Lall, S.B., Peshin, S.S., 1997. Role and functions of Poisons Information Centre. Indian J. Pediatr. 64 (4), 443–449.

Li, P., Feng, X.B., Qiu, G.L., Shang, L.H., Li, Z.G., 2009. Mercury pollution in Asia: a review of the contaminated sites. J. Hazard. Mater. 168 (2–3), 591–601.

Lochtefeld, J.G., 2002. The Illustrated Encyclopedia of Hinduism: A-M. The Rosen Publishing Group, New York.

McCally, Michael, Cassel, C., Kimball, D.G., 1994. U.S. Government-Sponsored Radiation Research on Humans 1945–1975. Med. Glob. Surviv. 1 (1), 4–17.

Mongillo, J.F., Warshaw, L.Z., 2000. Encyclopedia of Environmental Science. University Rochester Press, New York.

Nelson, D., 2012. Indian girl kills father and brother over love affair dispute. The Telegraph, July 2.

Radenkova-Saeva, J., 2008. Historical development of toxicology. Acta Medica Bulgarica 35, 47–52.

Rahman, M.M., Chowdhury, U.K., Mukherjee, S.C., Mondal, B.K., Paul, K., Lodh, D., Biswas, B.K., Chanda, C.R., Basu, G.K., Saha, K.C., Roy, S., Das, R., Palit, S.K., Quamruzzaman, Q., Chakraborti, D., 2001. Chronic arsenic toxicity in Bangladesh and West Bengal, India—a review and commentary. J. Toxicol. Clin. Toxicol. 39 (7), 683–700.

Ravichandran, R., Dudani, R.A., Almeida, A.F., Chawla, K.P., Acharya, V.N., 1984. Methyl alcohol poisoning. (Experience of an outbreak in Bombay). J. Postgrad. Med. 30, 69–74.

Rosner, F., 1968. Moses Maimonides' Treatise on poisons. J. Am. Med. Assoc. 205 (13), 914–916.

Sahu, R., Saxena, P., Johnson, S., Mathur, H.B., Agarwal, H.C., 2014. Mercury Pollution in Sonbhadra District of Uttar Pradesh and its Health Impacts. Toxicol. Environ. Chem. 96 (8), 1272–1283.

Sharma, Y.C., Prasad, G., Rupainwar, D., 1992. Heavy metal pollution of river Ganga in Mirzapur, India. Int. J. Environ. Studies 40 (1), 41–53.

Sharma, R.K., Agrawal, M., Marshall, F., 2007. Heavy metal contamination of soil and vegetables in suburban areas of Varanasi, India. Ecotox. Environ. Safe. 66 (2), 258–266.

Sideris, L.H., Moore, K.D. (Eds.), 2008. Rachel Carson: Legacy and Challenge. State University of New York Press, Albany.

Singh, S.P., Aggarwal, A.D., Oberoi, S.S., Aggarwal, K.K., Thind, A.S., Bhullar, D.S., Walia, D.S., Chahal, P.S., 2013. Study of poisoning trends in north India—a perspective in relation to world statistics. J. Forensic Leg. Med. 20, 14–18.

Singh, S.P., Kaur, S., Singh, D., Aggarwal, A., 2015. Lorazepam: a weapon of offence. J. Clin. Diagn. Res. 9 (3), HD 01-02.

Thakur, J.S., Prinja, S., Singh, D., Rajwanshi, A., Prasad, R., Parwana, H.K., Kumar, R., 2010. Adverse reproductive and child health outcomes among people living near highly toxic waste water drains in Punjab, India. J. Epidemiol. Community Health 64, 148–154.

The Hindu, 2006. Number of cases involving gang of 'biscuit bandits' on the rise. April 1.

The Times of India, 2013. Punjab's cancer cases exceed national average, Chandigarh, January 29.

Trestrail, III, J.H., 2007. Criminal Poisoning: Investigational Guide for Law Enforcement, Toxicologists, Forensic Scientists and Attorneys, second ed. Humana Press, Totowa, New Jersey.

United Nations Scientific Committee on the Effects of Atomic Radiation, 2014. UNSCEAR 2013 Report: Sources and Effects of Ionizing Radiation. Report to the General Assembly with scientific annexes United Nations, New York.

Vronsky, P., 2007. Female Serial Killers: How and Why Women Become Monsters. Berkley Books, New York.

Whitlock, F., 2013. Buchenwald: hell on a hilltop: murder, Torture & Medical Experiments in the Nazi's Worst Concentration Camp, first ed. Brule, Wisconsin.

Toxicological profile of Indian foods—ensuring food safety in India

S.P. Singh*, S. Kaur**, D. Singh[†]

*Department of Forensic Medicine and Toxicology, Government Medical College and Hospital, Chandigarh, India; **Department of Biochemistry, Government Medical College and Hospital, Chandigarh, India; [†]Department of Forensic Medicine, Post Graduate Institute of Medical Education and Research, Chandigarh, India

9.1 INTRODUCTION

Poisoning through food is an age-old affair in India. Our ancestors were aware of many plants and their adverse effects on the human body. They used these plants or their extracts to kill their prey while hunting and to poison their enemies. Human history has witnessed many plant- and animal food-related deaths. Most noted were those of Socrates (Greek scholar) who was forced to drink the poison Hemlock (*Conium maculatum*); Cleopatra (Egyptian queen) who killed herself by means of an asp bite, a poisonous snake; Durdhara (wife of an ancient Indian emperor, Chandragupta Maurya), who accidently got poisoned when she ate the poisoned food meant for her husband; and Charles VI (Roman emperor), who consumed poisonous mushrooms. Many ancient Indian kings used *vish kanyas* (poison damsels) as a part of their secret service to assassinate their enemies. These girls, supposedly, lured the victims for sexual intercourse and poisoned them. The Borgias of Italy killed many people, including Popes, with the help of arsenic for political and financial gains. As the times advanced, the knowledge of these poisons and their effects grew and they formed an inseparable part of modern toxicology. Today, the subject of toxicology has expanded into many subspecialties including biochemical toxicology, pathotoxicology, toxicogenomics, nanotoxicology, and food toxicology.

Food toxicology is the branch of science that deals with the study of nature, properties, effects, detection of toxins in food, and their disease manifestations in human beings. Although, the science has greatly supported toxicology in ensuring the delivery of safe foods for human consumption, it is still important to know the safety concerns of various food products and their recommended quantities. With the ever-increasing air, soil, and water pollution, it is imperative that the food that reaches the common man shall pass though stringent quality tests to ensure maximum safety and nutrition. In 1963, WHO and FAO established the Codex Alimentarius Commission to provide standards, guidelines, and codes of practice to ensure safe and quality food to the consumers. Today majority of countries, including India, are its members.

Every year, thousands of people fall sick and many die due to food poisoning. A number of factors contribute to this situation. *First*, the vast population of India has put a great burden on its resources. This has led to unrestricted and injudicious use of pesticides to increase the crop yield. Many of these toxic and banned pesticides are freely available in the market and, as they are quite inexpensive, they attract a lot of poor farmers. Most of these agriculturists are illiterate and possess limited knowledge about the correct methods of pesticide use. As a result, they harvest crops before the effects of pesticides mitigate. As a result, increased amount of pesticides remain deposited on crops, which then enter the food chains. *Second*, the widespread use of pesticides and industrial pollution of water bodies has now affected the water table at most of the places in country. So there is a greater probability that crops grown in these areas are inherently contaminated. *Third*, the dismal state of country's grain storage facilities plays an important role in the infestation of food grains by various insects, rodents, and microorganisms. Such spoiled grains are likely to cause food poisoning. According to an estimate, India loses about 12–16 million metric tons of food grains annually that costs more than Rs 50,000 crores. *Fourth*, the growing use of preserved foods is also contributing to increased food-related problems. *Fifth*, consumption of improperly cooked food, food prepared from rotten vegetables/grains, or food prepared under unhygienic conditions also contributes to the burden of disease. In the recent times, many food-poisoning deaths have been reported among schoolchildren from various parts of India. It is claimed to be due to the substandard food that is being served to schoolchildren under the Midday Meal Scheme.

As agriculture is one of the main industries, India faces a major challenge posed by the deleterious effects of pesticides on the environment and human health (Singh et al., 2013). The widespread use of these agents has not only contaminated the soil and water bodies across the country, but has also resulted in human and animal casualties over a long period of time. Ignorance about banned pesticides, injudicious use, poor adherence to instructions, and improper implementation of laws and poverty are some of the important factors governing the selection and usage of these agents.

Besides pesticides, metals and their compounds comprise another important group of pollutants. The metals in trace amount are essential for the adequate functioning of the enzymes in humans; however their increased ingestion through polluted water and foods produces adverse reactions and sometimes deaths. Cheaper methods to detect pollution of water bodies are available but underutilized in our country. It includes frogs that are capable of detecting heavy metals and pesticides in freshwater bodies. Studies conducted abroad have found them to be sensitive to heavy metals, such as, mercury, copper, and zinc and agricultural pesticides, such as, DDT, dieldrin, chlordane, herbicides, and fungicides.

The plants and animal species of such environments are likely to contain higher amounts of pollutants ultimately leading to their bioaccumulation and biomagnification in higher animals and humans. For example, say a large water body receives a lot of toxic material generated from industries. The planktons of this water body will contain a certain amount of toxicants. The small fish that feed on these planktons will contain a higher amount of toxins and the bigger fish that feed on such smaller fish will contain yet a higher amount of the toxins and so on. Ultimately if humans consume such fish they will have the highest concentration of toxicants in their bodies when compared to the planktons and the fish of that pond. The pesticides and heavy metals are discussed at length in the later sections of this chapter.

As far as the food storage facilities across the country are concerned, about 60–70% of India's food grains are still stored in indigenous storage structures, such as, *kanaja*, *kothi*, *sanduka*, *gummi*, *kacheri*, and *earthen pots*. These structures are not suitable for long-term storage of food grains and can lead to grain spoilage. Therefore the improved structures, such as, PAU bin, PUSA bin, Hapur tekka, CAP (Cover and Plinth) storage, and silos are being provided.

The microorganisms play a pivotal role in spoilage of stored/preserved foods. The list of such microorganisms involved in causing food poisoning is very long and includes a variety of bacteria, viruses, protozoa, and fungi (Ramanathan, 2010; Ramesh, 2007). Certain plant parasites, fish, and various chemicals, added to food for preservation and flavor, are also capable of producing ill health/death.

9.1.1 EFFECTS ON THE HUMAN BODY

The presence of toxicants in food produces variable effects in the human body, ranging from diarrhea, gastritis, enteritis, vomiting, hypersensitivity, anaphylaxis, and death. These ill effects are largely dependent on the type of toxicant present, its quantity, availability of antidotes, time taken for remedial measures, and any idiosyncratic reaction. A detailed knowledge of the subject is thus essential to understand the concept of food toxicology to produce an impact on public health.

9.2 CLASSIFICATION OF TOXINS/TOXICANTS

Toxins are the toxic substances of biologic origin, that is, they are produced by living organisms. Examples include venoms of snakes and insects.

Toxicants are the poisonous substances that are not of biologic origin and are produced by the human activities, for example, pesticides.

These can broadly be classified into the following:

1. natural toxins and
2. environmental toxicants.

9.2.1 **NATURAL TOXINS**

They occur in nature and are not the result of human activities. They are further classified as follows.

9.2.1.1 *Toxins inherently present in foods*

As the name implies, there is an inbuilt presence of toxins in plants, animals, and microbes. This category includes phytoalexins, glycosides (cyanogenic), alkaloids, carcinogens, enzyme inhibitors, and vitamin antagonists.

9.2.1.1.1 Phytoalexins

These compounds are produced in plants and possess antimicrobial activity for self-defense. They are particularly important in protection against fungal attacks. A few studies have suggested their possible role in inhibiting cell proliferation (including that of cancer cells) in humans. These studies have mainly focused on some particular phytoalexins and their effects in human colon cancer, prostate cancer, and their antiestrogenic potential.

9.2.1.1.2 Cyanogenic glycosides

Certain plants contain glycosides with a cyanide group. When attacked by pests, these plants release cyanide that results in the formation of hydrogen cyanide, which kills the invaders. Some of the important cyanogenic glycosides include amygdalin, sambunigrin, vicianin, lanimarin, and lotaustralin. Examples of some edible plants with cyanogenic glycosides include almond, apricot, sorghum cherry, plum, lima beans, cassava, prunasin, stone fruit, and bamboo shoots.

In humans, acute poisoning with cyanide has been reported with overconsumption of preparations made from bitter almonds and cyanide-rich apple seeds. The cyanide interferes with cellular respiration and may cause seizures, loss of consciousness, and even cardiac arrest, if taken in sufficiently large amounts. The enzyme rhodanase detoxifies cyanide to produce thiocyanate.

Some relevant medical conditions are discussed.

9.2.1.1.2.1 Goiter. The thiocyanate generated during detoxification of cyanide in the body is believed to interfere with iodine uptake because of its structural similarity. It increases the iodine requirement in iodine-deficient populations. The thiocyanate thus acts as an additional factor for the development of goiter in such populations.

9.2.1.1.2.2 Konzo. It is a neurologic condition associated with the intake of improperly processed (i.e., without peeling, slicing, and cooking) cassava fruit (Nzwalo and Cliff, 2011). The cyanogenic glycosides of the plant, namely linamarin and lotaustralin, are released by the plant to defend it against animals. These cyanide compounds produce konzo that is characterized by a sudden onset of spastic paraplegia. Konzo has been reported in the African continent.

9.2.1.1.2.3 Tropical ataxic neuropathy. Tropical ataxic neuropathy is another condition that is associated with chronic consumption of products derived from cassava and consequent exposure to cyanide. It manifests in the form of neurologic symptoms that mainly affect vision and hearing abilities. This condition has been reported in Kerala, India.

9.2.1.1.3 Alkaloids

Alkaloids are nitrogenous compounds of low molecular weight. They are mainly produced by plants and animals for defense. Examples of alkaloids include morphine, codeine, coniine, quinine, scopolamine, hyoscamine, atropine, caffeine, sangunarine, berberine, etc. About a fifth of total plant species are believed to produce them. Frogs, for example *Bufo marinus*, contain large quantities of morphine in their skin. Many of these alkaloids have been in use in modern medicine for various purposes. For example, morphine and codeine are used to produce analgesia, while caffeine is used as a central nervous stimulant.

9.2.1.1.4 Carcinogens

Here only those cancer causing agents will be discussed that are naturally present in plants and animals and are of dietary significance to man.

9.2.1.1.4.1 Plant carcinogens. The most important plant carcinogens include heterocyclic amines (HCAs) that occur naturally and hence cannot be totally avoided in diet. The HCAs are also present in meat. They possess the capacity to induce cancers of the colon, breast, prostate, lymphoid tissue, liver, and blood vessels in humans. However, their effect can be minimized by avoiding the direct exposure of meat to flames, using microwave ovens for cooking, and by wrapping the meat in aluminum foil before roasting in ovens.

Other carcinogens include *mycotoxins*, ptaquiloside (bracken fern) norsesquiterpene glucoside, cycasin (cycad nuts), hydrazine (mushrooms), pyrolysis products, nitrites, nitrates, and dioxins.

The ptaquiloside from bracken fern has been found to cause carcinomas of the intestines, mammary glands, and urinary bladder in rats. This fern is eaten even today in Japan after boiling, which removes a major portion of the toxin. Cycad nuts contain a carcinogen, methylazoxymethanol, known to cause colon cancer in rats but the ingestion of these nuts in now very limited. Heating biscuits and broiling of steaks results in formation of polycyclic aromatic hydrocarbons (PAHs) in the charred portions. Digoxin and related compounds are also toxic when they contaminate the foods but the toxicity is not uniform and depends on the offending compound and on different species.

9.2.1.1.4.2 Animal carcinogens. Excessive fat, particularly animal fat, is associated with an increased incidence of cancers of the colon, prostate, and breast. Increased fat deposition in the adipose tissue is also believed to prompt the enzyme aromatase that produces androgens and estrogens and contribute to the development of breast cancer. A diet rich in calories leads to greater deposits of fat and contributes to carcinogenesis.

9.2.1.1.5 Enzyme inhibitors

9.2.1.1.5.1 Botulinum toxin. It is produced by the bacterium, *Clostridium botulinum*, and is found in foods that are contaminated with its spores, improperly cooked, or preserved foods that have not been properly heated prior to packaging. The spores are resistant to many common preservation techniques. Botulinum toxin is one of the most potent toxins known to man. Botulism presents mainly with vomiting and paralysis of muscles (skeletal, ocular, pharyngeal, and respiratory muscles). Death from botulism is not uncommon. Good manufacturing practices, therefore, should be meticulously followed while handling, packaging, and storage of such food articles.

9.2.1.1.5.2 β-N-oxalyl-1-α-diamino propionic acid. Khesri dal (*Latyrus sativus*), and some other legumes of the genus *Lathyrus*, contain this toxin. The toxin is structurally similar to glutamate, which acts as a neurotransmitter. Consumption of this legume over prolonged periods of time has been linked with the development of a condition known as neurolathyrism, which is characterized by spastic paralysis of lower limbs. This disease commonly occurred during famines and has now virtually disappeared from India. More details on the subject are given in Chapter 2.

Due to a similar appearance and low cost involved, it is a common agent used to adulterate *arhar dal*. It is also used to adulterate gram flour and other foods products that are based on the expensive pulses.

This *dal* is nowadays making headlines again as a few groups are urging the government to lift ban on its cultivation. It is claimed that this *dal* alone is not responsible for neurolathyrism. Malnutrition, low immunity, and some other factors that are prevalent during famines, along with ingestion of *khesri dal* over a prolonged period of time collectively, play a vital role in the development of disease. These groups also claim that *dal* is cheap, nutritious, and safe if consumed in small quantities, hence it can be a good source of nutrition.

9.2.1.1.6 Vitamin antagonists

9.2.1.1.6.1 Thiaminase. Many plants and fish contain an enzyme thiaminase, which is capable of rendering the thiamine inactive. Thiaminases are of two types. *Type 1* is present in fish, shellfish, ferns, and some bacteria. *Type 2* has been reported in some bacteria. Both these types follow different mechanisms to inactivate thiamine and are capable of producing beri beri.

9.2.1.1.6.2 Chlorogenic acid. It is present in coffee and inactivates thiamine and can produce beri beri.

9.2.1.2 Toxins brought in food through spoilage

9.2.1.2.1 Bacterial contamination of food

There are a number of bacterial, fungal, protozoal, fish, and chemical agents responsible for food poisoning. The outbreaks of food poisoning are usually reported from common eating-places, such as, restaurants, hotels, picnics spots, dormitories, etc. The food poisoning may or may not be infectious in nature. Common bacteria involved in food poisoning include, *Staphyloccocus aureus*, *Escherichia coli*, *Bacillus cereus*, *Campylobacter jejuni*, *Salmonella cryptococcus*, *Clostridium botulinum*, and *Clostridium perfringens*. The noninfective sources of food poisoning involve certain fishes, fungi, strawberries, etc.

Generally, *S. aureus* gains access to food from unclean hands of the food handlers and, under suitable temperature conditions, produce an enterotoxin that is responsible for food poisoning. The enterotoxin attaches itself to neural receptors in the gastrointestinal tract and stimulates the vomiting center in the brain. Incubation period varies from about 1–6 h. The usual complaints are nausea, vomiting, and abdominal cramps. The enterotoxins are relatively resistant to heat and might not be inactivated during improper heating.

The gastroenteritis associated with *B. cereus* presents similar to staphylococcal food poisoning and is generally linked to consumption of contaminated rice.

Clostridial infection is usually seen in cases of meat and preserved fish. In addition to *C. botulinum* (as discussed earlier), *C. perfringens* is another species that is associated with food poisoning. It is part of the normal gut flora and the female genital tract. The bacillus is encapsulated and produces enterotoxins responsible for causing gastroenteritis. The spores remain viable for over 1 h at 100°C. The incubation period varies from 8 to 14 h. A typical episode of *C. perfringens* attack lasts for about a day or two and is characterized by nausea and abdominal cramps. Vomiting is infrequent and fever is rare. Without adequately heating the contaminated food at suitable temperature, the spores are likely to escape inactivation and cause food poisoning.

E. coli is the most common organism responsible for *Enterococcus*-related gastroenteritis. The effects are seen due to release of verocytoxin. Proper cooking of the food is likely to destroy the organism.

Further details of these and other related illnesses are elaborated in Chapters 2 and 8 of this book.

9.2.1.2.2 Fungal contamination of food

9.2.1.2.2.1 Alkaloids. Though a variety of fungal alkaloids have been reported, ergot alkaloids form the most important group. They are mainly produced by *Claviceps purpura* and include ergometrine, ergotamine, ergosine, ergocristine, ergocryptine, and ergocornine. Many outbreaks have been reported in the past involving this fungus. It is responsible for the release of ergot alkaloids and leads to a condition called as ergotism. The symptoms are variable and may include convulsive, gangrenous, or gastrointestinal features. In the past, outbreaks have resulted from the ingestion of baked bread that was made from ergot-contaminated wheat. Presently, this condition does not pose a significant health risk to humans because of the improved agricultural practices and milling procedures. The last outbreak of ergotism in India was in 1975.

9.2.1.2.2.2 Aflatoxins. The mycotoxins or fungal toxins are secondary metabolites of molds that are known to have an adverse effect on human health (Zain, 2011). Mycotoxicosis is the result of mycotoxins on human and animal health as a result of fungal toxins that mostly enter the body through ingestion. The adverse effects mainly depend on the toxicity of mycotoxin, duration of exposure, age, general health of the person, and interaction with other chemicals to which the person is exposed. *Aspergillus* is an important source of aflatoxins. Two species of this genus pose a significant health problem in humans, namely *Aspergillus flavus* and *Aspergillus parasiticus*. The fungus grows in a hot and humid environment. *A. flavus* produces aflatoxin B only, while *A. parasiticus* produces B and G aflatoxins. After consumption of the contaminated food, the aflatoxins appear in milk, urine, and feces. If mothers consume the contaminated food, it leads to an appearance of aflatoxins in the neonatal umbilical cord blood. Aflatoxins produce a condition known as toxic hepatitis. Other effects include carcinogenesis (hepatocellular carcinoma), immunosupression, and teratogenesis. Chapter 2 can be referred to for more details.

9.2.1.2.2.3 Ochratoxins. Many species belonging to *Penicillium* and *Aspergillus* genera secrete this toxin. These fungi grow on many food articles, such as, eggs, dried fruits, coffee, fish, certain cereals, and peanuts. Ochratoxins are known for their nephrotoxic and possible carcinogenic effects. The usual features include headache, fatigue, and loss of body weight. Ochratoxin A is the most lethal toxin in this group and is also the most frequently encountered.

9.2.1.2.2.4 3-Nitropropioninc acid. It is associated with the *Arthrinium* genus and causes acute food poisoning within 2–3 h of ingestion of contaminated moldy sugar cane. Clinical features involve vomiting, dystonia, carpopedal spasm, convulsions, and coma. This type of poisoning is most commonly seen in young adults and children.

9.2.1.2.2.5 Zearalenone. *Fusarium* presents as a plant pathogen and contaminates stored grains. Zolerone is one of the important mycotoxin associated with *Fusarium* genus besides trichothecenes. This toxin is known to produce estrogenic effects in mammals including humans (in males—testicular atrophy and enlargement of male breast; in females—vulval edema, vaginal prolapse, and infertility). It also might be involved in precocious puberty.

9.2.1.2.2.6 Trichothecenes. Trichothecenes are produced by the *Fusarium* genus and are infrequently associated with food poisoning. The most commonly incriminated tricothecene is deoxynevalinol (also known as vomitoxin). Others are nevalenol and diacetoxyscirphenol. Foodstuffs that are commonly affected include wheat, maize, oats, rice, barley, and other crops.

The toxin production is greatest when the temperature is in the range of 6–24°C and humidity is high. Clinical features are the result of toxic effects on the rapidly dividing cells of the alimentary tract, erythroid and lymphoid cells, and skin. There is necrosis of skin and the buccal mucosa. The bone marrow is depressed with consequent fall in immunity.

Other fungal toxins, such as, fumonsinins and moniliformins also sometimes cause food poisoning.

9.2.1.2.3 Protozoal contamination of food

Many protozoal species are involved with food poisoning in humans. Some of important ones are *Cryptosporidium, Giardia, and Cyclospora* (Dawson, 2005). Their spread is related to contaminated water that is used in the final process of production of foods or improper heating of ready-to-eat foods.

Giardia enters the body though water that is contaminated with its cyst form. Three stool samples after a gap of 2–3 days are required for the examination for cysts. It is responsible for causing diarrhea with steatorrhea, bloating, and flatulence.

Cryptosporidium enters the body through the oral–fecal route and after an incubation period of about a week, produces watery diarrhea and abdominal cramps. *Cyclospora* is responsible for causing nonbloody diarrhea, loss of weight, abdominal cramping, nausea, vomiting, and fever.

9.2.1.2.4 Noninfective food poisoning

Various fungi and fish have been incriminated in these cases. Fungi have already been discussed earlier in this chapter.

Some marine algae (primarily dinoflagellates) produce toxins that are responsible to produce food poisoning. These include ciguatoxins and saxitoxins associated with ciguatera and shellfish-induced posioning (Ramanathan, 2010).

About 50,000 cases of ciguatera poisoning are reported every year worldwide. The ciguatera toxin is heat-, cold-, and gastric-acid resistant. Within 2–6 h after ingesting the offending fish, there is an appearance of nausea, vomiting, and abdominal cramping. One of the characteristic findings includes reversal of hot and cold sensation within 3–5 days and may persist for months.

Globally, there are about 1600 cases of paralytic shellfish poisoning and about 300 of these are fatal. The paralytic shell poisoning was unknown in India until 1980. In 1981, 3 people died and 85 people were hospitalized due to paralytic shellfish poisoning in Tamil Nadu, India. In 1983, a boy died and a few people were hospitalized as they had ingested clam (*Meretrix casta*). The toxins accumulate in the digestive glands of the shellfish but do not harm it. The effects vary on the type of poisonous fish consumed, quantity ingested, age of the victim, and the presence of other intoxicants, such as alcohol.

The diarrheic shellfish poisoning mainly produces gastrointestinal effects comprising of diarrhea, nausea, vomiting, and abdominal pain. The symptoms of poisoning appear within a few hours of consumption of the fish and may last for a few days. The toxins affect metabolism, membrane transport, and cell division. They also posses mutagenic potential and may promote tumor growth.

9.2.1.2.5 Algal contamination of food

India has a coastline of more than 7500 km, which means that the sea serves as an important source of food for the people living near the coastline. In the Indian context, many terrible incidences related to seafood have been documented in the past and present centuries. In 1997, 7 people lost their lives and more than 500 fell sick in Vizhinjam (Kerela) after consuming mussel, *Perna indica* that was contaminated with an algal bloom (D'Silva et al., 2012; Karunasagar et al., 1998). In Sep. 2004, a lot of fish died in Kerala due to a massive algal bloom (*Noctiluca scintillans*). This generated a massive stench due to which 200 people mainly children had to be hospitalized because of nausea and breathlessness.

9.2.1.3 External factors influencing microbial growth in foods

9.2.1.3.1 Temperature

As India has an enormous landmass area, there is a considerable difference of temperature in different parts of the country. In general, Southern India experiences a tropical climate, Northern India experiences a temperate climate, and the Himalayan regions experience an alpine climate. The overall mean temperature in India is about 10°C during winters and 32°C during summers, which is ideal for microbial growth.

Microorganisms can be classified based on the temperature requirement for their growth:

1. Psychrophiles require temperatures less than 20°C for growth.
2. Mesophiles require a midrange temperature (20–40°C) for optimal growth.
3. Thermophiles require higher temperatures for growth (more than 40°C).
4. Hyperthermophiles require very high temperatures for growth (more than 60°C).

Almost all the microorganisms of public health significance are mesophiles.

The temperature is one of the most important factors that affects the growth of microorganisms. When the temperature rises, various chemical and enzymatic reactions progress at a much faster pace thus increasing the growth rate of microorganisms. However, above a certain temperature, denaturation of proteins occurs and the growth is arrested. The effect of minimum temperature on growth is understood to a lesser degree and it is believed that at this temperature the cytoplasmic membranes get "frozen" and its ability to form proton pumps and transport nutrients is restrained. Therefore, keeping foods at a very low or high temperature can only prevent growth of certain microorganisms. Viruses and some bacteria may still be able to produce poisoning through such foods (such as pathogenic *E. coli*).

9.2.1.3.2 Oxygen
Microorganisms have variable oxygen requirement for growth. Based on oxygen requirement, organisms are classified into four types:

1. Aerobes are capable of growing at full oxygen tension and many can tolerate elevated levels of oxygen (greater than 21%).
2. Microaerophiles are aerobes that can use oxygen if only it is present at reduced levels in air.
3. Facultative organisms, under appropriate nutrient and culture conditions, can grow in either aerobic or anaerobic conditions.
4. Anaerobes lack respiratory systems and thus cannot use oxygen as the final electron acceptor. There are two types of anaerobes:
    a. Aerotolerant anaerobes can tolerate oxygen and grow in its presence even though they cannot use it.
    b. Obligate anaerobes are killed by oxygen. Obligate anaerobes are unable to detoxify some of the by-products of oxygen metabolism. Anaerobes lack the enzymes, which aerobes have that decompose toxic oxygen products.

9.2.1.3.3 Osmotic pressure
The pressure exerted by the environment on the organism is called as the osmotic pressure. Water is one of the main agents contributing to osmotic pressure. This factor is sometimes employed in preservation of meat products by increasing the salt concentration and thus producing a hypertonic environment and plasmacytosis.

9.2.1.3.4 pH

Most of microorganisms grow maximally at neutral pH (i.e., pH = 7). Some organisms, however, grow at a lower pH and their growth is retarded if the pH is raised (e.g., *Helicobacter pylori*). Changes in the pH affect the protein structure and inhibit growth.

9.2.1.3.5 Processed foods (carcinogens and mutagens)

The global trade of processed foods runs in trillions of US dollars. However, the Indian food industry forms less than 1.5% of this trade. It is estimated that there are more than 300 million consumers. Many mutagens and carcinogens enter the foods (e.g., meats) thorough the process of preservation. These include *N*–nitroso compounds and high-temperature cooking (HCAs and PAHs). The problem of botulism has been already been discussed earlier in this chapter. Various preservatives, coloring-, and flavoring agents are discussed in the later sections.

9.2.2 ENVIRONMENTAL TOXICANTS

9.2.2.1 Pesticides

Pesticides are defined as the chemical or biologic agents used to protect crops from insects, weeds, pests, and fungal diseases and also to shield the harvest from rodents, flies, and insects. They act by targeting body systems or enzymes in the pests, which might be almost similar to systems or enzymes in humans and consequently they pose risks to human health as well. *Pesticide residues* are present in very minute quantities that are retained in the crop and even after harvest or storage and subsequently, gain entry into the food chain.

To increase the crop yield, pesticides are now widely used all over the world. According to a WHO estimate, only about a quarter of global production of pesticides occurs in developing countries while these countries report about 99% of the total pesticide-related deaths. Some pesticides resist degradation and hence persist for longer time in the environment (e.g., DDT, endrin, aldrin, heptachlor clordane, mirex, and hexachlorobenzene). Many of these persistant organic pollutants represent long-term hazards as they magnify up the food chain. Everyone in general and breast-fed babies particularly are vulnerable to their ill effects as these agents are lipophilic and thus deposit in adipose tissues and are also secreted in milk.

Maximum residue limit (MRL) is the maximum legally permitted concentration of a pesticide residue in food commodities considering good agricultural practices have been followed. It is expressed in mg/kg. The MRLs, however, do not represent the safety limits. MRLs are always set at a level far below the levels considered to be safe for humans. Consequently, a food having residues above the MRL may be safe for consumption. In India, MRLs are set under the Food Safety and Standards Act (FSSA), 2006 that was previously known as the Prevention of Food Adulteration Act, 1954. The *Accepted Daily Intake* represents the maximum pesticide intake that can be tolerated from all dietary sources in a day without posing any long-term risk to health. The *Theoretical Maximum Daily Intake* estimates the maximum pesticide intake with the existing MRLs for a person following a particular dietary practice.

The ill effects of pesticide on human health have been long debated. Today, there are numerous chemicals and biologic agents available in the market. Examples include insecticides, herbicides, fungicides, rodenticides, and fumigants. Their effects depend not only on the type of pesticide involved but also on its quantity, age, general health of the person, and the time period for which a person is exposed. Intake of pesticide residues through food and water is linked to neurotoxicity, reproduction defects, endocrine disruption, toxicity to fetus, birth defects, genetic defects, cancers, and blood disorders.

9.2.2.1.1 Organophosphorus compounds

They are the esters, amides, or thiol derivatives of phosphoric, phosphonic, and phosphorothioic acids. In addition to the gastrointestinal route, they also enter the body through the skin and respiratory pathways. Depending upon the agent, poisoning features may vary from mild dizziness, anorexia, anxiety, fascicultions to dyspnoea, respiratory depression, convulsions, cardiac ischemia, arrhythmias, and peripheral neuropathy. Examples include malathion, parathion, demeton, trichlorfon, and diazinon.

9.2.2.1.2 Organochlorines

They are very toxic, highly resistant to degradation, and persist for a longer time in the environment. These compounds contain carbon, chlorine, and hydrogen. They can cause both acute and chronic poisoning. Studies have reported their associations with cancers, neurotoxicity, respiratory problems, Parkinson's disease, and many birth defects. Some of these compounds also interfere with the reproductive and immune systems of the fetus. They accumulate in human tissues over time and are secreted in milk. Examples include DDT, aldrin, endrin, endosulfan, and heptachlor.

9.2.2.2 Heavy metals in foods

Some of the heavy metals, such as, copper, zinc, and manganese are naturally present in human body as micronutrients and play an important role in growth and development. Other heavy metals, such as, cadmium, lead, and mercury produce ill effects on health after a particular concentration (Singh et al., 2011). The main source of entry of these and other such heavy metals is mainly through the food (Liu et al., 2006).

Heavy metal contamination of the food mainly occurs at the time of crop production. Plants absorb heavy metals due to exposure to the polluted air or from contaminated soils. Due to a fall in the ground-water level that is reported from all corners of the country, the wastewater from industries is increasingly being utilized to raise the crops (Singh et al., 2010; Gupta et al., 2008). Some studies have reported that even after proper treatment of wastewater, the heavy metals still persist in such water. When such water is utilized for irrigation over a period of time, these metals enter the food chain and get accumulated in vital organs of the human body, such as, kidneys, liver, and bones. The metals, such as, mercury, lead, arsenic, copper, zinc, aluminum, etc., are capable of causing vomiting, stomatitis, diarrhea, tremor, depression, pneumonia, hemoglobinuria, ataxia, paralysis, and convulsions. Ingestion of nickel in the diet is associated with hand eczema. Chronic intake of cadmium is reportedly associated with ovarian, renal, and prostate cancers. Furthermore, every metal is capable of producing certain specific effects.

Budha nala of Punjab is a classic example of heavy metal contamination of water. The *Budha nala* is a seasonal water stream that runs through the Ludhiana city in the state of Punjab. The city is densely populated and heavily industrialized. As it passes through the city, this seasonal water stream turns into an open drain due to unchecked drainage of industrial and household waste. According to a study conducted in the villages that surround the *Budha nala*, its water was found to be contaminated with mercury, lead, fluoride, chloride, chromium, calcium, magnesium, arsenic, nickel, selenium, beta endosulfan, heptachlor, and many other pollutants. This poses a significant health hazard to the local population.

Similar cases of water contamination with heavy metals have been reported from other parts of the country. West Bengal, Uttar Pradesh, Bihar, Chhattisgarh, Jharkhand, Assam, and Andhra Pradesh are reporting chronic arsenic poisoning due to contaminated ground water (Chakraborti et al., 2009; Guha, 2009; Mukherjee et al., 2006). The urban areas of the country experience greater exposure to atmospheric pollutants that get deposited on various vegetables during their transportation and marketing (Sharma et al., 2009). Consequently, the vegetables are more contaminated with heavy metals in markets as compared to their sites of production.

Constant monitoring of their level in the environment and food products is thus essential to prevent the appearance and progression of adverse effects.

9.2.2.3 Food additives and preservatives

Any substance that is not a food but is either added to or used in food at any step to maintain its quality, taste, consistency, texture, and pH or to meet any other function is called a food additive (Tuormaa, 1994). Similarly, an agent added to food to prolong its shelf life is called a preservative. There is a plethora of such agents especially for preserved food articles (FAO-WHO report, 2012). The overwhelming usage of these agents has raised concerns about their ill effects on human health.

9.2.2.3.1 Coloring agents

These agents are used to produce the desired color in the food articles. They include erythrosine, tartazine, quinoline yellow, sunset yellow, brilliant blue, indigo carmine, carmosine, Ponceau 4R, allura red, amaranth, etc. The allura red, quinoline yellow, and tartazine are linked to hypersensitivity reactions including asthma and skin rashes. In contrast, the erythrosine, sunset yellow, Ponceau red, indigo carmine, and brilliant blue have a potential role in carcinogenesis. Most of these agents are banned in many developed countries or their use has been restricted to fixed levels.

9.2.2.3.2 Food preservatives

They have the quality of preserving foods for longer periods of time. Examples of such preservatives include agents, such as, potassium nitrate (E249), benzoic acid (E210), sodium benzoate (E211), sodium metabisulfite, sulfur dioxide (E220), calcium benzoate (E213), and calcium sulfite (E226).

Out of these, potassium nitrate, sodium benzoate, and sulfur dioxide have possible direct or indirect involvement in development of cancer. Some of the mentioned preservatives are involved in gastrointestinal tract problems (gastric irritation, nausea, and vomiting) while others in allergies and interference with cellular enzymatic machinery.

9.2.2.3.3 Flavoring agents

They are added either to alter the natural taste of the food or to create flavor for substances that do not have desired flavors. The members of this group include saccharine (E954), high fructose corn syrup (HFCS), aspartame (E951), monosodium gluatamate (E621), and acesulfame K (E950), etc. Sacchrine is associated with coagulation problems, carcinogenesis, and obesity while aspartame is linked with neurologic damage. Similarly, monosodium glutamate has been blamed for accelerating Alzheimer's disease and Parkinson's disease. Detection of these agents is discussed in Chapter 11 of this book.

9.3 ADULTERATION OF FOODS

As the Indian population continues to grow, the pressure on agriculture and dairy sectors has always been immense. The increased demand for higher yields from these sectors has led to a wide gap between the demand and supply of food products.

Most of the food products available today in the Indian markets contain some kind of adulterant (Bhatt et al., 2012). The FSSA 2006, describes an *adulterant* as any substance that is or can be used to make the food unsafe, substandard, misbranded, or contains inappropriate material. Milk is probably the most commonly adulterated food product. Reports of synthetic milk (that looks like natural milk but it is prepared by mixing caustic soda, table salt, urea, water, and refined oil) and spurious dairy products and sweets, etc., frequently make headlines in India. An alarming report appeared in a major newspaper, Times of India, in Jan. 2012 that about 70% of milk samples from Delhi did not conform to standards. The rest of country also presented a gloomy picture.

The demand–supply gap has attracted many unscrupulous elements that indulge in adulteration of foods and dairy products to make easy profits. An idea of monetary gain that these people make by selling synthetic milk can be obtained from the fact that a liter of synthetic milk costs around Rs 3 in preparation and is sold at a price of Rs 15–16 a liter after adding it to natural milk. Similarly, adulteration of ghee, sweets and other dairy products is rampant in India. The situation is particularly problematic during the festive seasons in India when the markets are flooded with sweets and other milk-based products. The problem of milk adulteration is so grave that Supreme Court of India, in Dec. 2013, asked state governments to make the necessary amendments in laws to make milk adulteration punishable with life imprisonment. Some of Indian states, such as, Uttar Pradesh, West Bengal, and Odisha have already made such provisions but other states are lagging behind in doing so.

A similar is the situation exists with other daily-use food articles used by the common man. This includes spices, honey, wheat flour, vegetable oils, and other foods. A list of some of the commonly adulterated food stuffs in India is summarized in Table 9.1.

Table 9.1 The Common Food Articles and the Adulterants in India

S. No	Food Item	Adulterant(s)
1	Milk	Water, glucose, detergents, urea, refined oil, formalin, skimmed milk powder, and synthetic milk
2	Ghee	Vanaspati/mashed potatoes
3	Butter	Vanaspati
4	Rabdi (a sweet dish)	Blotting paper
5	Ice cream	Washing powder, saccharin, and metanil yellow
6	Khoa	Starch
7	Sweet curd	Vanaspati
8	Jaggery	Washing soda, chalk powder, and metanil yellow color
9	Pulses	Lead chromate
10	Cereals	Dust, pebbles, and seeds of weeds
11	Wheat flour	Chalk powder, barn dust, and sand
12	Besan	Kesri dal and metanil yellow
13	Yellow dal	Kesri dal
14	Honey	Sugar
15	Whole spices	Dust, damaged seeds, other seeds, and straw
16	Red chilli powder	Brick powder, red color dye, and Sudan III color
17	Green chillies	Malachite green
18	Black pepper	Papaya seeds
19	Turmeric powder	Metanil yellow and aniline dye
20	Olive oil	Tea tree oil
21	Vinegar	Mineral oil
22	Asafetida (hing)	Soap stones, starch, and foreign resin
23	Saffron	Maize cob (dried tendons)
24	Green peas	Artificial color
25	Common salt	White powder

Note: One or more adulterant(s) may be present in the food article at a given time.

In 2008, infant-formula feeds affected nearly 290,000 infants in China, out of which more than 50,000 had to be hospitalized and 6 children died. These infant-formula feeds contained melamine levels up to 2500 ppm which is well above the safe limit. Many of the affected children presented with renal stones. The Chinese farmers had added this compound to raise the protein levels of milk. This incident adversely affected the image of China and consequently many countries imposed a ban on the import of food products from China. The Chinese government took a serious view of the situation. Criminal proceedings began and soon two offenders were executed while a few were sent to prison for life. This is an excellent example of the strict attitude that the government can take to check adulteration.

The authors believe that such sincere efforts are needed in India as well. All parties, including the general public, traders, food inspectors, and the governments are responsible for making adulteration go unabated. The public is responsible because of a disinterest in raising their voice against adulteration even after frequently hearing about the ill effects on health, the traders and food inspectors for adopting a wrong way to make profits, and the governments because of their apathetic attitude. Until and unless all the four segments are involved, the problem is likely to persist for a long time in India. Further details on detection of food adulterants are available in Chapters 10, 11, and 12.

9.3.1 **ADULTERATION AND LAW**

The Agricultural Product Standards Act 1990, regulations relating to dairy products and imitation dairy products (Regulation 2581 of Nov. 20, 1987) states that "milk" represents the normal secretion of the mammary glands of bovines, goats, or sheep.

Under the FSSA sale of spurious and adulterated food articles is punishable. The governments of various states and union territories of India have the duty to implement the FSSA, 2006 and Rules and Regulations, 2011 and take the necessary action against the culprits.

In Dec. 2013, the High Court of the Indian state of Jammu and Kashmir, penalized three manufacturing units involved in turmeric, milk, and *saunf* powder production. It imposed a fine of Rs 30 crores as their products were unfit for human consumption.

REFERENCES

Bhatt, S.R., Bhatt, S.M., Singh, Anita, 2012. Impact of media and education on food practices in urban area of Varanasi. Natl. J. Community Med. 3, 581–588.

Chakraborti, D., Das, B., Rahman, M.M., Chowdhury, U.K., Biswas, B., Goswami, A.B., Nayak, B., Pal, A., Sengupta, M.K., Ahamed, S., Hossain, A., Basu, G., Roychowdhury, T., Das, D., 2009. Status of groundwater arsenic contamination in the state of West Bengal, India: a 20-year study report. Mol. Nutr. Food Res. 53, 542–551.

D'Silva, M.S., Anil, A.C., Naik, R.K., D'Costa, P.M., 2012. Algal blooms: a perspective from the coasts of India. Nat. Hazards 63, 1225–1253.

Dawson, D., 2005. Foodborne protozoan parasites. Int. J. Food Microbiol. 103, 207–227.

FAO-WHO, Evaluation of certain food additives, 2012. Seventy-first report of the Joint FAO/WHO Expert Committee on Food Additives. World Health Organ. Tech. Rep. Ser. 974, 1–183.

Guha, M.D.N., 2009. Chronic arsenic toxicity and human health. Indian J. Med. Res. 128, 436–447.

Gupta, N., Khan, D.K., Santra, S.C., 2008. An assessment of heavy metal contamination in vegetables grown in wastewater-irrigated areas of Titagarh, West Bengal, India. B. Environ. Contam. Tox. 80, 115–118.

Karunasagar, I., Joseph, B., Philipose, K.K., Karunasagar, I., 1998. Another outbreak of PSP in India. Harmful Algae News 17, 1.

Liu, W.X., Li, H.H., Li, S.R., Wang, Y.W., 2006. Heavy metal accumulation of edible vegetable cultivated by People's Republic of China. B. Environ. Contam. Tox. 76, 163–170.

Mukherjee, A., Sengupta, M.K., Hossain, M.A., Ahamed, S., Das, B., Nayak, B., Lodh, D., Rahman, M., Chakraborti, D., 2006. Groundwater arsenic contamination: a global perspective with special emphasis to Asian scenario. J. Health Popul. Nutr. 24, 142–163.

Nzwalo, H., Cliff, J., 2011. Konzo: from poverty, cassava, and cyanogen intake to toxico-nutritional neurological disease. PLoS Negl. Trop. Dis. 5, 1051.

Ramanathan, H., 2010. Food poisoning. A threat to humans. Marsland Press, New York.

Ramesh, K.V., 2007. Food microbiology, first ed. MJP Publishers, Chennai.

Sharma, R.K., Agrawal, M., Marshall, F.M., 2009. Heavy metals in vegetables collected from production and market sites of a tropical urban area of India. Food Chem. Toxicol. 47, 583–591.

Singh, A., Sharma, R.K., Agrawal, M., Marshall, F.M., 2010. Health risk assessment of heavy metals via dietary intake of foodstuffs from the wastewater irrigated site of a dry tropical area of India. Food Chem. Toxicol. 48, 611–619.

Singh, R., Gautam, N., Mishra, A., Gupta, R., 2011. Heavy metals and living systems: an overview. Indian J. Pharmacol. 43, 246–253.

Singh, S.P., Aggarwal, A.D., Oberoi, S.S., Aggarwal, K.K., Thind, A.S., Bhullar, D.S., Walia, D.S., Chahal, P.S., 2013. Study of poisoning trends in North India—a perspective in relation to world statistics. J. Forensic Leg. Med. 20, 14–18.

Tuormaa, T.E., 1994. The adverse effects of food additives on health: a review of the literature with special emphasis on childhood hyperactivity. J. Orthomol. Med. 9, 225–243.

Zain, M.E., 2011. Impact of mycotoxins on humans and animals. J. Saudi Chem. Soc. 15, 129–144.

Detection of food adulterants/contaminants

10

D.P. Attrey

Central Military Veterinary Laboratory, Meerut, Uttar Pradesh, India; High Altitude Research, Defence Research and Development Organisation, Leh, Jammu and Kashmir, India; Amity Institute of Pharmacy, Amity University, Noida, Uttar Pradesh, India; Innovation and Research Food Technology, Amity University, Noida, Uttar Pradesh, India; Amity Institute of Seabuckthorn Research, Amity University, Noida, Uttar Pradesh, India; Lala Lajpat Rai University of Veterinary and Animal Sciences, Hisar, Haryana, India

10.1 INTRODUCTION

We must consume safe and nutritious food to remain healthy. It is the duty of a "Welfare State" to ensure that all its citizens have access to affordable, safe, and nutritious food. As per law, the food offered for sale for human or animal consumption must be uncontaminated/unadulterated and safe. Contamination and adulteration in food not only reduces its nutritional quality but may affect its safety also, which may in turn affect the health of the consumers adversely. Timely and proper detection of adulteration/contamination in food is an essential requirement of food safety controls in a country. Quick detection at household level is likely to render immense help in controlling/reducing the menace of adulteration in food in the society.

10.2 DEFINITION OF ADULTERANT/CONTAMINANT AND ADULTERATION

As per Indian Food Safety and Standards Act, 2006 (FSS, 2006), an "adulterant" means any material which is or could be employed for making the food unsafe, substandard, misbranded, or containing extraneous matter. A "contaminant" means any substance, whether or not added to food, which is present in such food as a result of the production (including operations carried out in crop husbandry, animal husbandry, or veterinary medicine), manufacture, processing, preparation, treatment, packing, packaging, transport, or holding of such food or as a result of environmental contamination and does not include insect fragments, rodent hair, and other extraneous matter.

In USA, a food is generally considered to be adulterated if it contains a poisonous or toxic substance that may cause injury to health. US FDA has, however, laid a clear distinction in food adulteration between those that are added and those that are naturally present. Substances that are added, "may render it injurious to health," whereas substances that are naturally present need only to be present at a level that "does not ordinarily render it injurious to health."

As per the Federal Food, Drug, and Cosmetic Act, 1938 of USA (FDA, 2002), food is "adulterated" if it bears or contains; any "poisonous or deleterious substance" which may render it injurious to health or if it bears or contains a pesticide chemical residue that is unsafe. US FDA also considers food to be adulterated if a substance has been added to increase the product's bulk or weight, reduce its quality or strength, or make it appear of greater value than it is (i.e., "economic adulteration").

10.3 ECONOMIC ADULTERATION

"Economic adulteration" is a huge temptation for making quick and easy money. Although India and China are among the largest producers and consumers of food, especially foods of animal origin, yet due to huge populations, the gap between their production and consumption is also large. The unscrupulous FBOs exploit this gap and adulterate the food fraudulently with cheaper or inferior commodities to increase quantity, with the intention to make quick profit. But such adulterants usually affect the health of the consumers adversely. Milk has often been found to be adulterated with water, synthetic milk, starch, cane sugar, skim milk powder, melamine, fat, ammonium sulfate, etc., which are used to increase its volume while maintaining its specific gravity, fat and protein content, etc. and to give it a natural look. Formalin, caustic soda, antibiotics, hydrogen peroxide, benzoic acid, salicylic acid, carbonates, bicarbonates, etc. are frequently used to increase shelf life of adulterated milk. Other additives such as urea and *vanaspati* are also used.

Like China, one of the biggest food safety issues faced by India is also the "adulteration of milk." But in India it is not melamine, but synthetic milk (prepared by mixing caustic soda, table salt, urea, water, refined oil, etc. and looks like natural milk). It is "economic adulteration." Preparation of synthetic milk may cost Rs 5 per liter, but it may fetch a price of Rs 20 a liter when added to natural milk. Times of India had reported in January 2012 that about 70% of milk samples from Delhi did not conform to standards (Sinha, 2012). Almost similar situation prevails in other parts of India also. Adulteration of sweets, ghee, and other dairy products with lard and animal fats, etc. is also rampant in India. The fraudsters almost have free run during festive seasons when the Indian markets are overstocked with sweets and other milk-based products. Problem of adulteration of milk is so grave that Supreme Court of India, in December 2013 asked State governments to amend laws to make milk adulteration punishable with life imprisonment. Some of Indian states such as Uttar Pradesh, West Bengal, and Odisha have already made such provisions.

While milk adulteration is so rampant in the developing countries, meat adulteration is also resorted to a great extent. According to Shears (2010), the unscrupulous FBOs feel that "why sell meat when you can sell water?" In India, injecting water in carcasses has been a persistent commercial problem and many unscrupulous FBOs inject water in the carcasses after slaughter to increase weight and cheat the consumers.

10.4 COMMON ADULTERANTS/CONTAMINANTS IN FOODS

According to Gahukar, nonpermitted colors are the most common additives to food. Presence of harmful chemicals or microorganisms including those unaffected by thermal processing is also common. Mycotoxins such as, aflatoxins, ochratoxins, fumonisins, zearalenone, patulin, and trichothecenes produced by moulds are found in food supply chain. In fact about 70% of deaths are likely to be of foodborne origin. Gahukar has also observed that illiterate consumers in India often get confused about the quality norms of permitted additives and become victims of irregularities or malpractices in the market. Examples of adulteration/contamination are given here:

Poor-quality cardamoms (from which essential oils have been extracted) are mixed with good-quality green cardamoms. Red pepper powder is adulterated with colored sawdust. Black pepper seeds are adulterated with papaya seeds. Grass seeds (coated with charcoal dust) or mineral oil, split grains or flour of pigeon pea (*Cajanus cajan*) or chickpea (*Cicer arietinum*) are adulterated with grass pea (*Lathyrus sativus*) while preparing snacks or meals. Wheat/millet grains are mixed with buck wheat (*Polygonatum fagopyrum*) flour, or seeds of *Crotalaria* spp. containing toxic alkaloids. Oils and fats containing butylated hydroxyanisole or butylated hydroxyl toluene are mixed with edible oils (rancid oil in edible oils destroys vitamins A and E). Oleomargarine (a product of beef fat) is mixed in butter and gelatin.

"Loose" food (i.e., without any branding or packaging) may contain extraneous matter and/or food mixed with moldy grains containing fumonicin toxin, etc. are also sold sometimes in the open market. Cereals and pulses, sold in the open market, many a times have been found to be adulterated with sand, gravels, stones, earth, or talc, etc. Healthy sorghum and corn grains are mixed with moldy grains containing fumonicin toxin.

Other forms of adulteration/contamination in food are use of formaldehyde as preservative in milk, application of a systemic fungicide (benomyl) to vegetables to inhibit the growth of microorganisms to prevent spoilage, coating of inferior quality, and quantity of wax (containing morpholine as a solvent and emulsifier) on fruits to retain moisture, prevent bursting and physical damage, enhance appearance, and extend storage period/shelf life, artificial ripening of unripe fruits with calcium carbide (produces acetylene, an analogue to ethylene) to retain firmness and to give ripening appearance, application of calcium carbonate containing traces of arsenic and phosphorus to fruits; injecting hormone "oxytocin" to fruits and vegetables to retain their freshness, injecting colored and sweetened water into water melon to impart redness and sweetness to the pulp. There are number of reports of use of nonpermitted colors (NPC) in foods, which cause deleterious effects on health.

10.5 IMPACT OF FOOD ADULTERATION ON HUMAN HEALTH

In 2004, water diluted milk was reported to cause death of 13 infants in China from protein malnutrition and nephrotoxicity. Raw milk, which was being adulterated regularly with water, used to have low protein content. This prompted the fraudsters to increase the protein content of diluted milk. Some unscrupulous FBOs in China found that melamine (which is a flame-retardant used in furniture industry as a plastic "melamine-formaldehyde resin") could be used to increase the nitrogen content of food and started adding it to water-diluted milk and powdered infant formula to increase the nitrogen and, in turn, the apparent protein content. This had disastrous results. An estimated 300,000 children were reported to suffer from kidney problems, of whom 54,000 babies were hospitalized and 6 infants died from kidney stones in China in July 2008. By November 2008, 16 more infants, who were fed with milk powder having melamine, were diagnosed with kidney stones (Wikipedia, 2016). The use of melamine in food is not approved by WHO or any national authorities.

10.6 ARTIFICIALLY RIPENED FRUITS AND MERCURY POLLUTION IN FISH

As per Kidula (2014), calcium carbide is used in the production of calcium cyanamide in fertilizer industry. When it comes into contact with moisture/water, it produces acetylene gas which is not only used in welding but also hastens the ripening of several fruits such as mangoes, bananas, and apples. This chemical is extremely hazardous because it contains traces of arsenic and phosphorus, both having dangerous effects on human health. Consumption of such artificially ripened fruits can cause mouth ulcers, gastric problems, diarrhea, and skin rashes. Free radicals from carbide play a major role in the ageing process as well as in the onset of cancer, heart disease, stroke, arthritis, and perhaps allergies. If pregnant women consume these artificially ripened fruits, it can cause miscarriages and developmental abnormalities if the child is born. Calcium carbide is in fact a bigger threat to us than we may think.

The artificially ripened fruits can be identified by placing it in a bucket of water. If the fruit sinks to the bottom, it means it is naturally mature and fine. However, if it floats, it means it has been harvested prematurely and you should not consume it as yet. Artificially ripened fruit can also be identified by their lack of a uniform color.

10.7 LIST OF ADULTERANTS/CONTAMINANTS IN FOODS AND THEIR HEALTH EFFECTS

Jaiswal (2008) has compiled a list of adulterants/contaminants in foods and their health effects (Table 10.1).

Table 10.1 List of Adulterants/Contaminants in Foods and Their Health Effects

S. No.	Adulterant/ Contaminant	Foods Commonly Involved	Diseases or Health Effects
1	Argemone seeds, argemone oil	Mustard seeds, edible oils, and fats	Epidemic dropsy, glaucoma, cardiac arrest
2	Artificially colored foreign seeds	As a substitute for cumin seed, poppy seed, black pepper	Injurious to health
3	Foreign leaves or exhausted tea leaves, saw dust artificially colored	Tea	Injurious to health, cancer
4	TCP	Oils	Paralysis
5	Rancid oil	Oils	Destroys vitamins A and E
6	Sand, marble chips, stones, filth	Food grains, pulses, etc.	Damage digestive tract
7	*L. sativus*	Khesari dal alone or mixed in other pulses	Lathyrism (crippling spastic paraplegia)
8	Mineral oil (white oil, petroleum fractions)	Edible oils and fats, black pepper	Cancer
9	Lead chromate	Turmeric whole and powdered, mixed spices	Anemia, abortion, paralysis, brain damage
10	Methanol	Alcoholic liquors	Blurred vision, blindness, death
11	Arsenic	Fruits such as apples sprayed over with lead arsenate	Dizziness, chills, cramps, paralysis, death
12	Barium	Foods contaminated by rat poisons (barium carbonate)	Violent peristalsis, arterial hypertension, muscular twitching, convulsions, cardiac disturbances
13	Cadmium	Fruit juices, soft drinks, etc. in contact with cadmium plated vessels or equipment. Cadmium contaminated water and shell-fish	"Itai-itai (ouch-ouch) disease," increased salivation, acute gastritis, liver and kidney damage, prostrate cancer

(Continued)

Table 10.1 List of Adulterants/Contaminants in Foods and Their Health Effects (*cont.*)

S. No.	Adulterant/ Contaminant	Foods Commonly Involved	Diseases or Health Effects
14	Cobalt	Water, liquors	Cardiac insufficiency and myocardial failure
15	Lead	Water, natural and processed food	Lead poisoning (foot-drop, insomnia, anemia, constipation, mental retardation, brain damage)
16	Copper	Food	Vomiting, diarrhea
17	Tin	Food	Colic, vomiting
18	Zinc	Food	Colic, vomiting
19	Mercury	Mercury fungicide treated seed grains or mercury contaminated fish	Brain damage, paralysis, death
20	*Bacillus cereus*	Cereal products, custards, puddings, sauces	Food infection (nausea, vomiting, abdominal pain, diarrhea)
21	*Salmonella* spp.	Meat and meat products, raw vegetables, salads, shell-fish, eggs and egg products, warmed-up leftovers	Salmonellosis (food infection usually with fever and chills)
22	*Shigella sonnei*	Milk, potato, beans, poultry, tuna, shrimp, moist mixed foods	Shigellosis (bacillary dysentery)
23	*Staphylococcus aureus* Enterotoxins–A, B, C, D, or E	Dairy products, baked foods especially custard or cream-filled foods, meat and meat products, low-acid frozen foods, salads, cream sauces, etc.	Increased salivation, vomiting, abdominal cramp, diarrhea, severe thirst, cold sweats, prostration
24	*Clostridium botulinum* toxins A, B, E, or F	Defectively canned low or medium-acid foods; meats, sausages, smoked vacuum-packed fish, fermented food, etc.	Botulism (double vision, muscular paralysis, death due to respiratory failure)
25	*Clostridium perfringens* (welchii) type A	Milk improperly processed or canned meats, fish and gravy stocks	Nausea, abdominal pains, diarrhea, gas formation
26	Diethylstilbestrol (additive in animal feed)	Meat	Sterlites, fibroid tumors, etc.

Table 10.1 List of Adulterants/Contaminants in Foods and Their Health Effects (*cont.*)

S. No.	Adulterant/ Contaminant	Foods Commonly Involved	Diseases or Health Effects
27	3,4-Benzopyrene	Smoked food	Cancer
28	Excessive solvent residue	Solvent extracted oil, oil cake, etc.	Carcinogenic effect
29	Nonfood grade or contaminated packing material	Food	Blood clot, angiosarcoma, cancer, etc.
30	Nonpermitted color or permitted food color beyond safe limit	Colored food	Mental retardation, cancer and other toxic effect.
31	BHA and BHT beyond safe limit	Oils and fats	Allergy, liver damage, increase in serum cholesterol, etc.
32	Monosodium glutamate (flour) (beyond safe limit)	Chinese food, meat and meat products	Brain damage, mental retardation in infants
33	Coumarin and dihydrocoumarin	Flavored food	Blood anticoagulant
34	Food flavors beyond safe limit	Flavored food	Chances of liver cancer
35	Brominated vegetable oils	Cold drinks	Anemia, enlargement of heart
36	Sulfur dioxide and sulfite beyond safe limit	In variety of food as preservative	Acute irritation of the gastro-intestinal tracts etc.
37	Artificial sweeteners beyond safe limit	Sweet foods	Chances of cancer
38	Aflatoxins	*Aspergillus flavus*—contaminated foods such as groundnuts, cottonseed, etc.	Liver damage and cancer
39	Ergot alkaloids from *Claviceps purpurea* toxic alkaloids, ergotamine, ergotoxin, and ergometrine groups	Ergot-infested bajra, rye meal or bread	Ergotism (St. Anthony's fire-burning sensation in extremities, itching of skin, peripheral gangrene)
40	Toxins from *Fusarium sporotrichioides*	Grains (millet, wheat, oats, rye, etc.)	Alimentary toxic aleukia (ATA) (epidemic pan-myelotoxicosis)
41	Toxins from *Fusarium sporotrichiella*	Moist grains	Urov disease (Kaschin–Beck disease)

(*Continued*)

Table 10.1 List of Adulterants/Contaminants in Foods and Their Health Effects (*cont.*)

S. No.	Adulterant/ Contaminant	Foods Commonly Involved	Diseases or Health Effects
42	Toxins from *Penicillium inslandicum, Penicillium atricum, Penicillium citreoviridin, Fusarium, Rhizopus, Aspergillus*	Yellow rice	Toxic moldy rice disease
43	Sterigmatocystin from *Aspergillus versicolor, Aspergillus nidulans,* and *bipolaris*	Food grains	Hepatitis
44	*Ascaris lumbricoides*	Any raw food or water contaminated by human faces containing eggs of the parasite	Ascariasis
45	*Entamoeba histolytica*	Raw vegetables and fruits	Amoebic dysentery
46	Virus of infectious hepatitis (virus A)	Shellfish, milk, unheated foods contaminated with feces, urine, and blood of infected human	Infectious hepatitis
47	Machupo virus	Foods contaminated with rodents urine, such as cereals	Bolivian hemorrhagic fever
48	Fluoride	Drinking water, sea foods, tea, etc.	Excess fluoride causes fluorosis (mottling of teeth, skeletal, and neurological disorders)
49	Oxalic acid	Spinach, amaranth, etc.	Renal calculi, cramps, failure of blood to clot
50	Gossypol	Cottonseed flour and cake	Cancer
51	Cyanogenic compounds	Bitter almonds, apple seeds, cassava, some beans, etc.	Gastro-intestinal disturbances
52	Polycyclic aromatic hydrocarbons (PAH)	Smoked fish, meat, mineral oil-contaminated water, oils, fats, and fish, especially shell-fish	Cancer
53	Phalloidin (alkaloid)	Toxic mushrooms	Mushroom poisoning (hypoglycemia, convulsions, profuse watery stools, severe necrosis of liver leading to hepatic failure and death)

Table 10.1 List of Adulterants/Contaminants in Foods and Their Health Effects (*cont.*)

S. No.	Adulterant/ Contaminant	Foods Commonly Involved	Diseases or Health Effects
54	Solanine	Potatoes	Solanine poisoning (vomiting, abdominal pain, diarrhea)
55	Nitrates and nitrites	Drinking water, spinach rhubarb, asparagus, etc. and meat products	Methaemoglobinemia especially in infants, cancer and tumors in the liver, kidney, trachea esophagus and lungs. The liver is the initial site but afterward tumors appear in other organs
56	Asbestos (may be present in talc, Kaolin, etc. and in processed foods)	Polished rice, pulses, processed foods containing anticaking agents, etc.	Absorption in particulate form by the body may produce cancer
57	Pesticide residues (beyond safe limit)	All types of food	Acute or chronic poisoning with damage to nerves and vital organs such as liver, kidney, etc.
58	Antibiotics (beyond safe limit)	Meats from antibiotic-fed animals	Multiple drug resistance hardening of arteries, heart disease

Source: Jaiswal, P.K., 2008. Common adulterants/contaminants in food—injurious and their health effects; and simple screening tests for their detection. Available from: http://agmarknet.nic.in/adulterants.htm

10.8 DETECTION OF ADULTERANTS/CONTAMINANTS

Food adulteration and contamination can be prevented to a great extent if people are made aware of health hazards and the concerned food safety officials are vigilant and active and work sincerely. Aware consumers of food may be able to prevent health hazards by quick scanning of doubtful food materials. Although estimation of all ingredients, present as adulterants/contaminants in the food may not be possible without an established laboratory (having sophisticated equipment), yet an aware consumer may be able to know whether the food procured by him/her for consumption contains any commonly added adulterant/contaminant. Adequate precautions taken by the consumer at the time of purchase of food can make him alert to avoid procurement of such food. It is equally important for the consumer to know the common adulterants and their effect on health. Initially it may seem to be a little cumbersome, but soon the consumers as well as the suppliers get accustomed and they may ensure supply of unadulterated food for fear of punishment.

Food adulteration, especially with the intention of fraud, is a constant struggle between the "science of deception" and the "science of detection" (Shears, 2010). Food science may develop to any extent, but the fraudsters will always find fraudulent ways and means to cheat public. They are like terrorists whose moves always remain ahead of security forces and who generally succeed in using terrorist devices by overmaneuvering the administrators and even the scientists. Although sophisticated lab techniques are accurate, precise, and reliable, yet they are costly and time consuming. We must find ways to detect adulteration quickly in the kitchen.

The official Indian Standard, that is, IS 15642-1 and -2 (BIS, 2006), provides quick methods for detection of adulterants/contaminants in common food products. IS 15733:2006 further prescribes the apparatus and reagents which may be conveniently kept in a portable kit for quick detection of adulterants/contaminants in common food products. Food Safety and Standards Authority of India has also suggested "quick tests" for detection of adulterants in food. However, many other authors/ agencies have also suggested quick home methods for detection of adulterants to make a broad assessment whether the food item is adulterated or not? Pujani (2014) has identified 10 common food products prone to adulteration. As per Sinha (2012), official machinery these days for food safety and prevention is being constantly maneuvered by the booming adulteration business. Almost every food item which we normally buy has a potential of being adulterated. As such, onus of safeguarding one's family against unsafe food falls on citizens themselves. For this, it is essential that common man should understand the basics about adulteration, that is, which foods are adulterated by which adulterants and simple methods to detect them. The Consumer Guidance Society of India has also listed commonly used food items and has suggested simple home tests using common tricks to check the most common adulterants. Government authorities with great efforts have succeeded in reducing the recurrent occurrences, but have not been able to eliminate it. Only an aware and well-informed consumer will be able to eliminate it by continuous routine monitoring (Dixit, 2016).

New scientific innovative products for quick screening of foods have also been claimed by various commercial organizations which need sophisticated instrumentation and laboratory setup. Examples are GM foods' testing kits, sensing and measurement techniques based on magnetoelastic biosensors to detect foodborne pathogens on-site, directly and in real time; food safety quick detection boxes. The 3M Microbial Luminescence System II (MLS II) is designed for rapid positive release of ultra high temperature (UHT) processed dairy and dairy-related products. It is able to detect the presence of microbial contamination in finished products; neogen food safety and allergen testing kits and various products such as planet-orbitrap, etc. have been claimed to screen the foods for safety quickly.

10.9 SIMPLE SCREENING TESTS FOR DETECTION OF ADULTERATION AT HOME

Although it is not possible to know about adulteration only on visual examination, especially when the toxic contaminants are present at ppm/ppb levels, yet visual examination of the food before purchase makes sure that it contains no insects, visual fungus, foreign matters, etc. Also careful reading of the label on packed food is essential for knowing the ingredients and nutritional value. It also helps in checking the freshness of the food and the period of best before use. The consumer should avoid taking food from an unhygienic place and food being prepared under unhygienic conditions. Jaiswal (2008) has provided simple screening tests (Table 10.2).

Table 10.2 List of Common Contaminants in Food Along With Simple Screening Tests for Their Detection

S. No.	Food Article	Adulteration	Test
1	Vegetable oil	Castor oil	Take 1 mL of oil in a clean dry test tube. Add 10 mL of acidified petroleum ether. Shake vigorously for 2 min. Add 1 drop of ammonium molybdate reagent. The formation of turbidity indicates presence of castor oil in the sample.
		Argemone oil	Add 5 mL, conc. HNO_3 to 5 mL sample. Shake carefully. Allow to separate yellow, orange yellow, crimson color in the lower acid layer which indicates adulteration.
2	Ghee	Mashed potato, sweet potato, etc.	Boil 5 mL sample in a test tube. Cool and add a drop of iodine solution. Blue color indicates presence of starch. "Color" disappears on boiling and reappears on cooling.
		Vanaspati	Take 5 mL of the sample in a test tube. Add 5 mL of hydrochloric acid and 0.4 mL of 2% furfural solution or sugar crystals. Insert the glass stopper and shake for 2 min. Development of a pink or red color indicates presence of Vanaspati in Ghee.
		Rancid stuff (old ghee)	Take one teaspoon of melted sample and 5 mL of HCl in a stoppered glass tube. Shake vigorously for 30 s. Add 5 mL of 0.1% of ether solution of phloroglucinol. Restopper and shake for 30 s and allow to stand for 10 min. A pink or red color in the lower part (acid layer) indicates rancidity.
		Synthetic coloring matter	Pour 2 g of filtered fat dissolved in ether. Divide into two portions. Add 1 mL of HCl to one tube. Add 1 mL of 10% NaOH to the other tube. Shake well and allow to stand. Presence of pink color in acidic solution or yellow color in alkaline solution indicates added coloring matter.

(Continued)

Table 10.2 List of Common Contaminants in Food Along With Simple Screening Tests for Their Detection (*cont.*)

S. No.	Food Article	Adulteration	Test
3	Honey	Invert sugar/ jaggery	Fiehe's test: Add 5 mL of solvent ether to 5 mL of honey. Shake well and decant the ether layer in a petri dish. Evaporate completely by blowing the ether layer. Add 2 to 3 mL of resorcinol (1 g of resorcinol resublimed in 5 mL of conc. HCl). Appearance of cherry red color indicates presence of sugar/jaggery.
			Aniline chloride test: Take 5 mL of honey in a porcelain dish. Add aniline chloride solution (3 mL of aniline and 7 mL of 1:3 HCl) and stir well. Orange red color indicates presence of sugar.
4.	Pulses/besan	Kesari dal (*L. sativus*)	Add 50 mL of dil. HCl to a small quantity of dal and keep on simmering water for about 15 min. If pink color develops, it indicates the presence of Kesari dal.
5	Pulses	Metanil yellow (dye)	Add conc. HCl to a small quantity of dal in a little amount of water. Immediate development of pink color indicates the presence of metanil yellow and similar color dyes.
		Lead chromate	Shake 5 g of pulse with 5 mL of water and add a few drops of HCl. Pink color indicates lead chromate.
6	Bajra	Ergot infested bajra	Swollen and black Ergot-infested grains will turn light in weight and will float also in water
7	Wheat flour	Excessive sand and dirt	Shake a little quantity of sample with about 10 mL of carbon tetra chloride and allow to stand. Grit and sandy matter will collect at the bottom.
		Excessive bran	Sprinkle on water surface. Bran will float on the surface.
		Chalk powder	Shake sample with dil. HCl. Effervescence indicates chalk.
8	Common spices such as turmeric, chilly, curry powder, etc.	Color	Extract the sample with Petroleum ether and add 13N H_2SO_4 to the extract. Appearance of red color (which persists even upon adding little distilled water) indicates the presence of added colors. However, if the color disappears upon adding distilled water the sample is not adulterated.
9	Black pepper	Papaya seeds/light berries, etc.	Pour the seeds in a beaker containing carbon tetrachloride. Black papaya seeds float on the top while the pure black pepper seeds settle down.

Table 10.2 List of Common Contaminants in Food Along With Simple Screening Tests for Their Detection (*cont.*)

S. No.	Food Article	Adulteration	Test
10	Spices (ground)	Powdered bran and saw dust	Sprinkle on water surface. Powdered bran and sawdust float on the surface.
11	Coriander powder	Dung powder	Soak in water. Dung will float and can be easily detected by its foul smell.
		Common salt	To 5 mL of sample add a few drops of silver nitrate. White precipitate indicates adulteration.
12	Chillies	Brick powder grit, sand, dirt, filth, etc.	Pour the sample in a beaker containing a mixture of chloroform and carbon tetrachloride. Brick powder and grit will settle at the bottom.
13	Badi elaichi seeds	Choti elaichi seeds	Separate out the seeds by physical examination. The seeds of badi elaichi have nearly plain surface without wrinkles or streaks while seeds of cardamom have pitted or wrinkled ends.
14	Turmeric powder	Starch of maize, wheat, tapioca, rice	A microscopic study reveals that only pure turmeric is yellow colored, big in size and has an angular structure. However, foreign/added starches are colorless and small in size as compared to pure turmeric starch.
15	Turmeric	Lead chromate	Ash the sample. Dissolve it in 1:7 sulfuric acid (H_2SO_4) and filter. Add 1 or 2 drops of 0.1% diphenyl carbazide. A pink color indicates presence of lead chromate.
		Metanil yellow	Add few drops of conc. Hydrochloric acid (HCl) to sample. Instant appearance of violet color, which disappears on dilution with water, indicates pure turmeric. If color persists metanil yellow is present.
16	Cumin seeds (black jeera)	Grass seeds colored with charcoal dust	Rub the cumin seeds on palms. If palms turn black adulteration in indicated.
17	Asafoetida (heeng)	Soap stone, other earthy matter	Shake a little quantity of powdered sample with water. Soapstone or other earthy matter will settle at the bottom.
		Chalk	Shake sample with carbon tetrachloride (CCl_4). Asafoetida will settle down. Decant the top layer and add dil. HCl to the residue. Effervescence shows presence of chalk.
18	Food grains	Hidden insect infestation	Take a filter paper impregnated with ninhydrin (1% in alcohol). Put some grains on it and then fold the filter paper and crush the grains with hammer. Spots of bluish purple color indicate presence of hidden insects infestation.

Source: Jaiswal, P.K., 2008. Common adulterants/contaminants in food—injurious and their health effects; and simple screening tests for their detection. Available from: http://agmarknet.nic.in/ adulterants.htm and also through http://agmarknet.nic.in_adulterants.htm

10.10 **CONCLUSIONS**

Food adulteration is a menace, which all of us face regularly. Detection of adulteration in food is an essential requirement for ensuring safety of foods we consume. Although sophisticated lab techniques are accurate, precise, and reliable, yet they are costly and time consuming. It is essential to develop reliable "quick screening tests" which a common person can perform at the level of household so as to have a broad picture of status of adulteration in his food in case of doubt. Although there is great scope for improvement and further development, some quick methods of detection of adulterants, developed by various government and private agencies for household application, have been presented.

REFERENCES

BIS, 2006. Indian Standard (IS 15642-1 and 2). Foodgrains, starches and ready to eat foods. Adopted by Bureau of Indian Standards. Available from: https://law.resource.org/pub/in/bis/S06/is.15642.1-2.2006.pdf

Dixit, 2016. Identifying common adulterants. Consumer Guidance Society of India. Available from: http://face-cii.in/sites/default/files/dr_sitaram_dixit_-_2.pdf

FDA, 2002. Federal Food, Drug, and Cosmetic Act (As Amended Through P.L. 107-377, Dec. 19, 2002); 21 U.S.C. 301. Document No. Q:\COMP\FDA\FDA.001. Available from: http://www.epw.senate.gov/FDA_001.pdf

FSS Act, 2006. Food Safety and Standards Authority of India (FSSAI), Ministry of Health and Family Welfare, Government of India. Available from: http://www.fssai.gov.in/portals/0/pdf/food-act.pdf

Jaiswal, P.K., 2008. Common adulterants/contaminants in food - Injurious and their health effects; and simple screening tests for their detection. Available from: http://agmarknet.nic.in/adulterants.htm; http://agmarknet.nic.in_adulterants.htm

Kidula, O., 2014. How to detect & avoid fruits ripened with the calcium carbide chemical. Available from: http://www.afromum.com/detect-avoid-fruits-ripened-calcium-carbide-chemical/

Pujani, S., 2014. 10 Common food products prone to adulteration. Available from: http://listdose.com/10-common-food-products-prone-to-adulteration/

Shears, P., 2010. Food fraud—a current issue but an old problem. Br. Food J. 112 (2), 198–213.

Sinha, D., 2012. Kitchen tricks to expose food adulteration. Mumbai Mirror.

Wikipedia, 2016. Wikipedia, the free encyclopedia. Available from: https://en.wikipedia.org/wiki/List_of_food_contamination_incidents

FURTHER READING

Links to new commercial developments in food safety screening.
http://www.abraxiskits.com/moreinfo/PN510001pub.pdf
http://www.bjzw.com/KuaiJianXiang/PeiZhiYingWen.htm

http://www.abraxiskits.com/moreinfo/PN510001pub.pdf

http://spie.org/newsroom/technical-articles/5836-a-fast-simple-food-safety-check?
ArticleID=x113130

http://planetorbitrap.com/targeted-screening-and-quantitation-of-food-contaminants#.Vqwe-
GtJ97IU

http://www.waters.com/webassets/cms/library/docs/720004066en.pdf

http://www.3m.com/3M/en_US/company-us/

http://www.3m.com/3M/en_US/company-us/all-3m-products/~/All-3M-Products/Food-
Safety-Microbiology/UHT-Testing/?N=5002385+8711017+8711414+8716593+329485
7497&rt=r3

Recent advances in detection of food adulteration

11

D. Banerjee*, S. Chowdhary*, S. Chakraborty*, R. Bhattacharyya**

*Experimental Medicine and Biotechnology Department, Postgraduate Institute of Medical Education and Research, Chandigarh, India; **Biotechnology Department, Maharishi Markandeshwar University, Mullana, Haryana, India

Economically motivated food adulteration or food fraud is currently recognized as a great threat to public health. It is high time to shift the focus of addressing this problem from intervention to prevention (Spink and Moyer, 2011). It is a widely occurring phenomenon practiced throughout the globe and has attracted the attention of the community since the last century (http://www.ncbi.nlm.nih.gov/pubmed/20756445). The public health related aspect of food adulteration is the matter of academic study (Wiley, 1899) and the malpractice is attempted to be controlled by systematic governance since the last century (Richards, 1889; Valade, 1898). A specialized database in this field is expected to encourage objective research for development of strategies that may prevent the occurrence of food fraud in the days to come (Moore et al., 2012). Almost all food is reported to be subjected to food adulteration including dairy products, fish, seafood, honey, oils, grains, alcoholic beverages, infant formula, etc. It appears that the traditional food safety strategies are not sufficient to control the issue and innovative means have to be developed that may be used by less trained personnel in field settings exploring less expensive, user-friendly tools to achieve the desired goal (Everstine et al., 2013). This chapter aims to review the available methods of detection of food fraud with a special focus on the detection of common adulterants and on recent advances.

Conceptually food fraud and food contamination are different. Food fraud is generally done for some economic gain. Food contamination may happen as a natural consequence of a process. For example, contamination of alcoholic beverages with biogenic amines is possible while production by means of fermentation that can happen without any ill motive of the producer (Romano et al., 2012). Another example of food contamination is by the growth of aflatoxin producing fungus on chillies or red pepper during storage without any motive of such a contamination (Khan et al., 2014; Alpsoy et al., 2013). The effect of food contamination can be deleterious. Aflatoxins can cause liver cancer (Farkas and Tannenbaum, 2005). Microorganisms may also be food contaminants that can pose a significant threat to public health.

Mankind is equipped with modern technology enforceable by regulation for appropriate governance for microorganism contamination (Food and Drug Administration, HHS, 2014). It is quite natural that food adulteration is done in such a manner so that the adulterant is not identified easily by technologies that are available in public laboratories.

Dairy products and pet foods are adulterated with melamine to falsely inflate the protein content. Chronic melamine exposure is expected to cause nephropathy and a wide range of abnormalities in the human body. Detection of melamine in food materials requires sophisticated instrumentation and highly trained personnel, which the common food detection laboratories of developing nations may not be equipped with. The most recent melamine scandal has been reported from China, which has caused deaths and had generated awareness. Numerous methods are developed for melamine estimation (Rai et al., 2014). Therefore, now traders use low-cost plant proteins to adulterate dairy products. To detect this, new technology has been developed (Scholl et al., 2014).

11.1 COLOR AS FOOD ADDITIVE: DETECTION METHODOLOGIES

Food color is known to affect perception of the food quality (Clydesdale, 1993). Therefore, scientific research has been diverted for the maintenance of food color (Zhou et al., 2014). For the same reason, prohibited coloring substances are common food adulterants in spite of the fact that they cause many types of health hazards. Even permitted food dyes are used in excess amounts to attract the customers (Dixit et al., 2013; Husain et al., 2006; Stevens et al., 2015). However, there have been efforts to develop techniques to detect food dyes. Initially, thin-layer chromatography (Oka et al., 1994), adsorptive voltammetry (Ni et al., 1996), and spectrophotometry-based methods (Nevado et al., 1994; Cruces Blanco et al., 1996; Vidotti et al., 2005) have been developed. Nonetheless, all these methods are time consuming and are not suitable for colorant mixtures because both water-soluble and fat-soluble adulterants are present in foods. Thus, capillary electrophoresis (Sádecká and Polonský, 2000), reversed-phase liquid chromatography (RPLC) (Hann and Gilkison, 1987; Greenway et al., 1992), and ion-pair RPLC (White and Harbin, 1989; Chen et al., 1998) have been tested for color adulterant detection. Among these methods, RPLC or high performance liquid chromatography (HPLC) and ultrahigh performance liquid chromatography (UPLC)-based techniques are predominant because of the usability of two phases; one stationary nonpolar or hydrophobic phase where fat-soluble adulterant will bind and another mobile phase for elution of water-soluble adulterants. In advance cases elution gradient is used to detect water-soluble adulterants more specifically. In this area of research, a number of methods are published in the scientific literature and an outline is available in Table 11.1. The published methodologies, which are in use mostly, rely on chromatography-based techniques with improved detection system. These methodologies are not yet possible to be used by less trained

Table 11.1 Coloring Dye Adulteration Detection Methods

Coloring Agent/Dye	Food	Technique	Limit of Detection (LOD)[a]	Reference
Allura red and sunset yellow	Strawberry jelly and wine	Liquid chromatography–mass spectrometry (LC/MS)	Less than 69%	Ates et al., 2011
Erythrosine, tartrazine, sudan I–IV, metanil yellow, rhodamine B, and red B, 7B, G, black B	Confectionery products, dried fruits, wines, bitter sodas, juices, sauces, pastes, and spices	LC/MS	69.6–116%	Ates et al., 2011
Eight permitted and nonpermitted orange II and metanil yellow dyes	Sugar, fat, and starch based	HPLC with reversed-phase C18 micro Bondapak column, ammonium acetate and acetonitrile gradient elution, λ_{max} detection	Recovery 82–104% LOD 0.01– 0.12 mg L^{-1}	Dixit et al., 2010
Anthocyanins	Saskatoon berry, black currant, blackberry, black raspberry, elderberry, cherry, plum, grape, bilberry, and red cabbage	HPLC method with reversed phase C18 stationary phase and water–methanol–formic acid mobile phase		Mazza, 1986
Carotinoids	Dark orange carrots in raw and frozen forms	HPLC with C18 ODS-3 column with a mobile phase of acetonitrile–methylene chloride–methanol		Simon and Wolff, 1987
Betalaines	Fermented red beetroot extract	HPLC with reverse phase C18 stationary phase and mobile phase methanol–water containing phosphate buffer with gradient elution		Pourrat et al., 1988
Lac color	Jellies	HPLC with reverse phase C18 column and acetonitrile–water as mobile phase with detection at 495 nm		Yamada et al., 1989
Turmeric	Extract from curcuma turmeric	HPLC with reverse phase C18 column and water–tetrahydrofuran as mobile phase with isocratic elution and spectrophotometric and fluorescence detection		Rouseff, 1988

(Continued)

Table 11.1 Coloring Dye Adulteration Detection Methods (*cont.*)

Coloring Agent/Dye	Food	Technique	Limit of Detection (LOD)[a]	Reference
Azo dyes	Boiled sweets, fruit gums, lemon curd, jelly, blancmange, and soft drinks	HPLC with ODS-2 Spherisorb (5 µm), mobile phase water–methanol containing ammonium acetate buffer, absorbance at 475 nm	Amaranth 0.64 mg mL^{-1}, brown FK 1.58 mg mL^{-1}, Ponceau 4R 0.5 mg mL^{-1}, and sunset yellow 0.5 mg mL^{-1}	Greenway et al., 1992
Xanthene (fluoresceine, 4',5'-dibromofluoresceine, eosine Y, ethyleosine, 2',7'-dichlorofluoresceine, tetrachlorofluoresceine, 4',5'-diiodofluoresceine, erythrosine B, and phloxine B	Sugar solution	HPLC with polystyrene–divinylbenzene column and water–acetonitrile mixture containing tetramethylammonium hydroxide and phosphoric acid mobile phase, with diode array spectrophotometric detection		Van Liedekerke and De Leenheer, 1990
Sudan I, Sudan II, Sudan III, Sudan IV, canthaxanthin, and astaxanthin	Animal feed	Sample extracted in acetonitrile, UPLC with C18 SPE column coupled with a diode array detector at 500 nm, acetonitrile–water mobile phase, eluted by acetonitrile–formic acid gradient	Recovery 62.7–91.0% LOD 0.006–0.02 mg kg^{-1}	Hou et al., 2010
Water-soluble tartrazine, amaranth, Ponceau 4R, sunset yellow FCF, and fat-soluble Sudan (I–IV)	Water-soluble dyes—soft drink and ginger Fat-soluble dyes—chilli powders and chilli spices	Extracted in DMSO (for solid sample), HPLC (C18 column) with diode array detector coupled with a Micrcmass ZQ2000 electrospray mass spectrometer, acetate buffer with methanol mobile phase with gradient elution	Recovery 93.2–108.3% LOD 0.01–4.0 ng	Ma et al., 2006

| Triphenylmethane dyes including malachite green, leucomalachite green, crystal violet, leucocrystal violet, and brilliant green | Fish muscle and skin | Sample extracted by acetonitrile containing 1% acetic acid, LC–MS/MS analysis by ODS-4 C18 column with ammonium acetate buffer and acetonitrile gradient, mass detection on a triple-quadrupole tandem mass spectrometer by multiple reaction monitoring mode via electrospray ionization | Recovery 96–106% LOD for malachite green 0.56 µg kg^{-1}, leucomalachite green 0.31 µg kg^{-1}, crystal violet 0.43 µg kg^{-1}, leucocrystal violet 0.37 µg kg^{-1}, and brilliant green 0.22 µg kg^{-1} | Kaplan et al., 2014 |

^{a}Detection limits are mentioned wherever available.

personnel on field visit; and without development of simple tests, it is difficult to control the widespread use of banned food colors. The matter has also been reviewed from an estimation point-of-view since considerable current interest has developed to control the hazards of food colorants (Gennaro et al., 1994; Kartheek et al., 2011). Details of some food color-detection methods are presented in Table 11.1.

11.2 PRESERVATIVES AS FOOD ADDITIVES

Every civilized society is using food preservatives but such practice can pose a threat to public health (Wiley, 1908; Boğa and Binokay, 2010). Safe and efficient preservative development for perishable food items is a matter of intensive research. For example, under refrigeration conditions, appropriate mixture of potassium lactate and sodium diacetate is observed to be an acceptable preservative (Fik et al., 2008). Salt is observed to be an efficacious meat preservative but it can induce hypertension (Uzan and Delaveau, 2009). Safety and efficiency of preservatives are the fundamental parameters that have to be taken into account for long-term food preservation. However, addition of harmful preservatives to food is often reported which is a malpractice (Williams et al., 2004). We should be equipped to control addition of harmful food preservatives but in the modern world banning of safe and efficacious food preservative is not possible. Therefore, appropriate methodologies have to be in place to screen food preservatives and the quality of food. In Table 11.2, various techniques of food preservative detection are discussed.

11.3 HARMFUL ASPECTS OF FOOD PRESERVATIVES

Formalin is misused for preservation or for processing of fish, meat, etc. (Lakshmi, 2012). It is particularly used to avoid the cost of refrigeration. The toxicity aspect of formalin in food items is proved beyond any doubt and its presence in the food has to be controlled strictly to avoid toxicological consequence.

Sodium nitrite is widely used as a meat preservative and it inhibits the growth of various microorganisms in meat products. However, at higher concentrations it is known to act as a cocarcinogen. Therefore if permitted dose is increased for food-preservation purposes, there is a high chance of toxicity to the public at large (Cammack et al., 1999; Walker, 1990). Nitrites in food items are reported from Finland (Penttilä, 1996).

Sulfite and benzoic acids that are commonly used as food preservatives are known to induce urticaria, asthma, and even an anaphylaxis reaction (Ortolani et al., 1999). Therefore, any concept of a universal food preservative has to be examined with caution and indepth toxicological analysis should be performed. From this discussion it is evident that food preservatives are necessary items but they have to be used in moderation following accepted guidelines. This can be only achieved if detection methodologies are available for each food preservative and are widely practiced

Table 11.2 Preservative Adulteration Detection Methods

Preservative	Food	Technique	Detection Limit[a]	References
Formaldehyde	Aquatic products, such as, shrimp and squid	Surface enhanced Raman Spectroscopy, Au/SiO$_2$ used as enhancer substrate	LOD 0.17 µg L^{-1}	Zhang et al., 2014
Benzoates, bronidox, 2-phenoxyethanol, parabens, butylated hydroxyanisole, butylated hydroxytoluene, and triclosan	Cosmetic products	Solid-phase microextraction followed by gas chromatography–tandem mass spectrometry	Recovery >85% LOD <0.000092%	Alvarez-Rivera et al., 2014
Sulfite, ascorbate, benzoate, and sorbate	Ground beef	Colorimetric estimation	Recovery 95% (sodium sulfite), 103% (sodium benzoate), 90% (potassium sorbate), 81% (sodium ascorbate) LOD 0.005% (potassium sorbate) and 0.001% (other preservatives)	Karasz et al., 1976
Nitrate and nitrite (simultaneous)	Cured meat	Sequential injection analysis, ammonium chloride + EDTA as carrier solution, copperized cadmium column for reduction of nitrate, sulfanilamide and N-(1-naphtyl)-ethylenediamine dihydrochloride used for diazotization reaction, absorbance at 538 nm		Oliveira et al., 2004
Benzoate and sorbate	Orange beverage and tomato concentrate	Gas chromatography, extracted by diethyl ether in acidic media, converted into methyl esters by thionyl chloride, methyl esters separated on column packed with 15% EGA coated on 80–100 mesh Chromosorb W AW DMCS	Recovery 96–97% (benzoic acid), 89–92% (sorbic acid) LOD 20 ng (benzoic acid), 16 ng (sorbic acid)	Giryn and Gruszczyn'ska, 1990

(Continued)

Table 11.2 Preservative Adulteration Detection Methods (*cont.*)

Preservative	Food	Technique	Detection Limit[a]	References
Dehydroacetic acid, benzoic acid, sorbic acid, and salicylic acid	Cosmetic products	Isocratic reversed-phase HPLC (TSK gel ODS-80TM column), tetra-*N*-butylammonium hydroxide as an ion-pair reagent, mobile phase as mixture of water, methanol, and TBA hydroxide	Dehydroacetic acid 2.5 ng, benzoic acid 4.0 ng, sorbic acid 2.0 ng, and salicylic acid 5.5 ng	Mikami et al., 2002
Sweeteners and preservatives (acesulfame K, alitame, aspartame, cyclamic acid, neotame, saccharin Na, *p*-hydroxybenzoic acid methyl, *p*-hydroxybenzoic acid ethyl, *p*-hydroxybenzoic acid isopropyl, *p*-hydroxybenzoic acid propyl, *p*-hydroxybenzoic acid isobutyl, *p*-hydroxybenzoic acid butyl, dulcin, glycyrrhizic acid, neohesperidin dihydrochalcone, rebaudioside A, stevioside, sucralose, benzoic acid, sorbic acid, and dehydroacetic acid	Solid and liquid samples	Liquid chromatography and tandem mass spectrometry XSelect CSH phenyl–hexyl column, mobile phase of acetate buffer–acetonitrile, MS detection with negative-ion electrospray ionization	Recovery 70.9–119.0% LOD 0.001 g kg⁻¹ (acesulfame K, alitame, aspartame, cyclamic acid, neotame, saccharin Na, *p*-hydroxybenzoic acid methyl, *p*-hydroxybenzoic acid ethyl, *p*-hydroxybenzoic acid isopropyl, *p*-hydroxybenzoic acid propyl, *p*-hydroxybenzoic acid isobutyl, and *p*-hydroxybenzoic acid butyl) LOD 0.005 g kg⁻¹ (dulcin, glycyrrhizic acid, neohesperidin dihydrochalcone, rebaudioside A, stevioside, sucralose, and benzoic acid) LOD 0.02 g kg⁻¹ (sorbic acid and dehydroacetic acid)	Tsuruda et al., 2013

Aspartame	Soft drinks and commercial pharmaceutical formulations	Flow-injection analysis with bienzymatic biosensor (alcohol oxidase, carboxyl esterase, bovine serum albumin, and glutaraldehyde) and cobalt–phthalocyanine screen-printed electrodes	Recovery 94.1–106.2% LOD 0.2 µM	Radulescu et al., 2014
Sinigrin and allyl isothiocyanate	Mustard samples (ground and cracked seeds, powders, and bran)	HPLC (Sphereclone ODS-2 column) with photodiode array detector (λ_{max} 228 nm for sinigrin and λ_{max} 242 nm for allyl isothiocyanate), mobile phase ammonium acetate and acetonitrile	LOD 0.1 µg mL^{-1} for both sinigrin and allyl isothiocyanate	Tsao et al., 2002

[a]Detection limits are mentioned wherever available.

to screen illegal addition of food preservatives. The developed nations are vigilant about the issue and estimation of amount of food preservatives is regularly done by government agencies to observe that food preservatives are used within limits (Ishiwata et al., 2001). Inadequate use of a food preservative is another issue that may contribute to food fraud and that is also often reported (Tfouni and Toledo, 2002). Several attempts have been documented to develop food preservative database prospectively, but till now no systematic effort is visible to create a real-time food additive/ preservative database (Jakszyn et al., 2004). To avoid the hazards of chemical preservatives, physical methods of food preservation are a matter of intense research (Heinz et al., 2001) but chemical preservation of food is still a prevalent practice throughout the globe.

11.4 OTHER ADULTERANTS: A CHALLENGE FOR DETECTION

Other than food coloring substances and preservatives some adulterants are reported to be deliberately mixed in food items for economic gains. Olive oil, milk, honey, and saffron are food items that are reported to be mostly adulterated and such adulterants are detectable by HPLC and infrared spectroscopy-based methodologies (Moore et al., 2012). Butter adulteration with margarine is detected with the help of Raman spectroscopy (Uysal et al., 2013). Butter is also reported to be adulterated with mutton fat and may be screened by FTIR spectroscopy (Fadzlillah et al., 2013). FTIR spectroscopy may also be explored to identify pork adulteration in beef meatballs (Rohman et al., 2011) and lard adulteration from meatball broth (Kurniawati et al., 2014). FTIR spectroscopy is gradually emerging as a tool for rapid detection of adulterants in a variety of food materials (Rodriguez-Saona and Allendorf, 2011). Recently, meat adulteration is also detected by PCR- and RFLP-based molecular technologies (Doosti et al., 2014; Rahman et al., 2014). Sunflower and corn oils are common adulterants for olive oil and can be detected by FTIR spectroscopy coupled with chemometrics. Trans fats are common adulterant of edible oils. Gas chromatography with flame ionization detection or ATR–FTIR spectroscopy are approved methods for detection of the same (Tyburczy et al., 2013). Although, gas chromatography-based methods are more sensitive for the detection trans fats (Delmonte and Rader, 2007) but the analytical methodology requires standards that may have to be prepared by published protocols due to lack of commercial availability of such standards (Delmonte et al., 2009). Edible oil is adulterated with argemone oil, which exposes the consumer to sanguinarine and dihydrosanguinarine, the toxic alkaloids, causing dropsy (Sarkar, 1948). Thin-layer chromatography based technology coupled with chemical identification tests is used for their identification (Assefa et al., 2013, Dhayal et al., 2013). Oxytocin is frequently used to enhance growth of vegetables and fruits or to enhance milk production from lactating animals. It is detected by HPTLC-based method (Rani et al., 2013). Monosodium-L-glutamate is used as an adulterant to enhance flavor and it may be detected by HPTLC-based methods (Krishna et al., 2010). Sugar syrups are common honey adulterants, which may be detected using one-dimensional and two-dimensional nuclear magnetic resonance (Bertelli et al., 2010). Near infrared spectroscopy is used as a tool to detect jaggery syrup adulteration of Indian honey (Mishra et al., 2010). Polysaccharide detection has emerged as a test for corn syrup adulteration for honey (Megherbi et al., 2009). Milk may be adulterated with anionic detergents and that may be detected by a paper chromatographic method (Barui et al., 2013). LC/MS-based protocol for the same purpose is reported from infant milk formulation (Tay et al., 2013). Liquid chromatography coupled with tandem mass spectrometry is explored to detect multiple nitrogen-containing milk adulterants (Abernethy and Higgs, 2013). Apart from the aforementioned methods, biosensors (Amine et al., 2006; Patel, 2006), nanosensors (Li and Sheng, 2014), and high-performance capillary electrophoresis-based (Dong et al., 2012) methods for food safety analysis have been published.

11.5 CONCLUSIONS

Since food adulteration is a huge concern in all parts of the globe, there is voluminous literature on various aspects of food adulteration including its detection. In fact, methodological procedures of food adulteration detection have been reviewed extensively (Druml and Cichna-Markl, 2014; Li and Sheng, 2014; Cheng et al., 2015; Qu et al., 2015; Feng and Sun, 2012).

Many of the methods for detection of food adulteration require elaborate steps of sample preparation prior analysis involving high-end technologies and that makes the whole process difficult to perform and time consuming.

Therefore, considerable interest has emerged in developing rapid methods for food-adulteration detection (Rodriguez-Saona and Allendorf, 2011). Thus, rapid on-line detection of food quality, in a nondestructive manner becomes even more relevant (Ruiz et al., 2008; Sun et al., 2009). The need of the hour is to develop composite in silico tools and computer-vision systems with minimal analytical technology for rapid, nondestructive, highly efficient, and economic food-adulteration detection maneuvers that may be used in the field/at point of use by less trained personnel to generate data with significant reproducibility (Huang et al., 2007; Ma et al., 2016).

REFERENCES

Abernethy, G., Higgs, K., 2013. Rapid detection of economic adulterants in fresh milk by liquid chromatography-tandem mass spectrometry. J. Chromatogr. A 1288, 10–20.

Alpsoy, L., Kiren, A., Can, S.N., Koprubasi, A., 2013. Assessment of total aflatoxin level in red pepper obtained from Istanbul. Toxicol. Ind. Health 29, 867–871.

Alvarez-Rivera, G., Vila, M., Lores, M., Garcia-Jares, C., Llompart, M., 2014. Development of a multi-preservative method based on solid-phase microextraction-gas chromatography-tandem mass spectrometry for cosmetic analysis. J. Chromatogr. A 1339, 13–25.

Amine, A., Mohammadi, H., Bourais, I., Palleschi, G., 2006. Enzyme inhibition-based biosensors for food safety and environmental monitoring. Biosens. Bioelectron. 21, 1405–1423.

Assefa, A., Teka, F., Guta, M., Melaku, D., Naser, E., Tesfaye, B., et al., 2013. Laboratory investigation of epidemic dropsy in Addis Ababa. Ethiop. Med. J. (Suppl. 2), 21–32.

Ates, E., Mittendorf, K., Senyuva, H., 2011. LC/MS method using cloud point extraction for the determination of permitted and illegal food colors in liquid, semiliquid, and solid food matrixes: single-laboratory validation. J. AOAC Int. 94, 1853–1862.

Barui, A.K., Sharma, R., Rajput, Y.S., Singh, S., 2013. A rapid paper chromatographic method for detection of anionic detergent in milk. J. Food Sci. Technol. 50, 826–829.

Bertelli, D., Lolli, M., Papotti, G., Bortolotti, L., Serra, G., Plessi, M., 2010. Detection of honey adulteration by sugar syrups using one-dimensional and two-dimensional high-resolution nuclear magnetic resonance. J. Agric. Food Chem. 58, 8495–8501.

Boğa, A., Binokay, S., 2010. Food additives and effects to human health. Archives Medical Review Journal 19, 141–154.

Cammack, R., Joannou, C.L., Cui, X.Y., Martinez, C.T., Maraj, S.R., Hughes, M.N., 1999. Nitrite and nitrosyl compounds in food preservation. Biochim. Biophys. Acta 1411, 475–488.

Chen, Q.C., Mou, S.F., Hou, X.P., Riviello, J.M., Ni, Z.M., 1998. Determination of eight synthetic food colorants in drinks by high-performance ion chromatography. J. Chromatogr. A 827, 73–81.

Cheng, J., Sun, D.W., Zeng, X.A., Liu, D., 2015. Recent advances in methods and techniques for freshness quality determination and evaluation of fish and fish fillets: a review. Crit. Rev. Food Sci. Nutr. 55 (7), 1012–1225.

Clydesdale, F.M., 1993. Color as a factor in food choice. Crit. Rev. Food Sci. Nutr. 33, 83–101.

Cruces Blanco, C., García Campaña, A.M., Alés Barrero, F., 1996. Derivative spectrophotometric resolution of mixtures of the food colourants Tartrazine, Amaranth and Curcumin in a micellar medium. Talanta 43, 1019–1027.

Delmonte, P., Rader, J.I., 2007. Evaluation of gas chromatographic methods for the determination of trans fat. Anal. Bioanal. Chem. 389, 77–85.

Delmonte, P., Kia, A.R., Hu, Q., Rader, J.I., 2009. Review of methods for preparation and gas chromatographic separation of trans and cis reference fatty acids. J. AOAC Int. 92, 1310–1326.

Dhayal, G.L., Agarwal, H., Mathur, A., Mathur, S., Kishoria, N., Jain, S., et al., 2013. Case report of a small outbreak of epidemic dropsy. J. Indian Med. Assoc. 111, 200–201.

Dixit, S., Khanna, S.K., Das, M., 2010. Simultaneous determination of eight synthetic permitted and five commonly encountered nonpermitted food colors in various food matrixes by high-performance liquid chromatography. J. AOAC Int. 93, 1503–1514.

Dixit, S., Khanna, S.K., Das, M., 2013. All India survey for analyses of colors in sweets and savories: exposure risk in Indian population. J. Food Sci. 78, 642–647.

Dong, Y., Chen, X., Hu, J., Chen, X., 2012. Recent advances in the application of high performance capillary electrophoresis for food safety. Se. Pu. 30, 1117–1126.

Doosti, A., Ghasemi Dehkordi, P., Rahimi, E., 2014. Molecular assay to fraud identification of meat products. J. Food Sci. Technol. 51, 148–152.

Druml, B., Cichna-Markl, M., 2014. High resolution melting (HRM) analysis of DNA—its role and potential in food analysis. Food Chem. 158, 245–254.

Everstine, K., Spink, J., Kennedy, S., 2013. Economically motivated adulteration (EMA) of food: common characteristics of EMA incidents. J. Food Prot. 76, 723–735.

Fadzlillah, N.A., Rohman, A., Ismail, A., Mustafa, S., Khatib, A., 2013. Application of FTIR-ATR spectroscopy coupled with multivariate analysis for rapid estimation of butter adulteration. J. Oleo Sci. 62, 555–562.

Farkas, D., Tannenbaum, S.R., 2005. In vitro methods to study chemically-induced hepatotoxicity: a literature review. Curr. Drug. Metab. 6, 111–125.

Feng, Y.Z., Sun, D.W., 2012. Application of hyperspectral imaging in food safety inspection and control: a review. Crit. Rev. Food Sci. Nutr. 52, 1039–1058.

Fik, M., Surówka, K., Firek, B., 2008. Properties of refrigerated ground beef treated with potassium lactate and sodium diacetate. J. Sci. Food Agric. 88, 91–99.

Food and Drug administration, HHS, 2014. Establishing a list of qualifying pathogens under the Food and Drug Administration Safety and Innovation Act. Final rule. Fed. Regist. 79, 32464–32481.

Gennaro, M.C., Abrigo, C., Cipolla, G., 1994. High performance liquid chromatography of food colours and its relevance in forensic chemistry. J. Chromatogr. A 674, 281–299.

Giryn, H., Gruszczyńska, Z., 1990. Use of gas chromatography for determining benzoic and sorbic acid levels in orange beverage and tomato concentrate. Rocz. Panstw. Zakl. Hig. 41, 217–222.

Greenway, G.M., Kometa, N., Macrae, R., 1992. The determination of food colours by HPLC with on-line dialysis for sample preparation. Food Chem. 43, 137–140.

Hann, J.T., Gilkison, I.S., 1987. Gradient liquid chromatographic method for the simultaneous determination of sweeteners, preservatives and colours in soft drinks. J. Chromatogr. A, 317–322.

Heinz, V., Alvarez, I., Angersbach, A., Knorr, D., 2001. Preservation of liquid foods by high intensity pulsed electric fields—basic concepts for process design. Trends Food Sci. Technol. 12, 103–111.

Hou, X., Li, Y., Wu, G., Wang, L., Hong, M., Wu, Y., 2010. Determination of para red, sudan dyes, canthaxanthin, and astaxanthin in animal feeds using UPLC. J. Chromatogr Sci. 48, 22–25.

Huang, Y., Kangas, L.J., Rasco, B.A., 2007. Applications of artificial neural networks (ANNs) in food science. Crit. Rev. Food Sci. Nutr. 47, 113–126.

Husain, A., Sawaya, W., Al-Omair, A., Al-Zenki, S., Al-Amiri, H., Ahmed, N., Al-Sinan, M., 2006. Estimates of dietary exposure of children to artificial food colours in Kuwait. Food Addit. Contam. 23, 245–251.

Ishiwata, H., Nishijima, M., Fukasawa, Y., 2001. Estimation of preservative concentrations in foods and their daily intake based on official inspection results in Japan in fiscal year 1998. Shokuhin Eisei. Zasshi. 42, 404–412.

Jakszyn, P., Agudo, A., Ibáñez, R., García-Closas, R., Pera, G., Amiano, P., González, C.A., 2004. Development of a Food Database of nitrosamines, heterocyclic amines, and polycyclic aromatic hydrocarbons. J. Nutr. 134, 2011–2014.

Kaplan, M., Olgun, E.O., Karaoglu, O., 2014. A rapid and simple method for simultaneous determination of triphenylmethane dye residues in rainbow trouts by liquid chromatography tandem mass spectrometry. J. Chromatogr. A 4 (1349), 37–43.

Karasz, A.B., Maxstadt, J.J., Reher, J., Decocco, F., 1976. Rapid screening procedure for the determination of preservatives in ground beef: sulfites, benzoates, sorbates, and ascorbates. J. Assoc. Off. Anal. Chem. 59, 766–769.

Kartheek, M., Smith, A.A., Muthu, A.K., Manavalan, R., 2011. Determination of adulterants in food: a review. J. Chem. Pharm. Res. 3, 629–636.

Khan, M.A., Asghar, M.A., Iqbal, J., Ahmed, A., Shamsuddin, Z.A., 2014. Aflatoxins contamination and prevention in red chillies (*Capsicum annuum* L.) in Pakistan. Food Addit. Contam. B 7, 1–6.

Krishna, V.N., Karthika, D., Surya, D.M., Rubini, M.F., Vishalini, M., Pradeepa, Y.J., 2010. Analysis of monosodium L-glutamate in food products by high-performance thin layer chromatography. J. Young Pharm. 2, 297–300.

Kurniawati, E., Rohman, A., Triyana, K., 2014. Analysis of lard in meatball broth using Fourier transform infrared spectroscopy and chemometrics. Meat Sci. 96, 94–98.

Lakshmi, V., 2012. Food adulteration. Int. J. Sci. Invent. Today 1, 106–113.

Li, Z., Sheng, C., 2014. Nanosensors for food safety. J. Nanosci. Nanotechnol. 14, 905–912.

Ma, M., Luo, X., Chen, B., Su, S., Yao, S., 2006. Simultaneous determination of water-soluble and fat-soluble synthetic colorants in foodstuff by high-performance liquid chromatography-diode array detection-electrospray mass spectrometry. J. Chromatogr. A 1103, 170–176.

Ma, J., Sun, D.W., Qu, J.H., Liu, D., Pu, H., Gao, W.H., Zeng, X.A., 2016. Applications of computer vision for assessing quality of agri-food products: a review of recent research advances. Crit. Rev. Food. Sci. Nutr. 56 (1), 113–127.

Mazza, G., 1986. Anthocyanins and other phenolic compounds of saskatoon berries *Amelanchier alnifolia* nut. J. Food Sci. 51, 1260–1264.

Megherbi, M., Herbreteau, B., Faure, R., Salvador, A., 2009. Polysaccharides as a marker for detection of corn sugar syrup addition in honey. J. Agric. Food Chem. 57, 2105–2111.

Mikami, E., Goto, T., Ohno, T., Matsumoto, H., Nishida, M., 2002. Simultaneous analysis of dehydroacetic acid, benzoic acid, sorbic acid and salicylic acid in cosmetic products by solid-phase extraction and high-performance liquid chromatography. J. Pharm. Biomed. Anal. 28, 261–267.

Mishra, S., Kamboj, U., Kaur, H., Kapur, P., 2010. Detection of jaggery syrup in honey using near-infrared spectroscopy. Int. J. Food Sci. Nutr. 61, 306–315.

Moore, J.C., Spink, J., Lipp, M., 2012. Development and application of a database of food ingredient fraud and economically motivated adulteration from 1980 to 2010. J. Food Sci. 77, R118–126.

Nevado, J.J.B., Cabanillas, C.G., Salcedo, A.M.C., 1994. Spectrophotometric resolution of ternary mixtures of amaranth, carmoisine and Ponceau 4R by the derivative ratio spectrum-zero crossing method. Fresenius J. Anal. Chem. 350, 606–609.

Ni, Y., Bai, J., Jin, L., 1996. Simultaneous adsorptive voltammetric analysis of mixed colorants by multivariate calibration approach. Anal. Chim. Acta 329, 65–72.

Oka, H., Ikai, Y., Ohno, T., Kawamura, N., Hayakawa, J., Harada, K., Suzuki, M., 1994. Identification of unlawful food dyes by thin-layer chromatography-fast atom bombardment mass spectrometry. J. Chromatogr. A 674, 301–307.

Oliveira, S.M., Lopes, T.I.M.S., Rangel, A.O.S.S., 2004. Spectrophotometric determination of nitrite and nitrate in cured meat by sequential injection analysis. J. Food Sci. 69, C690–C695.

Ortolani, C.C., Bruijnzeel-Koomen, C., Bengtsson, U., Bindslev-Jensen, C., Björkstén, B., Høst, A., Ispano, M., Jarish, R., Madsen, C., Nekam, K., Paganelli, R., Poulsen, L., Wüthrich, B., 1999. Controversial aspects of adverse reactions to food. Allergy 54, 27–45.

Patel, P.D., 2006. Overview of affinity biosensors in food analysis. J. AOAC Int. 89, 805–818.

Penttilä, P.L., 1996. Estimation of food additive intake. Nordic approach. Food Addit. Contam. 13, 421–426.

Pourrat, A., Lejeune, B., Grand, A., Pourrat, H., 1988. Betalains assay of fermented red beet root extract by high performance liquid chromatography. J. Food Sci. 53, 294–295.

Qu, J.H., Liu, D., Cheng, J.H., Sun, D.W., Ma, J., Pu, H., Zeng, X.A., 2015. Applications of near infrared spectroscopy in food safety evaluation and control: a review of recent research advances. Crit. Rev. Food Sci. Nutr. 55 (13), 1939–1954.

Radulescu, M.C., Bucur, B., Bucur, M.P., Radu, G.L., 2014. Bienzymatic biosensor for rapid detection of aspartame by flow injection analysis. Sensors 14, 1028–1038.

Rahman, M.M., Ali, M.E., Hamid, S.B., Mustafa, S., Hashim, U., Hanapi, U.K., 2014. Polymerase chain reaction assay targeting cytochrome b gene for the detection of dog meat adulteration in meatball formulation. Meat Sci. 97, 404–409.

Rai, N., Banerjee, D., Bhattacharyya, R., 2014. Urinary melamine: proposed parameter of melamine adulteration of food. Nutrition 30, 380–385.

Rani, R., Medhe, S., Raj, K.R., Srivastava, M., 2013. Standardization of HPTLC method for the estimation of oxytocin in edibles. J. Food. Sci. Technol. 50, 1222–1227.

Richards, E., 1889. Certain provisions of continenental legislation concerning food adulteration. Science 14, 308–310.

Rodriguez-Saona, L.E., Allendorf, M.E., 2011. Use of FTIR for rapid authentication and detection of adulteration of food. Annu. Rev. Food Sci. Technol. 2, 467–483.

Rohman, A., Sismindari, Erwanto, Y., Che Man, Y.B., 2011. Analysis of pork adulteration in beef meatball using Fourier transform infrared (FTIR) spectroscopy. Meat Sci. 88, 91–95.

Romano, A., Klebanowski, H., La Guerche, S., Beneduce, L., Spano, G., Murat, M.L., Lucas, P., 2012. Determination of biogenic amines in wine by thin-layer chromatography/ densitometry. Food Chem. 135, 1392–1396.

Rouseff, R.L., 1988. High performance liquid chromatographic separation and spectral characterization of the pigments in turmeric and annatto. J. Food Sci. 53, 1823–1826.

Ruiz, D., Reich, M., Bureau, S., Renard, C.M., Audergon, J.M., 2008. Application of reflectance colorimeter measurements and infrared spectroscopy methods to rapid and nondestructive evaluation of carotenoids content in apricot (*Prunus armeniaca* L.). J. Agric. Food Chem. 56, 4916–4922.

Sádecká, J., Polonský, J., 2000. Electrophoretic methods in the analysis of beverages. J. Chromatogr. A 880, 243–279.

Sarkar, S.M., 1948. Isolation from argemone oil of dihydrosanguinarine and sanguinarine; toxicity of sanguinarine. Nature 162, 265.

Scholl, P.F., Farris, S.M., Mossoba, M.M., 2014. Rapid turbidimetric detection of milk powder adulteration with plant proteins. J. Agric. Food Chem. 62, 1498–1505.

Simon, P.W., Wolff, X.Y., 1987. Carotenes in typical and dark orange carrots. J. Agric. Food Chem. 35, 1017–1022.

Spink, J., Moyer, D.C., 2011. Defining the public health threat of food fraud. J. Food Sci. 76, R157–R163.

Stevens, L.J., Burgess, J.R., Stochelski, M.A., Kuczek, T., 2015. Amounts of artificial food dyes and added sugars in foods and sweets commonly consumed by children. Clin. Pediatr. (Phila) 54 (4), 309–321.

Sun, T., Xu, H.R., Ying, Y.B., 2009. Progress in application of near infrared spectroscopy to nondestructive on-line detection of products/food quality. Guang Pu Xue Yu Guang Pu Fen Xi 29, 122–126.

Tay, M., Fang, G., Chia, P.L., Li, S.F., 2013. Rapid screening for detection and differentiation of detergent powder adulteration in infant milk formula by LC-MS. Forensic Sci. Int. 232, 32–39.

Tfouni, S.A.V., Toledo, M.C.F., 2002. Determination of benzoic and sorbic acids in Brazilian food. Food Control 13, 117–123.

Tsao, R., Yu, Q., Potter, J., Chiba, M., 2002. Direct and simultaneous analysis of sinigrin and allyl isothiocyanate in mustard samples by high-performance liquid chromatography. J. Agric. Food Chem. 50, 4749–4753.

Tsuruda, S., Sakamoto, T., Akaki, K., 2013. Simultaneous determination of twelve sweeteners and nine preservatives in foods by solid-phase extraction and LC-MS/MS. Shokuhin Eisei. Zasshi. 54, 204–212.

Tyburczy, C., Mossoba, M.M., Rader, J.I., 2013. Determination of trans fat in edible oils: current official methods and overview of recent developments. Anal. Bioanal. Chem. 405, 5759–5772.

Uysal, R.S., Boyaci, I.H., Genis, H.E., Tamer, U., 2013. Determination of butter adulteration with margarine using Raman spectroscopy. Food Chem. 141, 4397–4403.

Uzan, A., Delaveau, P., 2009. The salt content of food: a public health problem. Ann. Pharm. Fr. 67, 291–294.

Valade, F.X., 1898. On the working of the food and drug adulteration act in Canada. Public Health Pap. Rep. 24, 199–205.

Van Liedekerke, B.M., De Leenheer, A.P., 1990. Analysis of xanthene dyes by reversed-phase high-performance liquid chromatography on a polymeric column followed by characterization with a diode array detector. J. Chromatogr. A 528, 155–162.

Vidotti, E.C., Cancino, J.C., Oliveira, C.C., Rollemberg Mdo, C., 2005. Simultaneous determination of food dyes by first derivative spectrophotometry with sorption onto polyurethane foam. Anal. Sci. 21, 149–153.

Walker, R., 1990. Nitrates, nitrites and N-nitrosocompounds: a review of the occurrence in food and diet and the toxicological implications. Food Addit. Contam. 7, 717–768.

White, P.C., Harbin, A.M., 1989. High-performance liquid chromatography of acidic dyes on a dynamically modified polystyrene-divinylbenzene packing material with multi-wavelength detection and absorbance ratio characterization. Analyst 114, 877–882.

Wiley, H.W., 1899. Food adulteration in its relation to the public health. Public Health Pap. Rep. 25, 145–153.

Wiley, H.W., 1908. Preservatives in food and the effect thereof on the public health. Am. J. Public Hygiene 18, 27–30.

Williams, P., Stirling, E., Keynes, N., 2004. Food fears: a national survey on the attitudes of Australian adults about the safety and quality of food. Asia Pacific J. Clin. Nutr. 13, 32–39.

Yamada, S., Noda, N., Mikami, E., Hayakawa, J., Yamada, M., 1989. Analysis of natural coloring matters in food. III. Application of methylation with diazomethane for the detection of lac color. J. Assoc. Off. Anal. Chem. 72, 48–51.

Zhang, Z., Zhao, C., Ma, Y., Li, G., 2014. Rapid analysis of trace volatile formaldehyde in aquatic products by derivatization reaction-based surface enhanced Raman spectroscopy. Analyst 139 (14), 3614–3621.

Zhou, C., Tan, S., Li, J., Chu, X., Cai, K., 2014. A novel method to stabilize meat colour: ligand coordinating with hemin. J. Food Sci. Technol. 51, 1213–1217.

Role of public health food safety laboratories in detection of adulterants/contaminants

12

D.P. Attrey

Central Military Veterinary Laboratory, Meerut, Uttar Pradesh, India; High Altitude Research, Defence Research and Development Organisation, Leh, Jammu and Kashmir, India; Amity Institute of Pharmacy, Amity University, Noida, Uttar Pradesh, India; Innovation and Research Food Technology, Amity University, Noida, Uttar Pradesh, India; Amity Institute of Seabuckthorn Research, Amity University, Noida, Uttar Pradesh, India; Lala Lajpat Rai University of Veterinary and Animal Sciences, Hisar, Haryana, India

12.1 INTRODUCTION

Every human being wants to have good and safe food. Generally tasty food is considered as good food since common man does not differentiate much between tasty and nutritious food and considers the tasty food as nutritious food. Quality of food is affected not only by taste, texture, color, size, and shape, etc. but also by microbes, chemicals, and other physical factors which are generally hidden/invisible. Food-testing laboratories are essentially required to provide accurate and reliable information to consumers, regulators, researchers, and industry about the safety and quality parameters of food in question through an independent third-party testing of the food.

Food analysis for food safety and quality is carried out in a food-testing laboratory by detecting microbial, chemical, and other contaminants/adulterants, and by assessing quality/nutritional parameters for implementing labeling and other legal regulatory requirements; and also to conduct research and development. A food sample has to be characterized in a food-testing laboratory to assess its composition, structure, physicochemical properties, and sensory attributes. Characterization of food is essential to ensure safety and quality and also for producing the food, which is economical, consistently safe, nutritious, desirable and for helping consumers to make informed choices about their diet (McClements, 2003),

According to Westmoreland (2015), the global food testing market is projected to reach $4.63 billion in 2018 from $3.5 billion in 2013. Although chemical contaminants have long been a safety concern, primarily in the form of pesticides, the

2007 discovery of melamine—a toxic chemical typically used in plastics, fertilizers, and pet foods mainly in China aroused public awareness in China and worldwide. Chemical contaminants reach human body through processed foods (e.g., nitrosamines and acrylamide), environmental pollutants (e.g., pesticides, polychlorinated biphenyls, heavy metals, and dioxins), veterinary drugs' residues (e.g., coccidiostats, antibiotics, beta-agonists, and anabolic hormones), naturally occurring toxins (e.g., mycotoxins, plant toxins and marine biotoxins), and migration from packaging materials (e.g., bisphenol A and 4-methylbenzophenones). Chemicals are being introduced into the food chain at a disturbingly high rate. The food industry is also developing new ways to reduce the presence of naturally occurring toxins in products, for example, use of asparaginase, an enzyme used to mitigate acrylamide's natural formation in dough-based foods, is now used globally. In addition, ongoing studies are investigating the mitigating effects of various antioxidants on acrylamide formation. FSLs help in monitoring level of these contaminants in foods.

US Food and Drug Administration's (FDA) Food Safety Modernization Act (FSMA) became law in 2011, which shifted food safety focus from a reactive to preventive approach in USA. FDA, under FSMA, ensures testing for contaminants every 2 years and establishes science-based performance standards, as appropriate, for specific food matrices or products, through coordinated surveillance and response effort from all major stakeholders in food safety (Westmoreland, 2015).

Generally food safety is linked to microbial contamination only, since most people are either unaware of the other contaminants/adulterants or do not take them seriously. A glaring example is misuse of pesticides in India, which is not only affecting the soil and water but animal and human health also. An Indian Council of medical Research (ICMR) survey has reported that 51% of food in India is polluted. Out of this, 20% of the food has the residues above maximum residue limit (Tetsuo and Gupta, 2010). This persistent problem of pesticide residues in the environment raises severe health-related issues, threatens the marketing of many commodities such as food grain, wool, meat, fruits, vegetables, nursery plants, etc. This problem is being highlighted continuously by various public health groups and environmental agencies (Shodhganga, 2016). Most of the farmers/growers are ignorant about pesticide use. They resort to indiscriminate use of pesticides and chemicals, ignoring the recommended levels. Even veterinary drugs and antibiotics are being used indiscriminately. Such hidden contaminants can only be detected in foods/feeds through proper testing in a competent accredited laboratory. Similar is the case with many other chemicals, heavy metals, adulterants, etc.

Food analysis by Public Health Food Safety Laboratories (FSLs) not only helps in identifying whether the food is contaminated or not but also helps in assessing its quality. Food safety is the backbone of many national health programs and is also an essential part of any national public health system, whereas FSLs are the backbone of any food safety control system. Food control systems now focus mainly on prevention through adequate process controls rather than end-product testing (McKenzie and Hathaway, 2006). Food testing labs help in proving that products produced by food business operators (FBOs) are safe.

Implementation of food safety regulations depends upon elaborate testing of various types of foods and food products in an accredited Public Health FSL for controlling safety and quality of food (Hitech, 2016). The FBOs always try their best to ensure that no harmful substance is present in their product, or such substances are removed before the food is consumed. This can be achieved by following good manufacturing practices (GMP), implementation of government regulations and by regular laboratory (Lab), analysis of food in question which detects harmful substances. Very low levels of harmful chemicals, etc. can be detected only by using highly sensitivity and reliable analytical techniques and equipment. Since the food industry is highly competitive, FBOs continue to make efforts to ensure that their products are of high quality, safe, nutritious, cost effective, and are more acceptable than their competitors. One of the most important concerns of the FBO is to produce a final product that consistently has the same overall properties always, that is, appearance, texture, flavor, and shelf life. Biggest challenge for FBOs is how to control such variations in raw materials/ingredients to produce a product with consistent quality (McClements, 2003). Regular testing can mitigate this problem.

Continuous scientific innovations have made food production a highly specialized and technical field. Similarly global food trade with vast regulations covering different consumption patterns has become more complex. New health threats have also emerged, which require proper surveillance to estimate the burden of foodborne diseases, assess their impact on health and economy, evaluate disease prevention, and control program, along with rapid detection and quick response to outbreaks. Lab data is a major source of information for conducting risk assessment and to plan for risk management and communication.

Foodborne disease surveillance should be integrated with food-monitoring data and data from food animals along the entire food chain. Integrating such data would result in robust surveillance information and allow appropriate priority setting and public health interventions (Wong et al., 2005). In USA, the US Department of Health and Human Services' (HHS) Centers for Disease Control and Prevention (HHS/CDC), in close collaboration with State and Territorial Departments of Health, are responsible for conducting human disease surveillance. The US Department of Agriculture's (USDA) Food Safety and Inspection Service (FSIS), as well as the HHS' Food and Drug Administration (FDA), closely monitor this disease surveillance through a variety of human and technological liaison activities, for example, serotyping of clinical isolates of *Salmonella* at state public health laboratories is a critical part of this surveillance.

HHS/CDC also maintains a reporting system for foodborne disease outbreaks that are investigated and reported by local and state health departments. This is a web-based reporting system called the Electronic Foodborne Outbreak Reporting System (EFORS). EFORS collects standardized information on more than 1200 reports of outbreaks each year. HHS/CDC also conducts active surveillance for foodborne disease through a collaborative active surveillance network called FoodNet. The Foodborne Diseases Active Surveillance Network (FoodNet) is the principal food-borne disease component of HHS/CDC's Emerging Infections Program (EIP), which is a collaborative project of the HHS/CDC, 10 EIP sites in the United States, USDA, and HHS/FDA. FoodNet provides a network for responding to new and emerging foodborne diseases of national importance, monitoring the burden of foodborne diseases, and identifying the sources of specific foodborne diseases (Wong et al., 2005).

PulseNet, the national molecular subtyping network for foodborne disease surveillance in USA and established in 1996 by HHS/CDC, is facilitated by state health department laboratories in subtyping of bacterial foodborne pathogens for epidemiologic purposes. Public health laboratories in all 50 states of USA routinely determine the molecular fingerprints of *Escherichia coli* O157:H7, *Listeria monocytogenes*, and regularly subtype common serotypes of *Salmonella*. Standard protocols have also been developed for subtyping a growing number of other foodborne pathogens. Food regulatory laboratories at HHS/FDA and FSIS also participate, and HHS/CDC maintains the national database of patterns. Rapid electronic comparison of strain patterns in state and national databases provides early detection of clusters of related infections, guiding investigations, and verifying control. PulseNet identifies potential outbreaks that otherwise would have been missed, particularly those that are widely dispersed. Identifying and investigating such outbreaks can identify system problems in food safety, so that they can be corrected (Wong et al., 2005).

HHS/FDA and FSIS officials receive alerts from HHS/CDC's Epi-X electronic alert system. Epi-X is a web-based communications system operated by HHS/CDC. Distribution of information through the Epi-X network is to promote rapid communications of recent outbreaks and other health events among local, state, and federal health officials. Epi-X carries reports of disease events outside as well as inside the United States. Also HHS/FDA has the State Advisory FAX/Email system (SAFES) communication system, which allows HHS/FDA to broadcast FAX and email information to all 50 states on demand. It is regularly used to disseminate information by the Agency.

Wong et al. (2005) have also reported that FSIS and the USDA's Animal and Plant Health Inspection Service have also participated in the development of the "IHRs The Electronic Laboratory Exchange Network" (eLEXNET), which is a seamless, integrated, web-based data exchange system for food testing information that allows multiple agencies engaged in food safety activities to compare, communicate, and coordinate findings of laboratory analyses. eLEXNET is funded by HHS/FDA and supported by USDA and the Department of Defence (DOD). It enables health officials to assess risks and analyze trends, besides providing necessary infrastructure for an early warning system that identifies potentially hazardous foods. At present, there are 108 laboratories representing 49 states of USA that are part of the eLEXNET systems with 62 laboratories actively submitting data. Number of participating laboratories is being increased continuously.

The National Antimicrobial Resistance Monitoring System (NARMS) is an example of a well-coordinated surveillance program among HHS/FDA, HHS/CDC, and USDA. NARMS monitors antibiotic resistance of selected foodborne pathogens isolated from clinical settings (both human and animal) and the antibiotic resistance of isolates from foods. The system was initiated in 1996 in response to public health concerns associated with the approval of fluoroquinolone products for use in poultry. NARMS monitors changes in susceptibilities to 17 antimicrobial drugs of zoonotic enteric pathogens from human and animal clinical specimens, from healthy farm animals, from carcasses of food-producing animals at slaughter, and from isolates from samples of retail foods. The system includes a veterinary arm, a human arm, and a retail food-monitoring arm (Wong et al., 2005).

Foodborne disease surveillance within individual countries is important to track and to monitor domestic foodborne threats to public health. Existing national/regional systems such as that of the HHS/CDC, the European EnterNet, and that of the European Rapid Alert System for Food and Feed are examples of systems that may have applicability internationally. Collected information, including active and passive reporting from subjurisdictions (e.g., state and local public health officials), forms the basis of such systems and, when communicated to other countries, preferably though an international portal, is critical to global monitoring and surveillance.

Wong et al. (2005) have further observed that within individual countries, the surveillance arm of government must coordinate with the regulatory arm of government to enforce food safety standards. These internal food safety networks support global surveillance, communication, and coordination. The current structure for international/regional foodborne disease surveillance includes both formal and informal relationships between and among countries. Formal programs include Global Salm-Surv (a global network of laboratories and individuals involved in capacity building for surveillance, isolation, identification, and antimicrobial resistance testing of *Salmonella*) and the European Commission Health and Consumer Protection weekly reports from the Rapid Alert System for Food and Feed (RASFF). One goal of RASFF is to provide individual control authorities with an effective tool for exchanging information on food safety measures. Yet, formal international foodborne disease surveillance communication is limited.

Efforts are emerging to strengthen international foodborne disease surveillance. HHS/CDC works with other countries to assist in developing their version of Food-Net, such as OZFoodNet (Australia's program). In addition, a meeting (cochaired by HHS/CDC and WHO) at the last International Conference on Emerging Infectious Diseases, focused on the global effort to develop better foodborne disease reporting. Although not a foodborne disease surveillance program, the Global Environment Monitoring System/Food Contamination Monitoring and Assessment Programme, commonly known as GEMS/Food, is an example of a successful, internationally coordinated surveillance effort. GEMS began as a joint project between FAO, the United Nations Environment Programme (UNEP), and WHO in 1976. WHO is the implementing agency for the contributing institutions (located in over 70 countries around the world). GEMS' purpose is to compile data on food contamination and human exposure from different countries for global synthesis, evaluation, and presentation. Data may be submitted to GEMS/food using the compatible Operating Programmes for Analytical Laboratories (OPAL I and II). GEMS data are accessible at the WHO website. Uniform implementation and wide accessibility of the GEMS system make it a model for expanded, international food surveillance efforts (Wong et al., 2005).

Laboratory services must not work in isolation. They must be integrated with the national food control system to obtain the required scientific evidence on various food safety/quality issues affecting public health and to help the trade in solving these problems (FAO (Labs), 2016). United Nations Industrial Development Organization (UNIDO) has three objectives, namely, poverty reduction through productive activities, trade capacity building, and conservation of environment and energy (ITF, 2010). Compliance services are costly but should be considered a public good. Underdeveloped countries may require international assistance to establish the new standards infrastructure, which is costly.

UNIDO has many tools and guides to identify requirements for development of a national quality infrastructure that enables sustainable development and fulfils the technical requirements of the multilateral trading system for developing countries (ITF, 2010), such as "Fast Forward," a joint publication by the International Organization for Standardization (ISO) and UNIDO on the establishment and management of national standards bodies and "LabNetwork" Portal, a joint effort by UNIDO, and the World Association of Industrial and Technological Research Organizations, in partnership with the International Laboratory Accreditation Cooperation, ISO, International Weights and Measures Office (BIPM), and Vimta Labs Ltd (India), to create a global laboratory network in the field of testing and calibration.

Different analytic techniques are used in FSLs for estimating various components/parameters of a food/food product. According to McClements (2003), government agencies have specified a number of voluntary and mandatory standards regarding the composition, quality, inspection, and labeling of specific food products. There are certain "mandatory standards" as follows:

- "Standards of identity," which specify the type and amounts of ingredients that certain foods must contain if they are to be called by a particular name on the food label, for example, peanut butter must have less than 55% fat, ice-cream must have more than 10% milk fat, cheddar cheese must have more than 50% milk fat and less than 39% moisture, etc.;
- "Standards of quality" which set minimum requirements on the color, tenderness, mass, and freedom from defects, for example, in canned fruits and vegetables; or
- "Standards of fill-of-container," which state how full a container must be to avoid consumer deception, as well as specifying how the degree of fill is measured.

The "voluntary standards" are usually the "standards of grade," which grade the meat, dairy products, and eggs voluntarily according to their quality, for example, meats can be graded as prime, choice, select, standard, etc. according to their origin, tenderness, juiciness, flavor, and appearance because superior grade products can be sold for a higher price. The food producers send their products to government laboratories for testing and to get appropriate certification on payment.

According to ITC (2010), the two WTO agreements—Technical Barriers to Trade (TBT) and Sanitary and Phytosanitary Measures (SPS) quality certification followed by verification and assurance of conformance has become essential. It is extremely important to prove that the test results and the quality certificates are genuine. A country has to create infrastructure for standardized inspection sampling, testing certification, accreditation, and calibration that is internationally accepted. In the absence of this, importing country will not get assured quality of the product and required quality certification. To get success in international exports, laboratory services must be accredited at international level. The aim should be that "Once tested and certified, results should be accepted at international level."

Developing countries, in general, have weak infrastructure (including poor lab services) and are not able to take full advantage of the opportunities provided by international trade. They will have to comply with international technical standards to gain access to regional and global value chains in food. In the rules-based trading system, the agro-food sector provides immediate opportunity for food export to earn precious foreign exchange. The agro-processing can be developed into an industry for reducing rural poverty.

The technical regulations and standards applied in developing countries, including packaging, marking, and labeling requirements, are often incompatible with international standards. Laboratory capacity to test and certify goods for developed markets is also highly unsatisfactory. Steps taken by the concerned food safety authorities to nurture a quality culture will build client and consumer confidence not only in international markets but also in domestic markets. Developing countries must prove the reliability of their test data, maintain high-quality certification and inspection procedures, and establish conformity to international standards. Demonstrating a capacity to comply with the standards requires the establishment of efficient testing, certification, and accreditation mechanisms to meet the requirements of the SPS and TBT agreements. Compliance infrastructure will broadly include the following (ITC, 2010):

1. national standards institute;
2. microbiology and chemical testing laboratories;
3. national metrology institute; and
4. national accreditation certification capacity to certify enterprises for ISO 9001, ISO 14001, and ISO 22000 and to train internal auditors.

Although, India has aforementioned infrastructure, it is too inadequate to meet the requirement of such a large country having more than 1.2 billion people. The Indian Food Safety and Standards (Laboratory and Sample Analysis) Regulations, 2011, came into force on August 5, 2011. As per these regulations four Notified Labs for Import, that is, Central Food Labs, and four Referral Labs, that is, Referral Food Laboratories, have been established by Government of India at Kolkata, Ghaziabad, Mysore, and Pune (FSS (Labs), 2011). Their jurisdiction, functioning, and required quantity of sample have also been notified in aforementioned regulations. The functions of a referral laboratory have also been explained.

Broad objectives of establishment of FSL, especially in India, are as follows:

- To offer comprehensive and convenient laboratory services for the analysis and evaluation of raw materials, food ingredients, processed food products, and packaging materials, etc. to the food processing industries and other stakeholders.
- To reduce the time of analysis of samples by reducing the transportation time of samples.
- To aid and advise food-processing industries regarding food-quality parameters, nutritional labeling, and the certifications, etc.
- To develop reliable and specific procedures for food analysis and preparation of standard operating procedures for quality testing of food items.
- To established a surveillance system for monitoring the quality and composition of food especially for the residues and contaminants, etc.
- To generate scientific data on various aspects of food-quality standards and safety parameters.
- To facilitate compliance of national and international standards on various foods manufactured in the region and those being exported/imported.
- To provide training to industry personnel in analytical methods and food-quality management systems.

In India, the following *range of food products are commonly sent for lab tests.*

- *Food grains and milled products*: cereals, pulses, oilseeds and their flours, flour mixes, grits, flakes, puffed, and roasted products.
- *Bakery products*: biscuits, bread, buns, cookies, crackers, and other baked foods.
- *Fruits and vegetables*: fresh fruits and vegetables, minimally processed fruits and vegetables.
- *Processed fruits and vegetables*: canned and frozen fruits and vegetables, specialties, dried and dehydrated fruits and vegetables, jams, pickles, chutneys, ketchup, sauces, pastes, and other preserved items.
- *Convenience foods*: snack foods, wafers powders and mixes, instant mixes, noodles, pasta/vermicelli, etc.
- *Portable water and beverages*: packed drinking water, tea, coffee, soft drinks concentrates, fruit drink, fruit juices, fruit nectar, malted beverages, wines, and liquors.
- *Dairy products*: indigenous dairy products, milk, butter, cheese, ghee, milk powder, condensed milk, malted milk food, flavored milk, infant milk food, and milk-based sweets.
- *Meat products*: sausages and other prepared meats, fish, egg, poultry, and other meat products.
- *Sugar and confectionary products*: candy, caramels, chocolate, processed cocoa, chewing gum, traditional Indian confectionaries.
- *Fats and oils*: vegetable and animal oils and fats, shortenings and margarine.

- *Dietary supplements, nutraceuticals, and functional foods, etc.*: herbal products, breakfast cereals, snack and energy foods, health beverages, and nutrition bars, etc.
- *Spices and condiments*: all types of spices, basil, capsicum, chilies, cinnamon, clove, coriander, cumin, fennel, ginger, mace, mustard, nutmeg, paprika, pepper, sesame, thyme, turmeric, etc.
- *Packaging material*: films, pouches, boxes, laminates, etc.

Food parameters which are commonly tested:

- *Physical testing*: color, specific gravity, bulk density, true density, water activity, viscosity, total soluble solids, refractive index, acid value.
- *Chemical and nutritional analysis*
  - *Chemical analysis*: moisture, protein fat, ash, carbohydrate, reducing and nonreducing sugars, starch, total calories, pH, acidity, etc.
  - *Nutritional analysis*: minerals (iron, calcium, phosphorous, etc.), vitamins (Vit A, Vit C, thiamine, riboflavin, niacin folic acid, etc.), dietary fiber, beta-carotene, lycopene, chlorophyll, anthocyanins, tannins, trypsin inhibitor, phytates, etc.
- *Microbiological tests*: aerobic plate count (APC), yeasts and molds, total coliforms and *E. coli, Staphylococcus aureus, Salmonella, Listeria, E. coli* O157:H7, anaerobic plate count, aerobic spore former count, lactic acid bacterial count, anaerobic spore former count, thermophilic aerobic spore former count, thermophilic anaerobic spore former count, etc.
- *Contaminants and toxins*: aflatoxins, mycotoxins, veterinary drugs and antibiotic residues, pesticide residues, heavy metals, trace metals.
- *Water quality*: physical (clarity, color, odor, taste, turbidity), chemical (TS, hardness, alkalinity, acidity, pH, nitrates, nitrites, free ammonia, chlorides, sulfates, COD, BOD), microbiological (plate count, fecal coliforms).
- *Food ingredients*: food additives and preservatives (sodium chloride, sulfur dioxide, sodium benzoate, sorbic acid, etc.), antioxidants (gallates, BHA, etc.), synthetic colors (amaranth, erythrosine, sunset yellow, tartrazine NS, indigo carmine, brilliant black BN, annatto), artificial sweeteners (saccharin, aspartame, sucralose, etc.).
- *Subjective and objective analysis*: sensory and instrumental analysis of foods for color, flavor texture, and overall acceptability.

FBOs have to understand the role played by various ingredients of food and the processing operations in determining the final properties of the food product, so that they can rationally control the manufacturing process to produce a final product with consistent properties. FSLs help the FBOs in characterization of raw materials, monitoring of food properties during processing, characterization of final product with respect to composition, structure (molecular, microscopic, or macroscopic); physicochemical properties, and sensory properties (McClements, 2003).

Characterization of raw materials: By analyzing the raw materials, it is often possible to predict their subsequent behavior during processing so that the processing conditions can be altered to produce a final product with the desired properties of consistent quality. For example, the color of potato chips depends on the concentration of reducing sugars in the potatoes from which they are manufactured; the higher the concentration, the browner the potato chips shall be. Thus it is necessary to have an analytical technique to measure the concentration of reducing sugars in the potatoes so that the frying conditions can be altered to produce the optimum colored potato chip.

Monitoring of properties of food during processing: Estimation of properties of food during processing helps to solve problems arising during processing to improve overall quality of food and to reduce the amount of material and time wasted. For this reason, FBOs try to use analytical techniques which are capable of rapidly measuring the properties of foods on-line, without having to remove a sample from the process.

Characterization of final product: Once the product has been manufactured, it is important to analyze its properties to ensure that it meets the appropriate legal and labeling requirements, it is safe, of high quality, and retains its desirable properties till consumption (McClements, 2003).

Composition: The composition of a food largely determines its safety, nutrition, physicochemical properties, quality attributes, and sensory characteristics. Composition can be specified in a number of ways depending on the property of interest and type of analytical procedure used such as specific atoms (e.g., carbon, hydrogen, oxygen, nitrogen, sulfur, sodium, etc.); specific molecules (e.g., water, sucrose, tristearin, lactoglobulin, types of molecules (e.g., fats, proteins, carbohydrates, fiber, minerals), or specific substances (e.g., peas, flour, milk, peanuts, butter). Government regulations state that the concentration of certain food components must be stipulated on the nutritional label of most food products, and are usually reported as specific molecules (e.g., vitamin A) or types of molecules (e.g., proteins).

Structure: The structural organization of the components within a food also plays a large role in determining the physicochemical properties, quality attributes, and sensory characteristics of many foods. Hence, two foods that have the same composition can have very different quality attributes if their constituents are organized differently. For example, a carton of ice cream taken from a refrigerator has a pleasant appearance and good taste, but if it is allowed to melt and then is placed back in the refrigerator its appearance and texture will change, which is not likely to be accepted by consumers. Thus, there has been an adverse influence on its quality, even though its chemical composition is unchanged, because of an alteration in the structural organization of the constituents caused by the melting of ice and fat crystals. The structure of a food can be examined at a number of different levels:

> *Molecular structure (~1–100 nm)*: Ultimately, the overall physicochemical properties of a food depend on the type of molecules present, their

three-dimensional structure, and their interactions with each other. Therefore, it is important for food scientists to use analytical techniques to examine the structure and interactions of individual food molecules. *Microscopic structure (~10–100 μm)*: The microscopic structure of a food can be observed by microscopy (but not by the unaided eye) and consists of regions in a material where the molecules associate to form discrete phases, that is, emulsion droplets, fat crystals, protein aggregates, and small air cells. *Macroscopic structure (~>100 μm)*: This is the structure that can be observed by the unaided human eye, that is, sugar granules, large air cells, raisins, chocolate chips.

In order to design new foods, or to improve the properties of existing foods, it is important to understand the relationship between the structural and bulk properties of food. Therefore, analytical techniques are needed to characterize these different levels of structure.

Physicochemical properties: The physiochemical properties (optical, rheological, stability, flavor) of foods ultimately determine their perceived quality, sensory attributes and behavior during production, storage, and consumption.
The optical properties of foods are determined by the way they interact with electromagnetic radiation in the visible region of spectrum, that is, absorption, scattering, transmission, and reflection of light. For example, full fat milk has a whiter appearance than skim milk because a greater fraction of the light incident upon the surface of full fat milk is scattered due to the presence of the fat droplets.
The rheological properties of foods are determined by the way that the shape of the food changes, or the way that the food flows, in response to some applied force. For example, margarine should be spreadable when it comes out of a refrigerator, but it must not be so soft that it collapses under its own weight when it is left on a table.
The stability of a food is a measure of its ability to resist changes in its properties over time. These changes may be chemical, physical, or biological in origin. Chemical stability refers to the change in the type of molecules present in a food with time due to chemical or biochemical reactions, that is, fat rancidity, or nonenzymatic browning. Physical stability refers to the change in the spatial distribution of molecules present in a food with time due to movement of molecules from one location to another, for example, droplet creaming in milk. Biological stability refers to the change in number of microorganisms present in a food with time, for example, bacterial or fungal growth.
The flavor of a food is determined by the way that certain molecules in food interact with receptors in mouth (taste) and nose (smell) of human beings. Perceived flavor of a food product depends on the type and concentration of flavor constituents within it, the nature of food matrix, as well as how quickly

the flavor molecules can move from food to the sensors in mouth and nose. Analytically, flavor of a food is often characterized by measuring concentration, type, and release of flavor molecules within a food or in the headspace above the food.

Therefore, foods must be carefully designed so that they have the required physicochemical properties over the range of environmental conditions that they will experience during processing, storage, and consumption, for example, variations in temperature or mechanical stress. Consequently, analytical techniques are needed to test foods to ensure that they have appropriate physicochemical properties.

Sensory attributes: Ultimately, the quality and desirability of a food product is determined by its interaction with the sensory organs of human beings, for example, vision, taste, smell, feel, and hearing. For this reason the sensory properties of new or improved foods are usually tested by human beings to ensure that they have acceptable and desirable properties before they are launched onto the market. Although sensory analysis is often the ultimate test for acceptance or rejection of a particular food product, it is time consuming and expensive, besides being subjective and not objective. Moreover, these cannot be used on materials that contain poisons or toxins, and it cannot be used to provide information about the safety, composition, or nutritional value of a food. For these reasons, objective analytical tests which can be performed in laboratory using standardized equipment and procedures, are often preferred for testing food product properties that are related to specific sensory attributes. Moreover, FBOs generally try to correlate sensory attributes (such as chewiness, tenderness, or stickiness) to the results of objective analytical techniques, but with varying degrees of success (McClements, 2003).

Choosing an analytical technique: It has been observed that a number of analytical techniques are available to determine a particular property of a food material. Hence it is essential to select the most appropriate technique for the specific application, depending on the property to be measured, the type of food to be analyzed, and the reason for carrying out the analysis. Most foods contain a large number of components and it is necessary to distinguish the component being analyzed (the analyte) from other components surrounding it (the matrix). Food components can be distinguished from each other according to differences in their molecular characteristics, physical properties, and chemical reactions:

Molecular characteristics: Size, shape, polarity, electrical charge, interactions with radiation.

Physical properties: Density, rheology, optical properties, electrical properties, phase transitions (melting point, boiling point).

Chemical reactions: Specific chemical reactions between the component of interest and an added reagent.

Some of the criteria that are important in selecting a technique are as follows:

Precision: A measure of the ability to reproduce an answer between determinations performed by the same scientist (or group of scientists) using the same equipment and experimental approach.

Reproducibility: A measure of the ability to reproduce an answer by scientists using the same experimental approach but in different laboratories using different equipment.

Accuracy: A measure of how close one can actually measure the true value of the parameter being measured, for example, fat content, or sodium concentration.

Simplicity of operation: A measure of the ease with which relatively unskilled workers may carry out the analysis.

Cost: Total cost of analysis, including reagents, instrumentation, and salary of personnel required to carry it out.

Speed: The time needed to complete the analysis of a single sample or the number of samples that can be analyzed in a given time.

Sensitivity: A measure of lowest concentration of a component that can be detected by a given procedure.

Specificity: A measure of ability to detect and quantify specific components within a food material, even in the presence of other similar components, that is, fructose in the presence of sucrose or glucose.

Safety: Many reagents and procedures used in food analysis are potentially hazardous, that is, strong acids or bases, toxic chemicals, or flammable materials.

Destructive/nondestructive: In some analytical methods the sample is destroyed during the analysis, whereas in others it remains intact.

On-line/off-line: Some analytical methods can be used to measure the properties of a food during processing, whereas others can only be used after the sample has been taken from the production line.

Official approval of the method: Various international bodies have given official approval to methods that have been comprehensively studied by independent analysts and shown to be acceptable to various organizations involved, for example, ISO, AOAC, AOCS.

Nature of food matrix: The composition, structure, and physical properties of the matrix material surrounding the analytes often influences the type of method that can be used to carry out an analysis, for example, whether the matrix is solid or liquid, transparent or opaque, polar or nonpolar.

If there are a number of alternative methods available for measuring a certain property of a food, the choice of a particular method will depend on which of the aforementioned criteria is most important. For example, accuracy and use of an official method may be the most important criteria in a government laboratory which checks the validity of compositional or nutritional claims on food products, whereas speed and the ability to make nondestructive measurements may be more important for routine quality control in a factory where a large number of samples have to be analyzed rapidly (McClements, 2003).

12.2 CONCLUSIONS

Food safety is an essential part of any national public health system. In fact it may be considered as the base of any national health program. Public health FSLs play an important role in ensuring safety and quality of food through testing of foods/food products for adulterants/contaminants and for assessment of product quality and nutritive value. Implementation of labeling and other legal/regulatory requirements and conduct of research and development in the area of "foods" is unthinkable without the help from FSLs. It may not be an exaggeration to say that FSLs are the backbone of any food safety control system. Food testing laboratories are essentially required to conduct an independent third party testing to provide accurate and reliable information to the consumers, regulators, researchers, and industry about the quality of the food product in question. FSLs help by proving that the products produced by FBOs are safe and of requisite quality.

REFERENCES

FAO (Labs), 2016. Food safety and quality. Available from: http://www.fao.org/food/food-safety-quality/a-z-index/laboratory/en/

FSS (Labs), 2011. The Indian Food Safety and Standards (Laboratory and Sample Analysis) Regulations, 2011. Available from: http://www.fssai.gov.in/Portals/0/Pdf/5%20Food%20Safety%20and%20Standards%20Laboratory%20and%20Sample%20Analysis%20Regulation%202011%20and%20FAQ.pdf

Hitech, 2016. Food testing brochure. Hitech Testing and Research Centre Private Limited; Okhla Industrial Area, Delhi. Available from: http://www.htrcpl.in/profile.html

ITC, 2010. Challenges in agri-food exports: building the quality infrastructure. International Trade Forum (issue 3/2010). Available from: http://www.tradeforum.org/Challenges-in-Agro-Food-Exports-Building-the-Quality-Infrastructure/

ITF, 2010. The post-2015 development agenda—50 years of trade and development trade in services trade facilitation. International Trade Forum (issue 3/2010).

McClements, D.J., 2003. Analysis of food products, Food Science 581, Department of Food Physico-chemistry, University of Massachusetts Amherst, Chenoweth Laboratory; Holdsworth Way, Amherst, MA, extracted/reproduced with permission from Prof. McClements, D.J. Available from: http://people.umass.edu/~mcclemen/581Sampling.html

McKenzie, A.I., Hathaway, S.C., 2006. The role and functionality of veterinary services in food safety throughout the food chain, New Zealand Food Safety Authority, Wellington, New Zealand. Rev. Sci. Tech. Off. Int. Epiz. 25 (2), 837–848. Available from: http://citeseerx.ist.psu.edu/viewdoc/download?doi=10.1.1.118.746&rep=rep1&type=pdf

Shodhganga, 2016. Available from: http://shodhganga.inflibnet.ac.in/bitstream/10603/27236/11/11_chapter%20_1_%20introduction.pdf

Tetsuo, S., Gupta, R.C. (Eds.), 2010. Anticholinesterase Pesticides, Metabolism, Neurotoxicity, and Epidemiology. John Wiley & Sons, Inc., New Jersey, USA, p. 427.

Westmoreland, K.E., 2015. Lab outsourcing: what you need to know. Food Safety Magazine. Available from: http://www.foodsafetymagazine.com/magazine-archive1/december-2014january-2015/lab-outsourcing-what-you-need-to-know/

Lo Fo Wong, D.M.A., Andersen, J.K., Norrung, B., Wegener, H.C. 2005. Food contamination monitoring and food-borne disease surveillance at national level. Proceedings of Second FAO/WHO Global Forum of Food Safety Regulators; October 12–14, 2004, Bangkok, Thailand; Agenda Item 5.1, GF 02/10, extracted/reproduced with permission from Food and Agriculture Organization of the United Nations. Issued by the Joint Secretariat of the FAO/WHO Global Food of Food Safety Regulators, FAO, Rome; Secretariat of the FAO/WHO Global Food of Food Safety Regulators, FAO, Rome. Available from: http://www.fao.org/docrep/meeting/008/y5871e/y5871e00.htm#Contents

Food safety from farm-to-fork

4

Food-safety issues related to plant foods at farms

13

P. Dudeja*, A. Singh**

*Department of Community Medicine, Armed Forces Medical College, Pune, Maharashtra, India; **School of Public Health, Post Graduate Institute of Medical Education and Research, Chandigarh, India

13.1 INTRODUCTION

Food is vital for survival. Safe food is the essence of life. Foodborne diseases have been a constant concern of every society throughout the history of mankind. Though, fatality only occurs in a minority of cases, the morbidity associated with the cases of food-related illness has significant social and economic consequences. In ancient times, our ancestors would hunt, gather food, and consume it on the same day to satisfy the basic need of hunger. Early humans, probably by trial-and-error, also started to develop the art of recognition and avoidance of foods that were naturally toxic (Griffith, 2006). There is a faint evidence of avoidance of berries and mushrooms during specific seasons. With the discovery of fire, man started to cook food. This made the food tastier and also would eliminate threat from foodborne disease-causing microorganisms. With industrialization and globalization, there has been a lengthening of the food chain where food is stored for longer periods, transported across miles, and processed before it reaches the consumer. All these factors have caused a threat to safety of food.

13.2 FRAGMENTED APPROACH TO FOOD SAFETY IN PAST

Man started making efforts to keep food safe, even before the Biblical times. Various civilizations started basic forms of food preservation, which possibly also made food safer, for example, drying, salting, and fermentation. Ancient Egyptians developed a storage tank designed to store grain harvested from fields called the *silo*. Romans first recognized the importance of freshness of fruits and other foods. They also salted their foods for preservation—a practice that still holds true. Later, it was found that foods kept in cold lasted longer. That is how some people started keeping meat and fish in the waterfall to keep it fresh. Snow and ice were also identified as natural refrigerants.

Food Safety in the 21st Century. http://dx.doi.org/10.1016/B978-0-12-801773-9.00013-3

Our ancestors were also aware about the dangers of unsafe food. The Greeks have mentioned that food poisoning was so common that tasters were employed to check the food before it was served to the royal people. They also described about food adulteration as food became an item of trade. Nutmeg was used to hide the taste and smell of decomposed meat before serving. Numerous taboos existed in the past that indirectly ensured food safety. For example, in 1800 BC, Judea (present day Israel) prohibited consumption of pork. In 500 BC Confucius, a Chinese spiritual leader, warned against eating "sour rice." In India, a list of unclean foods was developed for the first time in 500 BC, which included food items, such as, meat cut with a sword, dog meat, human meat, etc.

The Middle Ages saw developments in food regulations. Ergot (*Claviceps purpura*) was responsible for a number of outbreaks of ergotism (St. Anthony's Fire). There were 40,000 deaths reported due to a single incidence in France in 944 AD. King John developed the first food law in 1202. During the period between 1300 and 1750, mold poisoning killed many people in England and other parts of Europe. This continued till the 19th century. Ergotism and alimentary toxic aleukia were responsible for significant morbidity in Europe. During World War II, thousands of Russians died of alimentary toxic aleukia as they were forced to eat infected grain. Food-related parasites, such as, *Ascaris*, *Trichura*, *Taenia*, and *Fasciola* were also discovered during this time (Motarjemi et al., 2014).

The period of industrial revolution between 1750 and 1900, witnessed a migration of lot of people to cities. The demand for food for people in cities grew. These led to the concept of mass production of food. There was a change in agricultural practices, processing, and trade patterns. There were many episodes of mass-food poisoning. This led to an increased demand for food, which would not get spoilt or cause illness. Canning came as a solution. At this stage, the scientific basis of food poisoning was not fully understood. Nevertheless, the advances in microbiology during the 19th century contributed to a great extent in understanding the concept of food spoilage (Caballero et al., 2003). Work by Antonie van Leeuwenhoek, Pasteur, Ferdinand Julius Cohn, and August Gartneu began to demonstrate that, although people could not see them, there were organisms in the air, soil, animals, and water that could make us sick (Hartman, 2001). Pasteur's work on pasteurization and fermentation had a mammoth impact on the science of food safety. The advances in microbiology complemented innovations in food technology.

Early in the 20th century, when food safety was a major concern for the public, two technologies, namely, milk pasteurization and retort canning, were developed, promoted, and virtually canonized as prevention measures against foodborne diseases. Social changes during this era included a rapid increase in the women-work force. The food-processing industry started growing and brought convenience to those cooking and consuming home-cooked food. This led to a decline in the consumption of home-cooked food. Along with this, the culture of eating-out started. Newer issues, such as, bovine spongiform encephalopathy also emerged and food safety emerged beyond the kitchen. These episodes of foodborne illnesses were paralleled by advances in food science and technology and formulation of food laws.

By and large, the two common tools for ensuring food safety were legislative methods and research in microbiology. The problem has been neatly summarized by Woolen (1999), who wrote *"Millions of words of advice and millions of pounds spent but the problem is getting worse."*

Industrialization, increased trade, and travel have enabled the transportation of food. Now, it frequently crosses international borders before consumption. As there are numerous possible routes for introduction and transmission of pathogens in food, ensuring food safety needs a coordinated multidisciplinary and intersectoral approach. Even in the 21st century, foodborne disease remains a major threat to the public health, as new foodborne pathogens have emerged. With rapid globalization, the forces spearheading the social changes in the food-science scenario have also changed. More and more people are moving to cities. There is a rapid increase in food trade. Globalization of the food supply has led to the rapid and widespread international distribution of foods. Changes in microorganisms have led to the constant evolution of new pathogens, development of antibiotic resistance, and changes in virulence of known pathogens. In many countries, as people increasingly consume food prepared outside the home, growing numbers are potentially exposed to the risks of poor hygiene in the commercial foodservice settings. All of these emerging challenges indicate that there is a strong link in the different stages of food chain and food safety. Preventive approaches in the form of Hazards Analysis and Critical Control Point (HACCP), training of food handlers, ISO certifications, Safe Quality Food, Food Safety Management Systems, Risk Assessments, Risk Analysis, Risk Communication, etc., have successfully been employed to ensure the safety of food at individual levels of the food chain (Bauman, 1994).

13.3 INTEGRATED APPROACH: FARM-TO-FORK

Food safety is a broad issue that requires an integrated response. The single answer to tackling the issue of foodborne hazards, which know no geographical boundaries, lies in the concept of the farm-to-fork approach or the food-chain approach. It has been well said that *"A chain is as strong as its weakest link."* The key here is to strengthen each and every link in the complex process of food reaching its consumer; from the way it is grown or raised, to how it is collected, processed, packaged, sold, and consumed. In the Indian culture, the highest grade of hygiene has been associated with kitchen. Elaborate codes of conduct for food handlers/cooks have been prescribed for ensuring food hygiene. But in the modern context, the concept of food safety is described by the "farm-to-fork" model. This is a well-structured, preventive approach that captures the essence of safe food and forms the backbone of food safety (Jones, 1998). It encompasses all stages of food distribution, storage, and handling from primary production to consumption, as food hazards can occur at any stage in the food chain. The main advantage of following this approach is the reduced risk of food contamination by applying the principles of prevention throughout the production, processing, and marketing chain. To achieve maximum consumer protection, it is essential that safety and quality be built into food products from production to consumption. This comprehensive and integrated approach is commonly known as the farm-to-fork, farm-to-table, stable-to-table approach in which the producer, processor, transporter, vendor, and consumer all play vital roles in ensuring food safety and quality. This concept is most acceptable to the consumer also because unsuitable products can be identified earlier along the chain. As the chain is long from farm-to-fork there are multiple stakeholders responsible to ensure food safety and quality at their respective place, who take the ownership of food safety and provide appropriate inputs to the decision-making process in the chain. The increasingly complex nature of hazards has triggered the need for this integrated approach. This approach places the primary responsibility of food safety on all contributors in the product–supply chain. Simultaneously, individual chain participants look for assurance of safety of products supplied by the preceding chain participant. These trends place greater attention on compliance with food-safety measures at the farm level, it being one of the important stages affecting the level of food safety of products consumed at the end of the chain. For example, the wholesalers believe that the product from the farm is safe. The retailers rely on wholesalers and so on. In case the safety of food cannot be ensured, responsibilities for failure can be attributed to stakeholders along the farm-to-fork continuum (Sargeant et al., 2007). Overall, the government is responsible for auditing performance of the food system through monitoring and surveillance.

There have been discussions about the paradigm of farm-to-fork model in the food-safety context. This concept implies that the responsibility of providing safe food to the consumer is shared among all those involved with food material/product at various levels, for example, production, processing, trade, cooking, serving, etc. It also portrays the journey/history of seeds from the time they are sown in the fields to their growth, harvest, storage, transportation, cooking, and the way they are served on the table (Fig. 13.1). For food from animal sources, the same concept can be explained as "stable-to-table." The benefit of this concept is that in case of a developing problem , such as a food-"poisoning" incident, the food, its processing, and

FIGURE 13.1 Farm-to-Fork

Plant sources	Animal sources
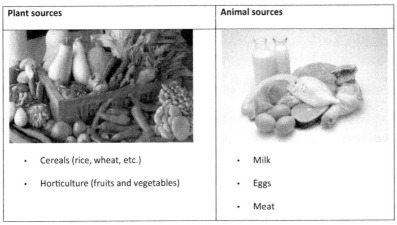	
• Cereals (rice, wheat, etc.) • Horticulture (fruits and vegetables)	• Milk • Eggs • Meat

FIGURE 13.2 Categorization of Food Sources

its source can be easily traced, investigated, and effectively corrected. The reverse "fork-to-farm" shall help in traceability and recall in case of an outbreak. In the long run, it benefits the consumers the most by raising the quality and standard of food. It translates into better marketing opportunities for all, namely, the farmer, wholesaler, retailer, FBO, etc. This umbrella concept takes care of all the hazards: biologic, chemical, and physical. It forms a platform where one of the basic principles of public health practice, that is, intersectoral coordination to prevent foodborne illnesses can be seen. The concept of food safety helps the regulators to follow an easy stepwise monitoring of the entire process. Above all, it plays a vital role in solving many issues related to global trade of various products. To understand the farm-to-fork concept we can broadly categorize food sources as in Fig. 13.2.

Today, we expect the food that we eat to be safe, regardless of its source. To achieve this public expectation and worthy objective, an enhanced integration and comprehensiveness in the food-safety system is necessary. First, this requires a greater clarity and acceptance of the role and responsibilities of each of the partners in this chain of farm-to-fork, including governments, industry, health professionals, educators, the mass media, and finally, the consumers. Second, it requires the complementary resolve and action of all stakeholders in making their necessary and timely contributions to the effort. It involves collaboration between departments of agriculture, medicine, veterinary science, food processing, transport industry, packaging industry, retail sector, government, research fraternity, and consumers. The government has a key role in setting and providing legislation that lays down minimum food-safety standards at all levels. Governments must ensure that these are implemented through training, inspections, and enforcement. The promulgation of the new Food Safety and Standards Act (FSSA) 2006, in India is an attempt by the government to achieve food safety. Although food-safety legislation affects everyone in the country, it is particularly relevant to anyone working in the production, processing, storage, distribution, and sale of food, no matter how large or small the business is. To understand and control all of the major factors that affect food safety from "farm-to-plate," it is necessary to integrate various disciplines more effectively (veterinarians, botanists, molecular biologists, microbiologists, inspectors, physicians, etc.) that are involved in the study, investigation, and control of foodborne illnesses. Today however, the situation is that we still have to worry about food safety. While we know enough about the *whys* and *hows* of food safety, the journey does not end here; there are many more challenges to be faced for keeping the fuel for our bodies from being our "last meal."

Let us begin our journey from the farm and understand which activities can compromise food safety at the farm level. The use of pesticides has led to increased food production globally (Chakravarti, 1973). Nevertheless, this method for increasing production, along with environmental pollutants due to industrial emission, has resulted in the presence of residues of these chemicals and their metabolites in food commodities, water, and soil. The general population is mainly exposed to organophosphorous pesticides through the ingestion of contaminated foods (such as, cereals, vegetables, and fruits), which are directly treated with these pesticides or are grown in contaminated fields. Therefore, contamination of food along with the environment by pesticide residues is a serious issue in many areas of the world. Food items, such as, cereals, grains, spices, herbs, etc., can also be contaminated by heavy metals, mycotoxins, and radionuclides at the farm. Even water used for irrigation, washing, and cooking of raw materials can be contaminated with pesticides, heavy metals (lead arsenic, radionucleides, etc.), surface active agents, environmental chemicals, and pathogens. Use of such water in turn can make food unsafe. Agrochemicals, such as, pesticides when used at the farm level are supposed to be degraded after their application on the crops but that does not happen in all the cases. Small amount of residues remain mixed with food and make it unsafe. Meat and poultry products can also have pesticides, heavy metal, dioxins, and environmental chemical residues in them (Jadhav and Vikas, 2011). They can also become unsafe with antibiotics, veterinary drugs, and microbes. Various processes during their travel from poultry farm, such as, quality of feed, veterinary treatments, hygiene, sanitary and phytosanitary conditions, processing of meat, preservatives used storage, handling, and packing affect food safety. Similarly, milk and milk products can compromise safety depending on the animal husbandry practices at the farm, quality of feeds, veterinary treatments, environmental conditions, storage, packing, and transportation. Veterinary drugs and antibiotics used for the animals to improve the output from animals continue their presence as residues in milk. Fruits and vegetables can also be contaminated with chemicals and microbes during their storage, transportation, and handling. Improper storage conditions with regard to temperature and humidity can cause infestation by the fungus *Aspergillus flauvus*, which can cause aspergillosis. Processed and packed food safety can be compromised with use of additives, banned dyes, toxic chemicals (residual solvents, acrylamisde, benzene, melamine, etc.), persistent agricultural pollutants, leechables from packaging material, microbes, and degraded products can also harm the food product during processing and packaging.

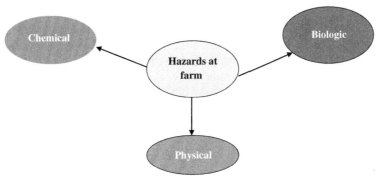

FIGURE 13.3 Food-Safety Hazards at the Farm Level

13.4 FOOD SAFETY AT THE FARM LEVEL

There are multiple factors that can affect food safety at the farm level, namely, quality of seeds, soil composition, preparation methods, crop management, water quality, irrigation techniques, management of pests, harvesting techniques, equipment, storage, and transport. Good Agricultural Practices (GAP) at the farm aim at ensuring food safety during preproduction, production, harvest, and postharvest stages. They also help protect the environment and the safety of workers. Farmers must apply GAP to produce safe food. Often, it is more difficult for small farmers to comply with GAP due to financial constraints. The hazards at the farm level can be classified as given in Fig. 13.3. Not all hazards are applicable to all the fields. Hazard identification specific to each field is an important activity that can prevent contamination later.

13.4.1 CHEMICAL HAZARDS

After green revolution in 1960s, India became self-sufficient in agrarian production. However, most of our farmers are less educated and follow a mix of both traditional and modern practices. The burden of meeting the increasing demand of grains and livestock products for the burgeoning population has been placed on them. This has led to practices, such as, indiscriminate use of chemicals, fertilizers, pesticides, and use of medicines for the farm animals and hence, the widespread contamination of food items. It is indeed a sad reality that the need to produce more, along with market-driven forces, has resulted in an excessive use of pesticides in our farms. This has also led to the deterioration of the nutritive value of food products. It has not only compromised the safety of food and affected health of the people, but has also harmed the environment and other living beings.

Nonpesticide chemicals that can make food unsafe at farm level are lubricants, cleaners, sanitizers, paints, fertilizers, etc. These can enter due to any oil leaks from nearby industries. Even farm equipment can contaminate produce with grease or paint. Similarly, picking containers, transportation vehicles, etc., that were previously used for chemicals can be contaminated and can harm the produce, if not cleaned properly. Heavy metal residues, such as, arsenic, lead, cadmium, etc., can be present in the food due to the continued use of fertilizers (including compost). In case the farm is near a busy road, lead contamination can also occur from car-exhaust fumes. The water used for irrigation too can have high levels of heavy metals, such as arsenic, in it. The soil at the farm may be naturally contaminated with heavy metals, from previous use, or leakage from industrial sites.

Natural toxins, various allergens, mycotoxins, alkaloids, enzyme inhibitors can occur in field due to unsuitable storage conditions and lead to mold on produce. Use of injectable chemicals in the produce to keep them fresh, intentionally added substances, such as, added color, mineral oil, carbides (for ripening), preservatives, etc., constitute additional hazards. It is a common practice to ripen mangoes by using calcium carbide to meet the increasing demand of fruit during summer season. These mangoes, called *masala mangoes*, in local areas are unsafe for consumption.

13.4.2 BIOLOGIC HAZARDS

Fruits and vegetables can get contaminated by microorganisms that often spoil them and bring undesirable changes in quality characteristics, such as, softening, bad odor, and flavor. The pathogenic ones can even cause foodborne illness. Some bacteria, such as, *Listeria* spp. and *Bacillus cereus*, are present in the soil and can contaminate crops directly or through dirty containers and equipment. Some bacteria pass through the intestinal tract of animals and humans and then contaminate fruits and vegetables through manure, contaminated water, and humans handling produce. One such bacterium is *Salmonella*.

Fruit and vegetables can act as vehicles to pass parasites from one host to another—animal-to-human or human-to-human. Cysts of *Giardia* can survive and remain infectious for up to 7 years in the soil. Water contaminated with fecal material, infected food handlers, animals in the field, or packing shed can be vehicles can contaminate produce with parasites.

Fungi/molds can affect food crops under suitable temperature and humidity conditions. For example, many species of the fungus *Aspergillus* produce aflatoxins. These are among the most carcinogenic substances known. Crops susceptible to *Aspergillus* infection include cereals, oilseeds (groundnut), and spices. The toxin can also be found in the milk of animals that are fed contaminated feed. Biologic hazards can harm the farm depending on various factors given in Table 13.1.

13.4.3 PHYSICAL HAZARDS

These are foreign objects that can cause illness or injury to consumers. These can occur during production and postharvest handling. Types of physical hazards include glass, wood, metal, plastic, soil, stones, personal items, such as, jewelry, hair clips, paint flakes, insulation, sticks, staples, weed seeds, and toxic weeds. These can enter

Table 13.1 Factors Affecting Hazards by Biologic Agents

Factor	Description	Example
Method of production of produce	Close to ground at higher risk than well above ground	Carrot > grapes
Contact with water	Frequent contact at higher risk	Rice crop
Surface of produce	Large uneven surface at higher risk than smooth	Lettuce > apple
Method of consumption	Raw at more risk than cooked	Leafy vegetables > potato
	Edible skin > inedible skin	Grapes > banana

from the environment during harvesting of ground crops during wet weather, dirty harvesting, packing equipment, picking containers, packaging materials, stacking of dirty containers on top of produce, broken lights above packing equipment and areas, inadequate cleaning after repairs, and maintenance.

13.4.4 OTHER HAZARDS

Some hazards are related to the application of newer technologies, the impact of which may not be fully understood (e.g., genetically modified plants or nanotechnology). Research is being carried out to fully understand the hazards associated with these.

13.5 METHODS TO REDUCE HAZARDS AT THE FARM LEVEL

Indian agriculture with a rich historic past has been described through hymns in *Rigveda*, a classic Hindu religious text, where ploughing, sowing, irrigation, and fruit and vegetable cultivation have been portrayed. An ancient Indian Sanskrit text, *Bhumivargaha*, classified agricultural land into 12 categories: *urvara* (fertile), *ushara* (barren), *pankikala* (muddy), *maru* (desert), *aprahata* (fallow), *jalaprayah* (watery), *kachchaha* (land contiguous to water), *sharkara* (full of pebbles and pieces of limestone), *shadvala* (grassy), *nadimatruka* (land watered from a river), *sharkaravati* (sandy), and *devamatruka* (rain-fed). Archaeological evidence suggests that rice was grown along the banks of the river Ganges in the 6th millennium BC. Indian farmers used to domesticate cattle, buffaloes, sheep, goats, pigs, and horses and most farmers, both big and small, followed traditional methods of cultivation. However, postindependence our population grew on a logarithmic scale and to ensure food security, an increase in farm production was the need of the hour. There was a paradigm shift in agricultural practices from traditional to advanced methods of irrigation infrastructure, use of hybridized seeds, and high-yielding varieties of crops. Our success in increasing the farm produce relied heavily on the use of synthetic fertilizers and pesticides at the farm. However, the use of modern methods had their own shortcomings. For example, high-yield crops were associated with land degradation, an

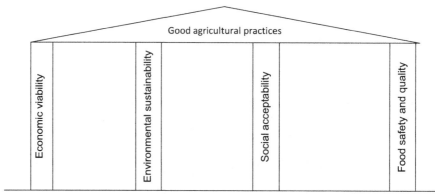

FIGURE 13.4 Four Pillars of GAP

increase in number of weeds, and pests at the farm. Even the water got contaminated with chemicals.

With this huge farm productivity, it becomes imperative that the safety and quality of farm produce is ensured at all stages of production. We need to balance the requirements of food security and safety both. The solution to this complex problem is by adopting GAP, which in combination with effective input use, are one of the best ways to increase productivity and improve quality. According to the Food and Agriculture Organization of the United Nations, GAP are practices that address environmental, economic, and social sustainability for on-farm processes, and result in safe and quality food and nonfood agricultural products (Fig. 13.4).

GAP enhance the production of safe and good quality food (Yuichrio, 2009). These practices are usually environmentally safe and ensure that the final product is appropriately handled, stored, and transported. When GAP are put in practice in true spirit it, the food will meet quality and safety standards at the time of harvest (Omore and Baker, 2011). GAP protects food at the primary stage of production from contamination as shown in Fig. 13.5.

The international market is becoming competitive. The economy of the country is benefited a lot by exporting food products to the developed world, which are becoming stringent in accepting export of food from developing countries. To have a good standing of our farm produce in the international market, *Indian Good Agricultural Practices* (INDGAP) have been formulated. Adopting these practices will ensure a safe and sustainable farm produce.

INDGAP defines certain minimum standards with a well-defined system of accreditation mechanism and implementation of GAP. These standards are voluntary and nondiscriminatory to the growers. INDGAP has different modules for all farm, crops, fruits and vegetables, combinable crops, green tea, and coffee. A broad outline of various aspects which needs to be managed are:

1. Site history and management
2. Soil management

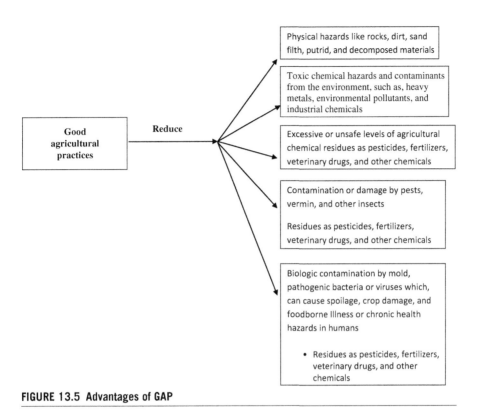

Good agricultural practices — **Reduce** →

- Physical hazards like rocks, dirt, sand filth, putrid, and decomposed materials

- Toxic chemical hazards and contaminants from the environment, such as, heavy metals, environmental pollutants, and industrial chemicals

- Excessive or unsafe levels of agricultural chemical residues as pesticides, fertilizers, veterinary drugs, and other chemicals

- Contamination or damage by pests, vermin, and other insects

  Residues as pesticides, fertilizers, veterinary drugs, and other chemicals

- Biologic contamination by mold, pathogenic bacteria or viruses which, can cause spoilage, crop damage, and foodborne Illness or chronic health hazards in humans

  - Residues as pesticides, fertilizers, veterinary drugs, and other chemicals

FIGURE 13.5 Advantages of GAP

3. Soil mapping
4. Plant nutrition management and fertilizers
5. Irrigation and fertigation
6. Integrated pest management
7. Plant-protection products
8. Traceability
9. Complaints management
10. Visitors safety
11. Record keeping
12. Health welfare and safety of workers
13. Environmental conservation
14. Waste and pollution management

The potential benefits of GAP are significant improvements in quality and safety of food and other agricultural products. There is a marked reduction in chances of noncompliance with national and international regulations regarding permitted pesticides, maximum levels of contaminants (including pesticides, veterinary drugs, radionuclide, and mycotoxins) in food and nonfood agricultural products, as well as, other chemical, microbiologic, and physical contamination hazards. The adoption

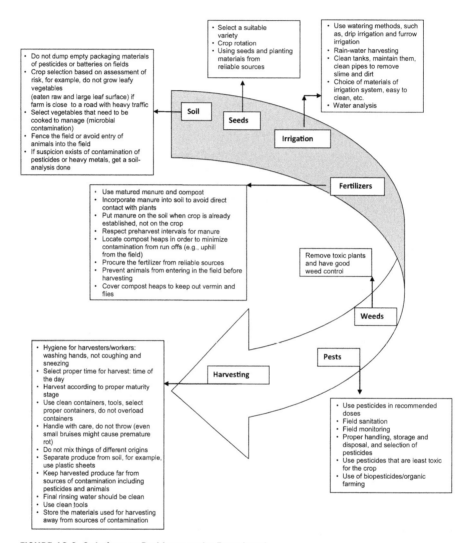

FIGURE 13.6 Solutions to Problems at the Farm Level

of GAP helps to promote sustainable agriculture and contributes to meeting national- and international, environment- and social development objectives (Grace et al., 2012).

However there are various challenges related to GAP. The most prominent is a definite increase in cost of production. There is lack of harmonization between existing GAP-related schemes and availability of affordable certification systems, which often leads to increased confusion, certification, and costs for farmers and exporters. There is a high risk that small-scale farmers will not be able to seize export-market opportunities, unless, they are adequately informed, technically prepared, and organized to meet this new challenge. The governments and public agencies must play a facilitating role in this aspect. However at times, compliance with GAP standards does not promote all the environmental and social benefits, which are claimed. Some solutions to problem at the farm level are given in Fig. 13.6.

REFERENCES

Bauman, H.E., 1994. The origin of the HACCP systems and subsequent evaluation. Food Sci. Technol. Today 8, 66–72.

Caballero, B., Trugo, L., Finglas (Eds.), 2003. Encyclopedia of Food Sciences and Nutrition. Elsevier Publications, Amsterdam.

Chakravarti, A.K., 1973. Green revolution in India. Ann. Assoc. Am. Geogr. 63, 319–330.

Grace, D., Dipeolu, M., Olawoye, J., Ojo, E., Odebode, S., Agbaje, M., Akindana, G., Randolph, T., 2012. Evaluating a group-based intervention to improve the safety of meat in Bodija Market, Ibadan, Nigeria. Trop. Anim. Health Prod. 44, S61–S66.

Griffith, J.C., 2006. Food safety: where from and where to? Brit. Food J. 108 (1), 6–15.

Hartman, P., 2001. The evolution of food microbiology, second ed. ASM Press, Washington DC, pp. 3–12.

Jadhav, V.J., Vikas, S.W., 2011. Public health implications of pesticide residues in meat veterinary world. Vet. World 4 (4), 178–182.

Jones, R., 1998. Food safety, farm to fork. American School Board Journal 185 (2), 34–38.

Motarjemi, Y., Moy, G., Todd, E., 2014. Encyclopedia of Food Safety, first ed. Elsevier, Missouri, USA.

Omore, A., Baker, D., 2011. Integrating informal actors into the formal dairy industry in Kenya through training and certification. International Livestock Research Institute. Proceedings of an international conference, Nairobi, Kenya, May 13–15, 2009.

Sargeant, J.M., Ramsingh, B., Wilkins, A., Travis, R.G., Gavrus, D., Snelgrove, J.W., 2007. Constraints to microbial food safety policy: opinions from stakeholder groups along the farm to fork continuum. Zoonoses Public Health 54 (5), 177–184.

Woolen, A., 1999. Safety and the Y2K. Food Processing, February, p. 20.

Yuichrio, A., 2009. Reflections on the growing influence of good agricultural practices in the Global South. J. Agric. Environ. Ethics 22, 531–557.

Food safety from farm-to-fork—food safety issues related to animal foods at farm

14

P. Dudeja*, A. Singh**

*Department of Community Medicine, Armed Forces Medical College, Pune, Maharashtra, India; **School of Public health, Post Graduate Institute of Medical Education and Research, Chandigarh, India

Man has placed himself at the top of food chain and mastered other animals. As per historical evidence, goat and sheep were the first species of animals which were domesticated for human use. This was followed by domestication of pig. About 8000 years ago, cow was the last major food animal that humans tamed. This led to introduction of milk as a useful foodstuff on the human platter. Goat, sheep, reindeer, and camel milk were also used. Other species used for food were avian, amphibian, fish, and various arthropods. In Indian mythology *Kamdhenu* is a mythological figure that symbolizes livestock as source of wealth. Indian agriculture in past was heavily dependent on livestock in terms of draught power and organic manure in our country. Food obtained from different animals is given in Table 14.1.

India owns the largest livestock in the world with 56.7% of world's buffaloes, 12.5% cattle, 20.4% small ruminants, 2.4% camel, 1.4% equine, 1.5% pigs, and 3.1% poultry. The Department of Animal Husbandry and Dairying (AH&D)—now renamed as Department of Animal Husbandry Dairying & Fisheries (DADF) is one of the departments in the Ministry of Agriculture. This department is responsible for matters relating to livestock production, preservation, protection from disease, and improvement of stocks and dairy development. It also looks after all matters pertaining to fishing and fisheries both inland and marine.

With modernization, Indian farms saw an increasing use of electrical and mechanical equipment mainly in form of tube wells and tractors. This reduced the use of livestock for crop production and they are now viewed as source of food. Apart from this, animal proteins are rich source of essential amino acids in diet of human being. They supply about 20–25% of total daily protein requirement. Milk is one of the most important sources of animal protein in the diet of predominately vegetarian population of India. The other protein-rich foods of animal origin are the meat of different animals as chicken, fish, and eggs.

Food Safety in the 21st Century. http://dx.doi.org/10.1016/B978-0-12-801773-9.00014-5

Table 14.1 List of Animals and Food Obtained From Them

Animal	Food
Dairy	Fluid and dried milk, butter, cheese and curd, casein, evaporated milk, cream, yogurt and other fermented milk, ice cream, whey
Cattle, buffalo, sheep	Meat (beef, mutton), edible tallow
Poultry	Meat, eggs, duck eggs (in India)
Pig	Meat
Fish (aquaculture)	Meat
Horse, other equines	Meat, blood, milk
Micro-livestock (rabbit, guinea pig), dog, cat, bulls	Meat
Insects and other invertebrates (e.g., vermiculture, apiculture)	Honey, 500 species (grubs, grasshoppers, ants, crickets, termites, locusts, beetle larvae, wasps and bees, moth caterpillars) are a regular diet among many non-western societies

In response to globalization, Indian society faced an increased demand of animal food products. The consumption of meat and meat production has increased radically throughout the world and is expected to increase in future also. There is changing trend that is occurring globally in how and what people eat. As the economic status of people changes, the food consumption pattern changes as well. Coupled with this increase in information technology, international trade, advertisements, and modern lifestyle have made the people to shift to nonvegetarian diet. Meat and fish consumption has increased remarkably since they are desirable, although expensive, food sources. The modern life style with high per capita purchasing power (PPP) has increased the meat production and consumption. Apart from this there has been increase in awareness about the quality and content of animal foods. Health advertisements bombard us with messages on importance of eggs and milk as complete protein foods of animal origin. Hence, there has been an increasing demand of protein-rich foods of animal origin in the society.

The food processing industry has grown well with the changing times and offered a long list of items in the menu of nonvegetarian platter. There are newer nonvegetarian products to pamper the taste buds especially in the frozen foods category. Better international trade has ensured the availability of these products all over world through the year and made them even more popular. International food chains like KFC, Mc Donalds, Subway, Al Kabir, Pizza Hut have modified the tastes of these nonvegetarian items as per the requirement/culture/traditional taste of the each country.

However, besides being a source of nutritious food items, animals are also source of zoonoses, that is, those diseases and infections the agents of which are naturally transmitted between vertebrate animals and man. Zoonoses make up more than 60% of all human infectious diseases and more than 70% of all emerging infectious diseases. The zoonoses can be classified as given in Table 14.2.

Table 14.2 Classification of Zoonoses

Anthropozoonoses	Zooanthroponoses	Amphixenoses
Infections which are transmitted to man from animals	Infections which are transmitted from man to lower vertebrate animals	Infections may be transmitted in either direction

Frequent outbreaks of various pandemics which were unknown previously such as SARS, avian flu, swine flu, bovine spongiform encephalopathy (BSE) has raised a question about the safety of nonvegetarian food or food of animal origin. This matter concerns us too. For the safety of the animal food in the farm to fork chain, it is important that only healthy animals enter the food chain. There have been instances of human suffering due to lapses in animal food safety; for example, in 2008 in Canada 28 people were affected by outbreak of listeriosis in meat products (Todd and Notermansb, 2011). This happened as most of these animals were reared in filthy and unhygienic conditions. These animals were being fed continuously with low doses of antibiotic and other growth promoters.

Another issue of concern is the indiscriminate use of veterinary drugs which are available everywhere without a prescription and their use is rampant in poultry and milk for extra profits. Good hygiene, proper feed, appropriate husbandry, and good management practices are by far the best strategies to be adopted for safety of food of animal origin. In this context, animal feeding and live stock products standards have been developed at the international level. Agencies such as codex, ISO, OIE, FAO, WHO have all made contribution for providing directions on safety of food of animal origin. With initiative of FAO, a website titled "International Portal on food Safety, Animal and Plant Health" has been developed (www.ipfsaph.org) to provide guidance on food safety. They have also developed a manual for good practices for meat industry.

Good animal husbandry (GAH) or good veterinary practices (GVP) have been established to assure consumers that foods derived from animals meet acceptable levels of quality and safety (Benavides Benavides and Rosenfeld Miranda, 2009). These practices are the guiding principles in professional veterinary practice for the care and treatment of animals, including animals used for human food production. The important components of GAH practices for safe food of animal origin can be understood better by using farm to fork model (Fig. 14.1).

14.1 ANIMALS AT FARM

Health of the animals at the farm plays a vital role in the quality and safety of food provided by them. By and large the four important aspects of GAH practices at farm comprise of safe animal feed, animal health and welfare, and healthy living conditions.

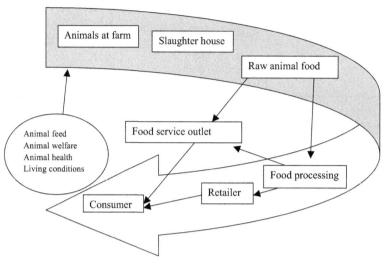

FIGURE 14.1 Farm-to-Fork Model for Food of Animal Origin

14.2 SAFE ANIMAL FEED

The first and foremost good practice at farm is provision of safe feed to the animals. Various threats in animal feed are given in Fig. 14.2. Safe animal feed will provide us with safe animal food. Adulteration of animal feed with subtherapeutic feeding

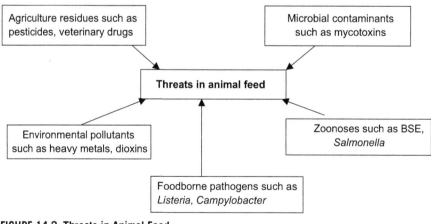

FIGURE 14.2 Threats in Animal Feed

of antibiotics residues is rampant. This has resulted in development of antibiotic resistance of zoonotic pathogens. Many antibiotics added to animal feed are also used in human medicine, and this can cause development of antibiotic-resistant bacteria which can later lead to infections in animals and humans.

Residues of antibiotics used in livestock or added to feed have been found in food-producing animals including dairy cows. Among these drugs are chloramphenicol and sulfamethazine. Alternatives to the prophylactic feeding use of antibiotics to maintain animal health include the modification of production systems. These modifications include reduced animal confinement, improved ventilation, and improved waste treatment. Various hormones such as oxytocin have been given to animals to improve the output of milk from the animal. Residues of such chemicals have also been detected in the milk.

The occurrence and detection of antibiotic residues in milk continue to be a concern for dairy industry, consumer, and the government with emerging issues of development of resistance to multiple antibiotics. Presence of antibiotic residues is known to cause allergic reactions, carcinogenicity, and spread of bacterial resistance. In India, many of these antibiotics are being used in an "extra label" fashion and furthermore, due to the malpractices new generation antibiotics recommended for human use are also being used in animals for disease control, for which no safe levels have been recommended in milk and are leading to development of antibiotic resistance (Paulson and Zaoutis, 2015).

Certain disease outbreaks such as bovine spongiform encephalopathy (BSE) in animals have raised serious public concern. A rare Creutzfeldt–Jakob disease (CJD) emerged among beef-importing nations in 1996 (Will et al., 1998). Eating beef infected with BSE, popularly known as mad cow disease, was the cause behind CJD infection. Although unproven, public perceptions include the proposition that the disease has entered cattle from feed containing bone meal and offal from sheep afflicted with the similar disease, scrapie. These diseases are fatal, their causes are unknown, and there are no tests to detect them.

Good feeding practices provide the herd with adequate feed and water, keep the animals healthy, preserve water supplies and animal feed materials from chemical contamination, and prevent microbiological or toxin contamination (FAO/WHO, 2007). Contamination of feed ingredients at times occurs due to improper storage, especially during warm and humid climate. Urea and/or ammonium salts are sometimes mixed in oilseed cakes to increase the protein contents. Various oil seed cakes and bran may contain fibrous material such as sawdust, husk, and hulls. Generally adulterants used in feed ingredients include excess moisture to increase weight which can lead to fungus infestations and low shelf life of feed. Good feed practices at farm can be classified as given in Fig. 14.3.

As per Food Safety and Standards Act in India, 2006 no article of food shall contain pesticides, veterinary drugs, antibiotic residues, and microbiological counts in excess of such tolerance limits as may be specified by regulations (Government of India, 2006).

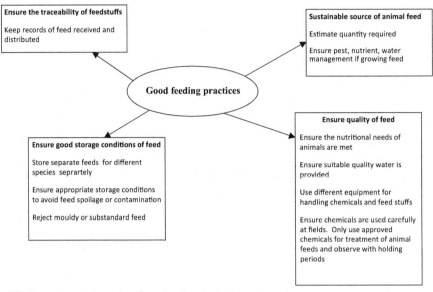

FIGURE 14.3 Good Feed Practices for Livestock at Farm Level

14.3 ANIMAL WELFARE

Looking after the welfare of animals is both an ethical and medical issue. Welfare activities will ensure good health of the animals at the farm. Animals should be reared in humane conditions. Animal welfare at the farm includes freedom from thirst, hunger, and malnutrition. They should be comfortable without any pain, injury, or disease. They should be capable of expressing normal behavior and free from fear and distress (Table 14.3).

Table 14.3 Animal Welfare Activities

Freedom From	Action to be Taken
Thirst, hunger, and malnutrition	Making sure animals have enough food and water of the right quantity and quality to meet their daily needs
Discomfort	Providing an environment that gives shelter and shade; sheds for newborn calves
Pain, injury, or disease	Vaccination program, drenching for internal parasites; pain relief for lame cows; safe facilities for handling animals
Freedom to express normal behavior	Providing natural conditions
Fear and distress	Not keeping animals way from the herd, handling animals calmly and gently during milking

14.4 ANIMAL HEALTH

Outbreaks of diseases such as foot-and-mouth disease, influenza, etc. have been documented in farm-reared animals. Many human foodborne illnesses result from pathogenic bacteria of animal origin, for example, *Listeria* and *Salmonellae* which are found in dairy products, and *Salmonellae* and *Campylobacter* found in meat and poultry (John et al., 2002). Dairy cattle in intensive production systems, commonly suffer from metabolic diseases with underlying nutritional etiology. The majority of human infectious diseases have their origins in animals, and many important zoonotic diseases not only adversely affect productivity and welfare of livestock but may also be transmitted to man either directly or via animal products (Mette et al., 2005). Sustainable future farming systems must mitigate risks such as avian influenza, Q-fever, *E. coli* O157, bovine tuberculosis, and various helminthiasis. It is of utmost importance that animal health is looked after. This will enhance herd immunity; reduce stress in animals; help in early disease detection; ensure food safety and traceability; and prevent occurrence of chemical residues in milk. Various practices to ensure good animal health at farm are given in Fig. 14.4.

Presently there are a total of 8,732 veterinary hospitals/polyclinics and 18,830 veterinary dispensaries in the country providing services for the large livestock population. These numbers are grossly inadequate. Whatever exists also has poor infrastructure in terms of wrecked buildings, lack of equipments, etc. The polyclinics, wherever established, lack the adequate infrastructure for surgical interventions and diagnostic imaging. Presently, the country is facing acute shortage of manpower to manage these institutions and provide required services. Diagnostic facilities too in terms of good clinical laboratories, equipments, quick and quality diagnostics, and the human resource having expertise in these areas are practically nonexistent.

Establish herd immunity	Prevent entry of disease onto the farm	Use all chemicals and veterinary medicines as directed	Have an effective herd health management program
• Choose breeds and animals well suited to the local environment and farming system • Determine herd size on the basis of local conditions, availability of land, infrastructure, feed, and other inputs • Vaccinate all animals as recommended or required by local animal health authorities	• Buy animals of known health status and control their introduction to the farm using quarantine if indicated • Ensure safe transport on and off the farm • Monitor risks from adjoining land and neighbours and have secure boundaries • Limit access of people and wildlife to the farm • Have a vermin control program in place • Only use clean equipment from a known source	• Only use chemicals approved for supply and use under relevant legislation • Use chemicals as per directions, calculate dosages carefully, and observe appropriate with holding periods • Only use veterinary medicines as prescribed by veterinarians • Store chemicals and veterinary medicines securely and dispose of them responsibly	• Use an identification system that allows all animals to be identified individually from birth to death • Regularly check animals for signs of disease and attend to sick ones quickly • Keep sick animals isolated • Separate milk from sick animals and animals under treatment • Keep written records of all treatments • Manage zoonoses

FIGURE 14.4 Good Animal Health Practices

Surveillance and monitoring of livestock diseases is a major component of good animal health. The present disease-reporting system is neither timely nor complete. Due to this delay, many a times animal disease outbreaks assume serious proportions before control and containment steps can be initiated. An authentic epidemiological data for realistic assessment of the prevalence and emergence of these diseases in different agroclimatic zones is essential not only for identification and prioritization of the most important diseases but also for their prevention and control. Recording the incidence of diseases is essential for estimating the economic loss, and conducting risk analysis.

National Disease Control Program involves the vaccination of all susceptible livestock against major infectious diseases. Except for foot-and-mouth disease, vaccine production of most other vaccines is with the state biological units. The biological production centers available in the government sector (both state and central) might be old and obsolete, and may not comply with the GMP requirements. They may not have technologies and infrastructure to meet the contemporary requirements. The availability of qualified manpower to run these institutions is also not available. Assistance is provided to state governments for control of economically important diseases of livestock and poultry by way of immunization. Foot and Mouth Disease Control Program (FMD-CP) has been implemented in 221 districts in Phase I with 100% central funding toward cost of vaccine, maintenance of cold chain, and other logistic support to undertake vaccination (Pattnaik et al., 2012). The program has lead to reduced incidence of disease. The ultimate objective of government is to eradicate this disease from the country in a time-bound manner on the lines of rinderpest eradication. Similarly, National Control Program on Brucellosis initiated in 2010 that envisages mass screening of cattle and buffaloes to ascertain exact incidence of the disease and vaccination of all female calves using S-19 vaccine is in progress.

14.5 ANIMAL LIVING CONDITIONS

Healthy living conditions at the farm will prevent the occurrence of various diseases in animals at the farm and will reduce the stress levels too in the animals. A rough checklist for ensuring healthy living conditions are given as follows:

- clean water
- clean feed
- clean housing
- good ventilation
- appropriate temperature
- consistent feeding
- vaccination
- quarantine sick animals
- balanced diet
- pest control
- sound control
- avoid overcrowding

The location of stable/coop/enclosure/pen/shed should be suitably selected with an appropriate drainage system, good lighting, and ventilation. There should be adequate space for each animal to avoid overcrowding. Good pest control is also a vital requirement to provide safe environment to animals. Some important GAH-related measures are as follows:

- Only healthy animals are slaughtered for the purpose of human food.
- Any drug used in the control of animal disease is safe for its intended use and used according to approved directions (i.e., appropriate amounts, frequency, and timing), and residues of such drugs do not remain in the edible tissues at unsafe levels when the food is made available for consumption.
- Chemicals utilized in animal husbandry (e.g., dips for insect pest control) are safe for their intended uses and used according to instructions (i.e., appropriate levels, frequency, and timing), and residues of such chemicals do not remain in the edible tissues at unsafe levels when the food is made available for humans.
- Live animal inspection and handling are properly conducted before slaughter, and carcass inspection and handling after slaughter.
- Appropriate temperature controls, storage conditions, handling and butchering techniques, and sanitary conditions are maintained during processing and butchering to prevent postslaughter contamination.
- Shipping and handling practices prevent any unnecessary exposure of the product to contamination.

REFERENCES

Benavides Benavides, B., Rosenfeld Miranda, C., 2009. Analysis of good animal husbandry practices and their epidemiological application. Rev. Sci Tech. 28 (3), 909–916.

FAO/WHO, 2007. Animal feed impact on food safety. Report of the FAO/WHO Expert Meeting, October 2007.

Government of India, 2006. The Food Safety and Standards Act, 2006, second ed. Commercial law publishers (India) Pvt. Ltd., India.

John, A., Patricia, M., Frederick, J., 2002. Bacterial contamination of animal feed and its relationship to human foodborne illness. Clin. Infect. Dis. 35 (7), 859–865.

Pattnaik, B., Subramaniam, S., Sanyal, A., Mohapatra, J.K., Dash, B.B., Ranjan, R., Rout, M., 2012. Foot-and-mouth disease: global status and future road map for control and prevention in India. Agric. Res. 1, 132–147.

Paulson, J.A., Zaoutis, T.E., 2015. Nontherapeutic use of antimicrobial agents in animal agriculture: implications for pediatrics. 136 (6), 1670–1677.

Todd, E.C.D., Notermansb, S., 2011. Surveillance of listeriosis and its causative pathogen, *Listeria monocytogenes*. Food Control. 22 (9), 1484–1490.

Vaarst, M., Padelb, S., Hovic, M., Younied, D., Sundrume, A., 2005. Sustaining animal health and food safety in European organic livestock farming. Livestock Product. Sci. 94 (1), 61–69.

Will, R.G., et al., 1998. Descriptive epidemiology of Creutzfeldt–Jakob disease in six European countries, 1993–1995. Ann. Neurol. 43 (6), 763–767.

Food safety from farm-to-fork—food-safety issues related to processing

15

P. Dudeja*, A. Singh**

*Department of Community Medicine, Armed Forces Medical College, Pune, Maharashtra, India; **School of Public Health, Post Graduate Institute of Medical Education and Research, Chandigarh, India

Food processing in not a recent phenomenon. It dates back to the prehistoric ages when raw food items were subjected to fermenting, sun drying, preserving with salt, and various types of cooking, such as roasting. There is archeologic evidence of salt-preservation for foods that constituted warriors' and sailors' diets until the introduction of canning. Processing methods, such as, making of pickles (using high salt solution), *morraba* (using high sugar solution), chutneys, drying grapes as *kishmish*, dried dates, etc., have been common practices in Indian households. These tried and tested processing techniques continued until the advent of the industrial revolution. Ready-to-eat meals, which date back to the preindustrial revolution period, are also considered as processed foods.

Modern food-processing technology was developed in the 19th and 20th centuries as a solution to the problem of food supply for military personnel. In 1809, Nicolas Appert, known as the "father of canning," a confectioner, invented a hermetic bottling technique that would preserve food for the French troops, which ultimately led to the development of tinning and later canning in 1810 (Pujol, 1985). Initially these foods were expensive. Also, the lead used in cans made the canned goods potentially harmful to the consumer. But in due course of time, these became popular around the world. In 1864, Louis Pasteur invented the method of pasteurization, which contributed to improved quality of preserved foods. This was the beginning for wine, beer, and milk preservation. After the Second World War, there were advances in food-processing areas, such as, spray drying, juice concentrates, freeze drying, and the introduction of artificial sweeteners, coloring agents, and preservatives, such as sodium benzoate. In the late 20th century, products, such as, dried instant soups, reconstituted fruits and juices, and self-cooking meals were also developed.

With the modernization forces in full swing and fast paced life in the West, there was a quest for convenience foods. The marketing strategies of food-processing companies targeted the middle-class working wives and mothers with frozen foods.

These had the benefits of being free from toxins, had ease of cooking, transportation, and less susceptibility to early spoilage than fresh foods. With long shelf lives these foods contributed to the success of long voyages.

The processing industry also contributed immensely to novel food items and brought a paradigm shift in the diet of people, from traditional to a modern one. Food consumption patterns changed from fresh, unprocessed, unbranded food products to processed, packaged, and branded products. Convenience foods found a major place in the modern diet. These are processed foods, which are so prepared and designed that they provide ease of preparation and consumption. Though meals served in a restaurant do meet this definition, nevertheless, the term is seldom applied to them. Convenience foods include prepared foods such as ready-to-eat foods, frozen foods, and prepared mixes, such as, *idli*, *upma*, *vada* mix, etc. The types of convenience foods can vary by country and geographic region. In simple words, they typically cost more money and less time compared to home cooking. Some examples of these foods, which have flooded the supermarkets in India are beverages, such as soft drinks; juices and milk; fastfood; nuts, fruits and vegetables in fresh or preserved states; processed meats and cheeses; and canned products, such as, soups and pasta dishes, frozen pizza, potato chips, and cookies. Even the spices/*masala* used in preparing traditional dishes are available in processed forms. With improved long-distance transportation, this modern diet was available all around the world. Transportation of more exotic foods gave the modern consumer easy access to a wide variety of food unimaginable to their ancestors.

Indian lifestyle has also undergone many changes in the present era. Today Indian households too welcome food with convenience in cooking and purchase. This is the result of both the working women culture present in the society coupled with increased and easy availability of processed and convenience foods. The increasing prevalence of nuclear families, rising disposable income, and more bachelors staying away from home for work have also contributed to the rise in consumption of processed foods in our country. More convenience has been added to such foods by offering home delivery/doorstep delivery by supermarkets for such products, reducing the gap between the shelves of supermarkets and our kitchens.

Revolution in the packaging industry, such as, retort packages have made it possible for Indians who go abroad for short-term assignments or those who are settled there to take these packages along. These provide them with Indian food at a lesser price, as the restaurants serving Indian foods are quite expensive abroad.

The Indian processed food industry has been divided into two main segments (Table 15.1).

The processed food industry is the fastest growing industry in India and is expected to be the world's largest food factory in the times to come (Umali-Deininger and Sur, 2007). The Ministry of Food Processing, Government of India, is responsible for managing the Food-Processing Sector. There are many food-safety concerns of

Table 15.1 Classification of Indian Processed Food Industry

Regional Processed Foods	Multinationals
• MTR, Kohinoor foods, ITC, Haldiram, Tasty Bites, Priya • Rice dishes, such as, *bisibele bhath, rajma chawal, sambar rice, jeera rice* • Ready-to-eat South Indian dishes, such as, a*vial, kesari bhath, khara bhath* • North Indian dishes, such as, *alu muttar, chana masala, dal fry, dal makhani* • Frozen foods, such as, *masala dosa, alu curry, rava idli, punjabi chole, paratha palak, paneer* • Instant sweet mixes, such as, *gulab jamun*, vermicelli • Instant snacks	• Quaker (oats) • Kellogs (cornflakes)

processed food, for example, the use of food additives. The health risks of any given additive vary greatly from person-to-person, for example, the use of sweeteners, preservatives, and stabilizers are permitted only at specified levels for their use in food products, which if not strictly adhered to can harm the consumer.

Food processing is typically a mechanical process that utilizes large mixing, grinding, chopping, and emulsifying equipment in the production process. These processes inherently introduce a number of contamination risks. This sector includes the food processors, suppliers of equipment, raw materials, ingredients, packing materials, processing aids, pesticides, fertilizers, and cleaning chemicals. However, *it does not include primary production, transportation, storage, and retail.* Various processing units and subsectors under the food-processing industry are given in Fig. 15.1.

To ensure safety of during processing *Good Manufacturing Practices* (GMPs) have been defined. These are referred to as practices and procedures performed by a food processor, which can ensure the safety of a food product. GMPs refer to the people, equipment, process, and the environment in the production process. It is a term that is recognized worldwide for the control and management of manufacturing, testing, and overall quality control of food and pharmaceutical products. The focus of GMPs is primarily diminishing the risks inherent in any food or pharmaceutical production process. These were first given by the US Food and Drug Administration. The various components of GMP are given in Fig. 15.2.

GMPs have been developed to reduce the risk of contamination in the food-processing sector. The design, documentation, and implementation of an organization's GMPs can be modified as per the specific needs of the products handled and the processes employed. GMPs are a set of procedures, which if adopted and implemented in true spirit, will lead to production of safe and wholesome foods. The various components of GMPs are described in succeeding paragraphs.

FIGURE 15.1 Subsectors of Food-Processing Industry in India

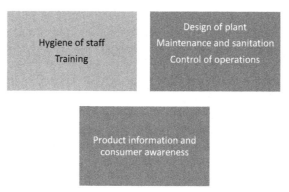

FIGURE 15.2 Components of GMPs

15.1 **HYGIENE OF STAFF**

Hygiene of all the personnel carrying out activities in the food-processing/manufacturing plant has an impact on the food safety. The plant owner has a direct responsibility to ensure that those who come directly or indirectly in contact with food do not contaminate the food product being manufactured in the plant. These handlers need to be instructed regarding their personal hygiene, method of operating, and handling of food items. A system is set up to ensure that all comply with the instructions issued to them.

Food handlers need to maintain a high degree of personal cleanliness and, where appropriate, wear suitable protective clothing, head covering, and footwear (Ansari-Lari et al., 2010). Cuts and wounds, where personnel are permitted to continue working, shall be covered by suitable waterproof dressings. Personnel shall always wash and disinfect their hands.

It is mandatory that an annual medical checkup of workers is carried out and records are maintained. People known, or suspected, to be suffering from, or to be a carrier of a disease or illness likely to be transmitted through food, shall not be allowed to enter any food-handling area. Any person so affected shall immediately report the illness to the management. In case a staff member in food operations is suffering from jaundice/diarrhea/vomiting/fever/sore throat with fever/visibly infected skin lesions (boils, cuts, etc.)/discharges from the ear or eye shall be reported to the management. The concerned officials may then take appropriate action and possible exclusion from food handling can be considered.

Vaccination of food handlers against typhoid is recommended as a good practice for the food handlers. People engaged in food-handling activities shall refrain from behavior that could result in contamination of food, for example, smoking, spitting, chewing or eating, and sneezing or coughing over unprotected food. Personal items, such as, jewelry, watches, pins, flowers, or other items shall not be worn or brought into food-handling areas. Visitors to food-manufacturing/processing or handling areas shall, where appropriate, wear protective clothing and adhere to the other personal hygiene provisions.

15.2 **TRAINING OF STAFF**

All the staff working in the unit shall have adequate and appropriate education, training, and skill about the plant operations with a special emphasis on critical points where food safety can be compromised. It is of utmost importance that extra attention is paid for training of those personnel who come directly or indirectly in contact with food.

To ensure that this aspect of GMPs is in place, the management must identify the training needs of personnel whose activities have an impact on food safety and quality. Apart from conducting training, periodic assessments of the effectiveness of training and instruction is essential. Routine supervision and checks during working hours can ensure that methods learnt during training are being strictly followed in practice.

It is necessary that the personnel are aware of the relevance and importance of their individual contributing activities to GMP system. All personnel shall be aware of their role and responsibility in protecting food from contamination or deterioration.

It is the responsibility of the management to ensure that training programs are routinely reviewed and updated where necessary. Records of all training and related actions should be maintained. The kind and level of training depends on following factors:

- The nature of the food and in particular its ability to sustain growth of pathogenic or spoilage microorganisms. For example, cooked rice is more susceptible than tinned fruits.
- The manner in which the food is handled and packed, including the probability of contamination, for example, whether direct contact with hands is required in the process or it is a fully automatic and a closed process.
- The extent and nature of processing or further preparation before final consumption, for example, using a mix for making *idli* will undergo cooking as compared to curd/ice cream, which are ready to be consumed.
- The conditions under which the food will be stored, for example, storage of ready-to-eat packets can be done in the storage area at room temperature as compared to frozen foods that require strict temperature control.
- The expected length of time before consumption, for example, storage of tetrapacks can be done for months at room temperature as compared to bread, which needs to be consumed within a specified duration.

15.2.1 GAPS IN THE IMPLEMENTATION OF GMPs

Laxity in hygienic practices of food handler in a food-processing plant can cause food contamination. The main challenge here is to motivate employees to comply with hygienic practices. Training alone is often not enough to ensure employee compliance. A carrot-and-stick model works well here. Various methodologies to ensure employee compliance, such as, using a CCTV to check hand-washing practices by employees, use of thumb sensors at hand-washing area to cross-check number of times hand washed by a particular employee can be used by managerial staff to ensure strict compliance in plants. Some indirect methods to measure compliance are counting the number of soaps/amount of liquid soap/number of towels dispensed by the end of the day, etc. In case sophisticated gadgets are not applicable, a strict monitoring of all workers by a senior employee can alone serve the purpose. To ensure compliance, a strict action against defaulters to set an example can make the employees understand that there is no scope of complacency in hygienic practices.

Another gap between training and implementation is that the plant managers outsource training of food handlers to an external agency. These outside agencies conducting training for employees are generic in approach. The content of training needs to be customized and tailor-made to suit a particular manufacturing unit. Other impediments in effective training might include training the wrong people, not training enough people, or not providing enough training.

15.3 **DESIGN OF PLANT**

Food-safety risks in a manufacturing unit depend on the nature of the food item and the process that is undertaken. The management needs to make certain that the premises, equipment, and facilities shall be located, designed, and constructed to ensure that contamination is minimized. The design and layout of equipment should permit appropriate maintenance and prevent cross-contamination. There shall be a unidirectional flow of process and materials. The surfaces and materials, in particular those in contact with food, should be nontoxic and durable. Appropriate facilities for light, temperature, humidity, and pest management ensure the smooth implementation of GMPs.

Various infrastructure requirements for GMPs are given in Table 15.2.

Table 15.2 Infrastructure Requirements for GMPs

Variable	Requirements
Location of manufacturing unit	Away from environmentally polluted areas and industrial activities
Premises and rooms: design and layout	Surfaces of walls, partitions and floors should be made of impervious materials; walls and partitions should have a smooth surface up to a height appropriate to the operation; floors should be constructed to allow adequate drainage and cleaning; ceilings and overhead fixtures should be constructed and finished to minimize the buildup of dirt and condensation; windows should be easy to clean; doors should have smooth, nonabsorbent surfaces
Equipments	Durable and movable or capable of being disassembled to allow for maintenance, cleaning, disinfection, and monitoring; equipment used to cook, heat, treat, cool, store, or freeze food shall be designed to achieve the required food temperatures; maintenance of equipment
Water	Potable in sufficient quantity
Waste-disposal sytem	Appropriate number and type of bins to prevent contamination and pest nuisance
Air quality and ventilation	Control of ambient temperatures and odors, which might affect the suitability of food; control of humidity, where necessary, to ensure the safety and suitability of food
Lighting	Adequate natural or artificial lighting shall be provided to enable the undertaking to operate in a hygienic manner; fixtures to prevent contamination from breakage; recommended lighting intensity in various areas: 1. 500 lx (minimum) in working area 2. 110 lx (minimum) in storage rooms 3. 600 lx (minimum) in inspection areas

15.4 MAINTENANCE AND SANITATION

The food-manufacturing unit shall establish effective systems to:

1. ensure adequate and appropriate maintenance and cleaning,
2. pest control,
3. waste management, and
4. monitoring the effectiveness of maintenance and sanitation procedures.

Establishments and equipment shall be kept in an appropriate state of repair and condition to facilitate all sanitation procedures. Cleaning and disinfection programs shall ensure that all parts of the establishment are appropriately clean and shall include the cleaning of cleaning equipment also. These procedures shall be continually and effectively monitored for their suitability and effectiveness and documented.

Biofilms are a serious food-safety hazard in any manufacturing units, which are formed when bacteria form a slime layer on a surface and provide an environment for pathogens to proliferate (Chmielewski and Frank, 2006). This biofilm can detach and become a significant source of food contamination. Biofilms are formed because attached bacteria can often survive conventional cleaning methods. Most of the times, the staff to take the cleaning step lightly thinking that the next step of sanitizing will take care of all pathogens. Hence, adequate cleaning, preferably scrubbing prior to sanitizing is of paramount importance to control this problem. Another method to tackle this issue is coating drains and equipment parts with antimicrobial material that can counteract biofilms. However, this does not eliminate the need for proper cleaning and sanitizing.

Pieces of equipment can break off and enter food products during processing if equipment is poorly maintained. Routine or preventive maintenance and other periodic checks of equipment can minimize the risk from this safety issue. Risk is further minimized with the use of metal detectors and X-ray machinery. Proper calibration of equipment and minimizing contact between pieces of machinery is also helpful.

15.4.1 LIGHTING FIXTURE/OTHER GLASS BREAKAGE

Glass can be controlled by having a glass-breakage policy, such as, throwing away all food within 10 ft. of the incident of any breakage. Light fixtures can be protected so that if they break, the glass does not spill out. Capping equipment should be properly calibrated and lines should be monitored for evidence of glass breakage. X-ray technology can also be helpful in identifying glass pieces in food.

15.4.2 PEST-CONTROL SYSTEMS

Good hygiene practices shall be employed to avoid creating an environment that is conducive to pests. Good sanitation, inspection of incoming materials, and good monitoring shall be implemented to minimize the likelihood of infestation and thereby limit the need for pesticides.

Buildings shall be kept in good repair and condition to prevent pest access and to eliminate potential breeding sites. Holes, drains, and other places where pests are likely to gain access shall be kept sealed. Animals shall, wherever possible, be excluded from the grounds of factories and food-processing plants. It is recommended to guard openings, such as, windows, exhaust fans with fly-proof mesh, and provide double doors, fix strip curtains, or air curtains at the entrance. Refuse shall be stored in covered, pest-proof containers.

Establishments and surrounding areas shall be regularly examined for evidence of infestation. Pest infestations shall be dealt with immediately without adversely affecting food safety or suitability. Treatment with chemical, physical, or biologic agents suitable for use in food and beverage industry shall be carried out without posing a threat to the safety or suitability of food. All chemicals used for pest control or cleaning/sanitation shall be supported by the "Material Safety Data Sheets."

15.4.3 WASTE MANAGEMENT

Suitable provision shall be made for the removal and storage of waste. Waste shall not be allowed to accumulate in food handling, food storage, other working areas, and the adjoining environment.

Sanitation systems shall be monitored for effectiveness, periodically verified by means, such as, audit preoperational inspections or, where appropriate, microbiologic sampling of environment and food-contact surfaces, and regularly reviewed and adapted to reflect changed circumstances.

15.4.4 PERSONAL HYGIENE FACILITIES AND TOILETS

Personal hygiene facilities shall be available to ensure that an appropriate degree of personal hygiene can be maintained and to avoid contaminating food. Such facilities shall be suitably located and should include:

- adequate means of hygienically washing and drying hands, including wash basins; and a supply of hot and cold (or suitably temperature controlled) water; and
- lavatories of appropriate hygienic design and adequate changing facilities for personnel.

15.5 CONTROL OF OPERATIONS

Food-business operators shall produce food as per its specification and reduce the risk of unsafe food by ensuring control of operations. Apart from building design and equipment process, controls needs to be established, for example, temperature controls. Based on the nature of food operations undertaken, adequate facilities shall be available for heating, cooling, refrigeration and freezing food, for storing refrigerated or frozen foods, monitoring food temperatures, and when necessary, controlling ambient temperatures to ensure safety and suitability of food. Temperature-recording devices shall be calibrated at stipulated intervals. Specific process steps, which may contribute to food hygiene are chilling, thermal processing, irradiation, drying, chemical preservation, and vacuum or modified-atmosphere packaging.

15.5.1 PREVENTION OF MICROBIOLOGIC CROSS-CONTAMINATION

Raw unprocessed foods shall be effectively separated either physically or by time, from ready-to-eat foods, with effective intermediate cleaning and where appropriate, disinfection. Access to processing areas may need to be restricted or controlled, where risks are particularly high; access to such processing areas shall be made through a changing facility. Personnel may need to put on protective clothing including footwear and wash their hands before entering. Surfaces, utensils, equipment, fixtures, and fitting as applicable, shall be thoroughly cleaned and where necessary disinfected after raw food, particularly meat and poultry, has been handled or processed. Certain guidelines specific to the plant shall be available, for example, in pickle and juice industry, best control practices—such as not using dropped fruit, removing damaged fruit, and washing/brushing fruit prior to processing or by pasteurization help in preventing contamination. Vegetables and fruits can be washed with chlorine water or can undergo ozone treatment. For dairy products, pasteurization is a foolproof method to eliminate any kind of contamination. In case of eggs shells pasteurization/washing/spraying of chicks/feed-ingredient control/use of *Salmonella*-free chicks are few options.

15.5.2 PHYSICAL AND CHEMICAL CONTAMINATION

Foreign matter in raw materials can be controlled with raw material inspections. X-ray technology is also available to examine incoming material. Adequate facilities for the storage of food, ingredients, and nonfood chemicals (e.g., cleaning materials, lubricants, and fuels) shall be provided. Where necessary, separate, secure storage facility for cleaning materials and hazardous substances shall be provided.

15.5.3 TRANSPORT

Products shall be adequately protected during transport. The type of conveyances or containers required depends on the nature of food and conditions under which it is to be transported. Where necessary, conveyances and bulk containers shall be designed and constructed so that they do not contaminate foods or packaging, can be effectively cleaned, and when necessary disinfected. It should permit effective separation of different food or food from nonfood items during transport; and provide effective protection from contamination including dust and dirt. There should be maintenance of effective temperature, humidity, and other conditions necessary to protect food from harmful or undesirable microorganisms. When the same conveyance is used for different food or nonfood items, effective cleaning and where necessary, disinfection shall take place between loadings.

15.5.4 **POSTPROCESSING CONTAMINATION**

Products can also be contaminated if the postprocessing environment, utensils, or equipment has been contaminated with a pathogen. This issue is especially relevant to the pathogen *Listeria monocytogenes*, due to its hardiness and pervasiveness in the environment. Effective controls against postprocess contamination include eliminating the pathogen from the postprocessing environment by using environmental sampling to eliminate niches, effective sanitation, and various in-package pasteurization methods (Krysinski et al., 1992). Contamination can be prevented by application of Hazard Analysis and Critical Control Point (HACCP) principles, regular environmental sampling and testing, in-package steam hot-water treatment, pasteurization, irradiation, or use of preservatives.

15.5.5 **TRACEABILITY**

The food business shall ensure that effective traceability procedures are in place from raw material to finished products and to the consumer as appropriate, so as to deal with any food-safety hazard and enable the complete, rapid recall of any implicated lot of product from the market. The organization shall identify product status with respect to its inspection and testing. Traceability records shall be maintained for a defined period for system assessment to enable the handling of potentially unsafe products and in the event of product withdrawal.

15.5.6 **MISTAKEN IDENTITY OF PESTICIDES**

The best way to control the risk of mistaken identity is to store pesticides away from food ingredients, keep an inventory of pesticides, and store the products in their original containers.

15.5.7 **CORROSION OF METAL CONTAINERS/EQUIPMENT/UTENSILS**

Metal poisoning can occur when heavy metals leach into food from equipment, containers, or utensils. When highly acidic foods (e.g., citrus fruits, fruit drinks, fruit fillings, tomato products, or carbonated beverages) come into contact with potentially corrosive materials, the metals can leach into the food. One solution to the problem is to use appropriate, noncorrosive materials in food processing.

15.5.8 **RESIDUE FROM CLEANING AND SANITIZING**

If an equipment and other food-handling materials are not rinsed well, then residue from detergents, cleaning compounds, drain cleaners, polishers, and sanitizers can contaminate a food product. This problem can best be controlled by properly training personnel about cleaning and sanitizing.

15.5.9 **ACCIDENTALLY ADDING TOO MUCH OF AN APPROVED INGREDIENT**

Some substances, such as, preservatives, nutritional additives, color additives, and flavor enhancers, are intentionally added to food products. But adding an approved ingredient in inordinate amounts by accident—such as adding too much nitrite to cured meat—can result in a toxic product. Nutritional safety issues can also arise when product labels' nutrition information is incorrect. Thus, it can be dangerous to public health when too little or too much of a specified nutrient is added. For example, malnutrition can occur if infant formula does not deliver the expected nutrient content during its shelf life. There are also many examples of nutritional food-safety issues arising when too much of a nutrient gets added to a product unintentionally. For example, some vitamins that are added to fortified foods (such as, vitamin A) are known to be toxic at high doses. Controlling chemicals by keeping an inventory of additives minimizes the occurrence of this type of contamination.

15.5.10 **NATURAL TOXINS**

Toxins, such as, mycotoxins and marine toxins are naturally produced under certain conditions. Given that these toxins generally occur in raw materials, especially crops and seafood, manufacturers should require suppliers to certify that the products they purchase are free from natural toxins.

15.6 **PRODUCT INFORMATION AND CONSUMER AWARENESS**

All food products shall be accompanied by or bear adequate information to enable the next person in the food chain to handle, display, store, prepare, and use the product safely and correctly. Flow diagrams shall be prepared for the products or process categories covered by the GMP system. Flow diagrams shall provide a basis for evaluating the possible occurrence, increase, or introduction of food-safety hazards.

The characteristics of end-food products shall be described in documents to the extent needed to ensure safe and quality food including information on the following as appropriate product name or similar identification, composition, biologic, chemical, and physical hazard specification and allergens relevant for food safety, intended shelf life and storage conditions.

The packaging design and materials shall provide adequate protection for products to minimize contamination, prevent damage, and accommodate appropriate labeling. The material shall be nontoxic so as not to pose a threat to safety and suitability of food under the specified conditions of storage and use. Where appropriate, reusable packaging shall be suitably durable and easy to clean and disinfect where necessary.

15.7 RECORD MAINTENANCE FOR GMPs

Good recording keeping is the backbone for effective GMPs (Min and Min, 2006). Each processing unit needs to have a GMPs policy of the plant along with the related objectives. This master document is expected to have the details of various processes, production system, and operation of the unit. It is required that any changes proposed in the food-manufacturing system is reviewed by the GMP experts of the unit prior to implementation to determine its effect on the food-safety system.

Records shall be established and maintained to provide evidence of conformity to requirements and evidence of the effective operation of the GMP system. Records shall remain legible, readily identifiable, and retrievable. A documented procedure shall be established to define the controls needed for the identification, storage, protection, retrieval, retention time, and disposition of records.

For example, for cleaning of the establishment and equipment following records shall be maintained:

1. areas, items of equipment and utensils to be cleaned
2. responsibility for particular tasks
3. method and frequency of cleaning
4. monitoring arrangements

15.7.1 FOOD-SAFETY IMPLICATIONS OF LACK OF CORRECT DOCUMENTATION

Documentation in every aspect of the process, activities, and operations involved with manufacture of the food product is a handy mechanism in ensuring safety of food. It not only works as a preventive tool but also helps in pinpointing where things went wrong and fixing accountability. In case a manufacturing unit undergoes plant renovations, it is vital that various *standard operating procedures* are revised and documented alongside. For example, if a new equipment is installed then the number of mops/time allotted for cleaning the plant should increase simultaneously.

15.8 GMP-SYSTEM VERIFICATION

The organization shall conduct internal audits at planned intervals to determine whether the GMP system conforms to the planned arrangements and requirements. An audit program shall be planned, taking into consideration the importance of the processes and areas to be audited, as well as, any updating actions resulting from previous audits. The audit criteria, scope, frequency, and methods, shall be defined.

Top management shall ensure that the GMP system is continually updated. In order to achieve this, the GMP team shall evaluate the GMP system at planned intervals.

15.8.1 **WITHDRAWALS**

A stable system for withdrawals helps to facilitate the complete and timely withdrawal of lots of end-products, which have been identified as unsafe. The management shall appoint personnel having the authority to initiate a withdrawal and personnel responsible for executing the withdrawal. There should be a documentation procedure in place for withdrawals and the sequence of actions to be taken.

Withdrawn products shall be secured or held under supervision until they are destroyed. GMPs aim at having a quality approach to manufacturing and reducing contamination in food processing. These are comprehensive and address issues, such as, record keeping, personnel qualifications, sanitation, cleanliness, equipment verification, etc. The requirements are open-ended and allow manufacturers to decide how best to implement necessary controls. They are not only flexible but also require that the manufacturer interpret the requirements suited to his business. The top management of the food business shall ensure that requirements, which are applicable to the nature of food operations, are complied with in addition to those applicable to the Statutory and Regulatory requirements.

REFERENCES

Ansari-Lari, M., Soodbakhsh, S., Lakzadeh, L., 2010. Knowledge, attitudes and practices of workers on food hygienic practices in meat processing plants in Fars, Iran. Food Control 21 (3), 260–263.

Chmielewski, R.A., Frank, J.F., 2006. Biofilm formation and control in food processing facilities. Compr. Rev. Food Sci. Food Saf. 2 (1), 22–32.

Krysinski, E.P., Brown, L.J., Marchisello, T.J., 1992. Effect of cleaners and sanitizers on *Listeria onocytogenes* attached to product contact surfaces. J. Food Protect. 55 (4), 246–251.

Min, S., Min, D.B., 2006. The Hazard Analysis and Critical Control Point (HACCP) system and its implementation in an aseptic thermal juice processing scheme: a review. Food Sci. Biotechnol. 15 (5), 1–3.

Pujol, R., 1985. Nicolas Appert: L'inventeur de la Conserve. Denoël, Paris, France.

Umali-Deininger, D., Sur, M., 2007. Food safety in a globalizing world: opportunities and challenges for India. Agric. Econ. 37 (S1), 135–147.

Food safety issues in production of foods of animal origin and from farm to plate

16

D.P. Attrey

Central Military Veterinary Laboratory, Meerut, Uttar Pradesh, India; High Altitude Research, Defence Research and Development Organisation, Leh, Jammu and Kashmir, India; Amity Institute of Pharmacy, Amity University, Noida, Uttar Pradesh, India; Innovation and Research Food Technology, Amity University, Noida, Uttar Pradesh, India; Amity Institute of Seabuckthorn Research, Amity University, Noida, Uttar Pradesh, India; Lala Lajpat Rai University of Veterinary and Animal Sciences, Hisar, Haryana, India

16.1 INTRODUCTION

It is well known that food must be nutritious and safe. As per World Health Organization (WHO, 1996), food security exists when all people have both physical and economic access to sufficient, safe, and nutritious food at all times to maintain a healthy and active life. Not only sufficient nutritious food should be available in the market, but people should also be able to purchase that food on sustainable basis, to meet their daily dietary needs. But what should that diet be and how to ensure that the diet people consume is healthy and safe. Foods of animal origin, that is, milk, meat, poultry, eggs, and fish are major source of class-one proteins in the human diet (containing all the essential amino acids required for overall development of body and brain, vitality, strength, and stamina), besides being a rich source of iron and vitamin B_{12} (which are scarce in foods of plant origin). They are more bioavailable than vegetable proteins and provide better nutrition. No doubt foods of animal origin are nutritious and healthy, but they are highly perishable and become unsafe if appropriate precautions are not taken in their primary production till consumption from farm to plate.

Quality and safety of foods of animal origin needs more attention. It is erroneously believed that this need will arise only when there is adequate production and availability of these foods of animal origin, since whatever is produced is generally consumed without much storage. Population tries to include one or the other food of animal origin in their daily diet. But poor people do not have economic access to such nutritious foods and cannot afford to purchase them. At the same time malnutrition cannot be removed without nutritious diet. Besides being nutritious, food must also be safe to consume. According to Kinsey (2004), safe food consumption focuses on simple but comprehensive performance standards. Safe food consumption makes a person feel good, that is, its eating or drinking facilitates the health and growth of human body. During Second International Conference on Nutrition at Rome, all participating nations, including India, had committed to "eradicate hunger and prevent all forms of malnutrition worldwide, particularly undernourishment, stunting, wasting, underweight and overweight in children under five years of age and anemia in women and children, among other micronutrient deficiencies,…, etc." (WHO, 2014).

Food safety is not only an important part of public health but it is regulatory requirement also, which works through inspection, education, and surveillance (Martin, 2009).

Safety of foods of animal origin is also linked to their availability and affordability. Poor people cannot afford diet containing animal proteins. Hence there is essential requirement of having not only sufficient availability but also accessibility at affordable rates.

16.2 FOOD SAFETY ISSUES DURING PRIMARY PRODUCTION

Farm produce, being highly perishable, cannot be stored even for few days at ambient temperature in India. Food safety involves reducing the risks in food to prevent infection and contamination during all stages of food production chain. WHO has given a slogan in 2015 for food safety, that is, "From farm to plate, make food safe" (WHO, 2015). Adequate production of foods of animal origin along with their adequate safety, are important. Pathogenic microorganisms (e.g., *Salmonella, Campylobacter, Listeria, Escherichia coli, Bacillus cereus,* etc.) and toxic chemicals (e.g., agrochemicals, pesticide, heavy metals, etc.), can enter meat food chain at various points, that is, at production site, from livestock feed, slaughterhouse, packing plant, during manufacturing, processing, retailing, catering, and during home preparation, and may make food unsafe. Chemical fertilizers and pesticides used in the farm can gain entry in to foods and animal feeds and indirectly enter in the human body. Similarly veterinary drugs, antibiotics, and growth hormones, etc. may also gain entry in the human body indirectly and cause harmful effects on health. *B. cereus* can form heat resistant spores and a heat resistant toxin at room temperature. Reheating or light cooking will not destroy this toxin.

Stale fish is often treated with carbon monoxide by many unscrupulous producers to give red color to the gills to project it as fresh. Formalin is also used illegally for fish preservation, which is not permitted. Seafood and fish can become contaminated with pathogens such as *Vibrio cholera, Salmonella, E. coli, Shigella, and Listeria*, etc. due to poor hygiene during production and processing. Methyl mercury is formed by bacterial action in an aquatic environment from dumping of industrial mercury as well as natural sources of elemental mercury. Heavy metal mercury is a known health hazard. Good agricultural and veterinary practices may help in reducing such risks (WHO, 2015).

Food authorities exert official food control through its various agencies to ensure food safety. The basic function of food control agencies is "inspection and control" besides coordination with food authorities and implementation of food laws. Farmers should focus on the safety and quality of their primary produce, for which it is essential to maintain detailed records of their raw materials, husbandry practices, animal movements, and customers to facilitate food quality and safety control as well as traceability (FAO/WHO, 2005).

16.3 PRESLAUGHTER SAFETY OF MEAT ANIMALS DURING TRANSPORTATION

It is essential that animals are reared, handled, transported, and slaughtered using humane practices. Only those animals which are disease free and in a condition to walk properly should be brought for slaughter. Safe transportation of meat animals is an important preslaughter handling issue which should be done meticulously to reduce stress to the animal before slaughter. The floor of handling areas should not cause slipping or falling of animals. Use of electric pods for moving animal should be discouraged and plastic wrapped sticks, etc. can be used in case of absolute requirement. All animals should have room to lie down simultaneously. The condition of animals arriving for slaughter should be closely monitored so that injured or diseased animals are not slaughtered.

Transportation of livestock exerts stress and may cause injuries to animals. Some may even die en route. Unnecessary stress during transportation may adversely affect the quality of meat. Muscles of stressed animals are low in water and glycogen affecting development of proper rigor mortis. Carcass pH fails to attain acidic values. Hence, animals must be rested adequately before slaughter with sufficient availability of feed and water. A well-fed and well-rested animal shall develop adequate rigor mortis which is an essential requirement for getting good quality meat (Attrey, 1987). No animal should be administered any chemical, drug, or hormone before slaughter except for its treatment as per the decision of the qualified veterinarian, who takes into consideration the withdrawal period of the particular drug.

Large animals should be transported in suitable partitions to protect the animals from infighting or crushing of young ones during transportation as per regulations. Before loading, animals should not be fed heavily. Only light feed may be allowed. For journeys less than 12 h no feed may be carried but for longer journeys sufficient feed shall be carried to last during the journey. Watering facilities shall be provided at regular intervals. All vehicles should be disinfected properly and must be inspected for safety, suitability, and cleanliness before loading the animals. A layer of clean sand to cover the floor to a thickness of not less than 6 cm shall be provided. This layer of sand shall be moistened with water during the summer months. During hot months, arrangements shall be made to sprinkle water on the animals at frequent intervals. In winter, a 2-cm layer of clean sand with another 6-cm layer of whole straw shall be provided.

Suitable ramp shall be provided for loading and unloading the animals. Overcrowding should be avoided. Each animal should have enough space to lie down. The speed of truck transporting animals should not exceed 40 km/h, avoiding jerks and jolts. The truck shall not load any other merchandise and shall avoid unnecessary stops on the road. For journeys exceeding 12 h, the animals shall be transported by railway, where possible. Loading shall be done by evening. Railway wagons shall not accommodate more animals than those specified in IS specifications.

16.4 THE INDIAN LEGISLATION ON MEAT BUSINESS

As per Food Safety and Standards (Licensing and Registration of Food Businesses) Regulations, 2011 [FSSR (L&R), 2011], all FBOs in India will be registered or licensed. Food business operators (FBOs) with a turnover greater than Rs 12 lakhs (approximately $6000) per annum shall be required to obtain license and petty FBOs, with a turnover greater than Rs 12 lakhs per annum shall be required to be registered with the concerned food authority of the state. Petty FBOs, who procure or handle and collect up to 500 L of milk per day or slaughter 2 large animals or 10 small animals or 50 poultry birds per day or less are required to be registered. The system of rearing and marketing of main meat animals (sheep and goats) is very resource intensive in India, involving up to 6 middlemen from farmer to the slaughterhouse, during which the animals are almost exhausted due to travel without adequate feeding and watering and there is no rest and recoupment for exhausted animals before slaughter (Attrey, 1988).

16.5 SLAUGHTERING OF ANIMALS

Part 4 of Schedule 4 in Food Safety and Standards Regulation [FSSR (L&R), 2011] provides the guidelines for hygienic and sanitary practices to be followed by FBOs in meat business, that is, slaughter of meat animals, processing, manufacture, storage, and sale of meat and meat products.

16.6 THE SLAUGHTERHOUSE

Persons/establishments intending to slaughter large and small animals including poultry birds, within their premises for production of meat/meat products for supply/ sale/distribution to the public, shall comply with the following requirements:

1. Must obtain "no objection certificate" (NOC) from local food authority before grant of license.
2. Locate the "slaughter house" at an elevated level in a sanitary place and should preferably be away from vegetable, fish, or other food markets with no undesirable odor, smoke, dust, or contaminants nearby.
3. Slaughterhouse should have slaughter hall for each species and for different methods of slaughter (e.g., halal, jewish, and jhatka). In addition, there should be adequately sized reception area/animal holding yard/resting yard, lairage, side halls for hide collection, paunch collection, offal's collection and separation rooms, holding room for suspected/condemned carcass, by-product harvesting, staff welfare inspector's office, refrigeration room/cold room, etc.

4. Slaughterhouse should have isolation pens for suspected animals and adequately sized partitions between clean and dirty sections and facilities for watering and examining animals before they are sent to holding pens/lairage with arrangements for watering and feeding

5. After introduction of a live animal into the slaughterhouse, there should be a continuous forward movement till the emergence of meat and offal.

6. Separate space shall be provided for stunning, collection of blood, and dressing of the carcasses. The slaughtering of an animal shall not be done in the sight of other animals. The dressing of the carcass shall not be done on the floor. Suitable hoists will be provided to hang the carcass before it is eviscerated.

7. Construction of all rooms and spaces in the slaughterhouse shall be of impervious and nonslippery material and as per standards, which must be cleaned properly every day. All processing operations must be carried out under strict hygienic conditions.

8. Every part of the internal surface above the floor or pavement of such slaughterhouse shall be washed thoroughly with hot lime wash within the first 10 days of Mar., Jun., Sept., and Dec. Every part of the floor or pavement of the slaughterhouse and every part of the internal surface of every wall shall be thoroughly cleaned, washed with water, wiped/dried, and disinfected within 3 h after the completion of slaughter.

9. Rooms and compartments in which animals are slaughtered or any product is processed or prepared shall be kept sufficiently free from steam, vapors, and moisture and obnoxious odors so as to ensure clean and hygienic operations. This will also apply to overhead structures in those rooms and compartments.

10. There shall be efficient drainage and plumbing systems and all drains and gutters shall be properly and permanently installed.

11. In case of slaughterhouses equipped to slaughter equal or less than 50 large animal, 150 small animals, and 1000 poultry birds, waste material should be composted which can be used for manure purpose and in case of slaughterhouses equipped to slaughter more than the aforementioned capacity, waste material should be rendered (cooked) in a rendering plant to produce meat, bone meal, and inedible fats.

12. Suitable and sufficient facilities shall be provided for isolation of meat requiring further examination by authorized veterinary officer in a suitable laboratory within the slaughterhouse, equipped and staffed with qualified chemist/analyst (a graduate with chemistry as one of subjects) and veterinary microbiologist (a qualified veterinarian), who are adequately trained. The licensing authority shall accord approval of the laboratory after inspection.

13. A constant and sufficient supply of clean potable cold water with pressure hose pipes and supply of hot water should be made available in the slaughter hall during working hours. The water quality shall comply with the standards prescribed by the licensing authority.

14. Suitable and sufficient facilities shall be provided for persons working in the slaughterhouse for changing their clothes, cleaning their footwear, and

cleaning their hands before entering rooms used for the preparation and storage of meat.

15. Provisions shall be made for sterilization of inspection equipment and toilets, etc. as per regulation.

16. Warm meat' meant for immediate sale need not be stored in cool conditions. It can be transported in suitable hygienic and sanitary condition in clean insulated containers with covers (lids) to the meat shops/selling units with precautions to ensure that no contamination/cross-contamination or deterioration takes place. Any surplus meat shall be stored in a refrigerated room maintained at 0–2°C.

17. Adequate natural or artificial lighting along with well-distributed artificial light should be provided throughout the abattoir/meat processing unit. As far as possible, meat inspection shall be carried out in the daylight.

18. *Equipment and machinery*: The equipment, fittings, and implements in slaughter hall (including chopping blocks, cutting boards, and brooms) shall be of such material and of such construction as to enable them to be kept clean and corrosion free. No iron or galvanized iron shall come in contact with meat food products.

19. *Personnel hygiene*: Every one working in the slaughterhouse must be neat, clean, tidy, and free from infectious or contagious diseases and should wear clean aprons and head wears.

16.7 SLAUGHTERING

Animals should be slaughtered after proper stunning followed by bleeding. Stunning can be done mechanically, electrically, or with gas. Mechanical stunning can be done by captive bolt, mushroom head percussive stunning or pneumatic percussive stunning. For cattle, pneumatic stunning should be preferred and the optimum position is to keep the center of the stunner in contact with the animal at a point of intersection of lines drawn from the medial corners of the eyes and the base of the ears. The best position for pigs is on the midline just above eye level, with the head directed down the line of the spinal cord and the optimum position for sheep and goat is behind the poll, aiming toward the angle of the jaw. If an animal shows signs of regaining consciousness after the initial stun, the animal must be immediately killed by the use of a captive bolt gun.

Electrical stunning consists of passing electricity through the brain to produce instantaneous insensibility. Electrical head stunners may be preferred for sheep and goat where both electrodes are placed on the head region. Water bath electrical stunning may be used for poultry birds. A low and controlled voltage must be maintained so that the stunning will not damage the heart and brain or cause physical disability and death to the animals. The minimum current levels recommended for stunning are indicated as follows:

cattle 1.5 A; calves (bovines of less than 6 month of age) 1.0 A;
pigs 1.25 A;
sheep and goats 1.0 A; lambs 0.7 A;
broilers 100 μA; turkeys 150 μA.

Gas stunning: Stunning by exposure to carbon dioxide is preferred for pigs. The concentration should be 90% by volume but shall not be less than 80% by volume. Ideally, pigs should be exposed for 3 min. Sticking should be done immediately after exit from the gas chamber. Overcrowding of animals should be avoided in the gas chamber.

16.8 **POULTRY MEAT**

Poultry suppliers and processors must have a documented program for rearing, transportation, marketing, and processing of poultry. They shall incorporate the same in their quality plans and manuals and carry out regular audit of the same. Poultry intended for slaughter should be clean and in good health. Every reasonable precaution should be taken to minimize injury to poultry. The catcher needs to be trained to this effect. Transport crates shall be in good condition, should not cause injury, must not open accidentally, should not be overfilled, and should have enough space to allow all poultry to lie down.

Poultry sheds should be provided adequate ventilation and climate control such as fans or curtains. Stunning equipment should be maintained to ensure that poultry are insensible prior to slaughter, and the time between stunning and slaughter should be kept minimum to avoid any possibility of the bird regaining consciousness prior to slaughter.

16.9 **VETERINARY INSPECTION**

Veterinary inspection envisages "antemortem inspection" of live animals/birds for suitability to slaughter for food/meat purpose and "postmortem inspection" after slaughter of animal/bird supported with subsequent laboratory investigation and testing for necessary confirmation as required, shall be carried out as per the manual for veterinary inspection and testing and necessary instructions/guidelines/procedures issued by the authority.

16.10 **ANTEMORTEM INSPECTION**

1. All animals must be rested before slaughter and shall be subjected to antemortem examination and inspection well in advance before slaughter.
2. No animal which has been received into a slaughter hall shall be removed before being slaughtered, except with the written consent of the qualified veterinarian.
3. An animal which, on antemortem inspection, is found to be unfit for slaughter shall be marked as "suspect" and kept separately.
4. An animal showing signs of any disease, or in febrile condition, at the time of antemortem inspection shall be marked as "condemned" and rejected.
5. No suspect animal shall be slaughtered until all other animals intended for slaughter on the same day have been slaughtered.

6. All animals which, on antemortem inspection show symptoms of railroad sickness, parturient paresis, rabies, tetanus, or any other communicable diseases shall be marked as "condemned" and disposed of as per procedure.

7. Animals presented for slaughter and found in a dying condition on the premises of a factory due to recent disease shall be marked as "condemned" and disposed of as provided for "condemned" animals.

8. Every animal which, upon examination, is found to show symptoms of or is suspected of being diseased or animals declared as "suspect" shall at once be removed for treatment to such special pen and kept there for observation for such period as may be considered necessary to ascertain whether the animal is diseased or not.

9. All animals declared as "condemned" on antemortem inspection shall be marked as "condemned" and killed if not already dead.

10. Such carcasses shall not be taken into the factory to be slaughtered or dressed, nor shall they be taken in the abattoir where edible products are handled, but shall be disposed of as per procedure.

16.11 POSTMORTEM INSPECTION

A careful and detailed postmortem examination and inspection of the carcass and its parts shall be conducted soon after slaughter by the designated veterinarian.

1. Postmortem inspection shall be a detailed one and shall cover all parts of carcass, viscera, lymph glands, and all organs and glands. Postmortem inspection shall be in accordance with the general rules laid down for such inspection in public slaughterhouses under the control of local bodies besides special instructions that may be issued from time to time by the licensing authority.

2. All organs and parts of carcass, including blood shall be retained in such a manner as to preserve their identity till completion of postmortem inspection so that they can be identified in the event of carcass being condemned.

3. Retained carcasses, detached parts, and organs thereof shall in no case be washed, trimmed, or mutilated in any manner unless otherwise authorized by the designated qualified veterinarian.

4. Every carcass or part thereof which has been found to be unfit for human consumption shall be marked as "inspected and condemned."

5. All such condemned carcasses, parts, and organs thereof, shall remain in the custody of the designated veterinarian pending disposal at or before the close of the day.

6. Carcasses, parts, and the organs thereof found to be sound, wholesome, healthy, and fit for human consumption shall be marked as "inspected and passed."

7. No air shall be blown by mouth into the tissues of any carcass or part of a carcass.

8. Carcasses found affected with anthrax or parts of such carcass shall be condemned and disposed of immediately through complete incineration or thorough denaturing with prescribed denaturant as per procedure.

9. Portion of slaughtering department including equipment, employees' boots, and aprons, contaminated by contact with anthrax material shall be cleaned and thoroughly disinfected immediately.
10. Bruised portion shall be removed immediately and disposed of as per procedure.
11. All condemned carcasses, organs, or parts thereof shall be completely destroyed in the presence of the designated veterinarian by incineration or denatured, after being slashed freely with a knife, with crude carbolic acid, cresylic disinfectant, or any other prescribed agent unless such carcasses, organs, or parts thereof are sterilized for the preparation of bone-cum-meat meal before leaving the slaughterhouse premises.
12. Destruction of condemned carcasses, organs, or parts thereof shall be carried out under the direct supervision of the designated veterinarian.
13. If in the opinion of designated veterinarian a carcass, organ, or part thereof is to be held back for further detailed examination the same shall be marked as "held." If on subsequent inspection, the carcass, organ, or part thereof is found to be unwholesome and unfit for human consumption, the designated veterinarian shall mark these as "condemned" and shall be disposed off as mentioned earlier.

16.12 MEAT FOOD SAFETY IN DOMESTIC KITCHENS

Now middle-class urban consumers in developing countries also buy food items at the weekend and store them in freezer or refrigerator. Microwave oven is often used for reheating of food. However, most of these consumers are still learning the scientific use of freezer, refrigerator, and microwave oven. Fat from meat, poultry, and fish should be trimmed to avoid chances of entry of pesticides and chemicals in the body. Avoid bigger fish, since smaller fish have less time to take up and concentrate pesticides and other harmful chemical residues and toxins. Meat must be cooked to the core properly to kill the pathogens and cysts.

Microwave ovens are often used for reheating, cooking, and defrosting. Cookware/products, specially manufactured for use in the microwave oven, should only be used. Dish should be covered with a lid or plastic wrap approved for microwave use. The moist heat that is created will help destroy harmful bacteria and ensure uniform cooking. Cooking bags also provide safe and even cooking. Stir, rotate, or turn food upside down (where possible) midway through the microwaving time to even the cooking and eliminate cold spots where harmful bacteria can survive. After removing food from the microwave, always allow a standing time of at least 3 min. This completes the cooking process. Remove food from its packaging, rotate, and turn food upside down periodically (2–3 times) during defrosting. Microwaves are absorbed preferentially by water rather than ice, and by the first layer that can absorb the radiation, so turning/stirring the food during the defrosting process improves the evenness of the heat.

Cooking whole, stuffed poultry in a microwave oven is not recommended. Because food in a microwave oven can cook faster on the outside than food on the inside, the stuffing might not have enough time to reach the temperature needed to destroy harmful bacteria. Raw meat, poultry, and seafood should be in sealed containers or wrapped securely to prevent raw juices from contaminating other foods. Review the contents of the refrigerator once a week and throw out perishable foods which are more than one week old.

16.13 FOOD SAFETY ISSUES DURING STORAGE

The topic of food preservation and storage has been discussed in Chapter 44 also. Meat preservation can be done either by altering environmental conditions in which spoilage microorganisms are unable to grow in the food or by removal of microorganisms and enzymes. High-fat foods do not store well in the frozen state. Oily fish are most suitable for canning which retains natural flavor of fish.

Botulinum spores are not always killed by high heat of canning process, but bacteria do not produce this toxin in acidic conditions. Therefore, canning or pickling of meat in airtight containers should contain enough acid to prevent toxin formation. Cans that have swelled or bulged should not be consumed and should be thrown away. Bacteria and fungi may grow inside refrigerator also if dryness and cleanliness are not maintained, for example, *Listeria monocytogenes*. Appropriate plastic bags may help in preventing cross-contamination by separating various items.

16.14 FOOD SAFETY CONTROLS

Primary producers and food industry are equally responsible for the quality and safety of their produce/products. Monitoring of certification programs, process control schemes, or hazard analysis critical control points (HACCP)–based control programs can provide a fair idea about status of safety of products produced. Ministries of Health and the Ministries of Agriculture/Food, along with their dedicated reference laboratories (which form the base of decision making by food control services) are responsible for legislative, technical, and practical implementation of food safety programs. But it has been observed that all these organizations work in isolation and independent of each other. For achieving comprehensive food safety at national level, the ministries, labs, their respective agencies, and private organizations must work closely together. Since food-control services function through inspection and control, it is necessary to have a uniform procedure for food inspection throughout the country and even internationally. The inspectors must follow a defined written procedure (FAO/WHO, 2005).

16.15 MEAT PROCESSING AND MARKETING

Like slaughterhouse, all the sanitary and hygienic requirements mentioned previously shall be satisfied in meat-processing unit also.

16.16 **FOOD RETAILING**

Food retailing involves sale of food to consumer. It is essential that food retailers such as processors also adopt a food safety management system to control the safety of their food products. Even the FBOs engaged in catering and distribution of prepackaged food must adhere to good hygienic practices and use HACCP approach to identify and manage food safety hazards. Street food is an important component of the food supply chain in developing countries. These foods are generally prepared and sold under unhygienic conditions, with limited access to safe water, sanitary services, or garbage disposal facilities. Hence street foods pose a risk of food poisoning due to microbial contamination, as well as improper use of food additives, adulteration, and environmental contamination. Specific guidance on the responsibilities of street-food vendors has been provided by WHO (1996).

16.17 **ROLE OF CONSUMER**

Consumers also have a responsibility to protect themselves from risks associated with the preparation and consumption of foods. Foodborne illness can occur as a result of foods being incorrectly stored, not fully cooked, or when cross-contamination between raw and cooked "ready-to-eat" food is allowed to happen. Adoption of WHO's five keys to food safety, that is, cook, clean, separate raw and cooked foods, store at correct temperature, and use safe water and materials can be very helpful in achieving food safety at the consumer level in the domestic kitchen (WHO, 2005).

16.18 **CONCLUSIONS**

Many farmers keep the safety and quality aspects aside and just want to sell whatever they produce. Government policies to help farmers in tackling these issues and creation of required infrastructure are necessary to encourage wholesome animal production in India. Food-control services should also encourage the farmers to get organized, and get appropriate training and expertise in modern agricultural practices and food hygiene to offer safe animal products.

REFERENCES

Attrey, D.P., 1987. An insight into the Rigor Mortis in sheep in India. Proceedings of Workshop on "Animal Reproduction, Farm Management and Associated Problems of Prolonged Hospitalization", August 28, 1987, Held at Headquarters, Uttar Pradesh area, Bareilly, India, pp. 82–87.

Attrey, D.P., 1988. Marketing of meat and meat products in India with special reference to production, procurement and distribution of meat and canned meat food products for defence services. In Proceedings of Seminar on "Quality Control Procedures of Meat in Army Butcheries and Veterinary Public Health Aspects in Relation to Wholesome Supply of Meat, Fish, Poultry and Eggs to the troops", held on 11 Jul 88 at Headquarters Uttar Pradesh Area, Bareilly Cantt., Uttar Pradesh, India, in collaboration with Indian Veterinary Research Institute, Izzatnagar, Bareilly, U.P., India, pp. 133–145.

FSSR, 2011. Food Safety and Standards (Licensing and Registration of Food Businesses) Regulations, 2011 from Food Safety and Standards Authority of India, Ministry of Health and family Welfare, Government of India. Available from: http://www.fssai.gov.in/ Portals/0/Pdf/Food%20safety%20and%20Standards%20(Licensing%20and%20Registration%20of%20Food%20businesses)%20regulation,%202011.pdf

Kinsey, J., 2004. Does food safety conflict with food security? The safe consumption of food. Working Paper 04-01, The Food Industry Center; University of Minnesota; USA, January 2004. Available from: http://ageconsearch.umn.edu/bitstream/14326/1/tr04-01.pdf

Martin, WRN, 2009. Exploring Food Safety & Food Security Tensions Report on the Canadian Public Health Association (CPHA) Annual Conference; Core Public Health Functions Research Initiative (CPHFRI), School of Nursing, University of Victoria, British Columbia, Canada. Available from: http://www.uvic.ca/research/groups/cphfri/assets/docs/cphfri_newsletter_vol1num2.pdf; http://www.uvic.ca/research/groups/cphfri/assets/docs/Food%20safety%20food%20security%20Wanda.pdf

WHO, 1996. Proceedings of the World Food Summit, 1996. Available from: http://www.who.int/trade/glossary/story028/en/

WHO, 2005. Five Keys to Safer Food Manual. Department of Food Safety, Zoonoses and Foodborne Diseases, World Health Organization (WHO). Food Handling Methods, Food Contamination-Prevention and Control, NLM Classification, WA, p. 695.

WHO, 2014. WHO. FAO/WHO Second International Conference on Nutrition (ICN2). Available from: http://www.who.int/nutrition/topics/WHO_FAO_announce_ICN2/en/ and http://www.who.int/foodsafety/Lancetfoodsafety_nov2014.pdf; http://www.fao.org/about/meetings/icn2/en/

WHO, 2015. Food safety: what you should know. World Health Day: April 7, 2015, SEA-NUT-196. World Health Organization. Available from: http://www.searo.who.int/entity/world_health_day/2015/whd-what-you-should-know/en/#intro

FURTHER READING

Attrey, D.P., 1990. Untapped potential of Indian agriculture. YOJNA, Planning Commission, New Delhi, 01–15 Jul 1990; vol. 34 (no. 12), pp. 10–14. Available from: http://yojana.gov.in/cms/(S(lslth1jkwp5ctt45wjkatjf3))/pdf/Yojana/English/1990/Jul_Vol34_No12.pdf

Attrey, D.P., 1995a. Livestock production and human nutrition—a future scenario. Key Note Address at the National Seminar on Animal, Man and Environment. December 21–23, 1995, Delhi Veterinary Association, IARI, Pusa, New Delhi, pp. 93–102.

Attrey, D.P., 1995b. Role of veterinary public health in maintenance of human health. Proceedings of National Seminar on Animal, Man and Environment. December 21–23, 1995, Delhi Veterinary Association, IARI, Pusa, New Delhi, pp. 75–80.

Attrey, D.P., 1997. Poverty alleviation through cattle development. Proceedings of National Seminar on Poverty Alleviation through Livestock Development. May 3, 1997, Delhi Veterinary Association, New Delhi, pp. 11–20.

FAO, 2014. Meat & meat products. Available from: http://www.fao.org/ag/againfo/themes/en/meat/home.html

Safe storage and cooking practices for foods of animal origin in home kitchen before consumption

17

D.P. Attrey

Central Military Veterinary Laboratory, Meerut, Uttar Pradesh, India; High Altitude Research, Defence Research and Development Organisation, Leh, Jammu and Kashmir, India; Amity Institute of Pharmacy, Amity University, Noida, Uttar Pradesh, India; Innovation and Research Food Technology, Amity University, Noida, Uttar Pradesh, India; Amity Institute of Seabuckthorn Research, Amity University, Noida, Uttar Pradesh, India; Lala Lajpat Rai University of Veterinary and Animal Sciences, Hisar, Haryana, India

17.1 INTRODUCTION

Safety of foods of animal origin encompasses production, handling, storage, and preparation/cooking of these foods in a way to prevent contamination in the food-production chain. Their quality depends on their correct origin, color, flavor, texture, and processing method. Poor quality and unsafe foods show signs of spoilage, discoloration, bad odors or tastes, and contamination with extraneous matter/filth. Good sanitary practices can minimize food contamination. Food, irrespective of its source, must be prepared and cooked properly. According to EU and WHO, food safety is a shared responsibility from "farm-to-fork" (Magkos et al., 2006) and now "farm-to-plate" (WHO, 2015).

Although the annual production of milk is in excess of 140 million tons (The Hindu, 2014) and meat is 220 million tons in India, (APEDA, 2014; Kandanuri, 2014), yet per capita availability and consumption is still far from satisfactory from a nutritional point-of-view. Kantor et al. (1997), had reported that food losses during retail, distribution, and consumption in United States were approximately 23, 24, 15, and 30% for fruits; vegetables; meat, poultry, fish products; and dairy products, respectively. Such losses seem to continue even now throughout the world. Therefore, for meeting the growing nutritional requirements of an ever-increasing population, it is essential to increase the yield of foods of animal origin and reduce food losses.

Proper preservation of food is essential to store food for a longer duration and to maintain its nutritional and culinary qualities. Most foods of animal origin are highly perishable, their safe handling, storage, and cooking practices are thus vital.

Food Safety in the 21st Century. http://dx.doi.org/10.1016/B978-0-12-801773-9.00017-0

Food safety issues at preslaughter stage, slaughtering, storage, transport, processing, sales, that is, food safety issues of foods of animal origin in the primary production and from farm-to-plate have been discussed in Chapter 17.

17.2 SAFE PRODUCTION AND STORAGE OF FOODS OF ANIMAL ORIGIN

Food safety starts with the primary production at a farm level where misuse of agrochemicals, including pesticides, growth hormones, and veterinary drugs that need to be monitored carefully. Proper storage will help in the retention of nutritional and functional properties of a food/food product by preventing microbial spoilage. Although proper food storage does not improve the original quality of food, it prevents further deterioration in the quality.

Besides the microbiologic deterioration of animal foods there are many other hazards to food safety. Certain fungi or molds produce mycotoxins on foods during primary production or during postharvest period due to poor storage or handling. Mycotoxins can also enter the food chain via meat or other animal products, such as, eggs, milk, and cheese as the result of livestock consuming contaminated feed. Continuous intake of aflatoxin may cause liver cancer. Other mycotoxins have been linked to kidney and liver damage (Fernández, 2016).

The use of antibiotics and growth hormones in livestock has been a controversial matter for many years. It has been shown that low residues of drugs may build up in fatty tissue, kidneys, and liver of livestock. These may not pose a direct risk to human health but may cause antibiotic-resistance, although this may also be due to poor drug management in the treatment of human health. EU had banned the use of growth hormones in 1998 and phased out use of antibiotics as growth-promoting agents in livestock by 2006. However, the use of hormones still continues in the United States, Canada, and Australia.

Industrial pollutants, such as, dioxins and heavy metals, affect the health of consumers adversely. Dioxins are by-products of manufacture of certain industrial chemicals and incineration or burning of farm and other waste matter. Dioxins are environmental contaminants that persist in the environment for many years and can find their way onto and into food. In fish, polluted water is the main cause of dioxin contamination, while animals are mostly exposed to dioxins through air. Dioxins settle on plants and feed, which are then eaten by animals. Dioxins concentrate in fatty tissues of livestock and fish. More than 90% of human exposures occur mainly through foodstuffs. Foods of animal origin normally account for approximately 80% of all exposures.

Other industrial pollutants include heavy metals, such as, mercury, lead, and cadmium. Fish are especially vulnerable to environmental pollutants because waters can become contaminated from industrial discharges or accidental spillage.

To produce safe and good quality foods of animal origin, careful planning has to be done with written operating procedures to ensure production of wholesome foods of animal origin. Monitoring of operating procedures is also essential to ensure that operations are carried out as intended to maintain safety and quality of such foods (Magkos et al., 2006). But in India, bovine meat production is incidental and secondary to production of milk. Bovines, which are not productive and are not linked with milk production due to age, sex, or other reasons, are sent for slaughter.

Storing of food to retain its natural flavor, texture, and nutritional qualities can be done as follows.

- Use appropriate containers for storage, such as, aluminium foil, tin, plastic, and wraps.
- Keep storage area clean and dry.
- Transfer nuts and sugar into airtight jars to protect them from pests.
- Keep raw foods away from ready-to-eat or contaminated foods.
- Choose containers that enable food to retain its quality and ensure safe storage.
- Label all containers for convenient handling.
- Food containers should be moisture resistant, odorless, easy to clean, and long lasting.
- Storage practices differ from food-to-food, depending on its composition, age from primary production, type of packaging material used, and shelf life.

17.3 STORAGE OF FOODS OF ANIMAL ORIGIN

A refrigerator with a temperature below 5°C will protect most foods from microbial spoilage, although not forever. In an efficient refrigerator, growth of most microbes slows down but it does not stop completely. Bacteria, for example, *Listeria* can grow well in refrigerator and cross-contamination can occur if raw and cooked food or salads come in direct or indirect contact. Food cannot be stored for an indefinite period even in a freezer. Eggs, cake, and some food items may absorb flavor from other food items. Refrigerators should be kept clean and dry all the time. Refrigerator should not be over-crowded as it reduces circulation of cool air. Doors of refrigerator/freezer should not be opened too often. Raw meat, poultry, and seafood should be stored in sealed containers or wrapped securely to prevent raw juices from contaminating other foods.

High-fat foods do not store well in the frozen state. Oily fish should be canned. Canning also retains natural flavor of fish. High temperatures of the canning process do not always kill *Botulinum* spores, but in acidic conditions toxins may not be produced. Thorough cooking inactivates toxins. Therefore, canning or pickling of vegetables or meat in airtight containers, where possible, should contain enough acid to prevent toxin formation. Cans that appear to have swelled or bulged should not be consumed. Pathogenic bacteria do not generally affect taste, smell, or appearance of food. They may cross-contaminate. Vinegar solution (about 15 mL per 500 mL of water) removes bad odors and interior walls of refrigerator and freezer should be washed with this solution. A container with vinegar in refrigerator or freezer should be kept for several hours. If odors persist, vinegar must be kept for 2–3 days, changed every 8 h.

Canned foods with low acidity may last for 2–4 years. Acidic canned foods, such as, fruits, fruit juices, and tomatoes will usually still be of good quality for 1–2 years. However, temperatures over 38°C for an extended time will speedup the loss of quality.

Appropriate plastic packaging plays a significant role in shelf life. Polyethylene terephthalate, which is strong, heat resistant, and resistant to gases and acidic foods, is used to make bottles for soft drinks, drinking water, sports drink, ketchup, etc. It can be transparent or opaque, not known to leach (chemicals that are suspected of causing cancer or disrupting hormones) and it can be recycled.

High-density polyethylene is used to make bottles for milk, water, juice, yogurt, and margarine tubs and grocery, and other retail bags. It is stiff and strong but is not heat stable. It is not known to leach any harmful chemicals and it can be recycled. Low-density polyethylene is used to make films of various sorts, some bread, frozen food bags, and squeezable bottles. It is relatively transparent and is not heat stable. Films may melt on contact with hot food. Polypropylene is more heat resistant, harder, denser, and more transparent than polyethylene. It is used for heat-resistant microwavable packaging and sauce or salad dressing bottles. Polycarbonate is clear, heat resistant, and durable and often used to make refillable water bottles and sterilizable baby bottles, microwave ovenware, eating utensils, and plastic coating for metal cans. Small amounts of "bisphenol A" are formed when polycarbonate bottles are washed with harsh detergents or bleach (sodium hypochlorite). At high levels of exposure, bisphenol A is potentially hazardous because it mimics the female hormone, estrogen. In addition, polystyrene and polyvinyl chloride are also used during food material transportation and handling in supermarkets. Modern food-safe plastic bags are plasticizer-free and do not release harmful chemicals into food while it is being cooked. However, all plastic is made from chemicals that have potential to harm a person's health. The following practices ensure the proper use of plastic packaging, which lowers chemical migration (WHO, 2015).

- Follow manufacturers' instructions when using household plastics, such as, cling films and bags.
- Follow recommendations for cleaning products to be used on containers, bottles, and lids.
- Use correct type of plastic, for example, use only the microwave-safe plastics in microwaves.
- Do not let cling film touch food during microwave cooking as it melts at a low temperature. It may be better to remove the film before cooking in a microwave.
- Leave a corner of dish uncovered to allow steam to escape. This reduces the risk of film being blown off and settling on the food.
- Reuse plastic containers that are food compatible only. For example, freeze food in ice-cream containers but do not heat them in microwave, as they were designed for use for cold food.

17.4 HANDLING AND COOKING OF FOODS OF ANIMAL ORIGIN IN KITCHEN

In India, people prefer hot and well-cooked food and milk is normally used only after boiling. These practices prevent foodborne infections. Street food is also very popular in urban India. But hygienic conditions during production, distribution, and consumption of street foods may be questionable. These can be improved if local authorities provide sufficient potable water supply, clean facilities, and adequate waste disposal. Increasing popularity of bottled drinking water in urban areas is also helping in the reduction of waterborne infections.

"WHO's five keys to safer food" should be implemented during handling of meat. In general, hands must be washed properly; preparation should be done on a separate surface than other cooking materials; separate cutting boards should be used for different items of food; vegetables; etc., should be kept away from meat; and cooking utensils must be cleaned after they come into contact with raw meat. Raw meat may be stored only for 3 days in a refrigerator at 1°C. For longer storage, it should be kept in freezer at −18°C in airtight packets.

According to Healthline (2016), meat can be stored in refrigerator for the following amount of time periods:

- uncooked poultry:1–2 days,
- uncooked ground meat: 1–2 days,
- uncooked steaks or chops: 3–4 days,
- uncooked fish: 1–2 days,
- cooked poultry, meat, or fish: 3–4 days, and
- hot dogs and lunchmeat: up to 1 week (open package) or 2 weeks (closed package).

Meat can be stored in the freezer for the following amount of time periods:

- uncooked ground beef: 3–4 months,
- uncooked steaks or chops: 4–12 months, depending on the item,
- uncooked fish: 6 months,
- cooked meat, poultry, or fish: 2–6 months, and
- hot dogs and lunchmeat: 1–2 months.

Cooking temperature affects both the taste and safety of food. Hotter temperatures at the core of the meat make it safer. Safe cooking temperatures at the core of meats are:

- 71°C for ground meats (beef, pork, and lamb),
- 63°C for fresh, whole meats, and the meat should be allowed to rest for at least 3 min before eating,
- 74°C for poultry, whole, or ground, and
- 63°C for finfish, or until the flesh is opaque and separates easily.

The resting time for cooking whole meats is important. Resting the food gives the heat more time to kill any bacteria. Rare to well-done cooking depends on the temperature at the core of the meat piece.

A meat thermometer is used to measure temperature.

- Rare and medium cooked meats may not have been cooked thoroughly enough to kill all bacteria and the risk may vary with the type of meat/cut.

- Poultry should never be eaten rare. It should always be cooked thoroughly. Undercooked poultry can spread *Salmonella* and other diseases.
- Pork should always be cooked to at least for the higher end range of medium. Pork can carry several potentially dangerous types of worms and parasitic larvae.

Beef has a wider safety range. However, rare meat lovers are safer if they stick to steaks, roasts, and chops. Ground meat needs to be cooked to a higher temperature. This is because whole cuts of meat typically have the most bacteria on their surfaces. Bacteria found in ground meats may get mixed throughout.

Fish also has a wider range of safe cooking, which depends on the type and quality of fish being cooked. The method of cooking fish is also important. Fish should generally be cooked to the well-done stage. For certain types of fish, medium-to-rare cooking may be acceptable. Raw sushi, should be eaten with great caution. It must be prepared carefully to reduce the risk of contamination. Foods may be classified as potentially hazardous and nonhazardous. Potentially hazardous foods permit growth of pathogens. Nonhazardous foods do not support the growth of microbes due to a low water activity (0.85 or less), pH being 4.6 or less, and *high temperature–short time*-treated products, etc. Foods from unapproved sources or which are unsafe, adulterated, or out of temperature should not be accepted. For this consumers should ensure the following at the domestic level (Idaho Food Code, 2008).

- Before purchase, all foods must be checked for damaged containers, leaks, off-odors, filth, and other signs that suggest food may not be wholesome.
- Packaged food without labels must not be accepted.
- Pasteurized milk and milk products should be accepted, packaged raw milk in retail can be accepted but must be boiled before storage in refrigerator.
- Cracked or dirty eggs must not be accepted.
- Shellfish must be obtained in containers bearing proper labeling with a certification number.
- All meat and meat products must be obtained from regulated meat processing establishments, which ensure inspection for wholesomeness by a qualified veterinarian.
- Care should be taken to ensure that produce from a local grower has not been mishandled or contaminated.
- Crustaceans, wild mushrooms, wildlife, and other foods not mentioned earlier must also be from approved sources.
- Foods canned or prepared in a private home or in an unregulated food establishment should not be accepted for use. Such foods may present a risk to the public health.
- Consumers should accept frozen foods only in their frozen state, with no signs of previous thawing.
- There should be no crosscontamination from and among raw foods of different species or foods that are already cooked.
- Raw meat of all types (beef, fish, lamb, pork, poultry, etc.) must be physically separated during transportation, storage, and processing. This is required because

different meats have different bacteria, parasite types, and numbers. Normally, beef and lamb have the least amount of microorganisms and poultry has the most. This requirement is particularly important considering different preparation methods and cooking temperatures for different products. Also, where custom meat processing is done, these meats must be stored and processed separately from inspected meats.

- Ready-to-eat food (including cooked food) must be physically separated from unwashed produce and uncooked food products during storage, preparation, holding, transportation, and/or service. Physical separation can be vertical with ready-to-eat food located above unwashed produce and uncooked food products, but not below.
- Separate storage areas must be provided for spoiled, returned, damaged, or unwholesome foods.
- Food, once served to the consumer, must not be served again (some exceptions, such as, crackers sealed in plastic, individual ketchup packets, etc.).
- Ready-to-eat foods must not be prepared in areas where raw meats are processed, except by scheduling and proper cleaning between operations.
- Foods must be protected against contamination resulting from the addition of unsafe or unapproved food, color additives, steam, gases, and air.
- Egg pooling and contamination must be avoided. Fresh eggs should not be cracked in quantity and pooled. Pasteurized eggs should be used. Raw eggs in ready-to-eat food products should not be used.
- Bulk foods must be protected. Prepared food, once removed from the original package or container, regardless of the amount, must not be returned. This also applies to consumer self-service displays, salad bars, etc.
- When using gloves, ready-to-eat products must always be handled in the order salad ingredients before raw meat. If necessary, raw foods should be handled in descending order of potential contamination as specified in Idaho Food Code. The food-handling procedure must not reversed. Gloves present no special protection against crosscontamination.

17.5 PREPARING FOOD SAFELY AT HOME

Foods of animal origin should be procured from reliable sources and must be inspected by an authorized qualified veterinarian. Fresh fish must be observed carefully for its freshness. It should have firm flesh, a stiff body, tight scales, no pitting on pressing, red gills, and bright eyes. It should not be slippery or slimy to touch and must be stored on ice or refrigerated. Seafood and fish can become contaminated with pathogens such as *Vibrio cholerae, Salmonella, Escherichia coli, Shigella, Listeria*, etc., due to poor hygiene and sanitation during handling or processing (WHO, 2015).

During cooking or reheating or for making ice, use potable water only. It is not a good habit to taste raw food during the cooking process. Oil should be put into pan for cooking at the beginning rather than later when pan is very hot, to prevent smoking of the oil. If possible avoid using used oil after frying and try to use fresh oil for each cooking, as reusing oil can create trans fatty acids, which may cause health problems, including cancer, atherosclerosis, etc. Sunflower, soybean, mustard, and canola oils have a high smoke point, that is, they do not break down at high temperatures and are suitable for frying. Oils which do not have a high-smoke point, such as, olive oil should only be used for sautéing and not for frying or tempering/seasoning. Cooking oil left after cooking or frying should be cooled down and then transferred into an airtight container through a strainer. If reused cooking oil is dark in color or is greasy/sticky, discard it (WHO, 2015). To reduce dietary exposure to pesticide residues, trim the fat from meat, poultry, and fish. Avoid bigger fish as the smaller fish have less time to take up and concentrate pesticides and other harmful residues.

17.6 COOKING OF MEAT OR FISH

Meat should be cooked properly to kill pathogens. Temperature and cooking time of meat varies according to the type of meat, type of cut, and method of cooking. Meat should be checked visually to see if it is cooked thoroughly. At home, one should make sure there is no pink meat left as the color of meat changes after cooking. On piercing the thickest part of the meat with a fork or skewer, the juices should run clear. Cut the meat open with a clean knife to confirm that it is steaming. Chicken is well cooked if the meat is white and there is no pink flesh. Fish is well done when it is opaque and flakes easily with a fork. Meat thermometer may be used to measure the internal temperature of cooked meat and poultry, or any other dishes, to assure that a safe temperature has been reached and that harmful bacteria have been destroyed.

Microwave ovens are used for reheating, cooking, and defrosting. Only microwaveable cookware should be used in a microwave. Metal paints on ceramics or glass may cause sparks and fire. Always cover the dish with a lid or plastic wrap approved for microwave use. Allow enough space between food and top of the dish so that plastic wrap does not touch the food. Loosen or vent the lid or wrap to allow steam to vent. Moist heat that is created will help to destroy harmful bacteria and ensure uniform cooking. Cooking bags also provide safe and even cooking. Stir, rotate, or turn food upside-down (where possible) midway through microwaving time to ensure even cooking. Even if the microwave oven has a turntable, it is still helpful to stir and turn food top-to-bottom. After removing food from the microwave, always allow a standing time of at least 3 min. This completes the cooking process (WHO, 2015).

Food pathogens are destroyed during microwave cooking if food is heated to >70°C. However food is not cooked evenly in a microwave oven and "cold spots" may remain where harmful bacteria can survive. To promote uniform cooking in a microwave, arrange food items evenly in a covered dish and add some liquid if needed.

17.7 METHOD OF THAWING OF FROZEN FOOD IN A MICROWAVE OVEN

1. Remove food from its packaging before defrosting.
2. Do not use foam trays and plastic wraps because they may melt from the heat of food and may cause harmful chemicals to leach in.
3. During microwave defrosting, rotate and turn food upside-down periodically (2–3 times during defrosting). Many microwaves have special settings for defrosting.
4. Microwaves are absorbed preferentially by water rather than ice and by first layer that can absorb the radiation, so turning/stirring food during defrosting process improves the evenness of heat.
5. Cooking whole, stuffed poultry in a microwave oven is not recommended, because food in a microwave oven can cook faster on the outside than on the inside, stuffing might not have enough time to reach the temperature needed to destroy harmful bacteria.

Microwaves cause water, fat, and sugar molecules to vibrate 2.5 million times per second, producing heat. After switching off, the molecules continue to generate heat as they come to a standstill. This additional cooking after microwaving stops is called "carry-over cooking time," "resting time," or "standing time." It occurs for a longer time in dense foods, such as a whole chicken roast, than in less-dense foods, such as, breads, small vegetables, and fruits. During this time, the temperature of a food can increase by several degrees. Additionally, microwave heating is often uneven with focal spots of intense heat near areas of cool (especially if originally frozen), so standing time allows for a more even heat dispersion. For this reason, directions usually advise to let a food "rest" for a few minutes after turning off the oven or removing food from the oven.

During pasteurization, milk is heated to a high enough temperature at which harmful bacteria are killed. Pasteurization does not kill all bacteria. Some are still left, although in very small quantities. But they may spoil the milk after a few days. Pasteurized milk must be kept in the refrigerator but it can be kept for a longer period in the freezer chamber.

Milk and cream can also be pasteurized at a higher temperature (132°C), but for a shorter time, which is known as ultrahigh temperature processing. Ultrahigh temperature milk passes through heating and cooling stages in quick succession, after which it is immediately put into a sterile tetrapack shelf-safe carton, in which it can be preserved for about 6 months without refrigeration or preservatives. Tetrapack products are sterile and safe to consume until they are opened. After opening it should be stored in a refrigerator. These products are labeled as ultrapasteurized (WHO, 2015).

According to FSSAI (2016), raw materials are of great importance as biologic, chemical, or physical hazards, which have been introduced in the initial stages, may persist through preparation and processing. Therefore raw materials of acceptable grade only (not low grade) should be accepted and these should be purchased in quantities that correspond to the adequate storage/preservation capacity. Packaged food products must be checked for "expiry date"/"best before"/"use by date," packaging integrity and storage conditions. Proper rotation of all raw materials and finished materials should be undertaken on *first in–first out*, *first expired–first out*, and *first manufactured–first out* basis. Receiving and storage temperature of potentially high-risk food should be at or below 5°C. Receiving and storage temperature of frozen food should be –18°C or below. Understand ingredients, indicative of possible allergens (milk, egg, fish, crustacean shellfish, nuts, wheat gluten, peanuts, and soya proteins), high sugar, etc., for health reasons.

If prepared food is to be seasoned with uncooked ingredients (e.g., fresh coriander, lettuce, etc.), do not expose food to excessive contact with these ingredients, in time or temperature, before consumption. Sandwiches should preferably be prepared at the moment of consumption. Raw eggs should not be used for preparation of food and beverages intended for direct consumption that are not to be cooked afterward.

Time required for transportation should be minimum, to avoid microbial proliferation. In case food is to be prepared and served in the near vicinity, it should be transported and served hot at temperature above 60°C and consumed within 4 h. In case food is prepared at long distances, which require transportation for a longer period, the food should be chilled to less than 5°C, transported and reheated at the time of service to a temperature of at least 70°C, served at temperature above 60°C, and consumed within 4 h. Reheating should be done once. Cooked food to be served hot should be kept at a temperature of at least 60°C and cooked food to be served cold should be kept below 5°C to prevent growth of pathogens. Otherwise time of holding should be limited. Only permitted food additives may be added within permissible limits. Vegetarian foods should always be stored above nonvegetarian foods and cooked foods above uncooked foods on separate racks in the refrigerator. All foods must be kept covered. Handling of food should be minimal. All surplus food and unused thawed food should be discarded. Food to be kept for cold storage should be distributed in small volumes to ensure uniform cooling. Even the dry, fermented, and acidified foods should be stored in a cool and dry place.

As per Fernández (2016), nonvegetarian products are cooked thoroughly (core temperature 75°C) for at least 15 s or an effective time/temperature control, for example, 65°C for 10 min or 70°C for 2 min. Nonvegetarian products should be stored covered in refrigerator below the vegetarian products. Raw and cooked products should be stored physically separated with cooked products at the top. All refuse/waste should be promptly removed from the preparation area. Milk should be received in clean and hygienic conditions at temperature below 5°C. Milk and milk products should be used immediately or pasteurized and refrigerated. Food must be reheated up to 70°C before consumption and consumed within 4 h of reheating. Food products are not stored at room temperature for more than 2 h during display or sale. For prolonged storage, foods are stored in refrigerators or kept for hot holding at or above 60°C.

RIU (2014) has presented "6 Consumer Control Points for Food Safety and 10 Steps to a Safer Kitchen." Application of Hazard Analysis Critical Control Point principles to prevent foodborne illness through "Consumer Control Point" in the kitchen is possible, which should be implemented. Consumer Control Points are critical points where food susceptibility to contamination from foodborne pathogens can be identified. The golden principle is to buy cold food, that is, meat, poultry, and dairy products from the market after finishing other purchases, get these home quickly and refrigerate them again as soon as possible. Select food packages with the longest period of time of expiry from the market.

If using reusable grocery bags, regularly wash them on a gentle cycle in a washing machine. Avoid storing reusable bags in the car or truck in warm weather because high temperatures can promote growth of harmful bacteria. Wash hands as soon as you return home with soap and water for 15–20 s before and after handling raw meat, poultry, or seafood products. Thaw frozen foods in the refrigerator, never at room temperature. A microwave oven can be used to thaw food. Thawed food should be cooked immediately. Keep pets away from food-preparation areas. Wash hands after touching your pets. Thorough cooking destroys harmful bacteria. Choose a serving style that will allow food to be served as quickly as possible, while maintaining temperatures below 41°F (4°C) or above 135°F (60°C). Keep hot foods above 135°F (60°C) and cold foods below 41°F (4°C). Divide leftovers into small units and store in shallow containers for quick cooling. Refrigerate within 2 h of cooking. Reheat leftovers to 74°C. Boil soups, sauces, and gravies before use. The most common food-handling mistake is cooling food too slowly. If in doubt, discard the food item.

Food should be procured from approved and trusted sources and then the following 10 points should be kept in mind for safe cooking (RIU, 2014).

1. Maintain refrigerator at 4°C or less to slow down the growth of most bacteria.
2. Refrigerate cooked food within 2 h after cooking. Leftovers should be used within 2–3 days. Discard the suspected food in case of doubt.
3. Sanitize your kitchen dishcloths and sponges regularly with a solution of 1 tsp chlorine bleach to one quart (about 1 L) of water, or use a commercial sanitizing agent, following product directions. Do not use the same cloth or sponge in other kitchen areas after use on dirty surfaces, such as, meat cutting boards. Sanitize them properly before using it on other surfaces in kitchen.
4. Wash cutting boards and utensils with soap and hot water after each use.
5. Cook ground meat to a safe internal temperature of at least 72°C.
6. Do not eat raw or lightly cooked eggs.
7. Clean kitchen counters and other surfaces that come in contact with food with hot water and detergent or a solution of bleach and water.
8. Allow dishes and utensils to air-dry in order to eliminate recontamination from hands or towels. When washing dishes by hand, it is best to wash them all within 2 h, before bacteria can grow.
9. Wash hands with soap and warm water immediately after handling raw meat, poultry, or fish. Wash for at least 20 s before and after handling food, especially raw meat. If you have an infection or cut on your hands, wear nonlatex rubber or plastic gloves, such as, vinyl, nitrile, and synthetics. Latex gloves may cause problem of allergies for certain people.

10. Defrost meat, poultry, and fish products in the refrigerator, microwave oven, or cold water that is changed every 30 min. Follow package directions for thawing foods in the microwave. Cook microwave-defrosted food immediately after thawing. Changing water every 30 min when thawing foods in cold water ensures that the food is kept cold.

According to FDA (2009), home kitchens, with their varieties of food and open entry to humans and pet animals, often become the source of microbial contamination of food if necessary precautions are not taken. The source of food is important because pathogenic microorganisms may be present in the farm animals, their feeds, in the farm environment, in waters used for raising and freezing aquatic foods, and in soils and fertilizers in which plant crops are grown. Chemical contaminants that may be present in field soils, fertilizers, irrigation water, and fishing waters can be incorporated into food plants and animals. As such food and food products must be obtained from approved sources, so that after harvesting and processing, they do not get adulterated.

REFERENCES

APEDA, 2014. Chapter 2, Red Meat Manual. Indian Meat Industry; Agricultural and Processed Food Products Export Development Authority (APEDA). Available from: http://apeda.gov.in/apedawebsite/MEAT_MANUAL/Chap2/Chap2.pdf

FDA, 2009. Chapter 3, FDA Food Code 2009 Annex 3–Public Health Reasons/Administrative Guidelines Food. US Food and Drug Administration. Available from: http://www.fda.gov/Food/GuidanceRegulation/RetailFoodProtection/FoodCode/ucm189169.htm

Fernández, L., 2016. Your guide to food safety and quality and health and nutrition for a balanced diet and healthy lifestyle. EUFIC. Available from: www.eufic.org

FSSAI, 2016. Guidelines for Food Safety (medium to small eating establishments). Available from: http://www.fssai.gov.in/Portals/0/Pdf/DOC_GUIDELINES.pdf

Healthline, 2016. Overview of meat safety: storing and handling meat, poultry, and fish. Available from: http://www.healthline.com/health/food-safety-meat#Fish5

Idaho Food Code, 2008. Available from: http://healthandwelfare.idaho.gov/Health/Food Protection/IdahoFoodCode/tabid/765/Default.aspx

Kandanuri, V., 2014. Comparative advantage of India in buffalo meat exports vis-à-vis major exporting countries. Res. J. Management Sci. 3 (2), 8–14.

Kantor, L.S., Lipton, K., Manchester, A., Oliveira, V., 1997. Estimating and addressing America's food losses. Food Rev. 20, 3–11.

Magkos, F., Arvaniti, F., Zampelas, A., 2006. Organic food: buying more safety or just peace of mind? A critical review of the literature. Crit. Rev. Food Sci. Nutr. 46, 23–56.

RIU, 2014. Six Consumer Control Points for Food Safety and 10 Steps to a Safer Kitchen. Available from: https://www.extension.iastate.edu/4hfiles/StateFair/SFDocuments/SF1Obtaining PermissionToUseCopyright.pdf

The Hindu, 2014. News report (The Hindu, Jun. 24, 2014). Available from: http://www.thehindu.com/business/Industry/milk-production-rises-to140-million-tonnes-in-201314/article6085584.ece

WHO, 2015. Food safety: what you should know. Available from http://www.searo.who.int/entity/world_health_day/2015/whd-what-you-should-know/en/

Food safety is a shared responsibility: role of various stakeholders in implementing food safety

Role of government authorities in food safety

18

P. Dudeja*, A. Singh**

**Department of Community Medicine, Armed Forces Medical College,
Pune, Maharashtra, India; **School of Public Health, Post Graduate Institute
of Medical Education and Research, Chandigarh, India*

Developed and developing countries both have shown an increasing trend in foodborne illnesses (FBI) over recent years. Although most of these infections cause mild illness, but death occurs due to severe infections and serious complications. The reports of melamine-contaminated milk powder in China and Bihar Midday Meal tragedy remind us that FBI can hit at anytime and anywhere. There will be no surprise if such incidents are reported in future also. Hence, it is the duty of the government to provide safe food to the citizens. From time to time, various laws have been enacted by the concerned ministries to ensure availability of safe food to all. Lately, the Government of India has given a new law to the country, that is, Food Safety and Standards Act (FSSA) 2006. This has been necessitated by various factors operating in the ever-changing contemporary society which affect food safety at the macro level (Fig. 18.1).

Over the last century, globalization has fuelled the transport of variety of food products across the countries. International chains of processed foods are increasing their business and India, because of its sheer population, is a business opportunity for them.

The advances in food-processing technology have undergone a paradigm shift. Restaurant industry has much to offer to clients. With the rise in economic growth and living standards, millions of Indians have now a richer and more varied diet than before. The media has influenced the food industry in a big way. The present day consumers want value for money. They are well aware about their rights. They demand better food safety from suppliers. At this juncture, the government authorities have to play their part and this scenario has necessitated a paradigm shift in the enactment of food-related laws in India. The requirement of food regulations in any society may be based on several factors. This depends on which food standards a country adopts. The international norms have been developed by the Codex Alimentarius Commission of the Food and Agriculture Organization (FAO) of the United Nations, and WHO. A country may also have its own set of food regulations. Usually more than one agency is involved in this, for example, health and agriculture. The

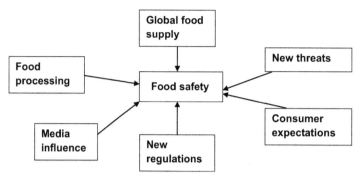

FIGURE 18.1 Factors Affecting Food Safety in Present Society

law may be centralized or regionally controlled. Different agencies may be involved in its enforcement.

18.1 HISTORY OF FOOD LAWS IN INDIA

Traditionally, food has always been worshipped in Indian culture. In traditional Indian society, kitchen was always given the place of a temple. However, in the same society incidents of food adulteration can also be traced back to the times of *Kautilya* (about 300 BC). Elaborate stringent laws, regulation, and procedures were evolved by *Kautilya* to ensure protection of king from any poisoning attempts through Royal Kitchen. The laws related to adulteration are as old as the crime itself. In the past, there were rules in Arthshastra. Kagle (1970) has translated his treatise into English: *"As to difference in weight or measure or difference in price or quality, for the weigher and measurer who by the trick of the hand brings about (difference to the extent of) one-eight part in (an article) priced at one panna, the fine is two hundred (pannas)... For mixing things of similar kind with objects such as grains, fats, sugar, salt, perfumes and medicines the fine is twelve pannas."* Adulteration was one of the gravest of socioeconomic crimes.

During the British era, the Indian Penal Code, 1860, came into force. The individual state laws, imposing strict liability started coming into force since 1912. But, there was considerable variance in rules and specifications of food which affected inter provincial trade. Until 1954, several states formulated their own food laws. Government of India appointed The Central Advisory Board and the Food Adulteration Committee in the years 1937 and 1943, respectively. A central legislation called the Prevention of Food Adulteration Act (PFA) in the year was framed 1954 which came into effect from June 15, 1955. The main objectives of PFA were to have a central legislation to bring uniformity in food laws, protection of public from poisonous and harmful foods, prevention of the sale of substandard foods, protection of the interests of the consumers by eliminating fraudulent practices and to prevent, curb and check

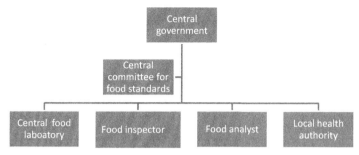

FIGURE 18.2 Composition of Authorities Under PFA

the rampant adulteration of food stuffs. The composition of authorities under PFA is given in Fig. 18.2.

18.2 AUTHORITIES UNDER PFA

Under the central government, there is a Central Committee for Food Standards. It plays an advisory role to the central and the state governments. Four Central Food Laboratories at Pune, Kolkata, Ghaziabad, and Mysore were established to analyze samples of food on payment. Posts of public analysts were created and they were appointed by the central or the state government for different local areas. Local health authority was made responsible for health administration. Food inspectors were appointed by the central or the state government for different local areas. They were entrusted with the responsibility to take food samples and send these to the public analyst. They could prohibit sale of the article of food with prior permission from local (health) authority, inspect any place where article of food is manufactured or stored, and seize any article or book of accounts or any other document. The act was amended in 1956, 1964, 1971, 1976, and 1986 to plug the loopholes and to make punishments more stringent. These were made to empower the consumers and voluntary organizations to play a more powerful and effective role in implementation. Various orders have been issued from time to time for regulating the licenses, permits, or otherwise the production or manufacturing of essential commodities (Fig. 18.3). Some of these orders are described subsequently.

18.2.1 FRUIT PRODUCTS ORDER, 1955

Fruit Products Order (FPO) is for nonfruit beverages, squashes, jams, jellies, tomato products, etc. Under the Central Government a Central Fruit Products Advisory Committee was constituted. There was appointment of a licensing officer (LO). Every manufacturer had to apply for a license. The LO were given the powers of refusal of a license. The manufacturer could appeal to the central government within

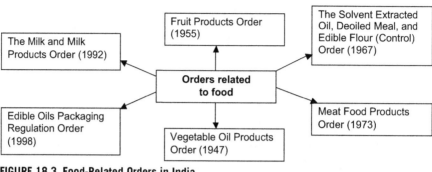

FIGURE 18.3 Food-Related Orders in India

30 days of such refusal. Under this order, license number was to be displayed or embossed prominently in case of bottle, tin, barrel, or any other container. It should specify the code number and date of manufacture. The code number was regulated to be given in English or Hindi numerals or alphabets or both. Label was not to be misleading or false. Any beverage not containing 25% of fruit juice was not to be described as a fruit syrup, fruit juice, or syrup. Non-fruit beverages, syrups, etc. should be labeled as "non-fruit." This order was not applicable to any syrup which contained fruit juices for medicinal use or are sold in bottles bearing a label with the words "for medicinal use only."

18.2.2 THE MEAT FOOD PRODUCTS ORDER, 1973

Meat food products (MFPs) meant any article of food, being used as a food which is derived from meat by means of drying, curing, smoking, cooking, seasoning, flavoring. It shall not include the following products unless the manufacturer himself desires to be covered under the provisions of the said order: meat extracts, meat consommé and stock, meat sauces; whole, broken, or crushed bones, animal gelatin, meat powder, bone extracts, and similar products; fats melted down from animal tissues; patties, puffs, rolls, samosas, cutlets, koftas, kababs, chops, tikkas, and soups made from mutton, chicken, etc. Under the Central Government a Meat Food Products Advisory Committee was constituted which appointed the licensing authority. It was empowered to refuse to grant the license to any applicant. Reasons were to be recorded in writing. The manufacturer could appeal against the refusal within 30 days. The validity of license was 1 year. Renewal of license was mandatory.

18.2.3 THE VEGETABLE OIL PRODUCTS (REGULATION) ORDER, 1998

Vegetable oil product (VOP) is any product obtained for edible purposes by subjecting one or more edible oil to any combination of processes such as blending, refining, etc. Under the central government, appointment of vegetable oil products commissioner was done. Under this order no producer was to be eligible for registration unless he has his own laboratory for testing of samples and the commissioner may refuse to grant registration and such reasons for rejection are to be recorded in writing. Any person could appeal against such a refusal/cancellation of the registration within 30 days of the receipt of the said order.

18.2.4 EDIBLE OILS PACKAGING REGULATION ORDER, 1998

Edible oils means vegetable oil and fats but does not include any margarine, vanaspati, bakery shortening, and fat spread. Under the Central Government appointment of edible oils commissioner was created. The registering authority was appointed by the state government and appointment of inspecting officers was created. Under the order any person who intended to carry on the business as a packer must be registered. The certificate of registration was valid for 3 years and could be further renewed for the same period.

18.2.5 THE MILK AND MILK PRODUCTS ORDER, 1992

As per the order milk means milk of cow, buffalo, sheep, goat, or a mixture thereof, either raw or processed. Milk product means cream, curd, yogurt, cheese and cheese spread, ice cream, milk ices, condensed milk (sweetened and unsweetened), condensed skimmed milk (sweetened and unsweetened), sweets made from khoya, etc. Under the Central Government Milk and Milk Product Advisory Board was constituted and a registering authority was constituted. The order entailed that no person or manufacturer should set up a new plant or expand the capacity of the existing plant without obtaining registration/permission. There was a provision of transfer of registration. The registering authority could suspend/cancel any certificate.

So far, there was a multiplicity of food related laws in India (Table 18.1). It created confusion in the minds of the consumers, traders, manufactures. Problems were faced as different products were governed by different ministries and orders. There were variations in specifications/standards in different orders. The food processing technology was getting advanced and there were emerging concerns for food safety. Table 18.2 describes different laws under their respective ministries. One of the loopholes in PFA was that the individual could not be punished if the item was not noxious. For example, mixing of water with milk. As per the need, the act was amended four times—in 1964, 1971, 1976, and 1986. The act did not provide for the mandatory standardization of food products. There was no requirement for training the food inspectors. Usually, they would not know how much sample is to be taken and in what quantity the preservative should be mixed in the sample for transportation to the laboratory. The minimum number of such inspectors required for the area was not given. In other words, the 'inspector to population' ratio norm was missing in the act.

The PFA gave right to any person to get the sample tested if he thought that it contained deleterious substance under Section 12. But for this he had to pass two hurdles. First, he had to inform the seller the purpose for which he was taking the

Table 18.1 Role of Different Ministries in Regulating Food-Related Laws

Ministry of Health and Family Welfare	Ministry of Agriculture	Ministry of Food and Consumer Affairs
Prevention of Food Adulteration Act, 1954 PFA Rules, 1955 Health Food Supplement Bill	Agriculture Produce Marketing Act Milk and Milk Product Order	Essential Com. Act, 1955 Standards of Weights & Measures Act, 1976 Packaged Commodities Rule, 1977 Consumer Protection Act, 1986 B.I.S. Act, 1986 VOP Control Order, 1947 VOP (Std. of Quality), 1975 SEO control (order), 1967
Ministry of Commerce	**Ministry of Food Processing Industries**	**Ministry of Rural Development**
Imports and Exports Regulations Export Inspection Agency Tea Board Coffee Board Coffee Act and Rules	Fruit Products Order, 1955	Agricultural Produce Grading & Marketing Act, 1937 Meat Food Products Order
Ministry of Forests & Environment	**Ministry of Science & Technology**	**Ministry of HRD (Development of Women & Child Welfare)**
Trade in Endangered Species Act Ecomark	Atomic Energy Act, 1962 Control of Irradiation of Foods Rules, 1991 G.M. & Organic Foods	Infant Milk Substitutes, Feeding Bottles & Infant Foods (Regulation of Production, Supply & Distribution) Act, 1992— Rules, 1993

sample and second that for analysis he had to pay the requisite fees. As far as the first issue was concerned, no trader who was really guilty would allow the consumer to take the sample. Second, though the fee was refundable if the analysis report was positive, it was not possible for all to afford it initially, as it was usually a costly affair. Moreover, it was always doubtful whether the analysis would be precise. Then nobody wanted to get entangled in legalities.

While sentencing, the judge had no discretion, as there was provision of minimum punishment. On the contrary, a burden was placed on him, to state in the judgment, the special and adequate reasons as to why a particular punishment was meted out. Lack of coordination was often witnessed between the food inspector and public

Table 18.2 Penalties Under FSSA 2006

Offences	Penalties
Selling food not of nature, substance of quality demanded	Penalty not exceeding Rs. 5 lakhs
Selling, manufacturing, or distributing sub standard food	Penalty not exceeding Rs. 5 lakhs
Selling, manufacturing, or distributing misbranded food	Penalty not exceeding Rs. 3 lakhs
Misleading advertisement	Penalty not exceeding Rs. 10 lakhs
Food containing extraneous matter	Penalty not exceeding Rs. 1 lakhs
Failure to comply with the directions of food safety officer	Penalty not exceeding Rs. 2 lakhs
Unhygienic or unsanitary processing or manufacturing of food	Penalty not exceeding Rs. 1 lakhs
Processing adulterant not injurious to health	Penalty not exceeding Rs. 2 lakhs
Processing adulterant injurious to health	Penalty not exceeding Rs. 10 lakhs
Unsafe food which does not result in injury	Imprisonment up to 6 months and fine up to Rs. 1 lakhs
Unsafe food which results in nongrievous injury	Imprisonment up to 1 year and fine up to Rs. 3 lakhs
Unsafe food which results in grievous injury	Imprisonment up to 6 years and fine up to Rs. 5 lakhs
If failure results in death	Imprisonment not less than 7 years but which may extend to imprisonment for life and fine not less than Rs. 10 lakhs

analyst who were not legal persons and the public prosecutor who was not a technical person. This benefited the accused. The magistrates usually handling criminal cases were not specialists in food adulteration matters. At the same time, they have the mindset of giving benefit of any doubt or any inordinate delay to the accused, which spoiled the prosecution case.

There was a major problem with procedural part of the act. The act failed to distinguish between the categories of adulteration and had same punishment for all kinds of adulteration. Coming to the practical side, under the present scenario the retailers were not in the position to press the manufacturers for giving guarantee. Moreover, there are no facilities available to the traders to test the purity of the articles at the time of purchase.

In India, the concept of food safety is now being looked into seriously. The felt need for an integrated food law has been met with by promulgation of FSSR, 2011. The archaic Prevention of Food Adulteration (PFA) Act, 1954, has been finally repealed. FSSA, 2006 consolidated the laws relating to food. The main intent of this endeavor was to ensure availability of safe and wholesome food for human consumption. The emphasis of Food Safety Act 2006 is on highlighting the responsibility of manufacturers, recall, risk analysis, good manufacturing practices, and process control, namely, hazard analysis and critical control point (HACCP).

18.3 NEW ERA OF FOOD LAWS IN INDIA

In 1988, the Prime Minister's Council on Trade and Industry appointed a subject group on food and agro-industries which recommended for comprehensive legislation on food with a food regulatory authority. The Standing Committee of Parliament in April 2005 in its 12th Report expressed that the much-needed legislation on integrated food law should be expedited. Therefore, the government introduced a bill called the Food Safety And Standard Bill 2005 in the Parliament (Lok Sabha) on August 25, 2005 which was finally passed in 2006.

The main objective of Food Safety and Standard bill was to bring out a single statute relating to food and to provide for a systematic and scientific development of a food processing industry. The bill incorporates the salient provisions of the Prevention of Food Adulteration Act, 1954 and is based on international legislations, instrumentalities and Codex Alimentarius Commission (which related to food safety norms). The bill was made keeping in view the international practices and envisaged an overreaching policy framework and provision of single window to guide and regulate persons engaged in food industry.

The Indian Parliament passed FSSA, 2006 that overrides all other food related laws. It specifically repealed eight laws, namely, The Prevention of Food Adulteration Act 1954, The Fruit Products Order (FPO) 1955, The Meat Food Products Order 1973, The Vegetable Oil Products (Control) Order 1947, The Edible Oils Packaging (Regulation) Order 1998, The Solvent Extracted Oil, De oiled Meal, and Edible Flour (Control) Order 1967, The Milk and Milk Products Order 1992, and the Essential Commodities Act 1955 relating to food.

The act established a new national regulatory body, Food Safety and Standards Authority of India (FSSAI), to develop science-based standards for food and to regulate and monitor the manufacture, processing, storage, distribution, sale, and import of food so as to ensure the availability of safe and wholesome food for countrymen.

On January 19, 2011, the Food Safety and Standards Authority of India (FSSAI) published a final draft of the Food Safety and Standards Rules, 2011. As previewed in earlier drafts, the final draft specifies the implementing rules of the Food Safety and Standards Act, 2006, which would replace the existing Prevention of Food Adulteration (PFA) Rules of 1955. On August 24, 2006, the Indian Government published the Food Safety and Standards Act, 2006 in the official Gazette, providing public notice of a set of consolidated and updated food safety and standards regulations and announcing plans to establish the FSSAI under the Ministry of Health and Family Welfare (Dhulia, 2010; Singh, 2011).

The FSSAI after its establishment drafted Food Safety and Standards Rules and Regulations. The Draft FSS Rules, 2010 contains qualifications of the enforcement agencies, sampling techniques, legal aspects, and other issues enumerated under Section 91 of the FSSA 2006. The Draft FSS Regulations, 2010 contain labeling requirements and standards for packaged food, permitted food additives, colors, microbiological requirements, etc.

However, these regulations are a re-titled version of the PFA Act, 1954, and Rules, 1955 and its amendments. FSSA 2006 officially repeals the regulatory framework established by the PFA, 1954, and Rules, 1955, 1955, MFPO 1973, VOP Order 1947, EOP Order 1988, the Solvent Extracted Oil, DeOiled Meal and Edible Flour (Control) Order, 1967, and the Milk and Milk Products Order, 1992.) The draft details the FSSAI's approach to risk analysis, risk assessment, risk communication, and risk management. The proposed draft also lays out FSSAI's policy on open stakeholders' consultation, processing of applications of food operators for getting registered with FSSAI and clearly does the functional separation of risk assessment and risk management.

The FSS Rules, 2010 contain qualifications of the enforcement agencies, sampling techniques, legal aspects, and other issues enumerated under of the FSSA 2006.

The FSS Regulations, 2011 contain labeling requirements and standards for packaged food, permitted food additives, colors, microbiological requirements, etc.

18.4 KEY PROVISIONS OF FSSA

1. Effective regulation, manufacture, storage, distribution, and sale of food to ensure consumer safety and promote global trade.
2. Single reference point for food safety and standards, regulations, and enforcement.
3. Prevention of sale of misbranded, unsafe/contaminated or sub-standard food
4. No article of food shall contain food additive, processing aid, contaminants or heavy metals, insecticides, or pesticides residue.

18.5 IMPORTANT DEFINITIONS AS PER FSSA

Food: Any substance, whether processed, partially processed, or unprocessed, intended for human consumption. It includes infant food, packaged drinking water, water used in food during its manufacture, chewing gum. It does not include live animals, plants prior to harvesting, cosmetics, drugs, and medicinal products.

Adulterant: Any material which renders the food unsafe or substandard, misbranded, or contains extraneous matter.

Fig. 18.4 describes *authorities under FSSA 2006*.

18.6 OTHER HIGHLIGHTS OF FSSA

FSSA 2006 has also attempted to define Nutraceuticals (Section 22) which are described in (Chapter 40). It has also laid restrictions on advertisements and prohibition of unfair trade practices (Section 24). It places the liability of the manufacturers, wholesalers, distributors, and sellers (Section 27) in the event any article of food supplied/sold after the date of expiry, unsafe or misbranded, unidentifiable of

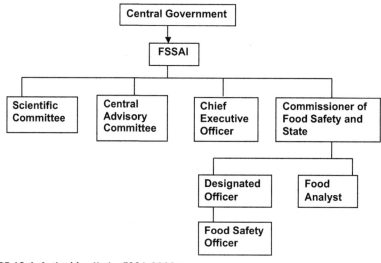

FIGURE 18.4 Authorities Under FSSA 2006

manufacturer, received with the knowledge of being unsafe, or stored in unhygienic conditions. It also consists of Food Recall Procedures (Section 28). As per the act, (Section 31) license is mandatory. However, petty manufacturers, hawkers, temporary stall owners, small-scale cottage industries, etc. are required to register with the concerned municipal authority. As per Section 32, improvement notices can be issued by the designated officer. Such notices include the following:

- Grounds stating the failure to comply with regulations.
- Matters which constitute FBO failure to comply.
- Measures to be taken by FBO in order to secure compliance within reasonable time.

Failure to comply the notice may lead to suspension/cancellation of the license. Under Section 40, rights of a purchaser to get food analyzed after paying the fees specified are mentioned. The purchaser has to inform the FBO at the time of purchase that it is for analysis. The fee will be refunded if sample fails and failing sample will lead to prosecution.

Penalties under FSSA are given in Table 18.2. *Comparison between PFA and FSSA* is given in Table 18.3.

18.7 CHALLENGES

Enormous workload of implementation of new law all across the country will be a slow and a long process. Lack of trained human resources is a definite bottleneck. The number of food safety officers in India is woefully small (approximately 2000).

Table 18.3 Comparison Between PFA and FSSA

PFA	FSSA
Definition of food	
"Food" means any article used as food or drink for human consumption other than drugs and water and includes 1. any article which ordinarily enters into, or is used in the composition or preparation of, human food; 2. any flavoring matter or condiments; and 3. any other article which the central government has declared by notification in the official gazette, as food with regard to its use, nature, substance, or quality.	"Food" means any substance, whether processed, partially processed or unprocessed, which is intended for human consumption and includes primary food to the extent defined in clause (ZK), genetically modified or engineered food or food containing such ingredients, infant food, packaged drinking water, alcoholic drink, chewing gum, and any substance, including water used into the food during its manufacture, preparation, or treatment but does not include any animal feed, live animals unless they are prepared or processed for placing on the market for human consumption, plants prior to harvesting, drugs and medicinal products, cosmetics, narcotic, or psychotropic substances: Provided that the Central Government may declare, by notification in the official gazette, any other article as food for the purposes of this act having regards to its use, nature, substance, or quality
Authority	
It consists of the Central Committee For Food Standards, central food laboratory, food inspector, etc.	It consists of the FSSA, scientific committee, central advisory committee, etc.
Prohibition of import	
Prohibition of import on any article of food which is adulterated, misbranded, or which is contravention of any provision of the PFA. Further, any officer of the custom may detain any package if he suspects it of being adulterated or misbranded.	Prohibition of import on any article of food which is unsafe, misbranded, substandard, or which is contravention of any provision of the FSSA. Further, the import of food articles is regulated under the Foreign Trade (Development and Regulation) Act, 1992 and the standards laid down by the food authority are to be followed. The central government may, from time to time, formulate and announce by notification in the official gazette, the export and import policy and may also, in the similar manner, amend that policy
Food recall	
No such provision	In order to remove unsafe food from the market and thus prevent injury to consumers food recall procedures can be initiated voluntarily by the manufacturers/distributors concerned or by the food authority

(Continued)

Table 18.3 Comparison Between PFA and FSSA (*cont.*)

PFA	FSSA
Improvement notice	
No such provision	If the designated officer has reasonable ground for believing that any food business operator has failed to comply with any regulations to which this section applies, he may, serve an improvement notice on that food business operator.
Licensing	
Licensing authority	All food business operators are required to get a license or registration that will be issued by the local authorities.
	Temporary stall holders are exempted from the same but are required to get their business registered with the panchyat or local municipality
Health supplements/foods for dietary uses/nutraceuticals	
PFA does not provide any limit on the vitamins, minerals, and other nutrients.	This has now been defined under Section 22 of the FSSA and may contain plants or botanicals in the form extract, powder, concentrate, etc., but does not include a narcotic drug or psychotropic substance and does not claim to cure any specific disease.

Manpower shortage is a main hurdle in implementation of this act. Moreover, the definition of "food" excludes the animal feed from its purview. It will be a mammoth task to educate the hawkers and street-food vendors about the concept of food safety.

The fact is that whatever extraneous harmful products like pesticides, insecticides, etc. get into the animal feed and are consumed by the animal (cow, goat, etc.) become a part of food chain. For example, if it is present in milk. Therefore, this should have been made a part of the definition of food contamination.

Food chain from farm to the products needs to be traced. But as the farmers are excluded from the purview of the act, the tracing is possible to the *sabzi mandi* only.

As there is lack of properly trained workforce, the Ministry of HRD can think about the role of universities, for organizing vocational training courses on food analysis and food testing.

A separate department in the concerned ministry must look after the matter of food adulteration since it is a serious matter that affects the health of millions of citizens. Ministry of Health as well as Ministry of Food Processing should deal with food safety and FSSA.

The act should have a compulsory provision for blacklisting of the companies or even suspension of their license when held guilty of the offence. It should be made a part of the punishment.

The Codex and the committees have suggested confidence building measures among the consumers for food safety. This can be done by attaching the logo displaying that products are safe. This logo that can be understood by literate or illiterate person should be made mandatory.

18.8 OTHER GOVERNMENT AGENCIES WORKING FOR FOOD SAFETY

There are two organizations that deal with voluntary standardization and certification systems in the food sector. The Bureau of Indian Standards (BIS) looks after standardization of processed foods and manufacturers complying with standards laid down by the BIS can obtain and "ISI" mark that can be exhibited on product packages. Standardization of raw agricultural produce is under the purview of the Directorate of Marketing and Inspection. It prescribes "Agmark" standards.

Management systems for quality and food safety are as per ISO 9000 quality management systems. The millennium standard (ISO 9000:2000) has changed the focus from procedure to process.

FSSR 2011 is more stringent with provisions like an enhanced penalty of up to Rs 10 lakhs and prosecution up to life-term for those ignoring hygiene standards. PFA, which had provisions for penalty along with prosecution, hardly saw any case reaching to its logical conclusions. The food manufacturers, suppliers, vendors, eateries, storage, distribution, imports and exports, food services, and other related businesses would now be governed by new rules under FSSR 2011. The 12th 5-year plan (2012–17) has also emphasized upon strengthening of food safety systems in states.

18.9 CONCLUSIONS

FSSA was born out of the need for an integrated food law, prioritizing consumer safety and harmonization of food standards with international regulations. FSSA 2006 is a new act that integrates eight different laws. It is a comprehensive enactment aimed at ensuring public health and safety. The implementation of this Act attempts to bring a paradigm shift in food regulatory scenario of India. However, considering the size of food industry in India, it will take lot of resources to implement the new food law. Awareness also needs to be created at all levels by the FSSAI authorities. However, all said and done, laws alone cannot solve the issue of food safety. The act only deals with prosecution and punishment. Commensurate infrastructure in terms of a network of public health laboratories with sufficient manpower is needed. Food catering and hospitality industry also need to raise their standards of production. People, on their part, also need to demand safe food products from providers (caterers/hoteliers).

REFERENCES

Dhulia, A., 2010. Laws on food adulteration: a critical study with special reference to food Safety and Standards Act 2006. ILI law review, vol. 1, pp. 164–188.

Kagle, R.P., 1970. The Kautaliya Arthashatra, vol. 2. Motilal Banarsidas, New Delhi, pp. 260–270.

Singh, R., 2011. India enforces the new Food Safety Law. Gain report, II1174. 2011.

Local governing bodies

19

R.K. Gupta

Department of Community Medicine, Army College of Medical Sciences, New Delhi, India

19.1 INTRODUCTION

Many developing countries, including India, are now arising and awakening to reality of "food-safety." Many food safety practices that were hitherto, considered acceptable and were "taken for granted" are now put under scanner. This scanner could be the vigilant public, the "story" hungry media, or the responsible government. This change with respect to food safety being "unimportant" a decade ago, has become so "vital" today, that the government has recrafted a law, in its honor. This goes with the fact that India has become a resurgent economy. The community now appreciates better and healthy living and is prepared to do all it can for safe food.

The central/union government can only formulate laws, guidelines, and statutes. But these regulations are only as good as the level of their implementation. The responsibility of their implementation lies with the local governing bodies (LGB). Hence the LGB assume extreme importance. Even though the LGB did exist earlier too, and had the statutory role under the Prevention of Food Adulteration Act 1954, now with the Food Safety and Standards Act (FSSA), 2006, they have been given a new lease of life with more specific roles and powers, with respect to food safety. This chapter deals with the issue of LGB at the state, district, city, and village levels.

19.2 LOCAL GOVERNING BODIES IN INDIA

India is a federal union constituted of 36 provinces (29 referred to as states, and 7 union territories) that are geographically and culturally diverse. These have their own democratically elected governments, independent to constitute and implement their laws in respective states. Although issues such as foreign and defense affairs are exclusively within the domain of the union government, matters such as law and order, education, and health are state subjects. This means that, on the basis of broad guidelines issued by the central government, the states are free to constitute their own laws and rules for effective implementation of regulations on health matters.

Food Safety in the 21st Century. http://dx.doi.org/10.1016/B978-0-12-801773-9.00019-4

The states are further divided into districts for administrative purposes, where the District Magistrate/Commissioner (DM/DC) is the overall administrator. The District Health Officer (DHO) is the overall health administrator for the district. The district is further divided into *talukas/tehsils*, governed by a SubDistrict Magistrate (SDM) with a chief medical/health officer (or equivalent) under him, to execute health administration duties. In the rural areas, many villages fall under a *tehsil/taluka*. The villages are run by a small LGB, the "*gram panchyat*," that is a congregation of few elected leaders from the village, under a village head (*sarpanch*). So far as cities are concerned, the administration is run by city municipalities, headed by a municipality commissioner, who would have a health department under him.

Further it is the duty of the LGB, within the state/district/city/village, to implement health regulations including the food policies. As elaborated previously, the local governing health bodies are constituted corresponding to the administrative LGBs.

The structure and process at the subdistrict—village levels, that is, *panchayati raj institutions* (local self government) is elaborated subsequently (Fig. 19.1).

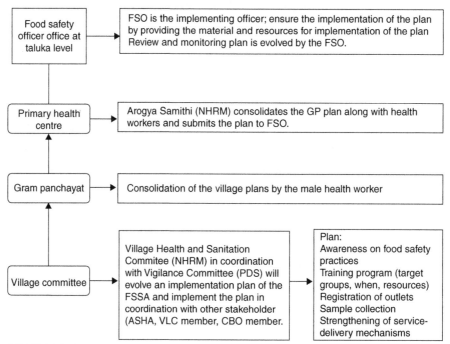

FIGURE 19.1 Structure and Process at Panchayati Raj Institutions (Government of India, 2010)

Responsibilities of various state and district LGB with respect to food safety (Government of India, 2010): Under the FSSA, 2006, each state is to have a food safety commissioner who will be the implementing agency for all food safety issues. The act also provides for appointment of "Designated Officer" (DO) in each district by the food safety commissioners of the states. The power to grant or cancel license of the Food Business Operator (FBO) is vested with the DO of the district/UT. Each DO has four to five Food Safety Officers (FSO) under him.

19.3 RESPONSIBILITIES OF THE STATE/DISTRICT FOOD SAFETY LGB

The LGB dealing with food safety have multifaceted responsibilities. These are elaborated as follows:

Inspection: FSOs shall monitor and inspect the general environment, sanitary and hygiene conditions, and preparation and processing facilities of the food-related establishment.

Food sample collection: Local FSOs of respective state/district LGBs shall collect the food samples from FBOs and send them to the food analyst for analysis. The same powers will also be vested with the respective *panchayat*/village authorities as well.

Training: DO of each district shall educate and organize training to food handlers, vendors, and others stake holders to create awareness using audio-visual aids, etc. in the municipal/*panchayat*'s areas under his/her jurisdiction.

Investigations: DO shall carry investigation of food poisoning incidents and shall take remedial measures to eliminate recurrence of such incidents in future.

Redressal of complaints: DO shall monitor health and safety complaints in a particular district/village/*panchayat* and take appropriate remedial measures.

Monitoring: DO shall supervise whether corrective actions are taken as per improvement notices issued or not.

19.4 RESPONSIBILITY OF MUNICIPAL COUNCIL/ CORPORATION

Besides ensuring the implementation of the aforementioned responsibilities, the LGB are accountable for the following as well:

Drinking water quality: Municipal corporation (or equivalent) of respective districts/cities/village/*panchayat* shall evaluate and ascertain the quality of drinking water and water used as an ingredient in food. Appropriate steps shall be taken to eliminate contaminants in the water.

Pest control: Municipal corporation (or equivalent) shall ascertain whether appropriate pest-control measures are followed.

Waste disposal/sewage disposal/drainage system: Municipal corporation (or equivalent) shall ascertain whether proper waste-disposal system/sewage/drainage system is in place.

Information, education and communication (IEC): Municipal corporation (or equivalent) will prepare and distribute to the general public pamphlets containing food safety measures to be taken by them. Such information may also be disseminated through appropriate (print or electronic) media.

19.5 **FOOD SAFETY PLAN**

The LGB are supposed to put in place a Food Safety Plan (FSP), for their area of jurisdiction. This plan puts in perspective all the components/functions of these LGB that have been enumerated in the previous paragraphs. This essentially means the adoption of good manufacturing practices, good hygiene practices, HACCP, and such other practices as may be specified by regulation, for the food business houses/stores within their jurisdiction. The FSP consists of programs, plans, policies, procedures, practices, processes, goals, objectives, methods, controls, roles, responsibilities, relationships, documents, records, and resources of all the installations dealing with food safety in that administrative zone. An FSP could often be part of a larger management system.

So, we will have an FSP for the village (*gram panchyat*) and city (municipality). In a nutshell, FSP at *gram panchayat*/municipality will envisage the following general parameters:

1. To identify and categorize the food business of the FBO in the area.
2. To inspect the premises of FBO at periodical intervals and on the basis of the inspection, issue necessary, improvement notices.
3. Food safety officer to take samples of food from FBO and send such samples for analysis.
4. To evaluate and ascertain the quality of drinking water used.
5. To review arrangements for disposal of waste by the public in general and the FBO in particular.
6. To investigate food-poisoning incidents in their area and send appropriate reports to the competent authorities and simultaneously take remedial measures to eliminate recurrence of such incidents in future.
7. To interact with industries and consumers and to create awareness among them for food safety.
8. To carry out IEC campaigns through various mechanisms.

19.6 **ROLE OF FOOD SAFETY OFFICER (FSO)**

The FSO is the key functionary and is responsible for implementation of the food-related regulations at the subdistrict (*taluka*) level. He will plan out strategies to ensure food safety. He would proactively notice hazards for food safety and take suitable action to overcome this. He will collect food samples for testing and ensure that food safety standards are followed in schools and *anganwadis*. He will also be responsible to network with other departments and various food laboratories at district level.

He reviews the food safety work done in the villages and *taluka* levels and ensures that various reports are documented and updated information is available at all the times. He also ensures smooth working of the resource center, grievance cell, and enforcement cell in compliance with the rules of the Food Safety Act. He is also responsible for registration and renewals of various licenses pertaining to FBO. The safety standards for FBO in the *panchayat* are also laid down under his supervision. He also builds awareness about food safety standards among *panchayat* members, teachers, school children, and other citizens.

19.7 **RESPONSIBILITY OF OTHER DEPARTMENTS/ MINISTRIES**

Besides the Ministry of Health & Family Welfare, many other ministries/departments have to contribute to make "food safety" a success at the national level. These could be ministries/departments as diverse as Ministry of Environment, to the Ministry of Sports & Youth Affairs! Role of some of these are discussed here.

a. Pollution Control Board (Ministry of Environment)
- To review arrangements for disposal of waste and waste water treatment facility being maintained in various food processing industries.

b. Department of Women and Child Department
- Train village level child/mother health workers (*anganwadi* workers) to spread awareness on food safety and counsel pregnant women for safe and nutritious food.
- Orient self-help groups on food safety issues and ask them to monitor school midday meals or areas in the village where food is prepared for community.
- Involve anganwadi workers in the planning process at the village level.

c. Department of Rural Development and *Panchayat Raj*
- Advocacy
- Issue instructions to *panchayats* to register the food business.
- Issue guidelines to *panchayats* to discuss food safety–related issues relevant to the village in village councils (*gram sabhas*) and other meetings.

  – Request *panchayats* with their own budget to allocate resources to supplement food sample collection cost.

d. Ministry of Youth Affairs & Sports
  – Conduct special campaigns/programs on safe food for rural youth.

e. Ministry of Education
  – Include food safety awareness in education programs.
  – Design project works for students.

19.8 CONCLUSIONS

The local governing bodies are vital to comprehend the local food safety issues faced by that particular region; translate the national laws into locally applicable terms and locally implement them for utmost efficacy. Thus, the LGB are not only the connecting link between the national government/laws with the community but also the engine to ensure propulsion and execution of the regulations on food safety, which would eventually make the consumer and the community safe.

REFERENCE

Government of India, 2010. Ministry of Health & Family Welfare. Training Manual for Food Safety Regulations. FSSAI, 2010. Food Safety Plan, General Parameters for Food Safety Plan for *Panchayat* and Municipalities, New Delhi, pp. 32–38.

Role of food business operators in food safety

20

P. Dudeja*, A. Singh**

*Department of Community Medicine, Armed Forces Medical College, Pune, Maharashtra, India; **School of Public Health, Post Graduate Institute of Medical Education and Research, Chandigarh, India

Indian society has witnessed a paradigm shift in the food culture along with industrialization and globalization. There has been a marked increase in eating out culture in all socioeconomic classes. The underlying driving force is the higher incomes, greater number of nuclear families and double-income households, and working women and urbanization. As per Indian Food Services Report 2013 "the continuously evolving economy, societal, and demographic changes have reshaped the Indian eating behavior and style." The promulgation of Food Safety and Standards Act (FSSA 2006) in India has been done to ensure supply of safe and wholesome food across the country. A food business operator (FBO) is defined as any undertaking, whether private or public, for profit or not, carrying out any of the activities related to any stage of manufacture, processing, packaging, storage, transportation, distribution of food, imports and including food services, sale of food or food ingredients. Examples of different categories are given in Box 20.1. Under the eating establishments, the law is applicable to all food business restaurants; small and medium eating joints, food served in retail outlets; fresh extracted juice and beverages shops/outlet; dispensing outlets, base kitchens, confectionary, bakery and sweet shop serving unlabelled prepackaged, or loose sweets/baked products/confectionary products and frozen desserts. Earlier the person manning these food businesses was termed as a vendor. However, under the FSSA 2006, any person or any firm or company which is involved with the business of food is an FBO.

20.1 RESPONSIBILITIES OF FBO

An FBO who is accountable for food safety in his business at any stage interacts with various agencies (Fig. 20.1). *As per FSSA 2006*, no food business can be carried out without obtaining a license or a registration. The FBOs have been categorized in three categories the one who are petty food business operators and whose annual turnover is less than Rs 12 lakh, need to obtain a registration certificate as they are covered under the registration provision. For others whose annual turnover is more

BOX 20.1 CATEGORIES OF FOOD BUSINESS UNDER FSSA 2006

1. Dairy units including milk-chilling units equipped to handle or process
2. Vegetable oil processing units and units producing vegetable oil by the process of solvent extraction and refineries including oil expeller unit
3. Slaughtering units
4. Meat-processing units
5. All food-processing units other than those mentioned previously including relabelers and repackers
6. 100% Export-oriented units
7. Importers importing food items including food ingredients and additives for commercial use
8. Food business operators manufacturing any article of food containing ingredients or substances or using technologies, processes, or combination thereof whose safety has not been established through these regulations or which do not have a history of safe use or food containing ingredients which are being introduced for the first time into the country.
9. Storage
10. Wholesaler
11. Retailer
12. Distributor
13. Supplier
14. Caterer
15. *Dhabha* or any other food-vending establishment
16. Club/canteen
17. Hotel
18. Restaurants
19. Transporter of food (having a number of specialized vehicles such as insulated refrigerated van/wagon, milk tankers, etc.)
20. Food ingredients
21. Marketer
22. Food catering services in establishments and units under central government agencies such as railways, air and airport, seaport, defense, etc.

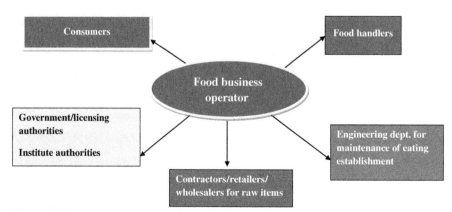

FIGURE 20.1 Interface of Food Business Operators With Stakeholders

than Rs 12 lakh, have to obtain a license either under the state licensing authority or from the central licensing authority, depending upon the criteria based on the nature and volume of the business. It is of utmost importance for the FBO to either register or obtain a license before initiating any food business.

On receipt of a license/registration it is the responsibility of FBO to get it displayed at a prominent place in the premises of eating establishment. At any point of time licensing authority can suspend or cancel license if the FBO is found not complying with the conditions of license. In case an FBO is carrying food business without valid license, he is liable with imprisonment up to 6 months and also with a fine up to Rs 500,000.

The law entrusts the responsibility on FBO to follow general hygienic and sanitary practices irrespective of the nature of business via food manufacturer/processor/handler/distributor/transporter and service provider of food articles. These encompass observation of certain norms regarding location of premises, general manufacturing, and hygiene practices, etc. Few guidelines to be followed (as given in FSSAI training manual, 2010) by FBOs are given here:

1. Location of business should be in a sanitary place and free from filthy surroundings; it should maintain overall hygienic environment. The premises should be clean, adequately lighted and ventilated and have sufficient free space for movement. The layout of the food establishment should be such that food preparation/manufacturing process are not subject to cross-contamination during processing of food (e.g., packaging, dishing/portioning of ready-to-eat food). The area occupied by machinery should not be more than 50% of the manufacturing area.

2. Use of potable water, efficient drainage system and adequate provisions for disposal of refuse.

3. Refrain persons suffering from infectious diseases to work and to prohibit eating, chewing, smoking, spitting, and nose blowing by food handlers within the premises. The FBO shall develop a system, whereby any person so affected, shall immediately report illness or symptoms of illness to the management. As a routine, medical examination of a food handler shall be carried out apart from the periodic checkups, if clinically or epidemiologically indicated. Arrangements shall be made to get the food handlers/employees of the establishment medically examined once in a year to ensure that they are free from any infectious, contagious, and other communicable diseases. A record of these examinations signed by a registered medical practitioner shall be maintained for inspection purpose. The staff shall be compulsorily inoculated against the enteric group of diseases once a year and a record toward that shall be kept for inspection.

4. Protect food from contamination during storage and transportation. Equipment and utensils used in the preparation of food shall be kept at all times in good order and repair and in a clean and sanitary condition.

5. All raw materials, food additives, and ingredients, wherever applicable should conform to all the regulations and standards laid down under the act.

6. No FBO shall manufacture, store, sell, or distribute any article of food which is unsafe/misbranded/substandard or contains extraneous matter or in violation of any other provision.

7. A detailed standard operating procedure (SOP) to be developed for proper management which in turn would help in identifying any problem at exact point, so the course of damage control would be faster. The FBOs shall ensure that technical managers and supervisors have appropriate qualifications, adequate knowledge and skills on food hygiene principles, and practices which shall enable them to ensure food safety and quality, judge food hazards, take appropriate preventive and corrective action, and ensure effective monitoring and supervision.

8. A system for food testing shall be ensured by the FBO either through an in-house laboratory or through an accredited laboratory.

9. A periodic audit of the whole system according to the SOP be done to find out any fault/gap in the system. Appropriate records of food processing/preparation, production/cooking, storage, distribution, service, food quality, laboratory test results, cleaning and sanitation, pest control, and product recall shall be kept and retained for a period of 1 year or the shelf-life of the product, whichever is more.

10. The FBOs shall ensure that visitors to its food manufacturing, cooking, preparation, storage, or handling areas must, wherever appropriate, wear protective clothing and footwear, and adhere to personal hygiene provisions envisaged in this section.

20.2 CHALLENGES

The food safety Act on issuing a license/registration imposes that it will be the duty of FBO to ensure that the articles of food satisfy the food safety requirements at all stages of production, distribution, and storage. The corollary to this is that in case of any problem the onus of responsibility rests with the person holding the licenses. Nevertheless, the FBOs are businessmen who run it for profit. Food businesses are frequently run from old constructions which need major repairs. This is often neglected by them to save on the cost. They fail to maintain the kitchens and functioning area as per the laid guidelines. The reasons for this are dual. Those FBOs who are not educated are often not aware of these guidelines or their relevance in ensuring safety and quality of food. Some of them feign ignorance to save money. They allow shortcuts when it comes to ensuring safe disposal of waste, pest management, cleaning of drains in premises, maintenance of equipment.

Another difficult task is management of food handlers. FBOs have to constantly mange issues related to food handlers, namely, salary, leave, addictions, welfare, etc. They have to handle food handlers who are often uneducated, amateur young boys of low socioeconomic status migrated from villages to cities in search of work and shelter. To handle this age group of boys/men is a difficult task. Most of the times, food handlers do not listen to/follow advice such as washing of hands, bathing, paring of nails, combing of hair, etc. They are smart enough to escape this being noticed by FBOs.

The FBOs value and respect their customers as the success of their business has a linear relation with customer satisfaction. However, their clients comprise of people of different socioeconomic status. They have to keep everyone happy which is not an easy task. Some want quality and are ready to pay for it whereas others want everything at minimal price even at the cost of food safety. With an endeavor to make all clients happy, the focus of FBOs is more on taste and presentation. Usually, food safety takes a back seat. Sometimes they sell leftover or stale unsafe food to avoid loss to business. As per the regulations it is important for FBOs to display their mobile numbers at the business place. This can help the clients as they can call and report to him directly about the problem if required. By displaying and sharing his contact number the FBO conveys his involvement and commitment toward the clients. However, even this practice is not adhered to, to avoid any complaints from the customers.

Some cost-saving measures used commonly by FBOs are to employ less number of workers or go for multitasking by deputing the food handlers for maintenance of premises. This compromises food safety due to opportunity cost principle. To save money they also buy substandard items at subsidized cost; they even adopt incorrect procedures for cleaning of utensils, overload refrigerators. They avoid expenditure on maintenance of premises. One way to save on expenses is avoiding major repairs in kitchen. This also affects food safety adversely.

20.3 **THE WAY FORWARD**

The FBOs in our country do not receive any formal training on food safety and hygiene. Their knowledge about the existence and implementation of new law is restricted only to getting licensed/registered. Some of them gain experience by working as an employee in a food business establishment and then start their own eating establishment. Most of the FBOs have some features in common—business motive, self-praise and a confrontationist approach toward the government enforcement authorities. Nevertheless, the objective of implementation of FSSA 2006 is dual—both to educate the FBOs and improve the food safety standards. Only penalizing the FBOs will defeat the purpose of food safety. Education and awareness about food safety to FBOs and food handlers is one main difference between this law and previous act (PFA). Training of FBOs can be effective in reducing food safety problem by implementation of realistic food safety practices within the workplace. If FBOs were trained to advanced levels, they could then provide basic training for food handlers in-house. Much of the noncompliance from the FBOs largely arises from a lack of awareness about the current rules and regulations and standards. Therefore, there is an urgent need to increase the awareness among FBOs to get their samples tested at least once in 6 months. With the ever-expanding food-processing industry, the government also needs to keep pace and increase the manpower to ensure implementation of regulations to cope up with the increasing demand. The laws and associated penalties need to be implemented stringently.

REFERENCE

FSSAI, 2010. Training Manual for Food Safety Regulators.

Food handlers

21

P. Dudeja*, A. Singh**

*Department of Community Medicine, Armed Forces Medical College, Pune, Maharashtra, India;
**School of Public Health, Post Graduate Institute of Medical Education and Research,
Chandigarh, India

India is presently the second largest producer of food in the world and has the potential of being the biggest, with the backing of a powerful and strong agricultural sector. The food-processing industry is one of the largest industries in India and globally it is ranked fifth, in terms of production, consumption, export, and expected growth in food processing. With globalization, a substantial increase in percentage of working women and a rise in eating-out, the role of a traditional homemaker in cooking food at home has been partly replaced by housemaids and food handlers working in various Eating Establishments (EEs). These food handlers are a missing link in food safety from the farm-to-fork chain.

A food handler is defined as any person who handles either *food or surfaces* that are likely to be in contact with food, such as, cutlery, plates, and bowls. They may be involved in preparation of raw material, cooking, packing, storage, displaying, serving of food, and cleaning of utensils. Examples of food handlers are waiter staff/service staff, chefs, head cooks, dishwashers, receiving and food storeroom staff, bartenders, and hosts/hostesses who handle food, street vendors who sell food items, and housemaids. They are employed at places where food is cooked and served, that is, at canteens, messes, restaurants, juice bars, street food vendors, snack bars, take away joints, etc. Even the housemaids working as help in the kitchen for chopping vegetables, kneading flour, etc., are food handlers.

The case of Mary Mallon (typhoid Mary) who as a food handler caused 7 outbreaks, 57 cases, and 3 deaths is a well-known example to the food industry (Leavitt, 2014). Poor standards of hygiene during food preparation and the lack of training in food safety by the food handlers are probably the most common causes of foodborne illnesses (FBI) (Greig et al., 2007). Mishandling of food and disregard of hygienic measures on the part of food handlers may enable pathogens to come in contact with food and, in some cases, to survive and multiply in sufficient numbers to cause illness.

The role of food handlers entails a lot of responsibility as they, on one hand, can provide us with food best suited to our taste buds, whereas, on other hand can also be

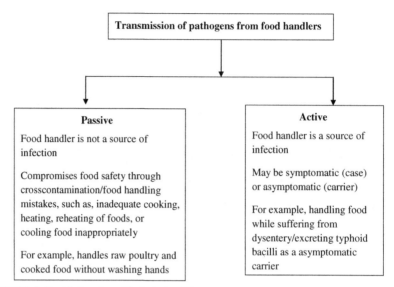

FIGURE 21.1 Restriction from Work during Illnesses

source of contamination and compromise food safety. They can transmit pathogens in two ways as shown in Fig. 21.1.

21.1 RESPONSIBILITIES OF FOOD HANDLER

A food handler has certain set of responsibilities while at work for ensuring safety of food. These are:

1. hand-washing practices
2. good personal hygiene
3. clean work attire
4. following food-hygiene practices during work
5. management of illnesses
6. regular training

21.1.1 HAND WASHING

Unclean hands are often the culprit in transmission of microorganisms to the food (Chin and APHA, 2000). The hand hygiene of a food handler at his/her work place is of utmost importance. Though it is the duty of food handlers to wash their hands as specified (for a particular job assigned to them), the bigger responsibility is that of Food Business Operator (FBO)/manager of EE to ensure availability of soap and clean water for washing hands at all times. The FBOs may focus on profit only in EEs and neglect these small issues or deliberately overlook to save money on soap. It is a common practice of food handlers in small EEs to use the kitchen sink for washing hands. This practice is not advisable. A separate sink exclusively for hand washing with a clean water source is required to maintain hygiene. In food manufacturing units with HACCP and ISO certifications, sanitizers are also provided near the sink. Here, use of sanitizer without washing hands should not be done. Such units also have a checklist where food handlers have to log entry of washing hands. This ensures an appropriate frequency of washing hands. It also acts as a tool to check handwashing by the supervisor.

It is often seen in small EEs, such as, *dhabhas*/small restaurants that they do not have appropriate hand-washing facility for customers and food handlers. Most often, the kitchen sink is used for this purpose. It is not uncommon for food handlers to wash hands without using soap. The food handlers just rinse their hands with plain water and then dry them by wiping with their dirty clothes or apron. This gives only a false sense of security to them (Hutin et al., 2003). They may or may not inform the FBO regarding nonavailability of soap. The onus of supplying soap lies with the FBO. These FBOs economize on soap use as their main motive is business and profit. They hardly supervise washing of hands by food handlers and neither punish or warn them frequently to ensure compliance of good hand-washing practices.

The other groups of food handlers are the street vendors who do not have any provision of washing hands. Usually they just wipe hands with the mop they are using to clean the food contact surface. These days some of them use gloves to attract customers. They feel as if wearing disposable gloves is a symbol of hygiene and cleanliness. To save on the cost they tend to reuse these gloves. Rather they keep gloves "on" for hours together and keep touching dirty surfaces of their cart, utensils, etc., with gloves on. A dirty glove is as bad as or even worse than an unclean hand.

Unclean hands can transmit germs from the hands to the food. Hands can get contaminated after handling raw food (e.g., after touching raw potatoes, carrot, onions, meat etc.), coughing, sneezing, visiting toilets, eating, drinking, smoking, handling money, using mobile phone, touching hair, scratching body, handling dustbin, and soiled equipments and utensils. Hands must be washed properly with soap and water after these activities. Hand washing is a must before the commencement of work in the kitchen.

21.1.2 GOOD PERSONAL HYGIENE AND WORK ATTIRE

Personal hygiene and cleanliness of food handlers is vital for ensuring safety of food. It is the moral and professional responsibility of food handler to maintain high standards of personal hygiene and cleanliness to protect the consumer from becoming ill following the consumption of their products. It includes the following practices.

21.1.2.1 Hair

Hair has also been known to crosscontaminate food. Bacteria cling to the hair and scalp, dust and perspiration collect in the hair. The hair of a food handler should be short and clean. Long hair is not only difficult to maintain but also might fall into food while cooking/handling. Hence, food handlers should preferably keep their hair short and trim every 4–8 weeks to keep them in shape.

In the course of handling food, hair should be covered with a clean cap or hair net. Long hair should be tied back. It should be ensured that combing of hair is not done in food-handling areas. Using a cap prevents food handlers from directly touching hair and scalp with fingers to move it out of their face and passing bacteria from the hair onto food. Hair can be partly covered by staff who are serving or handling food that is protected, for example, a person employed in a takeaway front counter, working in soft drink dispenser counter, ice cream shop, etc.

21.1.2.2 Face and neck

A food handler should not wear earrings or any necklace at the time of work. The reason behind this is that if the jewelry parts are loose then they can fall into the food and contaminate it. Beards, if any, must be trimmed and tidy. The use of beard nets is strongly recommended for bearded food handlers working in all food-preparation establishments.

21.1.2.3 Clothes

The clothes of a food handler should be neat and clean. Also, it is suggested that there should be no outer pockets on the shirts. In case there are any, they should be empty. Items, such as, pen/pencils/mobiles/medicine/paper, etc., if kept into the outer pocket can fall into food while working and may compromise food safety.

21.1.2.4 Apron

The next characteristic identifying feature of a food handler is the apron. It should be changed at least once during the day. Preferably it should be done in the middle of an 8-h shift. Apart from this, cooks should change aprons whenever these get soiled. Whenever a food handler changes workstations from raw food-preparation activities to ready-to-eat food-preparation activities, the apron needs to be changed.

Food handlers should not wear aprons outside food-preparation areas, such as, while going to washroom, etc. This procedure minimizes possible contamination of aprons by airborne pathogens, dirt, dust, possible soiling by washroom fixtures, and other unsanitary articles. So, they should remove aprons before going to the toilet. Buttons should preferably be avoided on the clothing as they may come off and fall in the food.

21.1.2.5 Hands and wrists

Wearing jewelry or other cosmetic items during food-handling activities is discouraged. These include, but are not limited to rings, nail polish, wristwatches, bracelets, clip-on earrings, false nails, false eye lashes, etc. Jewelry can hide microorganisms, may compromise hand washing and can also fall into food. In cases where rings are difficult to remove, clean gloves should be worn by the food handler while handling food.

21.1.2.6 Nail cutting

Food handlers should not maintain long nails. Nails of a food handler should be short and clean. Long nails accumulate dirt and bacteria, which can enter food while handling.

21.1.2.7 Foot wear

These must be clean and free of dirt and accumulated food particles on both the top and bottom. Street wear should preferably be avoided inside the kitchen. Accumulation of food particles and dirt on footwear may allow microorganisms to multiply. This may consequently affect the general sanitary conditions of the kitchen premises

21.1.2.8 Dressing on wounds

A food handler is exposed to various kinds of injuries through the use of sharp knives, grater, and other items while preparing food. There are frequent incidents of cuts, abrasions, and burns while handling food. It is imperative that cuts and wounds should not be left open. These should be covered with a colored dressing to prevent contamination of food. This is because in case it falls into the food it can be easily identified. A food handler may work if the cut has been bandaged and disposable gloves are worn.

21.1.3 MANAGEMENT DURING ILLNESSES

Food handlers suffering from infectious disease may transmit infection and render food unsafe (Olsen et al., 2001). A food handler must report to his senior/FBO/manager about his illness in case he is suffering from diarrhea, vomiting, fever, cough, skin lesions (including boils/cuts), and eye or nose discharge.

Such a person should not be engaged in handling food. In case it is unavoidable, then all measures must be taken to prevent food from being contaminated as a result of the disease. For example, an infected sore must be completely covered by bandage and clothing, or by a waterproof covering if it is on an area of bare skin. In case of cold/cough, a disposable tissue or a handkerchief must be used to handle the secretions. Various restrictions for food handlers during illnesses have been given by the Food Safety and Standards Authority of India in Table 21.1.

21.1.4 GOOD HYGIENIC PRACTICES AT WORK

Food handlers need to know how their work can affect the safety of food they handle (Walker et al., 2003). The following good hygiene practices must be observed by a food handler to ensure food safety.

- Wash and dry hands whenever they get contaminated.
- Avoid sitting on the food-preparation shelf.

Table 21.1 Restrictions for Food Handlers During Illnesses

Disease	Work Status	Duration of Work Restriction
Abscess, boils, etc.	Relieve from direct contact and food handling	Until drainage stops and lesion has healed or employee has negative culture
AIDS or AIDS-related complex	May work (as per CDC guidelines) No open lesions, upper respiratory diseases, or communicable diseases	Employee will be counseled and educated
Diarrhea		
Acute stage (etiology known)	Relieve from direct food handling	Until symptoms resolve and infection with *Salmonella*, *Shigella*, or *Campylobacter* is ruled out
Campylobacter		Until symptoms resolve or after appropriate antibiotic therapy for 48 h
Salmonella		Until stool is free of the infecting organism in two consecutive cultures not less than 24 h apart
Shigella		Until stool is free of the infecting organism in two consecutive cultures not less than 24 h apart
Hepatitis A		Until seven days after onset of jaundice; must bring note from physician on return.
Staphylococcus aureus		Until lesions have resolved and the employee has negative culture

Training manual for food safety regulators: Volumes 1 to 5. Food Safety and Standards Authority of India. 2010

- Avoid keeping personal belongings on the food-preparation shelf/cooking area.
- Cover exposed sores with waterproof dressing or disposable gloves.
- Wear clean outer clothing. Change aprons or other clothing if they are soiled.
- When sneezing or coughing inside the food-preparation area is unavoidable, turn away from food and cover noses and mouths with tissue paper or handkerchiefs. Hands should then be thoroughly cleaned at once.
- Never blow into a bag to open it that is used for storing food.
- Never blow on food for any reason.
- Do not spit, smoke, or use tobacco in areas where food is handled.

- Eat outside the food-preparation area only.
- Do not touch ready-to-eat food with bare hands.
- Do not taste food with fingers.
- Do not reuse a sampling spoon without washing.
- While cooking in kitchen do not touch hair or other parts of bodies, such as, noses, eyes, or ears.

Food handlers must tell their senior if they know or think they may have made any food unsafe or unsuitable for consumption. For example, jewelry or a band-aid worn by a food handler may have fallen into food or glass may have broken into or near exposed food.

21.2 FOOD-SAFETY REQUIREMENTS FOR FOOD HANDLERS UNDER FOOD SAFETY AND STANDARDS REGULATIONS (FSSR 2011)

According to the FSSR 2011 any food handler believed to be suffering from or to be a carrier of a disease or illness likely to be transmitted through food shall not be allowed to enter into the food-handling area. A food handler can transmit *Staphylococcus*, *Salmonella*, *Shigella*, *Escherichia coli*, *Entameoba histolytica*, *Campylobacter*, Hepatitis A, influenza, threadworm, and *Giardia* infections. A system should be developed in all EEs whereby any affected person shall immediately report illness or symptoms of illness to the management or FBO and medical examination is carried out apart from periodic checkups, if clinically or epidemiologically indicated. All arrangements should be made to get food handlers examined at least once in a year to ensure that they are free from any infectious, contagious, or communicable diseases. A record of these illnesses, duly authenticated by a qualified doctor should be maintained for inspection purposes. In case of a food-manufacturing unit, the staff shall be inoculated against enteric group of diseases and a record should be kept for inspection. At the time of recruitment also, recent history of illness along with a medical checkup must be done.

21.3 TRAINING OF FOOD HANDLERS

Review of studies on food-handler training has brought out that there is insufficient research evidence on the fact that food-handler training improves food-safety practices (Pajot and Aubin, 2011). Also, there is limited evidence that it may enhance knowledge and behavior. Implementation of various training courses, such as, "Hygienic Minimum" course to the food handlers has been studied and it was found that there was improvement in the most of examined parameters. To decrease the burden of FBI, WHO Department of Food Safety and Zoonoses actively promotes training as a means to improve the practices of food handlers for food safety. Many studies have documented that there is a wide gap between knowledge about various food-safety practices and their practical implication by these workers. Reasons are many. These involve time constraints, poor motivation levels, low wages, poor working conditions with high temperature and humidity levels, long working hours, ill-treatment by FBOs, lack of respect in the profession, poor supervision by the FBOs, unavailability of items as water and soap for hand washing, inadequate toilet facilities, nonconduct of medical examination of handlers, lack of training, etc.

21.4 OCCUPATIONAL HAZARDS OF FOOD HANDLERS

As per the International Labour Organization (ILO)/WHO 1950, occupational health is the promotion and maintenance of the highest degree of physical, mental, and social well-being of workers. It seeks to prevent unhealthy practices among the workers, as well as, control of job-related health risks. Every occupation has its own set of problems and hazards associated with it. What makes the food industry different from the rest is that maintaining occupational safety and looking after the health and welfare of employees will ensure safety of food. It has a direct bearing on the quality of product. Occupational hazards for food handlers can be broadly classified as physical, chemical, biologic, mechanical, and social (Fig. 21.2).

21.4.1 TEMPERATURE

Food handlers often have to work in environments with high temperatures. This is especially when they are employed near a cooking range/stove/*tandoor*. Lot of heat is generated in process, such as, deep-frying and cooking. The problem becomes worse in summer seasons. Often food handlers have to work in such areas for prolonged hours. Some EEs with star ratings do have a centralized cooling system to provide comfort to their workers from heat. However, such kitchens in any city are very few in number. In low-budget food-processing industries/EEs, least importance is given to creating a comfortable working zone for the food handlers. This often leads to prickly heat, dehydration, heat exhaustion, heat cramps, heat stroke, and burns in food handlers. For better productivity it is essential that temperature be maintained (corrected effective temperature 20–27°C).

Physical	Chemical	Biologic	Mechanical	Psychosocial
• Temperature • Lighting • Humidity • Noise • Fire • Lifting heavy loads	• Low-cost detergents	• Conatminated eggs • Unsafe food	• Equipment injuries • Cuts from sharps	• Lack of regonition • Alcoholism

FIGURE 21.2 Classification of Occupational Health Hazards in Food Handlers

21.4.2 **LIGHTING**

The workers may be exposed to the risk of low illumination or excessive brightness. This is seen at work places where either the owner's intentions are to save money by providing less number of lighting points or by not replacing the old one. Poor illumination can lead to eyestrain, headache, eye pain, lachrymal congestion, and eye fatigue. There should be sufficient and suitable lighting, natural or artificial, wherever persons are working. Inadequate lighting can compromise food safety.

21.4.3 **NOISE**

Various equipments in the kitchen produce noise, such as, chimney exhaust, exhaust fan, *chapatti maker*, food processor, etc. The effects of noise can be auditory or non-auditory leading to nervousness, fatigue, decreased efficiency, and annoyance. The degree of injury from exposure to noise depends on a number of factors, such as, duration of exposure and frequency range along with individual susceptibility. Among food handlers, annoyance due to noise can also affect food safety.

21.4.4 **FIRE**

Fire hazards/explosion of gas cylinders/stoves/pressure cookers can also take place in the kitchen. These can lead to burns and may be fatal also.

21.4.5 **LIFTING HEAVY LOADS**

At times food handlers have to lift heavy loads in carrying raw food items, such as, vegetables, fruits, grains, etc. This can lead to musculoskeletal problems, such as, backache, sprain, strain and pain in limbs, etc. They also have to stand for long hours and can develop varicose veins. Quite often ergonomics is not applied while designing the working shelves in the kitchens. Discordance between heights of food handlers and kitchen shelves may lead to lower back pain and easy fatigability. This can also affect the working of food handler in a negative way as (s)he may not be able to mop/clean/cut as per required instructions of food safety.

21.4.6 **CHEMICAL HAZARDS**

The use of dishwashers for washing utensils is restricted to high budgeted EEs only. FBOs prefer to employ staff for cleaning of dirty utensils and use low-cost detergents. In this way, food handlers come in contact with chemicals in the form of detergents and disinfectants. Prolonged exposure or increased duration of exposure can cause allergic contact dermatitis especially in hands. Such hands can be superinfected with bacterial infections and jeopardize food safety.

21.4.7 **BIOLOGIC HAZARDS**

Food handlers come in contact with organisms, such as, *Salmonella* while handling raw foods, particularly of animal origin, such as eggs. Food handlers may be carrying pathogens in/on their bodies and can be a source of infection to other coworkers in the same working area. Cases and carriers can also transfer pathogens to various foods and compromise food safety.

They need to taste the food prepared by them before it is served to the clients. In case the food has become unsafe, they themselves fall prey to FBI. Most of the low-budget EEs, messes, and canteens provide food to its workers after the lunch or dinner timings are over. This makes them the last ones to eat the meal, which most often has been lying in temperature-danger zone (5–60°C) for more than 4 h exposing them to risk of FBI.

21.4.8 MECHANICAL HAZARDS

The most common mechanical hazard is due to cuts with knives and other sharp equipments. Protruding parts of various machinery can also lead to injuries. It is known that 10% of accidents in any industry are probably due to mechanical causes.

21.4.9 PSYCHOSOCIAL HAZARDS

Food handlers do face monotony in their work environment. Most of the times they continue to work at same level for years together. There is very little scope of career progression. These people are always behind the curtain, do not get recognition, and appreciation for good work very often. Even the lady of the house waits for an appreciation from the family members after cooking meals. However, times are changing. Recently various live TV shows hosted by famous chefs have given due recognition to the skills and progression. The emerging hotel management industry, catering industry, eating-out culture in the changing society, and globalization have improved the social status of the occupation. The food handlers are often exploited with low wages and restricted leave. Most of them work in the unorganized sector. At times they are paid less for the skill they possess.

Easy availability of food and alcohol, results in obesity and alcoholism. They work hard to earn their livelihood but the job does not require strenuous activity. Often they are required to stand for long hours. Access to tasty and rich food at all times makes them consume more calories than required and hence the risk of noncommunicable diseases.

They also work under lot of stress as even ignorance on their part can dissatisfy the consumer/client and harm the reputation of the EE.

21.5 PREVENTION OF OCCUPATIONAL HAZARDS IN FOOD HANDLERS

21.5.1 ENGINEERING MEASURES

It is imperative that due care should be given to the ventilation and lighting area of work place/kitchens in EEs. Principles of ergonomics should be applied while designing the height of working shelves and food flow. Trolleys with wheels (similar to those in super markets) should be available to carry heavy loads. Equipments should be regularly serviced as per manufacturer's instructions. The quality of chemicals used should be checked before exposing food handlers to them.

21.5.2 **MEDICAL MEASURES**

Regular medical examination of food handlers will ensure good health. In case a worker is unwell, (s)he should be given leave and rest. However, this is not followed because of work-force constraints. In case leave is not possible, the manager should employ sick workers in an alternate job, not involving the direct handling of food. In case a food handler is returning to work on completion of a sick leave, medical examination of the employee should be done. It should be ensured that a policy on this issue is made in the industry based on the kind of risk involved to the food. A record of all medical examination needs to be maintained.

21.5.2.1 Preemployment screening of food handlers

The most important infections attributed to transmission from infected food handlers are norovirus, *Salmonella enteritidis*, and *Salmonella typhimurium*, which together account for the largest numbers of outbreaks and individual infections. Food handlers can be symptomatic or asymptomatic carriers of FBI—both the transmission of norovirus and *S. enteritidis* have been attributed to asymptomatic food handlers. All food handlers before employment should undergo a medical examination and a stool test by a registered doctor. Only those found fit should be allowed to handle food.

21.5.3 **LEGAL MEASURES**

Society has a responsibility to protect the health of the workers engaged in various occupations. Legislations on occupational health and safety have been in existence in India for over 50 years, for example, The Factories Act, 1948 and the Worker Workmen's Compensation Law. However, both these are not applicable to hotels and restaurants. The workers in hotels and restaurants are covered under the Employees Sate Insurance Act, which is a contributory social insurance scheme that protects the interests of workers in contingencies, such as, sickness, maternity, and employment injury causing temporary or permanent physical disability or death, loss of wages, or loss of earning capacity.

In spite of a good framework of this law, the implementation of this welfare act is far from satisfactory in case of food handlers working in EEs. The Indian government has expanded the coverage of Child Labour Prohibition and Regulation Act and banned the employment of children less than 14 years of age as domestic workers and as workers in restaurants, *dhabas*, hotels, spas, and resorts with effect from Oct. 10, 2006. The changes were necessitated after the Right to Education Act came into effect which promised free and compulsory education to all children aged between 6 and 14 years.

21.5.3.1 Workforce as food handlers in unorganized sector

A large number of food handlers mostly women in India work as maids and nannies in the unorganized sector. They are often involved in cooking along with other household chores. In Feb. 2014, the ILO stated: "Millions of maids working in middle-class Indian homes are part of an informal and 'invisible' workforce where they are abused and exploited due to a lack of legislation to protect them." According to the National Sample Survey of 2004–05, there are around 47.5 lakh domestic workers in the country. Out of these, 30 lakh are women working in urban areas. An ILO report suggests that the number of maids has surged by close to 70% from 2001 to 2010 in India. They number at least 10 million. There is a cyclical pattern of their daughters becoming maids to the same families. The rule of minimum wages is not maintained most of the time. In 2008, the government drafted a National Policy on Domestic Workers, the main points included were minimum wages, working hours and conditions, social security protection, the right to form trade unions, and develop their skills. The policy, however, has not been approved by the cabinet yet. Other laws which are applicable to domestic workers are the Minimum Wages Act, 1948; the Employees Compensation Act, 1923; the Equal Remuneration Act, 1976; and the Inter-State Migrant Workmen Act, 1976. These workers are often denied the benefits because they are generally illiterate and do not hold mutually agreed contracts, unlike say those working in factories.

In a nutshell, food handlers are an important cause of FBI and have a definite role in prevention. They are often neglected at their place of employment. Their food-handling practices can affect the well being of many people and hence they are a crucial link in food safety.

REFERENCES

Chin, J. and American Public Health Association, 2000. Control of communicable diseases manual (vol. 17). Washington, DC: American Public Health Association.

Greig, J.D., Todd, E.C.D., Bartleson, C.A., Michaels, B.S., 2007. Outbreaks where food workers have been implicated in the spread of foodborne disease. Part1, description of the problem, methods, and agents involved. J. Food Prot. 70 (7), 1752–1761.

Hutin, Y., Luby, S., Paquet, C., 2003. A large cholera outbreak in Kano City, Nigeria: the importance of hand washing with soap and the danger of street-vended water. J Water Health 1, 45–52.

Leavitt, J.W., 2014. Typhoid Mary: captive to the public's health. Beacon Press, Boston, USA.

Olsen, S.J., Hansen, G.R., Bartlett, L., Fitzgerald, C., Sonder, A., Manjrekar, R., Riggs, T., Kim, J., Flahart, R., Pezzino, G., Swerdlow, D.L., 2001. An outbreak of *Campylobacter jejuni* infections associated with food handler contamination: the use of pulsed-field gel electrophoresis. J. Infect. Dis. 183 (1), 164–167.

Pajot, M., Aubin, L., 2001. Does food handler training improve food safety? A critical appraisal of the literature. Available from: http://www.peelregion.ca/health/resources/pdf/mandatory-food-handler-training.pdf

Walker, E., Pritchard, C., Forsythe, S., 2003. Food handlers' hygiene knowledge in small food businesses. Food Control 14 (5), 339-L343.

Consumers

22

P. Dudeja*, R.K. Gupta**

*Department of Community Medicine, Armed Forces Medical College, Pune, Maharashtra, India; **Department of Community Medicine, Army College of Medical Sciences, New Delhi, India

22.1 INTRODUCTION

Every individual wants peace of mind that comes from knowing that the food which is put on the table is safe. There are multiple stakeholders in ensuring food safety, namely, government agencies (agriculture, animal husbandry, food processing, transport), food business operators (FBOs) and consumers (Fig. 22.1). Both government and FBOs shoulder the responsibility of providing safe food to the consumer. Globally, governments are working to meet consumer expectations for food safety. In the United States, consumers and the food industry came together to support passage of the FDA Food Safety Modernization Act. China is working to implement the comprehensive food safety modernization law it adopted in 2009. The Canadian Senate passed the Safe Food for Canadians Act. The enactment of Food Safety and Standards Act (FSSA) 2006, Food Safety and Standards Regulations 2011 in India is another example of commitment of the authorities toward the citizens.

A consumer/customer (also known as a client, buyer, or purchaser) is the recipient of a good, service, product, or idea, obtained from a seller, vendor, or supplier for a monetary or other valuable consideration. In the farm to fork model, the consumer is at the fork end. A small breach in food safety has great ramifications on the consumer. When an eating establishment manages to serve its customers well enough and makes them happy, it can create long-term customer relationships; and same applies to various brands of food items. Any major illness outbreaks and contamination incidents damage consumer trust in a particular food item leading to loss of sales and take a long time to recover. The setback suffered by a manufacturer of the famous instant noodles episode in India, in 2015 is an example how exaggerated can the consumer response be when it concerns safety issues and ultimately their health.

An effective food safety regulatory framework is imperative to ensure safe food for consumers in a country. However, considering the size of food industry in India, it will take lot of resources to implement the new food law. Laws alone cannot solve the issue of food safety. The need of the hour is not only to have an integrated law

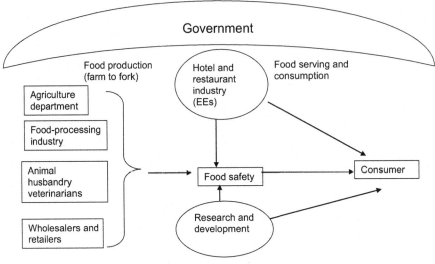

FIGURE 22.1 Key Stakeholders in Food Safety

but also to have an integrated approach to change the mindset of people regarding food safety. Consumers themselves need to stand up and demand for their right to safe food from the FBOs.

22.2 CONSUMERS' PERSPECTIVE ABOUT FOOD SAFETY

Most consumers understand that food is not risk free. But they do expect that everyone involved in producing, processing, transporting, and marketing food is doing everything they reasonably can to prevent problems and make food safe. In a huge country like India consumers belong to different cultural backgrounds, varied socioeconomic status, and wide environmental surroundings. Some of the perspectives of consumers are given subsequently.

22.2.1 CULTURE OF SILENCE

These groups of consumers are aware and want safe food but never complain about what is available or offered to them. Majority of the clients belong to this group. They overlook safety issues either as no other option is available to them or sometimes when cost, taste and variety may be their main concern over food safety. This specially applies to consumers of street foods in the country. This suits them as it is convenient too and they are often ready to compromise. At times the acceptance for substandard service or food item comes as they are not fully aware about their rights under the FSSA 2006 and how to exercise them. They hesitate in taking the responsibility to bring the change. When a consumer goes out to eat, there are many factors affecting

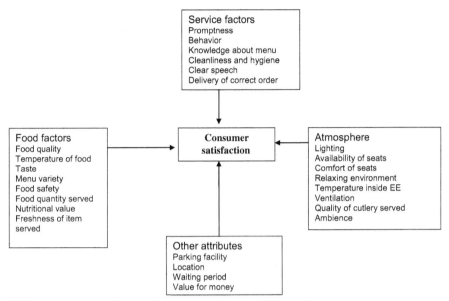

FIGURE 22.2 Factors Affecting Consumer Satisfaction While Eating Out

his satisfaction and experience (Fig. 22.2); and as we can see food safety is just one aspect for them as a part of satisfying experience!

22.2.2 LOW EXPECTATIONS

These consumers are themselves unaware and ignorant about issues of food safety. They accept what is offered to them as food security takes precedence over safety in these cases.

22.2.3 COST CUTTING

This group of consumers want food safety but at low cost. It is a known fact that food safety costs money or that safety is compromised by the FBO as a cost-cutting measure. This is acceptable to a section of consumers. This is the scenario while traveling, mass gatherings, religious occasions, wedding feasts in low socioeconomic strata, etc.

22.3 GOVERNMENT'S INITIATIVE IN CREATING CONSUMER AWARENESS

Food Safety and Standards Authority of India (FSSAI) has been entrusted with the responsibility of making the consumers aware about all aspects of food safety. Various initiatives taken by them are as given here.

22.3.1 **HOME-BASED DETECTION OF ADULTERATION**

A common man may not have sufficient knowledge about quality and purity of raw food items used at home. A look at them while purchasing is not sufficient to guarantee food safety. Sometimes food is bought loose also. FSSAI has taken a lead in preparing a manual to empower the consumers with home-based detection of adulteration of food items. Some states, (like Chandigarh) have recently taken the initiative to launch mobile vans in the city which will visit different parts of city as per a fixed schedule and offer free adulteration detection for the consumers. Apart from this, under the FSSA 2006, the consumer has been empowered to collect and submit any sample of food and submit the same to the food safety lab of the district/state along with requisite forms and get the sample tested. The result shall be available to him by the food analyst with 14 days of submission of sample.

22.3.2 **IEC CAMPAIGNS**

Various IEC campaigns are conducted to create awareness in masses. Resource material has been developed by experts under the government campaign "*Jago Grahak Jago*" meaning "wake up consumers" in collaboration with Ministry of Consumers Affairs, Food and Public Distribution and is freely available online from the website of FSSAI (Fig. 22.3) under title "IEC" on following subjects:

1. food safety for school children;
2. kitchen food safety;
3. how to read a label;
4. how to site a complaint against an FBO;
5. beware of food additives, etc.

22.3.3 **PROVIDING COMPENSATION TO CONSUMERS**

Under the FSSA, 2006 the eating establishments of the country can no longer dismiss complaints of food poisoning by simply blaming consumers' weak digestive system. They will have to serve up steep compensation or face police action. Cases of food poisoning can be registered under Section 65 (compensation in case of injury and death) of the act. However, the consumers have to follow a series of steps to seek compensation. They will have to file a complaint with the local police, provide a sample of their vomit and a copy of the bill. They will also have to submit doctor's certificate confirming the condition. The FDA and police officials need to complete the probe within 90 days of receiving the complaint. If the case of food poisoning is proved, the restaurant will have to pay up within 3 months. The amount of compensation will depend on the severity of food poisoning. There is also a provision of filing complaint to FSSAI against a misleading advertisement with a reward scheme for Rs 500 to consumer in case the claim is found correct by the experts.

FIGURE 22.3 Initiative by FSSAI and Ministry of Consumer Affairs, Food and Public Distribution

22.4 MANUFACTURER'S INITIATIVES TO PROTECT CONSUMER WELFARE

A manufacturer communicates with the consumer through labeling and advertisements. Both of them are double-edged swords. Their pros and cons have been dealt in separate chapters in the book (Chapters 14, 16, 20, and 27).

22.5 CONCLUSIONS

The government has done its bit by launching the new law. But for its implementation in true spirit, the consumers of the country need to be alert about their rights, particularly regarding adulteration in food products. Food safety education campaigns need to be strengthened to cover both the urban and rural sectors. The Internet and social media appear to be one of the best platforms to reach millions of consumers for food safety awareness. FSSAI has already taken a step forward in this direction. If we have good agricultural manufacturing, hygiene practices we need to promote good consumer practices to complete the chain from farm to fork!

Public health professionals and food safety

23

R.K. Gupta

Department of Community Medicine, Army College of Medical Sciences, New Delhi, India

In the game of cricket when a well established and accomplished batsman gets out repeatedly on poor balls, they advise him, *"whenever in doubt, go back to basics."* I was wondering whether this chapter might become monotonous; so I have decided to go to the very basics. As far as the disease causation and progression is concerned, the most basic is its "Natural History." It is felt that piggy-backing a Public Health Professional (PHP) on the natural history of disease would elicit, expose, and trace his/her role in the journey of food safety, to the fullest. So let us begin with the natural history of disease.

The natural history of disease is the course a disease takes, starting with a pathologic onset (inception) until its eventual resolution through complete recovery or death. As the inception of a disease is not a firmly defined concept, the natural history of disease is sometimes said to begin at the moment of exposure to causal agents. It progresses through various disease phases, namely, asymptomatic, early clinical disease, and classical disease phase. It ends with the termination phase concluding with one of these: recovery, chronic disease, disability, or death. The public health expert has a prominent role to play in each of these stages, which we shall see a little later.

If we peep into the history, it was that one episode of a food/waterborne disease, observed by one vigilant doctor that revolutionized medicine and laid the foundation of a new science. The episode was the 1854 cholera outbreak in London, when Dr. John Snow observed and identified a polluted public water well on Broad Street (now Broadwick Street). Although Snow's chemical and microscopic examination of a water sample from the Broad Street pump did not conclusively prove its danger, his studies on the pattern of the disease were convincing enough to persuade the local council to disable the well pump by removing its handle (Vinten-Johansen, 2003).

Snow later used a "spot map" to illustrate the cluster of cholera cases around the pump. He also used statistics to illustrate the connection between the quality of water source and cholera cases. He showed that the Southwark & Vauxhall Waterworks Company was taking water from sewage-polluted sections of the Thames river and delivering it to homes, leading to cholera. Snow's study was a landmark in the history of public health. It is regarded as the founding event of the science of epidemiology (Johnson, 2006).

Could we call John Snow, as the first PHP of modern medicine? This brings us to the question, who is a PHP? As per WHO, public health refers to all organized measures (whether public or private) taken to prevent disease, promote health, and prolong life among the population as a whole. It is concerned with threats to health, based on population-health analysis. The population in question can be as small as a handful of people, or as large as all the inhabitants of several continents (pandemics). The three main public health functions are as follows (WHO, 2015).

- The assessment and monitoring of the health of communities and populations at risk to identify health problems and priorities.
- The formulation of public policies designed to solve identified local and national health problems and priorities.
- Access to appropriate and cost-effective care, including health promotion and disease-prevention services.

Multidisciplinary teams of public health workers and professionals are required to execute these functions with respect to food safety as well. This includes, physicians specializing in public health/nutrition/community medicine/infectious disease, psychologists, epidemiologists, medical officers, public health nurses, medical microbiologists, health officers, public health inspectors, food inspectors, dental hygienists, dieticians, laboratory technicians, nutritionists, veterinarians, public health engineers, public health lawyers, sociologists, community development workers, communications experts, etc. In a broad sense each of these workers, is a PHP in his/her own capacity, exhibiting specialized skill sets (Public Health Agency of Canada, 2005).

In other words, it is not important, as to by what name do we know PHPs but rather what their job profile is. A PHP monitors and diagnoses the health concerns of communities and promotes healthy practices and behaviors to ensure that populations stay healthy. Food safety is an important public health priority for any PHP. The PHP is an important stakeholder in food safety from farm-to-fork. He/she is the link between the theory of public health and its practice; that is, from government policy to consumers/finance and business officers.

One way to illustrate the breadth of public health is to look at some notable food safety-related public health campaigns.

- Vaccination and control of foodborne infectious diseases.
- Safer workplaces for cooks.
- Safer and healthier foods.
- Safe drinking water.
- Healthier mothers and babies through access of healthy food.
- Policy and control of food adulteration and foodborne diseases.
- Various food-safety regulations, policies, and legislation.

It is evident that, the base of knowledge for public health evolves from a variety of disciplines. These range from social sciences to biologic sciences and business, brought together by a commitment to improve health. While physicians treat ills of individual patients, PHPs would address the community's food safety-related health problems—surveillance, monitoring, investigations, regulations, policies, research, etc.

The PHPs examine many questions listed as follows.

- Which foods are safe? (Recent cases of an international brand of noodles in India.)
- Who gets a particular disease and why? (Only few get food poisoning after a common community meal, consumed by many.)
- Can a common food-related element be changed to prevent illness? (e.g., reducing trans-fatty acids, consumption in a community to reduce cardiac risk.).
- Does that element require a change in behavior; technology, or the health care delivery system? (e.g., making it mandatory by law to endorse the t-FA content on product labels.)
- How can we motivate individuals to change their behavior? (E.g., through innovative education campaigns.)
- Can a new technology be developed? (e.g., air-cooking technology to minimize oil/trans-fatty acid for "frying" chips and wafers.)
- How the healthcare-delivery system be changed to improve situation? (e.g., through enforcing law.)

Resolving such questions often requires stepping into political arena where PHPs and advocates develop new policies and programs (USF, 2015).

Let us come back to the natural history of disease and the role of the PHP, over its stages (Table 23.1).

Table 23.1 emphasizes that a PHP journeys through the entire spectrum of the disease's natural history. This begins when the disease has not even started and there is perfect equilibrium between the host (individual/community) and the disease agent/determinants (microorganism/disease favoring conditions, e.g., poor sanitation of a cook house). But unlike a pure clinician, a PHP is active and (s)he is in the process of undertaking following activities.

23.1 STUDYING

During this "lull period, before the storm," a PHP dedicates himself/herself to study, and to be alert. He/she studies about the agent, host, and environment that are likely to be in the middle of causing a potential foodborne event. Their peculiarities and nuances are understood. We can illustrate this by studying a potential food-poisoning outbreak. Imagine a party is being hosted. There are 20 dishes to be served that include beverages, snacks, soups, salads, curds/yogurts, bread, vegetarian and non-vegetarian main course, rice, and desserts. It is the PHP's background study and knowledge about the circumstances and the actual potential of causing food poisoning of each one of these food articles that matters. A PHP has to be ever vigilant and sensitive to the surroundings, milieu, and the environment where the entire "act" is taking place. This includes having a background knowledge of each raw material, kitchen, processing room, dining room, the food handlers, and even the water used.

Table 23.1 Natural History of Disease and the Role of a PHP

Broad Stages of Natural History of Disease	I Positive Health/ No Disease	II Susceptible Community	III Asymptomatic	IV Early Clinical Disease	V Classical Disease/New	VI Termination (Recovery/Chronic Disease/Disability/ Death)
Role of a PHP in food safety	Study Teach Surveillance Information, Education, and Communication (IEC) Policy Research	Advise Vaccination Consultation	Suspicious Public health lab	Decides to detect—when, what, how. Interprets Compares Concludes. PHP as a clinician Early diagnosis Treatment	Investigation of outbreak PHP as a clinician, treats disease Research Critical review Tackling a new disease	Documentation Reporting Advisories Consultancy Advocacy Policy making

23.2 TEACHING AND TRAINING

Besides studying, another constant for a PHP is teaching. They continuously share their experiences and teach their students and other professionals. This could be an organized formal teaching at an institution/training institute or informal teaching through discussions in closed groups. Conferences, guest lectures, and seminars too are great forums for teaching and sharing experiences.

23.3 DISEASE SURVEILLANCE

Even though disease surveillance is a continuous process, its foundation is laid at this time of relative peace, when there is no disease. The protocols for surveillance could be worked out, stakeholders put on board and documents could be circulated. Mock drills and exercises could be planned involving all concerned (including the laboratory) and standard operating procedures may be worked out. PHPs can enrich the plan/document with their experience. Further details on surveillance are given in Chapters 3 and 4 of this book.

23.4 INFORMATION, EDUCATION, AND COMMUNICATION (IEC)

A major function of a PHP is to continuously engage in the IEC activities with respect to the health conditions/diseases under question to maintain food safety. While it has a much larger receptiveness when the disease episode has already occurred, but it has its own value during the nondisease period as well. In fact a crescendo is required to be initiated and maintained. For example, the community is to be continuously told about the advantages of iodized salt, low fat, the dangers of food adulterants, and what to look for on the food labels!

23.5 SPECIFIC ADVICE AND PROFESSIONAL CONSULTATIONS

In the second stage (Table 23.1), when the community is susceptible to disease, the functions of PHPs, elaborated in the preceeding paragraphs, continue; and besides these, the PHPs' role expands to specific advice, professional consultations, vaccinations, etc. Take a disaster situation for example. It might be an occasion for the PHP to render specific advise on food safety. This may range from how to manage the food aid, what to eat, how to purify water, manage a refrigerator when the current goes out, to the nutrition of children and relevance of cholera/typhoid vaccine in the given situation.

23.6 PHP IN PUBLIC HEALTH LABORATORY (PHL)

When there is a lurking fear of unscrupulous activity (e.g., food adulteration) or a likelihood of asymptomatic cases occurring in the community (that are not clinically manifested), PHPs have to be "suspicious," think of relevant investigations, and the role of PHL comes in. To cite some examples; there is an unconfirmed news of likely adulteration of milk in a community. The PHP has to take all actions to involve the local administration, collect samples as per procedures, get them analyzed in PHL, collect reports, and follow it up with definitive action. To quote another example, a study is undertaken in a town neighboring the sub-Himalayan region for likely iodine deficiency disorders, where manifestation of goiter might not be present. Various samples including, blood and urine of suspected cases in the community and salt and soil samples for iodine might have to be analyzed in the PHL before reaching a conclusion. More importantly these "numerical" results have to be interpreted and acted on by the PHP.

23.7 EARLY DIAGNOSIS OF DISEASE

Referring back to Table 23.1, (stage IV), in the phase of early clinical disease, PHPs have a major public health role. They have to be alert to the health/disease-related developments in the community. They have to decide on the "what, when, and how" of early detection of disease like a detective looking for clues. They have to be vigilant and observe, collate, compare, conclude, and interpret data intelligently for best results in the interest of the community. Based on their expertise, experience, and scientific interpretations, they would be able to advise the community and administrators for earliest possible control of the disease.

23.8 INVESTIGATION OF FOODBORNE OUTBREAK

When the disease manifests overtly and takes the form of an outbreak, it may call for the PHPs' attention, as a clinician, for treatment; and more importantly for "investigation of the outbreak." All their public health skills are tested in this singular situation. He/she has to be a clinician, a scientist, a PHP, a communicator, person with excellent soft skills, and most importantly a role that no other doctor/health professional performs, a health detective.

This job becomes exceptionally challenging because, more often than not, in this situation the PHP has to work in the most adverse and hostile environments. The administrators of the site where one is "investigating" the outbreak want to guard the reputation, more than their own lives. So, they are most unwilling to divulge the actual facts; the kitchen/food-related evidence would have been destroyed and they may not be cooperative at all.

To cite a real example, the author was called to investigate a food-poisoning outbreak, 3 days after an episode had occurred leading to 46 cases and 1 death. It had ensued following a dinner party organized at the residence of a senior executive. On arrival at the house, when I went in to seek permission of the host to carry out the investigation, I was ushered into the bedroom. The typical scene was like this: the host, a stout middle-aged gentleman with a mild American accent was in bed, down with gastroenteritis. The hostess, who had just recovered from mild gastroenteritis, quickly prepared tea for me. After the pleasantries, I sought their permission to visit the kitchen to carry out the "investigation." The kitchen was spic-and-span, no traces of food; everything including the basins and drains were sparkling clean. I carried out a general hygiene inspection. There seemed to be no point in taking any food samples, as no 3-day-old food was available.

Rather disappointed, I moved back to their bedroom to thank them. This is when I noticed a huge two-door (large almirah-like) fridge located in their bedroom. Such refrigerators were not to be seen in India, two decades back, as we used to have smaller single-door ones. I curiously asked about this fridge, when beaming with pride, the lady of the house announced that she had got it imported from "*Canada*." The gentleman then interjected, "*it can keep food fresh for 6 months*." Quite intrigued I nodded, but thought the fridge could be the best piece of technology from Canada, but it is housed in India, where it was routine for the electricity to be out for many hours each day; how could it maintain the temperature?

Suddenly, I saw a photograph hung on the wall. The gentleman standing was holding a large fish vertically, with its snout resting on the ground, the fish being "taller" than him. I could not stop asking him about the strange photo. Very proudly, he declared that he had "fished" it in New Zealand, 6 months back, and "*we are still eating it and it is right there in this Canadian fridge*," he added. I could not believe it, for two reasons, first, we Indians have thrived for millennia eating fresh food, and second, given the electricity situation, how can it still be edible? I promptly asked him, if this was the fish that was prepared for dinner that night. The lady proudly claimed, "*Yes*." Immediately I took the fish samples and the source of the outbreak was confirmed after lab tests!

Chapters 3 and 4 are dedicated to investigation of outbreaks in this book that can be referred for technical details on the subject.

23.9 TACKLING A NEW FOODBORNE DISEASE

A patient with a new syndrome or disease that does not "fit" the classical disease picture always perplexes the clinicians. They examine, investigate, and treat with their best intentions (albeit hit-and-trial), yet are unable to cure. They might give up, as probably the new disease is beyond their domain. This difficult situation now falls in the PHP's court. The PHP uses the entire armamentarium available at his/her disposal to assess the disease from every facet possible, employing the principles of epidemiology. Various interactions of agent–host–environment are worked out, signs and symptoms of disease studied, attack rates and fatality rates calculated, charts and figures are drawn, Epi Info and SPSS used, p values and Chi-square values calculated, mathematical modeling and global positioning done, and probably a new disease is born.

Many examples can be cited. Prior to the 19th century, it was believed that pellagra was either an infectious or a protein-deficiency disease. It took many experiments by Goldberger including on humans and dogs, with dietary supplementation, initially with milk, meat, and eggs and later with yeast, that he "established" pellagra to be a deficiency disease. Later he concluded that it was the deficiency of a "heat stable" factor present in yeast that caused pellagra.

Finally, natural history of disease concludes with "termination," when the patient either recovers, lapses into chronic disease, or ends with disability or death. Here too, the PHP has major roles to play.

23.10 DOCUMENTATION AND REPORTING

Documentation and reporting are important functions of the PHP. It may be part of a basic surveillance process, epidemic investigation, collection of food samples, or enumeration of children undergoing hemoglobin test during a research project. The PHP is always at the forefront and he/she is responsible for correct documentation, as the results depend on that.

23.11 ADVISORIES, CONSULTANCY, AND ADVOCACY

Foodborne diseases may present in varying forms, in different communities, and at different times. Their prevention, control, and management also differs. Likewise in the situation of a new, lesser known or reemergence of a forgotten disease, analysis of situation by PHPs becomes mandatory. Consequently, guidelines, advisories, standard operating procedures, and instructions are required to be issued from time-to-time for their best possible contemporary management. The PHP is the key to such advisories. Public health consultancy and advocacy are ongoing processes.

23.12 POLICY MAKING AND ADMINISTRATION

PHPs might be professional people but their inputs are vital in administration, governing institutions related to food safety, and in policy making. While policy making might be the holy grail and the final frontier of what the system, organization, or the nation wants, its valid and correct execution too is as important. This cascade of policy implementation trickling down to the consumer and the common masses is to be intelligently inferred and executed through the PHP, who is a key link between the administration and the community.

23.13 RESEARCH

Research is another important function of the PHP. It, in fact, is an ongoing process and has a multitude of facets. They are as varied as the basic fundamental research in nutrition safety, laboratory research, animal research, human research, community research, clinical trials, behavioral research, policy research, etc.

23.14 **LEADERSHIP**

The PHP is a leader in situations when others seem to be bewildered. They include, investigation of a foodborne epidemic, dealing with a new disease, tackling reemergence of a disease (Ebola), in creating international/national or regional guidelines on disease (prevention), execution of these guidelines, their monitoring and evaluation; in research, and in policy making. The PHP not only provides leadership but also assists in the development and strengthening of risk-based, integrated systems (national/regional) for food safety.

23.15 **CONCLUSIONS**

However, as vital the task of a PHP might be, there are inherent challenges that he/she faces with regards to food safety. The ideas and concepts of food and of diet and nutrition are deeply ingrained in our psyche and culture. It is not really easy to alter the behavior regarding food habits. Thus it calls for professional competence, conviction, confidence, and perseverance to carry forward the functions of a PHP in a desirable manner. Further, in addition to public roles, the role of a PHP spills over to the domains of government, nongovernmental organizations, legal bodies, policy makers, bureaucracy, civil service organizations, etc. These agencies might have their own inertia to an advice or a change that make the task of a PHP that much more challenging.

PHPs provide the scientific base for measures along the entire food chain to decrease foodborne health risks, through mediating the improvement of international and national cross-sectoral collaboration and enhance food-safety communication and advocacy. PHPs have always been the driving force behind the provision of healthy food. It is their multifaceted expertise, skills, and teamwork that facilitates the conduct of research, laboratory work, formulation of policies, and helps the community solve the jigsaw of outbreaks, with amazing dexterity, thus allowing us to live in the world of food safety.

REFERENCES

College of Public Health, University of South Florida, 2015. What is public health? Available from: http://health.usf.edu/publichealth/definition.html

Johnson, S., 2006. The Ghost Map The Story of London's Most Terrifying Epidemic—and How it Changed Science, Cities and the Modern World. Riverhead Books, New York City, USA.

Public Health Agency of Canada, 2005. Building the public health workforce for the 21st century. Joint Task Group on Public Health Human Resources; Advisory Committee on Health Delivery and Human Resources; Advisory Committee on Population Health and Health Security, Ottawa, Canada. Available from: http://publications.gc.ca/collections/collection_2008/phac-aspc/HP5-12-2005E.pdf

Vinten-Johansen, P., 2003. Cholera, Chloroform, and the Science of Medicine: A Life of John Snow. Oxford University Press, Oxford, UK.

WHO, 2015. Available from: http://www.who.int/trade/glossary/story076/en/

Role of veterinary experts in food safety

24

D.P. Attrey

Central Military Veterinary Laboratory, Meerut, Uttar Pradesh, India; High Altitude Research, Defence Research and Development Organisation, Leh, Jammu and Kashmir, India; Amity Institute of Pharmacy, Amity University, Noida, Uttar Pradesh, India; Innovation and Research Food Technology, Amity University, Noida, Uttar Pradesh, India; Amity Institute of Seabuckthorn Research, Amity University, Noida, Uttar Pradesh, India; Lala Lajpat Rai University of Veterinary and Animal Sciences, Hisar, Haryana, India

24.1 INTRODUCTION

A veterinarian (vet) is generally considered as a person, who commonly treats sick animals. Most people, due to their ignorance, perceive the role of a vet only as a clinician for sick animals and do not know that vets not only help to control or reduce the zoonoses but they are also closely involved in the safety of human foods for protection of public pealth. Since prevention and control of zoonotic and other diseases of livestock used to provide increased production and better quality of food products, it had indirectly brought veterinary and animal health services in to food production and food safety and had also increased their role in improving the public health services through reduced incidence of epizootic or zoonotic diseases. Food and Agriculture Organization of United Nations Organization (FAO) gives high priority to programs and activities dealing with food quality, safety, and consumer protection.

While discussing the role of vets in prevention of zoonoses, Dikid et al. (2013) observed that over 30 new infectious agents have been detected worldwide in the last three decades; 60% of these are of zoonotic origin. In the recent past, India has seen outbreaks of many organisms of emerging and re-emerging diseases in various parts of the country, six of these are of zoonotic origin. Developing countries such as India are more prone to such emerging foodborne infections due to social, economic, environmental, and other factors. Nipah virus, seen among pig farmers in Malaysia in 1999, and which has a high fatality rate (60–70%), has also been categorized as a foodborne risk, since it occurs from eating dates contaminated with urine or saliva of infected bats. Vets have an important role to play in reduction or control of such emerging infectious diseases. In fact as per World Organisation for Animal Health (OIE), control and/or reduction of biological hazards of animal and public health importance is a core responsibility of veterinary services (OIE, 2012). WHO also encourages evidence-based strategies for the control of foodborne diseases and to provide guidance in prioritizing such strategies (WHO, 2000).

24.2 ROLE OF VETS IN FOOD SAFETY CONTROLS AND FOOD SAFETY PLAN

Production of safe food depends upon adequate food safety controls for preparation and implementation of appropriate "food safety plan" for all processes of food production across the food chain, that is, primary production, raising of healthy animals and birds, slaughtering, primary and secondary processing, packaging, storage, distribution, sale, cooking, serving, and even eating of food in hygienic manner as per all relevant good practices, following the required food safety standards and statutory instructions. Vets provide invaluable help in preparation and implementation of food safety plan for almost all processes of food chain. WHO has also resolved to make food safety as an essential public health function, with the goal of developing sustainable, integrated food safety systems for the reduction of health risks from the entire food chain. WHO has also a continuing commitment to the fundamental principle that ensuring food safety is an essential activity and an integral part of any public health program (WHO, 2000).

Vets play a major role in implementation of risk analysis and "risk-based controls" in food safety control systems. Application of a vet's knowledge of comparative medicine, helps in clinical understanding of risks from foodborne infections, which is essential for maintaining and improving public health, especially in the rural settings, like in India, where majority of human population stays in close contact with animals. Most cases of human Salmonellosis are foodborne (Hoelzer et al., 2011), which are generally acquired through direct or indirect animal contact.

As per FAO/WHO (2005), during the 2nd Global Forum meeting, Hazard Analysis Critical Control Point (HACCP) was considered by many delegates as one of the important risk management tools. However, it alone cannot resolve food safety problems, and should be complemented by other control measures such as monitoring programs at primary production for agriculture chemicals, pollutants, contaminants and natural toxins, traceability and labelling.

According to McKenzie and Hathaway (2006), the food safety controls, which were based on science and good hygienic practice (GHP) in early 1990s, shifted to more targeted food safety control system, that is, HACCP during mid-1990s. However, during late 1990s, the food safety approach for hygienic production of food was shifted from the routine "food safety controls" to the "risk-based controls." Now the USFDA under its new "Food Safety Modernization Act" (FSMA), has adopted a new rule, that is, "Current Good Manufacturing Practice & Hazard Analysis and Risk-Based Preventive Controls (HARPC)" (FDA, 2013). Although "food safety controls" are based on current good manufacturing practice (cGMP) and hazard analysis, the risk-based preventive controls are based on "risk assessment."

Achievement of risk-based prevention through successful implementation of FSMA, depends on close partnership and full integration of all stakeholders across the Foods and Veterinary Medicine Program in USA, at both the planning and operational levels to protect public health (FDA, 2013). This integrated and collaborative approach is essential between the Offices of Foods and Veterinary Medicine and other stakeholders, such as Centre for Food Safety and Applied Nutrition (CFSAN), Centre for Veterinary Medicine, (CVM), Office of Regulatory Affairs (ORA), and the Office of Global Regulatory Operations and Policy, etc., to protect public health.

In the European Union, the responsibility for production and control of safe food is shared between operators, national authorities, and the European Commission (EC). Operators are responsible for compliance with legislative provisions, and for minimizing risk on their own initiative. National authorities are responsible for ensuring that operators respect food safety standards. They need to establish control systems to ensure that rules are respected and enforced. For this the EC, through its Food and Veterinary Office (FVO), carries out a program of audits and inspections. These controls evaluate the performance of national authorities against their ability to deliver and operate effective food control systems, and are supported by visits of FVO officials to individual premises to verify that acceptable standards are actually being met (FAO/WHO, 2005). Tuominen (2009) has observed that the new role in risk-based controls, through risk assessment in food safety, has placed new responsibilities on veterinary services. Veterinarian is most competent and suitable person to assess risks and implement quality assurance throughout the food chain. They play an important role in food safety controls.

As per Bousfield and Brown (2011), veterinarians have the knowledge and expertise to understand and audit the standards of animal health, animal welfare, and public health from "farm to fork," the areas in which veterinarians are involved have gradually been extended from animal production to all levels of the food production chain. All these are included in veterinary training and education.

Public Health, Animal Health and Animal Welfare are always interrelated. These should be dealt with holistically e.g. good animal welfare is the backbone of food safety, which in turn is the backbone of public health, since stressed animals are more prone to develop diseases, which will require maintenance of adequate health of food animals through proper veterinary care. For this, each farm should have effective "animal health plan," prepared by a qualified vet. Vets are also involved in checking health and welfare of animals at different stages of transport as per requirement through antemortem inspection. Adequate rest and recoupment (for about 3 days, with proper feeding and watering) is essential after transportation of all food animals to restore the glycogen reserves for development of adequate rigor mortis after slaughter. In India, most animals are slaughtered immediately after transportation and without any rest and recoupment of their base glycogen levels, which results in production of poor quality of meat. Only adequately rested healthy animals can provide safe and wholesome food and those only should be sent for slaughter. Veterinarians have scientific and medical training as well as statutory accountability to ensure that animal welfare expectations and standards are met and they play an important role in drafting animal welfare legislations/standards and in implementation of animal welfare programs.

24.3 SCOPE OF VETERINARY EDUCATION IN INDIA

The veterinary education aims not only at mitigating the ailments in animals, but in improving the public health also through prevention of zoonoses and food safety for the people. Further, this education not only addresses the health issues of animals (and to a great extent of human beings) but also the economic issues of the people and the country through safer and enhanced animal production. In fact economic issues influence all stakeholders more than any other issue. By catalyzing production of increased quantity and better quality of food, the vets help in improving the economy of the producers directly and of the entire country indirectly.

The base of veterinary education is basic degree of Bachelor of Veterinary Science and Animal Husbandry, that is, BVSc & AH. It has a vast curriculum covering broadly the fields of veterinary science and animal husbandry (VCI, 2008). Aim of veterinary education is to provide knowledge, training, and skills to the students in the areas of breeding, feeding and rearing, and health control of domestic animals kept for food, work or pleasure. Veterinary science component encompasses all activities related to clinical services for animals through veterinary clinics, that is, diagnosis and treatment of diseases, using the knowledge of the subjects such as Clinical Medicine, Surgery, Gynaecology & Obstetrics, and allied subjects like Pharmacology, Physiology, Pathology, Nutrition, Microbiology, Toxicology, and Parasitology, etc. Preventive medicine in veterinary education falls partly in veterinary science domain for prevention of epizootic diseases in animals and partly in the veterinary public health domain, where the veterinary knowledge and skills are applied for the promotion of health and prevention of zoonotic diseases among human beings.

According to Saunders (2007), veterinary epidemiology deals with distribution and determinants of animals such as health, welfare, production, related states, or events in specified populations, besides application of this study to control the health problems in animals as well as human population. Major thrust of study of this subject is to describe the animal health-related events in terms of its distribution in time, place, and animals. Similarly descriptive epidemiology, analytical epidemiology, observational epidemiology, and interventional epidemiology have been defined by Saunders (2007). A qualified vet is competent to undertake such studies and interpret their results.

The animal husbandry component of veterinary education encompasses all aspects of animal rearing and animal production. Animal rearing component deals with breeding and genetics, feeding and nutrition, and care and welfare of animals, etc. The animal production component deals with all aspects of animal production, including hygienic production of foods of animal origin such as milk, meat, poultry, eggs, etc. and their safety aspects along with animal care and welfare. Safety aspects of marine products are also handled by vets.

The main aim of veterinary education and training is to provide a minimum assurance that the qualified veterinarian has acquired at least following knowledge and skills adequately (Kechrid, 2014):

- Knowledge of all subjects of veterinary science on which activities of veterinarians are based, that is, knowledge of structure, functions, behavior, and physiological needs of animals.
- Skills of clinical, epidemiological, analytical competences required for prevention, diagnosis, and treatment of animal diseases, including zoonoses.
- Competences for preventive medicine, including those dealing with inquiries and certifications.
- Knowledge and skills of hygiene and technology involved in production of animal foodstuffs and foodstuffs of animal origin.
- Competences required for responsible use of veterinary medicinal products.

While describing the background of appointment of Official Veterinarians (OVs), the Animal & Plant Health Agency (APHA) announced plans to improve the quality and accessibility of training for OVs in July 2013 through introduction of a modern, flexible approach to training which reflects current business needs and changes to the veterinary sector (EU-FVO, 2015a,b). Previous system for training of OVs reflected a historic concept of veterinary practice structure and, with a variety of business models for veterinary practices now developing, APHA has reviewed the system of training and authorization to ensure it is relevant to this changing landscape. In addition, all state veterinary functions are now subject to audit by international organizations such as the FVO of EC.

A dedicated institution only for veterinary public health (VPH) was established in 2006 in Sweden, that is, European College of Veterinary Public Health (ECVPH), which has mainly two subspecialties, namely, population medicine and food science. As in the case of ECVPH, India and developing countries should now move up the ladder and start VPH education and training on a large scale, on the lines of ECVPH. As presented by Documenting Secretary (2015) in the case of ECVPH, developing countries should also aim to improve and promote the following:

1. The quality of animal health care and welfare by making available specialized knowledge and skills in the subspecialties of VPH, that is, "population medicine" and "food science" to the benefit of the animals.
2. The quality of veterinary practice through contacts of general practitioners with registered specialists.
3. The structure of animal health care through enhancing the application of formal "risk-assessment" procedures, quantitative problem analysis methods, systems of monitoring and surveillance at population level, food safety, and process quality management systems.
4. The structure of population medicine, risk management, and risk communication by improving the knowledge and perception of veterinarians, livestock owners, food processing industries, and the general public.

5. Consumer protection with regard to prevention and control of foodborne hazards and to food hygiene procedures.
6. Further development of VPH and its subspecialties, population medicine, and food science.
7. Integrated, multidisciplinary approach toward analysis, control, and prevention of hazards related to human and animal health.

24.4 ROLE OF VETS IN PUBLIC HEALTH THROUGH VETERINARY PUBLIC HEALTH IN IMPLEMENTATION OF PRINCIPLES OF RISK ANALYSIS IN FOOD SAFETY

As mentioned earlier, food safety is an essential public health function for the reduction of health risks from the entire food chain and ensuring food safety is an essential activity and an integral part of any public health program (WHO, 2000). Veterinary public health is a very important functional area of veterinarians being an important partner in most of the public health programs, which includes food safety activities through inspection of foods of animal origin and through appropriate lab tests, where possible. But earlier, almost entire food inspection in India used to be controlled by public health component of medical profession. Since there is commonality of interest between veterinary public health and public health and as the awareness has grown in policy makers as well as the general public, the vets are now increasingly being assigned with this responsibility and their role in food inspection, especially of animal origin, is being expanded.

Roles of veterinary experts in public health, as identified by McKenzie and Hathaway (2006), are as follows:

- development of public health policy for the competent authority;
- scientific evaluation of foodborne hazards and risk assessment;
- design, implementation, and verification of food controls at appropriate points in the food chain, including primary production;
- monitoring biological and chemical hazards at appropriate points in food chain;
- specialized veterinary inputs, for example, evaluation and control of antimicrobial-resistant zoonotic bacteria that may be transmitted by food;
- risk communication.

Food Safety and Inspection Service (FSIS) of US Department of Agriculture (USDA) is ensuring supply of safe foods of animal origin in USA for over a century. FSIS has a responsibility for preventing public health risks and managing the risks based on science by verifying systems. It has jurisdiction over products which are generating more than $120 billion in sales annually (Bousfield and Brown 2011). FSIS public health veterinarians perform various functions on day-to-day basis such as environmental sanitation, extended public health duties, antemortem inspection, humane handling of animals, postmortem examination, viscera disposal supervision/leading work unit teams, etc.

In many countries the role of the veterinary services has been extended to include entire food chain from "farm to fork." Food safety and quality are best assured by an integrated, multidisciplinary approach, considering the whole of the food chain. Eliminating or controlling food hazards at source, that is, a preventive approach, is more effective in reducing or eliminating the risk of unwanted health effects than relying on control of the final product, traditionally applied via a final "quality check" approach. The development of risk-based systems has been heavily influenced by the World Trade Organization Agreement on the Application of Sanitary and Phytosanitary Measures (SPS Agreement). This agreement stipulates that signatories shall ensure that their sanitary and phytosanitary measures are based on an assessment of the risks to human, animal, or plant life or health, taking into account risk-assessment techniques developed by relevant international organizations. The SPS Agreement specifically recognizes as the international benchmarks the standards developed by the OIE for animal health and zoonoses and for food safety by Codex Alimentarius Commission (CAC). The veterinary services play an essential role in the application of the risk-analysis process and in the implementation of risk-based recommendations for regulatory systems, including the extent and nature of veterinary involvement in food safety activities throughout the food chain (FAO/WHO, 2005). These services also contribute through auditing of animal and public health activities conducted by other government agencies, private sector veterinarians, and other stakeholders. Where veterinary or other professional tasks are delegated to individuals or enterprises outside the veterinary authority, clear information on regulatory requirements and a system of checks should be established to monitor and verify performance of the delegated activities. The veterinary authority retains the final responsibility for satisfactory performance of delegated activities.

Through their presence on farms and appropriate collaboration with farmers, the veterinary services play a key role in ensuring that animals are kept under hygienic conditions and in the early detection, surveillance, and treatment of animal diseases, including conditions of public health significance. The veterinary services may also provide livestock producers with information, advice, and training on how to avoid, eliminate, or control food safety hazards (e.g., drug and pesticide residues, mycotoxins, and environmental contaminants) in primary production, including through animal feed. The veterinary services play a central role in ensuring a responsible use of biological products and veterinary drugs, including antimicrobials, in animal husbandry. This helps to minimize the risk of developing antimicrobial resistance and unsafe levels of veterinary drug residues in foods of animal origin. Animal identification and animal traceability systems should be integrated in order to be able to trace slaughtered animals back to their place of origin, and products derived from them in the meat production chain. Vets play a major role in tracing back the root cause of food poisoning.

Another important role of the veterinary services is to provide health certification to international trading partners attesting that exported products meet both animal health and food safety standards. Certification in relation to animal diseases, including zoonoses, and meat hygiene is the responsibility of the veterinary authority.

McKenzie and Hathaway (2006), identified the key legislative responsibilities of veterinary services in the area of food hygiene as follows:

- establishment of policies and standards;
- design and management of inspection programs;
- scientific evaluation and risk assessment;
- assurance and certification that inspection and compliance activities are appropriately delivered;
- dissemination of information throughout the food chain—conformance with WTO obligations;
- negotiation of mutual recognition and equivalence agreements with trading partners.

Veterinary services play an important role in implementation of risk analysis (McKenzie and Hathaway, 2006) as follows:

1. Identification of food safety issue.
2. Prioritizing different food safety issues.
3. Identification and selection of risk management options
4. Balancing expectations in terms of minimizing risks against available food control measures, which may include reaching a decision about an appropriate level of protection. This process is facilitated mainly by veterinary services, with the help of industry and consumers.

According to Bellemain (2013), "The control of animal health and food safety has undergone profound changes and is now seen in terms of a global approach, 'from stable to table'. In terms of official controls, targeted control of the final food product has gradually been replaced by control of the production processes and an integrated approach to hazards throughout the production chain. This, in turn, has resulted in a new division of responsibilities among the producers (farmers), the manufacturers and the administration- namely the Veterinary Services. The areas in which veterinarians are involved have gradually been extended from animal production to all levels of the food production chain. Animal health interventions on farms are comparable to interventions in agri-food companies, where knowledge and competence in appropriate Quality Management System (QMS) should be part of Veterinary Education and training. To meet new challenges, the current trend is for Veterinary Services to be responsible for, or coordinate, sanitary interventions from the stable to the table. Coordination between Veterinary Services and other relevant authorities is a key component of good public governance, especially for effective action and optimal management of the resources available."

This is a very logical statement and needs to be taken up by the Indian Government seriously so as to achieve realistic results in the area of food safety in India, without which the implementation of Indian Food Safety and Standards Act (FSSA) 2006 and its associated regulations may not be possible in letter and spirit. Due to multifarious controls and unsystematic organization both at the national as well at the level of states, where veterinary public health and the public health are working in complete isolation at present, the food safety has become merely eyewash in India.

FSSA indicates the legal role of veterinarians in the area of food safety. Veterinary services play a key role in investigation and tracing back outbreaks of foodborne diseases up to farm level, using their professional knowledge and lab support by developing food safety plans for implementing the good agricultural practices (GAP), good manufacturing practices (GMP), and GHP, as per food laws and relevant food standards and to formulate remedial measures once the source of outbreak has been identified.

Bousfield and Brown (2011), had observed that efficient surveillance systems and risk-analysis system should be established to fight against foodborne zoonoses. In the "whole food chain" approach to food safety, appropriate controls should be established to prevent introduction of unacceptable levels of chemical hazards such as residues of veterinary drugs and pesticides at the time of primary production. Veterinary services are involved in ensuring good hygiene and good veterinary practices (GVP) in the use of veterinary drugs and food monitoring programs. Final responsibility for verification of the food safety program on an ongoing basis lies with veterinary services to control regulatory limits or procedures derived from risk assessment.

Veterinary professionals have to work in close coordination with primary producers and the government to overcome gaps due to lack of documentation to ensure safety of animal as well as public health (Birhanu et al., 2015). Veterinarians are present at every link in food chain since they have the knowledge and expertise to understand and audit the standards of animal health, animal welfare, and public health from stable to table (Bousfield and Brown, 2011). Bellemain (2013) studied the role of veterinary services in animal health and food safety surveillance and coordination with other services. Implementation of risk-based food hygiene program present many challenges in developing countries, due to which regulatory systems and scientific principles are not fully implemented. Development of risk-based standards on the basis of an integrated production-to-consumption approach to food hygiene ideally requires application of a well-designed risk-management framework (RMF). This, however, is difficult to achieve where there is limited interaction between veterinary public health, public health and animal health, leading to poor monitoring and feedback of information about zoonoses and other foodborne diseases.

Since veterinary services have technical capability to assess risks, they also carry out other processes of risk analysis, that is, risk management and risk communication for protection of human health. Emerging "risk-based" approach to food control demands has increased the involvement of veterinary services throughout food chain and systematic application of risk-management strategy when making decisions and taking regulatory action. In addition, vets are increasingly developing multidisciplinary skills that extend their activities well beyond the farm and initial processing of food such as prevention of environment from degradation by contaminating animal wastes and animal products and collection of new epidemiological information. The SPS Agreement (Rappard and Lausanne, 2010) and standards developed by CAC and OIE, all refer to the need for a systematic process to collect, evaluate, and document scientific and other information as the basis for public health and animal health controls.

A commitment to risk assessment as the basis for establishing food safety controls has placed new responsibilities and accountabilities on veterinary service components of competent authorities. Emerging "risk-based" approach to food control, demands increased involvement of veterinary services throughout the food chain and systematic application of risk management framework when making decisions and taking regulatory action (McKenzie and Hathaway, 2006). Vets play an irreplaceable role in these areas for effective food safety through animal health, animal welfare, preventing environmental degradation, collection, collation, evaluation, and proper documentation of scientific and other information for adequate public health and animal health controls; animal identification and trace-back systems; and prevention of bioterrorism (McKenzie and Hathaway, 2006). Hence, to achieve the objectives of food safety in developing countries like India, major responsibility of central government is to employ veterinary services properly to carry out risk analysis work to reduce foodborne public health problems.

As reported by Rue Defacqz (2012), the European Food Safety Authority (EFSA) was asked by EC in July 2009 to prepare concrete proposals which would allow the effective implementation of a modernized meat inspection system while making full use of risk-based principles. Federation of Veterinarians of Europe (FVE) has listed the basic principles for a modernized meat inspection system for evolving a new policy for this new modernized meat inspection system. Major thrust of these principles is to maintain consumer safety and their confidence in the controls.

Proposed modernized system requires checks of food safety systems on farm during primary production, to discover, eliminate, or reduce the potential hazards. This is best achieved through integrated herd or flock health planning and through a system of veterinary lead checks on farm which leads to important "Food Chain Information" (FCI) communication on farm standards to official veterinarian in slaughterhouse. As per Rue Defacqz (2011), Federation of Veterinarians of Europe (FVE) has suggested basic principles of modernized meat inspection system as follows:

- Providing professional advice, and carrying out inspection, audit, and enforcement tasks, since the veterinary services have four pillars of strength, namely, animal health, animal welfare, public health, and the environment; they play a key role in ensuring the safety of foods of animal origin from farm to consumer for which veterinarians have the education and necessary resources to enable them to fulfill their tasks.
- FVE offers leadership which focuses on ensuring consumer safety and confidence in the "official food controls."
- All veterinarians engage fully with the modernization process and make necessary changes where required.
- There is no single global answer to modernization. Each country or region, may have its own way to implement the equivalent "food safety objectives.";
- The new "official controls" should be based on peer reviewed science and should be risk based.

- There are differences in the ability of FBOs to take full responsibility for food hygiene. This needs to be acknowledged.
- FVE considers veterinary antemortem inspection of every animal, or group of animals, to be essential.
- FCI must be linked to a herd health plan and confirmed by veterinary checks on farm. This involvement of the veterinarian from "farm to fork," especially preharvest, is central to an integrated process control approach to food hygiene.
- Inspections required for third country trade should be seen as being additional to those required by the European Union for safety.
- Decisions on modernization should not be driven by cost reduction.
- Food hygiene controls are at least in part, for the public good. The charging system should reflect this.
- The consumer expects an independent body to ensure food safety.

The role of veterinary antemortem in the slaughterhouse is recognized as pivotal in interpretation of "FCI" from farm and in ensuring that risk assessment of status of the consignment of animals is correctly made and implemented. Therefore, it is unacceptable for EC to propose any dilution of the principle that the official veterinarian in slaughterhouse must see every animal or group of animals between its arrival at slaughterhouse and its slaughter. This inspection is not only for public health but also for the animal health and welfare. Possibility of a role for veterinarians on farm should be strengthened.

Old meat inspection system, that is, online inspection by official auxiliaries to the veterinarians, needs to be changed. New modernized meat inspection system is science based and for consumer safety. Consumer confidence can be built with integrated production systems, robust herd health programs, together with reliable FCI. For this reason, veterinary involvement in assurance of food safety in earlier stages of production chain needs to be strengthened. This can be best achieved through veterinary involvement in integrated herd health planning and through veterinary audit and verification of FCI. EFSA's proposal, if correctly implemented for modernization of meat inspection, has great potential to improve consumer safety. However, if is are not implemented in the integrated way, that is, from farm to slaughterhouse in the food safety assurance system, then the consequences for the health of the consumer could be serious.

24.5 ROLE OF VETS IN PREVENTING ANTIMICROBIAL RESISTANCE IN PUBLIC HEALTH

Illegal substances (such as clenbuterol, which is a "beta-2-agonist") are fed to livestock to assist growth and increases proportion of lean meat in the animal. By consuming residues of such drugs in meat, people may suffer from the symptoms of nervousness, fast heart rate, muscle tremors, etc. (Bousfield and Brown, 2011).

Borg (2014) (European Commissioner for Health), while assessing the emerging and pressing health challenges in Europe, discussed EU's food safety policy, which aims to ensure that European Union citizens enjoy safe and nutritious food produced from healthy plants and animals, while enabling the food industry to operate in the best possible conditions. One of the major challenges of European public health and food safety policies is antimicrobial resistance, since excessive and inappropriate use of antimicrobials in both the veterinary sector and in human healthcare is making many microorganisms resistant to these agents. As a result European Union is facing a growing problem of infections that cannot be treated. Another important challenge is tackling food waste to strengthen resource efficiency and reduce environmental impact of food and feed chain without compromising on human or animal health.

As per ECDC/EFSA/EMA (2015), the European Centre for Disease Prevention and Control (ECDC), the EFSA, and the European Medicines Agency (EMA) the consumption of several antimicrobials extensively used in animal husbandry was higher in animals than in humans, while consumption of antimicrobials critically important for human medicine (such as fluoroquinolones and 3rd- and 4th-generation cephalosporins) was higher in humans. Vets can and should play leading role in tackling these challenges for adequate food and public health safety.

24.6 CONCLUSIONS

Veterinary public health and public health departments are supposed to complement each other for achieving their common goal of promoting health and prevention of zoonotic and parasitic diseases among the human population through the utilization of veterinarians in food safety. However, at present, this role of veterinary public health (VPH) is quite limited and the Public Health and Veterinary Public Health departments are functioning in isolation with each other. "Food Safety Controls" are now being replaced by "Risk-Based Controls" in all stages of food production. Veterinarians are major stakeholders in undertaking "Risk-Based Controls" in Food Safety in developed countries. They are ideally placed in society to undertake the role of "Risk Based Controls." Since the role of Veterinarians in the context of Food Safety in India is vast, it needs to be strengthened and implemented for the sake of safety of public health of the food consumers in India, as discussed above.

REFERENCES

Bellemain, V., 2013. The role of veterinary services in animal health and food safety surveillance, and coordination with other services. Rev. Sci. Tech. 32 (2), 371–381, Available from: http://www.ncbi.nlm.nih.gov/pubmed/24547643.

Birhanu, T., Tesfaye, M., Ejeta, E., 2015. Review on roles of veterinary services in food safety of animal origin in Ethiopia. Nat. Sci. 13 (6), 93–99, Available from: http://www.sciencepub.net/nature.

Borg, T. (Ed.), 2014. Taking stock of EU public health, food safety, animal and plant health policy achievements 2010–2014. European Commissioner for Health. Available from: http://ec.europa.eu/health/docs/2010_2014_policy_achievements_en.pdf.

Bousfield, B., Brown, R., 2011. The veterinarian's role in food safety—Agriculture, Fisheries and Conservation Department, Chinese Academy of Agricultural Science. Vet. Bull. 1 (6), 1–16, Available from: http://www.afcd.gov.hk/english/quarantine/qua_vb/files/vfs_final.pdf.

Defacqz, R., 2011. The official veterinarian's role in food hygiene—an essential public good. Document No. FVE/11/PR/005. Federation of Veterinarians of Europe Secretariat, 1, 1000 Brussels, Belgium. Available from: http://www.fve.org/uploads/publications/docs/2011_12_005%20the%20role%20of%20ov_%20adding%20value%20v1.5_press%20release.pdf

Defacqz, R., 2012. Modernisation of meat inspection. Document No. FVE/12/docs/047c. Federation of Veterinarians of Europe AISBL General Assembly. November 16–17, 2012, Brussels. Available from: http://www.fve.org/veterinary/welfare.php

Dikid, T., Jain, S.K., Sharma, A., Kumar, A., Narain, J.P., 2013. Emerging & re-emerging infections in India: an overview. Indian J. Med. Res. 138, 19–31, Centenary Review Article, Available from: http://icmr.nic.in/ijmr/2013/july/0704.pdf.

Documenting Secretary, 2015. The European College of Veterinary Public Health (ECVPH), Sweden. www.ecvph.org/about-ecvph/the-council; Jeroen Dewulf, Council President; Ghent University, Belgium.

ECDC/EFSA/EMA, 2015. ECDC/EFSA/EMA first joint report on the integrated analysis of the consumption of antimicrobial agents and occurrence of antimicrobial resistance in bacteria from humans and food producing animals. EFSA J. 13(1), 114. Available from: http://ecdc.europa.eu/en/publications/Publications/antimicrobial-resistance-JIACRA-report.pdf

EU-FVO, 2015a. Report of EFSA Panel on Biological Hazards (BIOHAZ) EFSA Panel on Contaminants in the Food Chain (CONTAM) EFSA Panel on Animal Health and Welfare (AHAW). Against Question Number: EFSA-Q-2010-01469, EFSA-Q-2011-00110, EFSA-Q-2011-00019. EFSA J. 10(6), 179.

EU-FVO, 2015b. Creation of FVO. Available from: https://www.improve-ov.com/about/

FAO/WHO; 2005. Building effective food safety systems—strengthening official food safety control services (Agenda Item 4). Proceedings of Second FAO/WHO Global Forum of food safety regulators. Paper prepared by the Ministry of Agriculture, Nature and Food Quality and the Food and Consumer Product Safety Authority of the Netherlands; Agenda Item 4.2; GF 02/5b October 12–14, 2004, Bangkok, Thailand. Available from: http://www.fao.org/docrep/meeting/008/y5871e/y5871e00.htm#Contents

FDA 2013. FDA Food Safety Modernization Act. Under the provision of the Federal Food, Drug, and Cosmetic Act (21 U.S.C. 301 et seq.). U.S. Department of Health and Human Services, U.S. Food and Drug Administration (FDA). Available from: http://www.fda.gov/Food/GuidanceRegulation/FSMA/ucm247548.htm and http://www.fda.gov/Food/GuidanceRegulation/FSMA/ucm359436.htm

Hoelzer, K., Switt, A.I.M., Wiedmann, M., 2011. Animal contact as a source of human non-typhoidal salmonellosis. Vet. Res. 42, 34, Available from: http://www.veterinaryresearch.org/content/42/1/34.

Kechrid, F., 2014. The Role of Veterinarians in Animal Welfare and Inter-sectoral Collaboration. Conference of Federation of Asian Veterinary Associations. President, World Veterinary Association (WVA). November 26–30, 2014, Singapore. Available from: www.worldvet.org; ftp://ftp.fao.org/docrep/fao/meeting/008/y5871e/y5871e00.pdf

McKenzie, A.I., Hathaway, S.C., 2006. The role and functionality of veterinary services in food safety throughout the food chain, New Zealand Food Safety Authority, Wellington, New Zealand. Rev. Sci. Tech. Off. Int. Epiz. 25 (2), 837–848, Available from: http://citeseerx.ist. psu.edu/viewdoc/download?doi=10.1.1.118.746&rep=rep1&type=pdf.

OIE, 2012. Manual of Diagnostic Tests and Vaccines for Terrestrial Animals; Terrestrial Manual, seventh ed. vols. 1 and 2, 1404 pages. Available from: http://www.oie.int/ international-standard-setting/terrestrial-manual/; DL 8-8-15

Rappard, W., Lausanne, R., 2010. SPS Measures, WTO Agreement Series, 2010. World Trade Organization Centre, Geneva; Switzerland, Chapter 1211.

Saunders, 2007. Saunders Comprehensive Veterinary Dictionary, third ed. Saunders Elsevier, St. Louis, Missouri. Available from: http://medical-dictionary.thefreedictionary.com/ epidemiology.

Tuominen, P., 2009. Developing risk based food safety management. January 9, 2009. Academic Dissertation. Department of Food and Environmental Hygiene Faculty of Veterinary Medicine, University of Helsinki, Helsinki, Finland and Finnish Food Safety Authority, Evira Risk Assessment Unit Helsinki, Finland.

VCI, 2008. Veterinary Education Syllabus in India. Available from: http://www.pondiuni.edu. in/sites/default/files/downloads/51.pdf

WHO, 2000. Safer food for better health: Food Safety Programme—2002, World Health Organization. Eighth plenary meeting, May 20, 2000, Committee A, second report. Available from: http://www.who.int/fsf; http://apps.who.int/iris/bitstream/10665/42559/1/9241545747.pdf

Researchers and food safety

25

R.K. Gupta*, P. Dudeja**

*Department of Community Medicine, Army College of Medical Sciences, New Delhi, India;
**Department of Community Medicine, Armed Forces Medical College, Pune, Maharashtra, India

"The reasonable man adapts himself to the world; the unreasonable man persists to adapt the world to himself. Therefore, all progress depends on the unreasonable."
~ George Bernard Shaw

25.1 INTRODUCTION

Traditionally when we talk about research, a complex image of laboratories with burettes, chemicals, and microscopes emerges. But the diversity in the facets of research in the field of public health are simply mind-boggling; and research in food safety is no exception. Here the research that is carried out in the classical laboratories is merely the tip of the iceberg. The research on various aspects of food safety encompasses issues as varied as inventing newer, easier, and faster techniques for detecting food adulterants, developing information technology (IT) apps for effective inventory management, or developing "smart" environment-friendly self-destructive packaging material. Smart packaging is used to enhance/facilitate traceability of foods that confess their origin, traveling history, and temperature fluctuations endured through the use of a radio frequency identification device (RFID) sensor. Academic and office-based research on various policy, advocacy, legal, and statutory issues is also crucial in this regard. In fact, the idea of this very book had emerged as a direct off-shoot of a major food-safety research work undertaken by one of the editors of this book, Dr. Puja!

25.2 WHO IS A RESEARCHER?

If one believes the George Bernard Shaw statement quoted right in the beginning of this chapter, the "unreasonable man" is the researcher. But besides being unreasonable, a researcher needs to possess certain definite qualities.

One of my fellow participants at an International Conference, a Professor in Plant Pathology from the University of Copenhagen, presented an innovative e-poster on "Applying the health-system approach to plant healthcare," wherein she discussed, how the human primary healthcare approach could be used for the care of crops! While it seemed a "crazy" idea to me, she has already been working on it in Uganda for 7 years!

Even though Boudoulas puts it as, *"There is no substitute for talent,"* but those with moderate talent too, can think of research. Everyone saw the apple fall but it was only Newton with the keenest sense of observation who made laws out of it, which dictated mechanics forever. Ada Yonath, a Max Planck research scholar, did not care for a shelter and slept in her car for 7 long years while pursuing research on the chemical mapping of ribosomes. She was forced to find a place "to live," only when her mother visited her! This obsession paid off when she became the first Israeli woman to be awarded the Nobel Prize in 2009. She personified personal attributes of a researcher—single-minded devotion, passion, hard work, and dedication; amounting to obsession. In the mid-1800s the structure of benzene was an unsolved mystery, as the arrangement of six carbon atoms did not seem to fit in the then existing system of aliphatic organic compounds that were known to have only straight chains. Friedrich Kekule (1829–96), a German chemist was constantly intrigued while working for years on unraveling the mystery of the structure of benzene, discovered its ring shape after having a daydream of a snake seizing its own tail with its mouth, thus making a ring! This laid the foundation of study of aromatic hydrocarbons, an important branch of organic chemistry. This vision came to him after years of studying the carbon–carbon bonds, bringing to the forefront, the researcher's virtue of intense focus and constant meditation on the idea besides committed devotion.

Maria Curie (1867–1934) after decades of research, painstakingly separated 1 g of radium chloride emitting "pretty blue–green light" from a ton of pitchblende (ore) in 1902. She even died of aplastic anemia, contracted from the exposure to radiation. This talks of her diligence, dedication, and utter disregard to personal comforts and safety, besides of course discipline and extreme hard work.

While curiosity, persistence, and an open mind are the basic traits of a researcher, new understanding and challenging old ways of thinking is the crux of good research. A researcher has to be an avid reader, with sound knowledge of research methodology and statistics, fully aware of the latest work, open to suggestions, analytic, and critical. He cannot crave for instant results and has to be ruthlessly honest.

He has to be a good team member first and later, a good team leader, as leading a research team is equally important. Our ex-president Dr. APJ Abdul Kalam, the great scientific leader of the Indian Space Research Organization's rocket program and the brain behind Indian missiles, is a testimony to this fact. Those who know him closely, vouch that he was a greater "research-leader" than "researcher."

A curious and inquisitive mind along with all the personal attributes enumerated earlier might be good enough for the germination of a brilliant idea and pursuing it thereafter; but it is the environment that needs to nurture it through fertile soil so that it flourishes and bears fruit. Ample infrastructure, funding, and simple procedures are the basic requirements for a healthy research environment (Gupta, 2011).

25.3 NEED OF RESEARCH IN FOOD SAFETY

The world population has been rising and so has been the requirement of food. Concurrently, the pressure on farmland for urbanization has led to its shrinkage for agricultural use. This has led to a demand–supply gap in food produce and mankind has been exploring ways and means to "expand" the food production through whatever methods possible. Logically, the problem can be met through more production [better seeds, genetically modified (GM foods), and use of pesticides/fertilizers]; conserving food (better storage and packaging); tightening wastages (in harvesting, storage, transport, and distribution); and efficient management (inventory control and marketing). These mechanisms could be as straightforward as the use of current technology (which may be controversial at times, e.g., GM foods) or adopting unscrupulous ways to enhance profits (such as, food adulteration). Most of these modern processes struggle for more food availability in the shortest time, with maximum profitability entail a hit at food safety, if not regulated well. This calls for an ongoing need of monitoring these systems and research into various facets of food production, packaging, distribution, transport, marketing, and use, with an implication on food safety.

25.4 RESEARCH IN FOOD SAFETY

As elaborated, the canvas of food-safety research is so vast that it has to be painted in multiple colors with paintbrushes ranging from laboratory research, computer applications, and nanotechnology on one hand, to research into simple behavior-modification practices for the population to not get carried away by advertisements and select the right food to eat! The subsequent section tries to sum up the aspects of research in the major areas of food safety. By no means, does it claims to be comprehensive. It merely gives an indication of the diverse flavors that food research has to offer.

From planting of a crop until its consumption, there are many chances of contamination with harmful microorganisms, pesticides, and other toxic substances. On the farm, soil, manure, water, animals, equipment, and workers may spread such contaminants. Produce may be harvested on a farm, processed in one plant, repackaged in another, then stored, displayed, or served commercially or in the home. Each of these steps opens up a chance for harmful contamination of the food supply. Seeds, cultivation practices, and harvesting are the essential functions of food production. To a great extent food safety can be enhanced through appropriate research and the use of newer methods.

25.5 FOOD-SAFETY RESEARCH DURING FOOD PRODUCTION PROCESSES

25.5.1 GENETICALLY MODIFIED SEEDS AND FOODS

There has been a perception that GM seeds and foods are deleterious to health. However, there is a broad scientific consensus that the currently marketed GM seeds/

foods pose no greater risk than conventional foods. No reports of ill effects have been documented in the human population but the technique is new and concerns about its safety remain (USIMNRC, 2004). Yet, 20 out of 28 European Countries (including Switzerland and France) have said "no" to GMOs. A European Commission-funded scientist group chartered to set a research program to address public concerns about the safety and value of agricultural biotechnology indicated that *"the combination of existing test methods provides a sound test-regime to assess the safety of GM crops"* (IDEA, 2014). However, the consensus among scientists and regulators pointed to the need for improved testing technologies and protocols. This brings us to the fact that no matter how safe the GM foods are claimed to be, active research in GM foods is the order of the day.

25.5.2 FERTILIZERS

Much scientific research reports that manufactured fertilizers are not harmful directly to human/animal health. However, the environmental impact of over fertilization, eutrophication, acidification, nitrate pollution, and accumulation of chemicals (cadmium, fluorine, and radioactive elements) does have indirect health effects, which require research.

25.5.3 PLANT BREEDING

New and novel plant varieties of various fruits, vegetables (e.g., kiwi, berries, potato, and peas) are being bred not only for good health of food produce in terms of color, texture, and flavor but also for storage, pest, and disease resistance, which contribute to food safety. Modern research through conventional and advanced breeding techniques and using functional genomics techniques of molecular biology, genetic mapping, marker-assisted selection, bioinformatics, metabolic profiling, etc., is at play for best results.

25.5.4 PESTICIDES

Crop protection is the most limiting factor in crop production. The development of synthetic organic pesticides in the mid-1940s was a turning point in pest control but it brought health hazards along with it. Pesticide exposure can cause a variety of effects, such as, irritation of the skin and eyes and severe effects, such as, nervous system problems, reproductive problems, autism, and cancers. Many pesticides were banned (starting with DDT in 1973). Standards for pesticide residues were developed, a particularly important element of which, concerned the *maximum residue limit* of pesticides permitted on specific foods. These issues call for constant research (Bashour, 2010).

25.5.5 HARVESTING

While processes in food/crop harvesting are replete with compromised food safety (in terms of microbial, physical, chemical, and mechanical hazards), research in the field can substantially counter that. For example, research for molecular, immunologic, and epidemiologic approaches to identify new strains, genetic diversity, and antimicrobial profile of foodborne pathogens in fruits and vegetables, is currently underway. This enables testing of bioactive compounds from fruits and vegetables and their action against pathogens. Further, novel techniques are being developed to inactivate foodborne pathogens on the surfaces of fruits (and other produce).

25.6 POSTHARVEST

Many postharvest technological advances contribute to improved food safety. Smart packaging, in terms of biosensors and consumer-friendly packaging including ripe sensors, has been developed. Research on certain environment-friendly pest and disease treatment technologies, such as, "heat treatment" (water or air) for insect disinfestation or pathogen reduction can go a long way in food safety.

25.6.1 HI-TECH STORAGE

Composition of the internal atmosphere of a package can be modified to improve shelf life. It may involve lowering the amount of oxygen, slowing down the growth of organisms, and the rate of oxidation reactions. The removed oxygen can be replaced with nitrogen or carbon dioxide, which can lower the pH to inhibit growth of microorganisms.

25.6.2 TEMPERATURE-TRACKING DEVICES

Research has brought in temperature-tracking recorders to monitor products shipped in a cold chain and to validate the cold chain. Digital temperature-data loggers record the temperature history of food shipments. The data can be downloaded (through cable, RFID, etc.) to a computer for further analysis. These help in identifying any possible temperature abuse of products and in determining the remaining safe shelf life. They can also help in determining the time of temperature extremes during shipment so corrective measures can be taken.

Time-tracking indicators integrate the time and temperature experienced by the indicator and adjacent foods. Some use chemical reactions that result in a color change while others use the migration of a dye through a filter media. The degree of physical changes in the indicator match the degradation rate of the food and can indicate probable food degradation. The inks can also signal a desired temperature for consumers. There are certain dyes/inks put in the bottle of beer that change color when it reaches the right temperature to drink!

It is well known that cold chain/temperature is maintained for vaccines, meat, ice-creams, etc. But sky is the limit in the temperature-regulation technology. For example, gel packs are often used to actively keep the temperature of the contents within specified acceptable temperature ranges. Research to develop newer and more efficient packages with the ability to heat or cool the product for the consumer is being carried out. These will have segregated compartments where exothermic or endothermic reactions provide the desired effect. Self-heating cans are being made available for several products.

25.6.3 RADIO-FREQUENCY IDENTIFICATION TECHNOLOGY

Radio-frequency identification is applied to food packages for supply-chain control and creates a full real-time visibility of their supply chain. Radio-frequency identification chips are becoming more common as smart labels that are used to track and trace packages and unit loads throughout distribution. Newer developments include recording the temperature history of shipments and other intelligent packaging functions.

Special kinds of barcodes can be used for better tagging and identification of many products. Two-dimensional barcodes used in autocoding are now applied to food packaging to ensure products are correctly packaged and date coded.

25.7 ACTIVE PACKAGING

Terms, such as, active, intelligent, and smart packaging refer to packaging systems used with foods, pharmaceuticals, etc. They help extend shelf life, monitor freshness, display information on quality, and improve safety and convenience. Many innovations are being carried out by active packaging.

25.7.1 CONTAINERS FOR VISCOUS FLUIDS

The search for the best material with the ability of a package to fully empty a viscous food (e.g., oil/honey) is being carried out. Superhydrophobic material is useful for its ability to "repel" the viscous contents. The material can be further improved by using new lubricant-impregnated surfaces (Dainelli et al., 2008).

25.7.2 ANTIFOOD-DEGRADING PACKAGING

Some engineered packaging films contain antibacterial agents, enzymes, scavengers, and other active components to help control food degradation and extend shelf life.

25.7.3 ANTICORROSIVES PACKING

Anticorrosives can be applied to items to help prevent rust and corrosion. These can be provided inside a package or can be incorporated in a special paper. These organic salts condense on the metal to resist corrosion. Films are also available with copper ions in the polymer structure, which neutralize the corrosive gas in a package and deter rust. Oxygen controllers help remove oxygen from a closed package. Some may contain powdered iron, as iron rusts; oxygen is removed from the surrounding atmosphere. Newer systems can be built into package films or molded structures.

25.7.4 **MODIFIED ATMOSPHERE**

Some products, such as, cheese package is flushed with nitrogen prior to sealing; the inert nitrogen is absorbed into the cheese, allowing a tight shrink-film package. The nitrogen removes oxygen and "actively" interacts with the cheese to make the package functional. Similarly, other mixtures of gas may be used inside different food packages to extend the shelf life. The gas mixture depends on the specific product and its degradation mechanisms. Oxygen scavengers, carbon dioxide generators, ethanol generators, etc., can keep the atmosphere in a package at specified conditions.

25.7.5 **SHOCK DETECTORS**

Shock detectors have been available for many years. These are attached to the package or to the product in the package to determine if an excessive shock has been encountered. The mechanisms of these *shock-overload* devices have been spring-mass systems, magnets, drops of red dye, and several others. Recently, digital shock- and vibration-data loggers have been made available to accurately record the shocks and vibrations of shipment. These are used to monitor critical shipments to determine if extra inspection and calibration is required. They are also used to monitor the types of shocks and vibrations encountered in transit for the use in package testing in a laboratory.

25.7.6 **SECURITY**

A variety of security holograms are being developed that confirm that the product in the package is original. RFID chips are being used in this application also.

25.7.7 **BIODEGRADABLE PLASTICS**

Plastic packaging presently used is usually nonbiodegradable. Many biodegradable plastics are now being made, which include biodegradable films and coatings synthesized from organic materials and microbial polymers. Some package materials are even edible, for example, capsules made of natural substances, such as, gelatin, potatostarch, or other similar materials. Research is being carried out to develop newer bioplastics, films, and similar packaging products.

25.8 **CONTRIBUTION OF SPECIFIC BRANCHES OF SCIENCE TO FOOD-SAFETY RESEARCH**

We have discussed the application-based research innovations that have/are in the process of transforming the food-safety scenario. There are certain branches of science that either already have improved food safety or have a huge potential to improve food safety in the times to come.

25.8.1 MICROBIOLOGY

Probably no other science has contributed as much to the understanding and practice of food safety as microbiology has. From the effective production of wine since the times of early researchers, such as Louis Pasteur, to the control of food-poisoning epidemics, microbiology has been at the forefront of research. Production and safety of curd, yogurt, cheese, wine, bread, beer, pickles, fermented sausages, *idli–dosa*, are all possible because of microbiology. Food microbiology has been a major discipline that deals with the science of food spoilage and preservation. Now there is focus of using microorganisms and their products to combat pathogenic and food-spoilage microbes. Certain probiotic bacteria, including those that produce bacteriocins, can kill and inhibit pathogens. Purified bacteriocins, such as, nisin can be added directly to food products for maximum efficacy. Bacteriophage viruses that only infect bacteria are used to destroy bacteria.

25.8.2 LABORATORY MICROBIOLOGY

There have been major advances in the laboratory science of microbiology. From the basic bacterial culture and diagnostic techniques, which used to broadly indicate presence of a particular bacterium, the subsequent enzyme-linked immunosorbent assay technique could identify a particular antibody. Cutting edge advances, followed with the capability to even pick up antigens (of viruses/bacteria) conclusively recognize the organism DNA level, through the polymerase-chain reaction techniques. Many rapid tests have also evolved that can detect probable bacteria through rapid color indicators, thus giving valuable time to the clinician to act promptly in the eventuality of a foodborne-disease outbreak. Testing of food products and ingredients is also important through the entire food supply chain as possible flaws can occur at any stage of production. Apart from detecting spoilage, microbiologic tests can also determine germ content, identify yeasts and molds, bacteria, etc.

25.8.3 MEDICINE

The disciplines of medicine and pathology contribute to food safety through treatment and early diagnosis modalities. Advanced treatment techniques and newer drugs and antibiotics try to keep pace with the changing, mutating, and emerging bacteria and help effectively in the treatment of foodborne illnesses.

25.8.4 PUBLIC HEALTH

Public health has always been at the forefront in anticipating, trend forecasting, managing, controlling, investigating, and preventing foodborne-disease outbreaks. Public health specialists also help in awareness generation, information–education–communication, food-handler training, inspections, intervention tools for research, policy, and planning. Research in all these extensive fields is on going.

25.8.5 FORENSIC EXPERTS

The research in the disciplines of toxicology and forensic medicine deal with the intricacies of poisonous foods and plants.

25.8.6 VETERINARY SCIENCE

Veterinary medicine has a major stake in food safety with respect to animals, not only those which are consumed as food but also those that threaten the mankind from a distance, for example, fruit bats in Ebola! Diseases, such as, mad cow disease, avian influenza, etc., have of course been a subject of research and in future the emerging and reemerging diseases will be tackled through veterinary research.

25.8.7 INFORMATION TECHNOLOGY

One discipline of science that is not directly related to food safety or medicine, yet is all pervading and probably has the largest impact on food safety, is the science of IT. Anything and everything from the sowing and harvesting machine control consoles to the advances in RFIDs are an off-shoot of IT. Chapter 47 is dedicated to IT in food safety.

25.8.8 NANOTECHNOLOGY

Nanotechnology involves characterization, fabrication, and/or manipulation of structures, devices, or materials that are very small, in the range of 1–100 nm. This enhances polymer-barrier properties, making the material stronger, more flame resistant, with better thermal properties, and having favorable surface wettability and hydrophobicity. Nanotechnological innovations could produce remarkably new packaging concepts for barrier and mechanical properties, pathogen detection, and active and intelligent packaging.

25.8.9 BEHAVIORAL SCIENCES

While the major content of food safety is science based, there is a major stake of behavioral and social sciences. They contribute to food safety through research into understanding of issues, risk communication, information dissemination, education, changing/modifying public opinion, transforming ideas of stakeholders and opinion leaders, and influencing advocacy. Management of print/electronic and social media needs major understanding with constant supervision and research. This may also include consumer surveys on food preferences, behavior of finance and business operators/food handlers, and development of assessment tools pertaining to food safety.

25.9 **CONCLUSIONS**

Any discipline without a contemporary flavor becomes redundant and risks extinction. The same is true for food safety. With the emerging food-safety threats, a constant vigil has to be maintained to keep pace with the rapid march of microorganisms into mutations and drug resistance. Technologies for rapid detection and elimination of microorganisms, improving shelf life of products, improved smart packaging, minimizing environmental impact, and identifying consumer practices that compromise food safety remain some of the research priorities. Whether it was the prehistoric time with invention of wheel or the industrial revolution giving us trains or automobiles, "necessity" still remains the mother of invention.

REFERENCES

Bashour, I., 2010. Pesticides, fertilizers and food safety. Arab Environment: Future Challenges, 137–144.

Dainelli, D., Gontard, N., Spyropoulos, D., Zondervan-van den Beuken, E., Tobback, P., 2008. Active and intelligent food packaging: legal aspects and safety concerns. Trends Food Sci. Technol. 19 (1), 167–177.

Gupta, R.K., 2011. The Idea of research! Med. J. Armed Forces India 67, 215–216.

Irish Doctors' Environmental Association, 2014. IDEA position on genetically modified foods. Available from: http://ideaireland.org/library/idea-position-on-genetically-modified-foods/

United States Institute of Medicine and National Research Council, 2004. Safety of genetically engineered foods: approaches to assessing unintended health effects. National Academies Press, Washington, DC, pp. R9–10.

Role of hotel management and catering technology institutes in ensuring food safety

26

I. Kaur

Post Graduate Institute of Medical Education and Research (PGIMER), Community Medicine (School of Public Health), Chandigarh, India

Twenty-first century is an era of rapid globalization. There is a dominance of industrialization, service economy, and outsourcing in job markets. Standard of living and per capita income have increased across the globe. Even leisure-time availability has increased with technology advancement and mechanization. Modern lifestyle and improved transportation have led to a boom in hospitality and tourism industry. Human beings' natural urge and desire for new experiences, knowledge, and entertainment have also widened the tourism market. The demand for this sector is on a steep rise as more and more people are traveling across the world for trade, events, health care, meetings, education, recreation, adventure, sports, or religion-related outings. Expenditure on tourism has also induced demand for goods and services from allied sectors. It has also generated many new employment opportunities.

Today, India is one of the world's most dynamic economies. The hospitality and tourism sector is a contributor to gross domestic product and foreign exchange earnings (FEE) to India. This is quite evident from the latest statistical data on tourism. Number of foreign tourist arrivals (FTAs) in India during 2011 was 6.31 million which increased to 6.58 million in 2012. Tourism continues to play a major role for the country as the foreign exchange earner. FEE from tourism were US$ 17.74 billion (2012) as compared to US$ 16.56 billion (2011), registering a growth of 7.1%. The affluent population in India is rising. The domestic tourist visits in India during 2012 was numbered at 1036.3 million (provisional) as compared to 864.53 million in 2011, with a growth rate of 19.9%. During 2012, the number of Indian national departures from India was 14.92 million as compared to 13.99 million in 2011, reporting a growth rate of 6.7% (India Tourism Statistics, 2012). Foreign tourism arrivals registered a growth of 2.5% during the month of June 2013 as compared to June 2012. From January to June 2013, India received 3.3 million foreign tourists. It is expected that by 2017 this sector will generate around US$ 42.8 billion. India is currently ranked 68th in the list of the world's attractive destinations and 12th in the Asia Pacific region (*Statistics of hospitality industry in India*). During

Table 26.1 Evolution of Indian Hospitality and Tourism Sector

Before 1990	1990–2000	2000–05	2005 Onward
1. The National Tourism Policy was announced in 1982 2. In 1988, the Government of India formulated a detailed plan to promote tourism in the country	1. Government laid emphasis on private–public partnership in the sector 2. Government policies gave boost to the hotel industry	1. In 2002, a national policy on tourism was announced. It focused on developing robust infrastructure 2. Online travel portals and low-cost carrier airlines fillip domestic tourism	1. The government undertook various marketing initiatives to attract tourists 2. Domestic expenditure on tourism made up over 82% of the total tourism revenue in 2011 3. Foreign tourists visiting India per year rose to 6.3 million in 2011 from 3.9 in 2005

(Tourism and Hospitality, 2011)

travel, irrespective of the reason or destination, everyone needs basic amenities such as a safe and comfortable place to stay and good food. As the clientele of this industry is increasing so is the demand for hospitality sector. These growth drivers have led to increased hotel and catering market opportunities.

Hospitality and tourism is a multi billion dollar industry. It is one of the fastest-growing sectors in India (Table 26.1 depicts evolution of Indian hospitality and tourism sector). Hospitality means serving the guests to give them the best facilities, comfort, and a feel-good factor. The growth in tourism has increased the hospitality business. The subsegments of this industry are hotel, catering, restaurant, resorts, theme park, cruise, etc.

26.1 HOTEL AND CATERING INDUSTRY

The term "hotel" was first used in England in about 1760. A hotel is as an establishment that is expected to provide comfort to their guests. It is also referred to as "a house of entertainment for travelers." Their primary business is to provide lodging and boarding. Hotels also furnish food/catering services including restaurant, bars, coffee shops, buffet, and catering services.

Hotel sector has a wide range of establishments from guesthouses to five-star luxury hotels. They include independent, owner-run, government-run chains of hotels. Worldwide hotels are classified by different countries in many ways for different purposes. They are classified according to the location, ownership, types of client, standard facilities, star rating, and size (as depicted in Fig. 26.1). Many hotels fit into two or more categories to attract different types of guests.

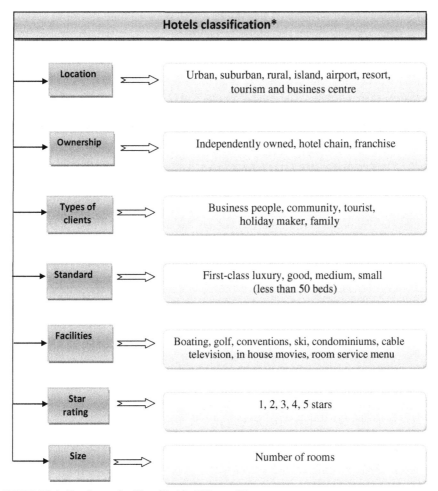

FIGURE 26.1 Hotels can be Classified in Different Ways

*A hotel may fit into more than one category.

Catering is defined as, "an art of providing food and beverages, aesthetically and scientifically, to a large number of people in a satisfactory and cost effective manner" (Sethi, 2008). This industry caters to people of all ages, and in varied situations and encompasses restaurants, cafeteria, catering firms, hotel caterers, etc. It is broadly of following two types:

1. *Off-premises catering*: The food is served away from the facility where food is cooked. The food and services are provided to clients hosting functions, parties, events, etc. The caterers have the following job duties and responsibilities:
   a. *Food preparation*: It is one of the most important duties of caterer to prepare the food aesthetically and hygienically. The food must not only be appealing and appetizing but also safe to consume.

   b. *Transporting the prepared food items*: The prepared food should be delivered in vehicles that are fully equipped to prevent spillage and spoilage.

   c. *Setting up and food service*: Once the food reaches its destination, the setup is established for serving food efficiently and hygienically.

   d. *Cleaning up*: The premises are cleared and clean.

2. *On-premises catering*: The caterer executes his duties at his own facility, for example, hotel caterer, hospital caterers, college, or school caterers. The food is prepared and served at the same place.

26.1.1 CATERING INSTITUTIONS ARE ALSO CLASSIFIED ON THE BASIS OF THE OBJECTIVES

1. *Commercial catering*: Their main goal is consumer satisfaction and profit making. The services must meet customers' requirement. They can accept or reject the service being provided. The establishments that come under this head are following:

   a. *Hotel catering*: These provide accommodation and food services. Food services/catering services are the lifeline of any hotel. Poor catering negatively affects hotel's image and hence its earnings. Food and beverage facility in hotels include all the restaurants, bars and lounges, coffee shops, banquets, etc.

   b. *Restaurants*: These are smaller variants of a hotel where lodging is not provided. These only provide meals. Banqueting is an important part in restaurant service. A new trend in restaurant business is the specialty restaurants. They have specific style menu. The décor is also according to the menu. Food specialty varies from Indian, Italian, French, Chinese, Thai, Mexican, or others.

   c. *Coffee shops*: These are established either as a separate entity or as a part of star luxury hotels. These days coffee shops are mushrooming with their growing popularity. They are casual eating and drinking places. They are the most happening places among the Indian youth.

   d. *Clubs*: They also offer foods and drinks. The services are open only for club members and their guests.

   e. *Transport catering*: In modern society people are on the move. This change has led to an explosive demand for catering on wheels.

     – *Train catering*: The food services provided depend upon length/time of the journey and the types of people traveling. They serve meals, beverages, and refreshments. They often follow cyclic menu. The focus is low-budget catering to retain popularity among majority of the passengers.

     – *Airways catering*: The airlines serve meals to passengers. Such meals are prepared, cooked, and supplied by caterers. Food is prepared on the ground and loaded into the aircraft. These caterers can be independent or linked with some hotels.

- *Marine catering*: The food prepared in ships varies with space and number of passengers, fuel engine, stores, etc. Food type, quality, and service also depend on the ticket price and passengers' willingness to pay.

f. *Catering functions*: They offer services for special parties/occasions such as weddings, receptions, birthdays, kitty parties, hen parties, events, meetings, etc. They set up services at various places such as open grounds, parks, community halls, and ballroom of the hotel. They provide the service of food and drink to the guests at these events. They may use generic label foods at such functions.

g. *Motels*: These are generally on roadsides. They serve simple foods that are quick to prepare and serve.

h. *Pubs*: The concept is based on public house in England. They serve all types of alcohol. Snacks may also be served. Music is played to create the ambience.

i. *Fast foods and take-away*: These are establishments to serve foods at affordable price with little or no waiting time. This section of catering industry aims at preparing food and beverages in least possible time. Mechanization of processes is done to save time. Many popular fast food joints work as chains through franchising.

j. *Small-scale eating establishments (dhabas) and street vendors*: These serve simple food. Generally few selected food items are prepared and served. Due to its affordability and location, the food is consumed by people from all walks of life. Often these are popular because of their distinctive taste patronized by customers. Easy access on road side/street is also their unique selling proposition.

2. *Noncommercial catering*: These catering establishments work for social welfare. They provide meals at subsidized rates. These include the following:

a. *Voluntary organizations*: In many social-welfare institutions voluntary organizations provide food and catering services for old age homes, orphanage, child-care centers, etc.

b. *Religious catering*: Religious belief in India is diverse. In most religions, "*Prasad*" is an edible religious offering consumed by worshippers. It is food that is considered sacred. It is first offered to a deity and then distributed among the followers. In fact people have great reverence for *prasad* offered at famous temples such as Vaishno Devi temple, Golden Temple, Balaji temple, etc. Apart from this *langar* (free kitchen) is a form of religious food service run in many temples. This form of food service is open for all people and without any discrimination. The food is prepared within the temple premises, usually by volunteers.

c. *Hospital catering*: The food prepared in hospital kitchen is for the patients. For many diseases food is prepared as per patients' requirement. Other than normal diets, therapeutic diets are provided to patients. These include diabetic diet, semisolid diet, liquid diet, renal diet, low sodium diet, low cholesterol diet, etc. The aim here is to promote healing and prevent complications related to some specific food items intake. Apart from this mess/canteen facilities are also there in hospital premises for doctors, nurses, and for relatives of patients.

d. *Industrial catering*: Industrialists have now realized that industry's product/service output is related to the welfare of the employees. So, most industries have separate canteens for employees and managers. These are not only a place to eat and drink but also act as social centers. A great sum of money is spent by industry owners in providing first-class kitchens and dining room to its employees.

Industrial catering is often a 24-h service depending on the duty timings. The manpower and investment is huge. Industrial caterers normally issue a contract to a company. A separate base kitchen is usually there for providing catering services.

e. *School/college catering*: Institutional catering may or may not be a commercial venture. In many schools these are run by private owners to earn profits, but they have to regulate their prices. Students are provided with healthy foods at controlled cost. The food served is expected to be palatable and nutritious at the same time.

The best example for absolute noncommercial catering is government's mid-day meal scheme in schools. This program aims to achieve enhanced nutritional status, as well as to improved enrolment, retention, and attendance of students in government schools. Other schools in India that are direct beneficiaries under this program are government-aided primary and upper primary schools, Education Guarantee Scheme (EGS)/Alternative and Innovative Education (AIE), and National Children Labour Project (NCLP) schools. However, there have been some reports of food poisoning in this scheme (Bihar school meal poisoning, 2013). Even fire incidents have been reported. In many cities (such as Chandigarh) the mid-day meal scheme has been outsourced to hotel management and catering technology institutes and government-run hotels.

In colleges, usually some private contractors run the catering services. These services are offered during college hours. The food items range from meals, snacks to various refreshments.

f. *Child-care-center meal scheme*: Another such nutrition supplementation services are given under the Integrated Child Development Scheme (ICDS) of department of Women and Child Development of Government of India. Meal is provided at *anganwadi* free of cost to children between 0 and 6 years of age and to pregnant and lactating women (*MDMS 2015*).

A separate Chapter is dedicated to food safety in canteens, messes, Mid Day Meal programme and ICDS in this book (Chapter 30).

All the aforementioned food services have thousands of job opportunities for hotel and catering work force of varying grades of skill.

All this has led to emergence of diverse needs of human resource in this sector (Fig. 26.2).

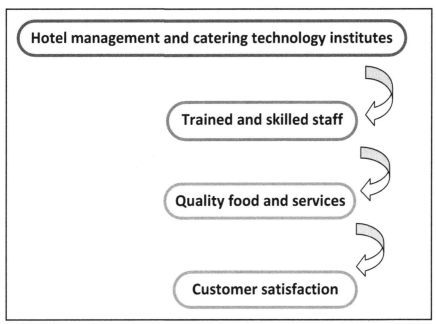

FIGURE 26.2 Emerging Needs of Hotel and Catering Industry

26.1.2 CUSTOMER SATISFACTION

The primary expectation from hospitality industry is customer satisfaction. This industry survives and flourishes on human beings' love for taste and efficiency in food/catering service. The end users of this industry are the clients. Their contentment with the food and services is directly related to the profit earnings. To achieve this goal, the food should not only taste good but also be of good quality.

26.1.3 FOOD SAFETY AND QUALITY CONTROL

Food safety and hygiene hold great value in hotel and catering industry where food is the prime focus. Food is highly vulnerable to contamination and adulteration due to both human interference and environmental/natural factors. Food is consumed not just for satisfying hunger, but also for taste and enjoyment. All the same, food safety is always a concern for food service provider. It is the first level of defense against food and waterborne diseases. If at any level of food handling, preparation, or storage this aspect is ignored or compromised it can result in major health problems among the customers apart from wastage and loses. The changing food pattern, international trade, public awareness, and expectation have further necessitated the demand for safe and high-quality food in hospitality sector.

Standardized product manufacturing in other industries is a simple task. The quality of product is generally made under controlled conditions and major work is mechanized and computerized. Each finished item is allowed to leave the premises only after inspection check. However, service businesses such as hotels and catering operate under an entirely different set of circumstances. It is not just the food but the total experience that guest pays for. Provision of standardized food products and services is difficult to achieve through mechanization due to involvement of various factors. High standard in quality of product and service is of primary importance. This can only be achieved through following:

- Stringent quality controls.
- Setting standards that match the needs and expectations of guests.
- Selecting skilled and motivated employees who are well trained and capable of achieving those standards.
- Conducting continual training and certification programs for the employees.
- Involving employees in structuring job descriptions, responsibilities, and setting standards.
- Having a feedback system will keep a check on the services provided by the managers and employees. This will help achieve what they intend to do and satisfy the guest.
- Rewarding employees and managers for achieving quality goals.

26.1.4 SKILLED MANPOWER

The human resource needs of hotel and catering industry have escalated the demand of skilled manpower. This industry now employs more people than any other industry in the world. There are ample employment opportunities that help in making this sector a viable and lucrative career option. In hotel, catering and allied sectors trained staff are required in departments such as food and beverage, housekeeping, accounting, marketing, and recreation among others.

Government and private units under this sector appoint qualified staff to carry out various operations smoothly such as cooking, serving, and housekeeping. This needs special manpower such as managers, chef, servers, housekeepers, kitchen workers, bartenders, and hospitality executives.

Trained and skilled personnel can play a great role in improving food safety and quality. Thus, it is vital to educate and train the hotel and catering industry personnel regarding healthy habits such as personal hygiene, food hygiene, and kitchen hygiene practices. It is also imperative to train them in food safety and related laws.

26.2 EDUCATION AND TRAINING OF HOTEL AND CATERING INDUSTRY PERSONNEL

In India, the supply of skilled and professionally trained workforce is insufficient, especially in hotel and catering sector. Everyone in this sector needs to have requisite knowledge and competence to provide standard and safe products and services. This has necessitated the opening of more hotel management and catering technology institutes in India for training requisite manpower.

There are many approved government and private institutes throughout the country that impart education and training in the field of hospitality and tourism sector. These are Institutes of Hotel Management & Catering Technology & Applied Nutrition (IHMs), The Indian Institute of Tourism & Travel Management (IITTM), Food Craft Institutes (FCIs), industrial training institutes, polytechnic institutes, universities, government colleges, government vocational schools, and private institutes.

26.3 GOVERNMENT INITIATIVE IN CREATING INFRASTRUCTURE FOR EDUCATION IN THIS SECTOR

From time immemorial people would eat food prepared by non-family members/cooks, for example, in India, kings had meals prepared in royal kitchen. Common man also consumed food outside homes like in temples, small eateries (during travel), sweet shops, etc. Unlike today, earlier eating establishments were not as diversified and formalized.

26.3.1 INDIAN INSTITUTE OF TOURISM AND TRAVEL MANAGEMENT (IITTM)

In 1993, IITTM was established as a registered society at New Delhi under the Ministry of Tourism. Its objective is to develop and promote education, training, and research in the field of travel and tourism. The institute has various Regional Centres at different places in India. The institute offers 2 years Post Graduate Diploma in Management (PGDM-Services/International Business/Tourism and Travel/Tourism and Leisure/Tourism and Cargo) programs.

26.3.2 NATIONAL COUNCIL FOR HOTEL MANAGEMENT AND CATERING TECHNOLOGY SOCIETY

In India, most of education and training in the field of hotel management and catering technology comes under the purview of National Council for Hotel Management and Catering Technology (NCHMCT) society, The Ministry of Tourism. There are more than 70 institutes registered under this council. All these Institutes follow a standardized course curriculum prescribed by the council for various professional programs. Food safety is an important part of this curriculum (A detailed list of institutes and food safety syllabus for various courses is given in Appendix 26.1 and 26.2, respectively).

Sensing the gap between "high-end" human resource development in hotel education and "run of the mill," routine working hand (at *dhabas*/"low end" restaurants), who do not have access to organized teaching and training, Government of India in 2009 started a new skill development program, namely, *hunar se* rozgar (skill to job).

It was launched for people who wish to work within hotel and catering industry but do not have formal education or experience in order to get jobs. This initiative has added immensely to food safety.

26.4 IMPORTANCE OF FOOD SAFETY EDUCATION AND TRAINING IN HOTEL AND CATERING INDUSTRY

Mastering the right skills is very important for smooth and safe functioning of this industry. The food safety concerns have become a major health issue. This aspect is taught as an important component in these institutes. The importance of these training institutes has been further enhanced due to recent changes in food safety laws in India.

To avoid ambiguity and for strict implementation of food laws, the Government of India passed Food Safety and Standards Act, 2006, which consolidates all the existing food regulatory laws under one act. Under this act and Food Safety and Standards Rules (FSSR), 2011 all the eating establishments are required to follow and fulfill the recommended guidelines.

As per the new FSSR 2011, all eating establishment where food is handled, processed, prepared, packed, stored, or even distributed should follow strict hygiene and sanitary measures, and other food safety laws/standards of their respective countries.

It is the responsibility of the FBOs to ensure that all food safety standards are being adhered to by all personnel of his establishment. It is also their duty to identify the steps followed in food business and also the critical control points that can help ensure better management, and implementation of food safety measures.

There is also a provision of punishment/penalty if the provisions of FSSR 2011 are violated. All eating establishments and potential job seekers in food and catering line must give due importance to implementing food safety measures and standards. Hence, requisite training of food handlers of these establishment in food safety is desirable to ensure food quality and thereby customers' health.

The recent changes in food safety laws have also forced FBOs to follow Hazard Analysis Critical Control Point (HACCP) system or a food safety system in their establishment. Implementation of HACCP requires lot of investment and expenditure on many resources (including monetary and human). In India, only few big business houses can afford to follow this expensive system. To maintain profit margin HACCP following establishments generally provide food and services at very high price. Such luxuries and high cost can only be afforded by elites or by upper middle class section of the society.

HACCP in India is followed only by four- and five-star hotels or heritage resorts and not by other hotels, catering and eating establishments, roadside vendors, etc.

To overcome the issue of maintaining safe and basic hygiene standards in absence of trained professionals, small-scale eating establishments can appoint professionals who are formally trained in food hygiene for training unskilled employees to in-house food safety–related aspects. The food establishment employees can be trained at different levels, depending on their job and type of food they handle. Staff handling "high-risk" foods requires more training than those who handle "low-risk" foods. All food handlers and staff should be trained and must receive written or verbal instruction in "the essentials of food hygiene" and "hygiene awareness instruction." In addition, staff must be instilled with the knowledge of how to do their particular job hygienically.

Thus, high standards of hygiene and food safety in all establishments can be ensured by imparting nonformal (in-house) training and periodic refresher courses in basics of food safety and hygiene.

APPENDIX

Appendix 26.1 A detailed List of Institutes (Indian tourism statistics 2012, NCHMCT Annual Report 2012-13))

A. List of Academic Courses/Programmes Offered Under NCHMCT Registered Institutes

S. No.	Academic Programmes	Duration
1.	M.Sc. in Hotel Administration	2 years
2.	Post Graduate Diploma in Accommodation Operations & Management	1½ years
3.	B.Sc. in Hospitality and Hotel Administration - Generic	3 years
4.	B.Sc. in Hospitality and Hotel Administration - Specialization	1½ years
5.	Diploma in Food Production	1½ years
6.	Diploma in Food & Beverage Service	1½ years
7.	Diploma in Bakery & Confectionery	1½ years
8.	Diploma in Front Office Operation	1½ years
9.	Diploma in Housekeeping	1½ years
10.	Craftsmanship Certificate Course in Food Production & Patisserie	1½ years
11.	Craftsmanship Certificate Course in Food & Beverage Service	24 weeks

Indian tourism statistics 2012, NCHMCT Annual Report 2012–13.

B1. List of Functional Central Institutes of Hotel Management

S. No.	Name of the Institute	Place
1.	Institute of Hotel Management & Catering Technology	Bengaluru
2.	Institute of Hotel Management & Catering Technology	Bhopal
3.	Institute of Hotel Management & Catering Technology	Bhubaneswar
4.	Dr. Ambedkar Institute of Hotel Management	Chandigarh
5.	Institute of Hotel Management & Catering Technology	Chennai
6.	Institute of Hotel Management & Catering Technology	Delhi (Pusa)
7.	Institute of Hotel Management & Catering Technology	Gandhinagar
8.	Institute of Hotel Management & Catering Technology	Goa
9.	Institute of Hotel Management & Catering Technology	Gurdaspur
10.	Institute of Hotel Management & Catering Technology	Guwahati
11.	Institute of Hotel Management & Catering Technology	Gwalior
12.	Institute of Hotel Management & Catering Technology	Hazipur
13.	Institute of Hotel Management & Catering Technology	Hyderabad

B1. List of Functional Central Institutes of Hotel Management (cont.)

S. No.	Name of the Institute	Place
14.	Institute of Hotel Management & Catering Technology	Jaipur
15.	Institute of Hotel Management & Catering Technology	Kolkata
16.	Institute of Hotel Management & Catering Technology	Lucknow
17.	Institute of Hotel Management & Catering Technology	Mumbai
18.	Institute of Hotel Management & Catering Technology	Shillong
19.	Institute of Hotel Management & Catering Technology	Shimla
20.	Institute of Hotel Management & Catering Technology	Srinagar
21.	Institute of Hotel Management & Catering Technology	Thiruvananthapuram

B2. List of Functional State Institutes of Hotel Management

S. No.	Name of the Institute	Place
1.	Institute of Hotel Management & Catering Technology	Dehradun
2.	Institute of Hotel Management & Catering Technology	Gangtok
3.	Institute of Hotel Management & Catering Technology	Jodhpur
4.	Institute of Hotel Management & Catering Technology	Delhi (Lajpat Nagar)
5.	Institute of Hotel Management & Catering Technology	Chandigarh
6.	Institute of Hotel Management & Catering Technology	Kurukshetra
7.	Institute of Hotel Management & Catering Technology	Kozhikode
8.	Institute of Hotel Management & Catering Technology	Faridabad
9.	Institute of Hotel Management & Catering Technology	Trichirapalli
10.	Institute of Hotel Management & Catering Technology	Bhatinda
11.	Institute of Hotel Management & Catering Technology	Sylvassa
12.	Institute of Hotel Management & Catering Technology	Hamirpur
13.	Institute of Hotel Management & Catering Technology	Puducherry
14.	Institute of Hotel Management & Catering Technology	Rohtak
15.	Institute of Hotel Management & Catering Technology	NITHM (Andhra)

B3. List of Functional Food Craft Institutes

S. No.	Name of the Institute	Place
1.	Food Craft Institute	Ajmer
2.	Food Craft Institute	Aligarh
3.	Food Craft Institute	Balangir
4.	Food Craft Institute	Darjeeling
5.	Food Craft Institute	Udaipur
6.	Food Craft Institute	Hoshiarpur
7.	Food Craft Institute	Nawgaon

Appendix 26.2 Food Safety Syllabus for Various Courses (Details of other courses and their curriculums are available at: http://nchm.nic.in/)

A. M.Sc. In Hotel Administration

Semester I	MHA-1	Management Functions & Behaviour in Hospitality
	MHA-4	Information Management Systems & Hospitality
	MHA-2	Hospitality Management
	MHA-3	Properties Development & Planning

Semester II	MHA-5	Revenue/Yield Management
	MHA-6	Market Research
	MHA-7	Equipment & Materials Management
	MHA-8	Managing Entrepreneurship : Small & Medium Business Properties
Semester III (Sales & Marketing)	MHA-9	Sales Management
	MHA-10	Principles of Marketing Management
	MHA-11	Marketing of Services and Consumer Behaviour
	MHA-12	International Marketing
	MHA-21	Mentorship & Research Project
Semester III (Human Resource Management)	MHA-13	Human Resource Planning
	MHA-14	Union Management Relations
	MHA-15	Managing Change in Organizations
	MHA-16	Social Processes and Behavioural Issues
	MHA-21	Mentorship & Research Project
Semester IV	MHA-17	Production & Operations Management
	MHA-18	Managerial Economics
	MHA-19/20	Sales & Marketing or Labour Laws
	MHA-21	Dissertation

Details of other courses and their curriculum is available from: http://nchm.nic.in/

B. B.Sc. degree in Hospitality & Hotel Administration

Semester I	BHM-111	Foundation Course in Food Production – I
	BHM-112	Foundation Course in Food & Beverage Service – I
	BHM-113	Foundation Course in Front Office Operations – I
	BHM-114	Foundation Course in Accommodation Operations – I
	BHM-105	Application of Computers
	BHM-106	Hotel Engineering
	BHM-116	Nutrition
Semester II	BHM-151	Foundation Course in Food Production – II
	BHM-152	Foundation Course in Food & Beverage Service – II
	BHM-153	Foundation Course in Front Office Operations – II
	BHM-154	Foundation Course in Accommodation Operations – II
	BHM-117	Principles of Food Science
	BHM-108	Accountancy
	BHM-109	Communication
	TS-01 (IGNOU)	Foundation Course in Tourism
Semester III & IV	BHM-201	Food Production Operations
	BHM-202	Food & Beverage Service Operations
	BHM-203	Front Office Operations
	BHM-204	Accommodation Operations
	BHM-205	Food & Beverage Controls
	BHM-206	Hotel Accountancy
	BHM-207	Food Safety & Quality
	-	Research Methodology
	TS-03 (IGNOU)	Management in Tourism
	BEGE-103 (IGNOU)	Communication Skills in English
	TS-07 (IGNOU)	Human Resource Development

B. B.Sc. degree in Hospitality & Hotel Administration (cont.)

Semester V	BHM-311	Advance Food Production Operations – I
	BHM-312	Advance Food & Beverage Operations – I
	BHM-313	Front Office Management – I
	BHM-314	Accommodation Management – I
	BHM-307	Financial Management
	BHM-308	Strategic Management
	BHM-309	Research Project
		Special topics/Guest speakers
	TS-06 (IGNOU)	Tourism Marketing
Semester VI	BHM-351	Advance Food Production Operations – II
	BHM-352	Advance F&B Operations – II
	BHM-353	Front Office Management – II
	BHM-354	Accommodation Management – II
	BHM-305	Food & Beverage Management
	BHM-306	Facility Planning
	BHM-309	Research Project
		Special topics/Guest speakers

C. Diploma in Food Production

Semester I & II	Computer Awareness
	Commodities and Costing
	Hygiene and Nutrition
	Larder
	Cookery

D. Post Graduate Diploma in Accomodation Operations & Management

Semester I	AOM-11	Accommodation Operations
	AOM-12	Front Office Operations
	AOM-13	Supervisory Management
	AOM-14	Accountancy
	AOM-15	Communication
		Guest Speaker/Field Visits
Semester II	AOM-21	Accommodation Operations
	AOM-22	Front Office Operations
	AOM-23	Interior Decoration
	AOM-24	Hotel Accountancy & Costing
	AOM-25	Business Communication
	AOM-31	Industrial Training (16 weeks × 6 days × 8 hours)
		Guest Speaker/Field Visits

FURTHER READING

ASA, 2013. A brief report on tourism in India—ASA (2013). Available from: www.asa.in/pdfs/surveys-reports/Tourism-in-India.pdf

FDA, 2014. Managing Food Safety: A Manual for the Voluntary Use of HACCP Principles for Operators of Food Service and Retail Establishments. Available from: http://www.fda.gov/Food/GuidanceRegulation/HACCP/ucm2006811.htm

FSSAI, 2006. FAQ's on FSSA 2006. Available from: http://fssai.gov.in/Portals/0/Pdf/FAQ.pdf

FSSAI, 2010. The Training Manual for Food Safety Regulators Who are Involved in Implementing Food Safety and Standards Act 2006 Across the Country. Available from: www.fssai.gov.in

Hospitality and catering. The UK hospitality industry. Available from: http://www.pearson-schoolsandfecolleges.co.uk/Secondary/Vocational/HospitalityandCatering/WJECGCSE-HospitalityandCatering/Samples/Samplepages/WJECGCSEHospitalitySamplePages.pdf

Hospitalityindia, 2012. Statistics of hospitality industry in India. Available from: http://www.hospitalityindia.com/hospitality-industry-in-india.htm

Hotel definition. Available from: http://dictionary.reference.com/browse/hotel

IBEF, 2011. Tourism and hospitality. Available from: www.ibef.org

ICHM, in press. Introduction to hospitality industry. Introduction: Scope & Nature of Hotel Management. Available from: http://thecareersguide.com/download/samples/Hotel_Management_BTR202/lesson%201.pdf

India Tourism Statistics, 2012. Government of India, Ministry of Tourism, Market Research Division. (2012). Available from: http://www.tourism.gov.in/TourismDivision/AboutDivision.aspx?Name=Market%20Research%20and%20Statistics.htm

MDMS, 2015. National Programme of Mid Day Meal in Schools (MDMS). Annual Work Plan & Budget 2015-16. Available from: http://mdm.nic.in/Files/PAB/PAB-2015-16/Tripura/1_Annexure-II_AWPB_ Writeup_2015-16_for_State_Plan.pdf

National Council for Hotel Management & Catering Technology, Noida. Syllabus. Available from: http://nchm.nic.in/coursedetails/index/1

NCHMCT, 2012. NCHMCT brochure 2012. National Council for Hotel Management and Catering Technology. Available from: http://specialtest.in/nchmct2012/NCHMCT2012-Brochure.pdf

NCHMCT, 2013. Annual report for the year 2012–2013. Available from: http://www.nchm.nic.in/nchmct_adm/writereaddata/upload/annualreports/Annual%20Report%202012-13.pdf

NCHMCT, in press. List of affiliated private institutes of hotel management. Available from: http://www.nchmct.org/affiliates.htm

Operational Guidelines for Development of Skilled Work Force By Classified Hotels under Hunar Se Rozgar. (2012) http://tourism.gov.in/writereaddata/CMSPagePicture/file/Primary%20Content/HRD/rozgarsena.pdf

Reimers, F., 1994. HACCP in retail food stores. Food Control 5,176–180. Available from: http://www.sciencedirect.com/science/article/pii/0956713594900795

RIHMCT, in press. List of affiliated state govt. sponsored institutes of hotel management. Available from: http://www.rihmct.edu.in/pdf/Affilated_Govt.Private_Instititue.pdf

Seth, M., 2006. Catering Management: An Integrated Approach, second ed. New Age International (P) Ltd, New Delhi.

Sethi, S. Land of opportunities—food industry in India. Available from: http://www.academia.edu/5608094/Land_of_Opportunities_-_Food_Industry_in_India_Report_by_Sanjay_Sethi

Whitelaw, P.A., Barron, P., Buultjens, J., Caircross, G., Davidson, M., in press. Training needs of the hospitality industry. Available from: works.bepress.com/jeremy_buultjens/138

Food safety in large eating establishments

Food safety in large organized eating establishments

27

P. Dudeja*, A. Singh**

**Department of Community Medicine, Armed Forces Medical College,
Pune, Maharashtra, India; **School of Public Health, Post Graduate
Institute of Medical Education and Research, Chandigarh, India*

27.1 INTRODUCTION

With globalization there has been a substantial rise in income of the Indian middle class. This increase in purchasing power has had a significant impact on the eating patterns of the people. The age-old tradition of eating home-cooked food together in the family has been replaced by eating out culture. Eating out which was earlier an infrequent activity has become a regular feature of present lifestyle. Eating out is defined as the "consumption of all foods, taking place outside one's own household." Reasons are many; ranging from lack of option for a home-cooked meal for migrants to having an enjoyable experience with friends/family. Factors such as staying away from home, working women, and increase in food varieties have contributed to this phenomenon of eating out. Advancements in food technology are fast contributing to this growth. However, the practice of eating out has many food safety implications as often people experience Food Borne Illness (FBI) as a result of eating contaminated food in these Eating Establishments (EEs). Food Safety and Standards Authority of India (FSSAI) is the statutory regulatory authority having the mandate for laying down standards for articles of food and to regulate their manufacture, storage, distribution, sale and import, to ensure availability of safe and wholesome food for human consumption. EEs can be classified based on their size and type of items served (Fig. 27.1) or services provided (Fig. 27.2).

27.2 FOOD SAFETY REQUIREMENTS IN AN EE

Food safety requirements in an EE comprise the last segment of farm to fork chain (kitchen to fork). They can be broadly categorized into requirements directly related to cooking (like preparation of raw material, cooking, hot and cold holding, cooling and storing, serving and reheating of leftover food) and those indirectly related to

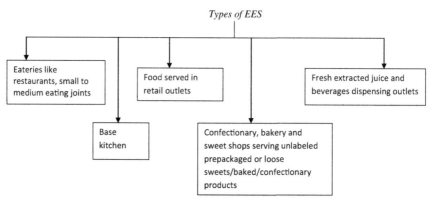

FIGURE 27.1 Types of EEs Based on Size and Type of Items Served

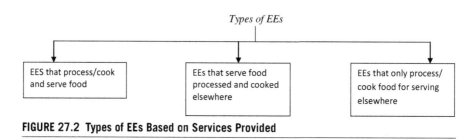

FIGURE 27.2 Types of EEs Based on Services Provided

cooking (hygiene of water and ice, transportation of prepared food, equipment and utensils, cleaning and sanitizing, waste disposal pest control) and hygiene and training of food handlers. Food safety precautions related to cooking and food handlers have been discussed in separate chapters. The current chapter describes issues indirectly related to cooking.

27.2.1 SAFE WATER

Use of safe water is critical to ensure safety of food. Water is required during cooking, processing, washing of equipment, utensils, containers, kiosks, hand washing and preparation of ice. Hence, it is imperative that all EEs should have access to sufficient quantity of safe water at close distance or at one's own premises. In case water is stored it needs to be in sanitary state in close containers with taps for withdrawal. The water containers including water dispensers should be regularly emptied and cleaned regularly and dried (by turning upside down) at the end of day's sale wherever possible. In case nonpotable water systems/containers exist, for cleaning and maintenance they should be identified and labeled. It should not connect with, or allow to reflux into potable water supply under any circumstances. All precautions to ensure safety of ice during preparation/transportation/storage must be taken. No food or beverages should be stored in a container that is intended to store ice.

27.2.2 **TRANSPORTATION AND HANDLING OF PREPARED FOOD**

The broad category of EEs that only process/cook food for serving elsewhere actually come under the broad umbrella of catering business. These provide cooked food services at strategic places such as hotels, public houses, events, weddings and parties, etc. Apart from hygiene and safety measures which are to be in place at the base kitchen, it is crucial to ensure safety of cooked food during transportation. Precautions to be taken for safe transportation of cooked food by these EEs can be classified into two groups: general and specific (Fig. 27.3). Furthermore, proper documentation accompanying each load (i.e., time of dispatch, time in transit and time of receiving, temperature readings, etc.) will ensure safety of food. Even documentation of the food received esp. food spoiled during transportation needs to be maintained. Last but not the least awareness and training of drivers will further help in transporting food safely.

27.2.3 **VENDING/SELLING UNITS**

These units commonly serve beverages like coffee, aerated beverages like fountain Pepsi/Cola, etc. or ready-to-eat items like sandwiches/*vadapav, pakoras, idlis, chaanakulchas,* etc. Design, construction and maintenance of all vending units should be

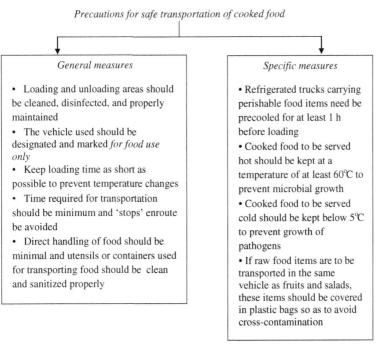

Precautions for safe transportation of cooked food

General measures	*Specific measures*
• Loading and unloading areas should be cleaned, disinfected, and properly maintained • The vehicle used should be designated and marked *for food use only* • Keep loading time as short as possible to prevent temperature changes • Time required for transportation should be minimum and 'stops' enroute be avoided • Direct handling of food should be minimal and utensils or containers used for transporting food should be clean and sanitized properly	• Refrigerated trucks carrying perishable food items need be precooled for at least 1 h before loading • Cooked food to be served hot should be kept at a temperature of at least 60°C to prevent microbial growth • Cooked food to be served cold should be kept below 5°C to prevent growth of pathogens • If raw food items are to be transported in the same vehicle as fruits and salads, these items should be covered in plastic bags so as to avoid cross-contamination

FIGURE 27.3 Precautions for Safe Transportation of Cooked Food

such that they can be cleaned and sanitized totally (to ensure no buildup of residues). Various precautions which needs to be taken at these vending units are:

- Vending stalls should have cover, namely, canopy, umbrella, etc.
- Personal belongings, such as clothes, footwear, etc. should not be kept at the vending unit, preparation, and storage and sale area.
- Vending units/sale/preparation area should be at least 30 cm off the ground and suitable for the quantity of food prepared, handled, and stored.
- Utensils used for raw food should be separated from that of prepared foods and should be sanitized properly before each use.
- Single-use/disposable items such as straws, paper towel, disposable cups and plates shall not be reused.
- Reusable serving utensils/items should be in good condition and should be washed, cleaned, and disinfected after each use.
- Utensils/equipment should be air dried or clean and sanitized cloth should be used for wiping, wherever required.

27.2.4 REQUIREMENTS AT THE POINT OF SALE

This is the last and final stage of the chain operation. At the point of sale it must be ensured that food is protected from dust, wind, rain, strong sun and flies and insects. The counter display of cold foods should be at 5°C or below and hot foods at 60°C or above (Garden-Robinson, 2009). Any item left at ambient temperature for more than 2 h should be discarded. In case the vendor is using disposable gloves these shall be discarded after every use. Many a times as a cost cutting measure these vendors don't change gloves and this gives them and the consumer a false sense of security. A dirty glove is even worse than a dirty hand. It is a common practice to use old newspapers for storing and packing food items. This must be discouraged and aluminum foil be used instead for display and packaging of cooked items.

27.2.5 ENVIRONMENT AND SURROUNDINGS

It should be ensured that food procured and prepared hygienically do not get contaminated due to an unclean and unhygienic environment, hence, all areas used for preparation and display and sale of food should have clean surroundings. Various precautions which need to be addressed are:

- The food preparation and selling areas are clean, dry, well-lit, and hygienic with proper ventilation system in place and should be in an airy environment and not in a damp and wet place.
- All wastes should be taken from these areas regularly.
- There should be no open drains, garbage stacks, or public latrines near the area.
- In the food establishment, the toilet should not open directly into the food processing, display, or selling area.
- No one should be allowed to spit or wash hands/face/body near the area.

- No wastewater should accumulate or run through the area.
- Cooked food selling areas should be separate from raw food selling areas.
- Adornments, such as vases with or without flowers or plants and other items should be maintained in such a way that they do not represent a source for food contamination.

27.2.6 WASTE DISPOSAL AND PEST CONTROL

Waste disposal (organic and other) is critical to keep food and beverage safe at every point of the chain, waste at no point should come in contact with the food directly or indirectly (through flies, insects, etc.). Hence, it has to be ensured that biodegradable and nonbiodegradable wastes; liquid and solid wastes should be separated right at the point of putting them into the bins. All garbage cans should be covered, cleaned daily, sanitized and collected at an assigned collection point at a public garbage collection system, and placed at sufficient distance to avoid food contamination. Adequate pest control measures should be in place.

27.2.7 CLEANING AND SANITIZING

Cleaning and sanitizing at every point of the preparation chain has to be ensured by using proper cleaning agents, methods and cleaning schedules. Cleaning of EE involves removal of soil, food residue, dirt, grease, or other objectionable matter. Separate cleaning materials, including cloths, sponges, and mops should be used for the designated clean area. Use of disposable, single-use cloths is recommended wherever possible. Effective cleaning is essential to get rid of harmful bacteria and stop them spreading to food (Edrees, 2014). Water alone is not a very efficient cleaning agent because of its high surface tension. Adding of detergent to water facilitates the contact between water and surface soil, allowing better penetration into soil by lowering surface tension.

Disinfection is the killing of infectious agents outside the body by direct exposure to chemical or physical agents. However, chemical disinfectants only work if surfaces have been thoroughly cleaned first to remove grease and other dirt. For effective disinfection, it is important to first clean the surface and remove visible dirt, food particles and debris, and then rinse to remove any residue. After this step, application of a disinfectant is done using the correct dilution and contact time, according to the manufacturer's instructions, and then rinsing with drinking water. Sanitizers have both cleaning and disinfection properties in a single product. But cleaning and disinfecting process must still be carried out as above, to ensure that the sanitizer works effectively, that is, to first provide a clean surface and then again to disinfect. Cleaning and disinfection programs shall be continually and effectively monitored for their suitability and effectiveness once in 6 months and records maintained. Cleaning and disinfection schedule in an eating establishment as given by FSSAI are given in Tables 27.1–27.5. Disinfection methods are of two types:

Table 27.1 Cleaning of Structure

Component	Minimum Frequency	Equipment and Chemicals	Method
Floors except washroom and store	End of each day or as required	Brooms, damp mops, brushes, detergents, sanitizers	1. Sweep the area and remove debris 2. Apply detergent and mop the area 3. Use scrub for extra soil 4. Rinse thoroughly with water 5. Remove water with mop
Walls, doors, ceiling, ventilators, fans, and exhaust fans	Fortnightly or as required	Clean wiping clothes (one time use) brushes and detergents	1. Remove dry soil 2. Rub with wet cloth or rinse with water 3. Apply detergent and wash 4. Wipe with wet cloth or rinse with water 5. Air dry
Air conditioners	As per manufacturers maintenance manual		
Desert coolers	Fortnightly or as required	Water, mop	1. Remove water 2. Rub with cloth or rinse with water 3. When not in use remove water and keep dry
Washroom	Once every 4 h	Brooms, damp mops, brushes, detergents, sanitizers	1. Sweep the area 2. Apply detergent and mop the area 3. Use scrub for extra soil 4. Rinse thoroughly with water 5. Remove water with mop

Component	Least Frequency	Equipment and Chemicals	Method
Store	End of each day or as required	Brooms and camp mops	1. Sweep the area 2. Mop the area 3. Use scrub for extra soil 4. Air dry
Water storage tank	Once in six month	Clean wiping clothes (one time use), detergents, sanitizers	1. Remove foreign matter and soil 2. Rub with wet cloth or rinse with water 3. Apply detergent and wash 4. Rinse with water and sterilizer 5. Air dry
Insect Electrocuting devices	Once a week or as required	Clean wiping clothes (one time use)	1. Remove insects and other foreign matter 2. Rub with wet cloth 3. Reinstall insectocutors
Waste bins and waste areas	End of each day or as required	Water, clean wiping clothes (one time use), detergents	1. Remove foreign material and soil 2. Rub with wet cloth and rinse with water 3. Apply detergent and wash 4. Air dry
Parking and open spaces	End of each day or as required	Water	1. Sweep the area and remove debris 2. Wash parking space thoroughly with water
Street lanes and other public places or the common part of building which are adhering and/or nearby food premises	End of each day or as required	Water or mop	1. Sweep the area and remove debris 2. Wash thoroughly with water

FSSAI

Table 27.2 Cleaning of Food Contact Surfaces

Component	Least Frequency	Equipment and Chemicals	Method
Work tables	After use	Clean wiping clothes (one time use), detergents, sanitizers	1. Remove food debris and soil 2. Rub with wet cloth or rinse with water 3. Apply detergent and wash 4. Wipe with wet cloth or rinse with water 5. Apply sanitizer 6. Air dry
Sinks	After each use	Running water, detergents	1. Remove food debris and soil 2. Rinse with water and or detergent

FSSAI

Table 27.3 Cleaning of Equipment

Component	Least Frequency	Equipment and Chemicals	Method
Utensils, cutting boards, knives, other cooking equipment, service ware, crockery, and cutlery	After use	Clean wiping cloths (one time use), brushes, detergents and sanitizers	1. Remove food debris and soil 2. Rinse with water 3. Apply detergent and wash 4. Rinse with water 5. Apply sanitizer 6. Air dry
Food processing equipment, vending machines	As per manufacturers cleaning and maintenance manual		
Refrigerators, freezers and storage areas, refrigerated display counters	Weekly or as required	Clean wiping cloths (one time use), brushes, and detergents	1. Remove food debris and soil 2. Rub with wet cloth or rinse with water 3. Apply detergent and wash 4. Wipe with wet cloth or rinse with water 5. Dry with clean cloths/ air dry

FSSAI

Table 27.4 Cleaning of Hand Contact Surfaces

Component	Least Frequency	Equipment and Chemicals	Method
Doors and Door knobs	Daily	Damp cloths and detergents	1. Remove debris 2. Apply detergent 3. Rinse or wipe with damp cloths 4. Dry with paper towels/ air dry
Upholstery	Daily	Clean wiping cloths (one time use)	1. Remove food debris and soil 2. Wipe with dry cloth
	Fortnightly or as and when required	Steam/chemicals	1. Remove debris 2. Apply chemicals 3. Vacuum dry

FSSAI

Table 27.5 Cleaning of Furniture and Decorative Items

Component	Least Frequency	Equipment and Chemicals	Method
Chairs and tables, reception and cash counters, display counters held at ambient temperature	Fortnightly or as required	Clean wiping cloths (one time use), brushes, and detergents	1. Remove dry soil 2. Rub with wet cloth or rinse with water 3. Apply detergent and wash 4. Wipe with wet cloth or rinse with water 5. Air dry
Paintings, artificial plants, and decorations	Fortnightly or as required	Clean wiping cloths (one time use), brushes, and detergents	1. Remove dry soil 2. Wipe with wet cloth 3. Air dry
Plants	Daily	Washing	Water

FSSAI

- nonchemical disinfection methods like heat/steam: expensive, impractical
- chemical disinfection methods: commonly used disinfectants [chlorine and chlorine-releasing, quaternary ammonium, amphoteric (ampholytic) and phenolic compounds, peracetic acid].

Chlorine is the most effective disinfectant available and sodium (or calcium) hypochlorite is a cheap disinfectant commonly in use (Rusin et al., 1998). A practical disadvantage of sodium hypochlorite is the risk of corrosion to all common metals (especially aluminum and galvanized iron), except perhaps high quality stainless steel. As per the Food Safety and Standards Act (FSSA) 2006, food premises, their fixtures, fittings, equipment and utensils shall be maintained clean, and in a good state of repair and working condition. A well-planned, well-executed, and controlled cleaning and sanitation program for eating establishment (service area, kitchen, equipment, utensils) is very important to achieve a high hygienic standard. Cleaning and sanitation alone, however, will not fully ensure hygienic standard in production; since process hygiene and personal hygiene are also equally important factors.

27.3 KITCHEN EQUIPMENT AS FOOD SAFETY HAZARD

Requirement of size and number of kitchen equipment in an EE is based on type of food production, convenience to user, and ease of cleaning and maintenance. Equipment in a typical kitchen of an EE are classified as in Fig. 27.4.

Equipment, fixtures, utensils, and food contact surfaces are a source of trouble, if the surfaces are not smooth, easy to clean, durable, nontoxic, nonabsorbent, and corrosion resistant. Accumulation of dust and dirt, and harborage of pests, or microorganisms can contaminate food when they come in contact with food. Many a times food particles get lodged in the blades, corners, internal edges of these equipment. The bottom tray of toasters, grills, *tandoors* often contains old bread crumbs or residues of burnt *chapattis*, etc. Cleaning and maintenance are easily put off until tomorrow in the busy work hours of the EE. But if they are not cleaned properly, tomorrow may be too late! Accumulation of dirt and food residues can jeopardize food safety. The first and foremost requirement before installation of any equipment is to create enough space for it. The equipment can be a large one which is either permanently placed on the floor or is mobile, for example, deep freezers, and refrigerators. Mobile equipment with casters are preferable as they are easy to move which allows cleaning of walls and floor. Adequate toe space (6 in.) from floor and gap between wall and equipment is maintained to clean and prevent pest nuisance.

Worktables and shelves used for preparation of food can also accumulate dust, dirt, and food particles in case they are not smooth and have developed cracks and crevices. This applies to chipped crockery also. Wooden surfaces are difficult to maintain and hence are not recommended. Wood also absorbs water, get chipped easily and encourages the growth of microorganisms. Aluminum tops are also difficult to maintain as they dent easily and durability is poor. Stainless steel, granite, marble tops are easy to clean and maintain and hence advised to be used. They are impervious to grease, food particles, or water. The edges and supporting stands of these shelves/tops need to be designed in a way that there is no place for accumulation of dirt and dust and these are easy to clean.

Maintenance implies keeping the equipment in good operating order, in clean, and usable state. Neglecting this may affect the efficiency of equipment and jeopardize food safety also. Cleanliness means to keep equipment and kitchenware safe and bacteria free that comes in direct contact with food. There are three general ways to establish a maintenance and sanitation program: corrective, regular, and preventive.

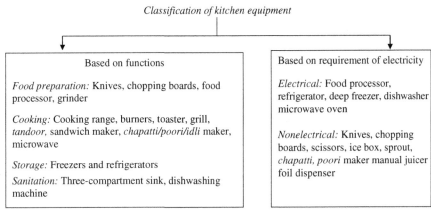

Classification of kitchen equipment

Based on functions

Food preparation: Knives, chopping boards, food processor, grinder

Cooking: Cooking range, burners, toaster, grill, *tandoor,* sandwich maker, *chapatti/poori/idli* maker, microwave

Storage: Freezers and refrigerators

Sanitation: Three-compartment sink, dishwashing machine

Based on requirement of electricity

Electrical: Food processor, refrigerator, deep freezer, dishwasher microwave oven

Nonelectrical: Knives, chopping boards, scissors, ice box, sprout, *chapatti, poori* maker manual juicer foil dispenser

FIGURE 27.4 Classification of Kitchen Equipment

- *Corrective maintenance* is the type used by most FBOs who wait until something breaks down and then summon an employee and try to fix it. If he fails, then to cut down on cost of repair from the manufacturing company he often calls a local repair service. As is often the case, the repairman is not familiar with the equipment and has to send out (again) for the repair parts. It is obvious that this corrective maintenance approach is not the best approach to be taken.
- *Regular maintenance* requires that major pieces of equipment be inspected and serviced at regular fixed intervals. When repairs are necessary, they can be scheduled and anticipated before there is a total break down, and don't generally disrupt normal operations. Some operators employ their own staff for this purpose while others may prefer to establish a regular maintenance contract with a service company to keep their equipment running at peak operating efficiency.
- *Preventive maintenance* often referred to as "PM," is more of a systematic approach and is based upon schedules of inspection and replacement that reflect the requirements of each piece of equipment. While failures can occur, but less frequently. While this approach may initially be more expensive than the other two systems mentioned, PM will more than pay for itself by increasing overall equipment life, maintaining top operating efficiency and with an overall lower frequency of failure (Oke and Charles-Owaba, 2006).

27.3.1 SPECIAL PROCEDURES FOR CLEANING AND MAINTENANCE OF SPECIFIC EQUIPMENT

Cleaning in place equipment (CIP): This practice is used for cleaning equipment that are difficult to dismantle and reassemble. In this method detergent and sanitizer are circulated for specific period of time at specific speed and in a specific sequence so that the food contact surfaces are cleaned and sanitized. These equipment need to be self-draining and leave no residue of cleaning material. For example, the tea/coffee/ juice/soup/crush dispensers/dairy milking machines are cleaned this way at the end of the day.

Chopping boards and meat blocks: Irrespective of the material (wood/plas- tic/ steel) they need to be cleaned immediately after use. The surface should be wiped, scrubbed, and cleaned with a cleaning solution. The sides and under sur- face should be cleaned with a wet cloth. Sanitizer solution should be poured over the board and kept for drying. Rubbing common salt on meat blocks helps in keeping them dry.

Tandoors: For *tandoor* maintenance, it is important to clear the soot from col- lecting inside. All accumulated ash should also be removed daily. The cracks and crevices develop need to be filled with clay. For those using electric *tandoor,* the tray is removed cleaned, scrubbed, sanitized and dried for next use.

Microwave: Microwave, a gadget known for its convenience for cooking and re- heating, has a linked controversy regarding compromised food safety. It is worth mentioning here that microwave energy does not penetrate well in thicker pieces of food, and may produce uneven cooking. This can lead to a health risk if parts of the food are not heated sufficiently to kill potentially dangerous microorganisms. Due to the potential for uneven distribution of cooking, food heated in a microwave oven should rest for several minutes after cooking is completed to allow the heat to distrib- ute throughout the food. In addition to this plastic containers should not be used for cooking and reheating purposes inside the microwave. Splatters of food from micro- waving too decrease the efficiency of microwave. Some methods used for cleaning microwave are given in Table 27.6.

27.3.2 FIXTURES

All lighting and light fixtures should be designed to avoid accumulation of dirt and be easily cleaned. Structures within food establishments shall be soundly built of du- rable materials and be easy to maintain, clean and where appropriate, able to be dis- infected. Other fixtures as door knobs, handles, and switches should also be cleaned to prevent accumulation of dust and dirt.

27.3.3 USE OF MOPS AND ISSUE OF FOOD SAFETY

In our set up often mops are used to wipe dirt from surfaces of equipment. It is ex- tremely important to note here that dirty mops should never be used on a food contact surface. For example, a mop which has been used to clean the microwave should not be used to clean the work table which is to be used for chopping of vegetables. Adequate number of clean dried mops should be available in the kitchen to prevent cross contamination. All used mops should be washed well and dried before reuse. In case fresh clean mops are not available disposable mops may be used. A separate mop for each worktable and equipment per day can ensure safety.

Table 27.6 Methods of Cleaning Microwave

Using Vinegar	Using Lemon	Using Dish Soap	Using Window Cleaner

a. Place a microwave-safe bowl half filled with water and a tablespoon of white vinegar inside the microwave
b. Turn on for 5 min. This will steam up the walls of microwave and loosen the dried-on gunk
c. Remove the container. Wipe the inside of microwave with paper towel or a clean rag
	a. Cut a lemon in half. Place both halves cut-side down on microwave plate with a tablespoon of water
	b. Microwave for 1 min or until the lemon is hot and inside of the microwave is steamy
	c. Wipe the inside of the microwave with kitchen paper and wash the plate
		a. Add dish washing liquid in a microwave-safe bowl and fill it with warm water. Place the bowl in microwave and turn it on for 1 min or until it starts to steam
		b. Take the bowl out. Wipe the inside of microwave with damp sponge
		c. Baking soda can also be added to the bowl to serve as a deodorizer
			a. Mix cleaning solution in a bowl (two parts window cleaner with one part warm water)
			b. Dip a sponge in the cleaning solution and use it to wipe the inside of the microwave. Remove the spin tray and wipe the microwave until all spots and stains have been removed. Also wipe the vents of the microwave oven
			  - Make sure microwave is unplugged
			  - Soak tough stains and spots in the window cleaner solution for 5 min before scrubbing them away
			c. Wipe it down with a clean rag. Finally clean with a rag soaked in freshwater. If you still smell the window cleaner, wipe again with a clean rag soaked in fresh water
			  - If tough spots remain, clean with a cloth soaked in olive oil
			  - Never use unsafe chemicals that can cause the unit to catch fire or cause safety hazards
			d. Let the microwave dry

FSSAI

27.3.4 REFRIGERATOR MAINTENANCE AND FOOD SAFETY

Perishable foods like cooked, ready to eat and high-risk foods are generally refrigerated or are frozen to preserve and keep them safe for longer time. A refrigerator maintains temperature from 3 to 5°C (37–41°F). It is a myth that no bacteria can grow in a refrigerator. The environment inside a refrigerator usually creates an inhospitable environment for many bacteria. But some bacteria are able to grow at cold temperatures, these are called psychrophiles (Toule and Murphy, 1978). This explains why food products still go bad in refrigerators. Bacteria such as some *Coliforms, Pseudomonas* sp., *Vibrio* sp., *Listeria* sp. and moulds such as *Penicillium* and *Cladosporium* sp., are all known to survive low temperatures and can cause illness.

Routinely verifying inside temperature of the refrigerator can prevent any lapse in food quality. Some refrigerators have built-in thermometers to measure their internal temperature. For those without this feature one can keep an appliance thermometer within the refrigerator to monitor the temperature. This holds special importance at the time of power outage. The "down time" period and temperature fluctuation is very crucial from food safety point of view. Once the electricity is back, the temperature of refrigerator must be checked. The food is safe if the refrigerator is still 5°C. Foods held above 5°C for more than 2 h should not be consumed and must be discarded. In the event of a power outage foods can be kept cool for several hours by refraining from adding foods to the appliance and opening the doors. If one knows there will be a power outage, produce more ice cubes and place them in the top section of the refrigerator. Don't keep the refrigerator door open unnecessarily and close as soon as possible. Never overload the refrigerator.

The temperature requirement varies with type of food and other items. The frozen food or highly perishable food items are kept best in freezer at −20 to −10°C. Other foodstuff requires temperature between 1 and 5°C. Another misconception is that hot food cannot be placed directly in the refrigerator. For this reason people keep hot foods at room temperature for long to lower their temperature. This habit should be disowned and instead food must be rapidly chilled in an ice or cold water bath before refrigerating. Always remember to place it in refrigerator within 2 h of heating/cooking.

Many unsafe refrigeration practices are followed in India at grocery shops, stores and EEs. Business operators rarely give any importance to cleaning and maintenance of refrigerators. Often, due to space constraints these are kept in direct sunlight and also to attract customers or advertise the food brands. To achieve safety of foods during refrigeration it is imperative to follow the basic cleaning and maintenance steps. Routine inspection by Food Safety Officers for temperature and training in this aspect can help in maintaining safety of food requiring cold temperature conditions.

REFERENCES

Edrees, N.O., 2014. Study on impact of household environment factors regarding milk storage and wheat powder prepared for feeding infants and some other regular storage flour infested with *Suidasia nesbetti*. J. Am. Sci. 10 (10).

Garden-Robinson J., 2009. Keep hot foods hot and cold foods cold: a consumer guide to thermometers and safe temperatures. Available from: library.ndsu.edu http://hdl.handle.net/10365/5212

Oke, S.A., Charles-Owaba, O.E., 2006. An approach for evaluating preventive maintenance scheduling cost. IJQRM 23 (7), 847–879.

Rusin, P., Orosz-Coughlin, P., Gerba, C., 1998. Reduction of faecal coliform, coliform and heterotrophic plate count bacteria in the household kitchen and bathroom by disinfection with hypochlorite cleaners. J. Appl. Microbiol. 85 (5), 819–828.

Toule, G., Murphy, O., 1978. A study of bacteria contaminating refrigerated cooked chicken; their spoilage potential and possible origin. J. Hyg. 81 (2), 161–169.

Design and construction of eating establishments for ensuring food safety

28

P.K. Ahuja

Postgraduate Institute of Medical Education and Research,
Chandigarh, India

"Food service establishment" is any place where food is prepared or provided in individual proportions for consumption on or off the premises and includes restaurants, delis, take-out food premises, etc.

28.1 OBTAINING APPROVAL AND PERMIT TO OPERATE A FOOD SERVICE

This is the first requirement for which there is a procedure for obtaining approval and permit to operate a food service establishment from registering authority. This may be a designated officer/food safety officer or any official in the "local government set-up" (*panchayat*, Municipal Corporation), or that notified by the State Food Safety Commissioner for the purpose of registration. This has been laid in Food Safety and Standards (Licensing and Registration of Food Businesses) Regulations which came into force on or after August 5, 2011. The registration form must be accompanied with a clear and complete detailed drawing/plan preferably to scale, with the information listed subsequently besides any other information that may be pertinent to the review of the proposal.

Prior to construction of any food establishment, various aspects such as form; space planning; aesthetics; fire and life safety; structural adequacy; plumbing services; lighting and natural ventilation; electrical and allied installations; air conditioning, heating and mechanical ventilation; acoustics, sound insulation and noise control; installation of lifts and escalators; building automation; data and voice communication; other utility services installations; landscape planning and design; energy efficiency measures, etc. need to be kept in view right at the concept stage. The project requiring such multidisciplinary inputs needs a coordinated approach among the professionals for proper integration of various design inputs. Here, it is desirable that the multidisciplinary integration is initiated right in the beginning.

Food Safety in the 21st Century. http://dx.doi.org/10.1016/B978-0-12-801773-9.00028-5

Ideally, the issuance of license to operate should imply that the concerned food service establishment meets all the basic hygiene requirements. This is in essence the implementation of health promotion principle, that is, if our policy and planning are right the results will be good.

28.2 CHARACTERISTICS OF GOOD KITCHEN DESIGN

Design and layout of the eating establishment (EE) should ensure good food hygiene practices. It should protect against cross-contamination between and during operations. Ideally, the food preparation area should be visible and/or easily accessible to the consumers. The layout of EE should facilitate the food flow in one direction as far as possible (i.e., receiving, storage, preparation, and serving). There should be adequate space for food preparation and storage. Also there should be space allocation for storage of equipment/utensils and installation of sanitary fitments, parking of vehicles; seating of customers, restroom for the food handlers, etc. Toilets should be completely segregated from food rooms. Customers need not have to pass through a food room for going to the toilet.

28.3 INTERNAL STRUCTURES AND FITTINGS

Structures within EEs should be built with durable materials. These should be easy to maintain, clean, and disinfect. It is preferable that in nonvisible areas of the kitchen plain tiles are used. The counter tops should be of granite because of its hardness. The following aspects of design has important bearing on its efficiency:

- *Ergonomic kitchen design*: It minimizes movement of kitchen staff while they work in the kitchen.
- *Energy efficiency*: Kitchen design directly affects the energy consumption. For example, placing the cookers in one location reduces energy costs for range hoods.
- *Appropriate size of the commercial kitchen*: Size of restaurant kitchen should be proportional to the size of the restaurant or better say the number of seats in the restaurant. The general rule is that for every seat in the restaurant it is necessary to provide at least 5 ft.^{2} of kitchen space.
- *Good ventilation*: Working in the kitchen is not possible without good ventilation. The presence of steam and smoke in the restaurant kitchen is unacceptable, if not dangerous and unhealthy for employees.
- *Equipment standards*: Installation of good quality equipment must be ensured.
- *Easy to maintain*: How easy is it to maintain the entire kitchen depends on the material of which the kitchen is made, the arrangement of elements, and the way the cookware in the kitchen has been stored.

A well-planned kitchen should provide adequate storage and space for raw materials, food being prepared, food awaiting service, equipment, utensils, crockery, and cutlery. It must be efficient and effective in terms of movement of staff, equipment, materials, and waste management system in place [food, oil, and grease (FOG)]. It should have a janitorial store with janitorial sink in place and a chemical store.

Commercial kitchen: For commercial kitchens, the following checklist is a must for good construction:

- Site/building evaluation to verify suitability
- Equipment schedule based on menu requirements
- Physical space requirements
- Process map of food handling and preparation techniques
- Layout of kitchen equipment and workstations
- Dry storage and refrigeration requirements
- Preliminary regulatory review
- Heating, ventilation, air conditioning (HVAC) requirements
- Lighting
- Ventilation hood, make up air and fire suppression requirements
- Sizing and installation for grease interceptor
- Hot water requirements
- Gas and electrical service

28.4 **WORKSPACE DIMENSIONS**

28.4.1 **CLEARANCES**

- Up to 1.2 m clearance in front of storage areas with a sliding door.
- 1.2 × 1.2 m clearance in front of other rooms with swinging or folding doors (i.e. dining room/service doors).
- Between work surfaces, about 900 mm.

To maintain effective movement through spaces, the area per person according to use of the equipment has been established by the Building Act 1984/Workplace (Health, Safety and Welfare) Regulations 1992. *It is recommended that in a kitchen each person needs a floor space of more than 10 m².*

28.4.2 **LAYOUT AND DESIGN OF FOOD ESTABLISHMENT PREMISES**

Layout should ensure that food preparation/manufacturing processes are not amenable to cross-contamination.

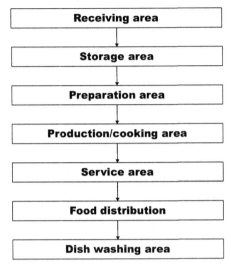

FIGURE 28.1 Logical Work Flow in a Kitchen

There should be a logical workflow in the facilities required as depicted in Fig. 28.1.

Receiving area: It should preferably be located at the ground floor for easy access for receiving of supplies, loading and unloading platforms. It should be equipped with weights and measuring instruments to weigh supplies and materials.

Storage area: It should have sufficient storage area for utensils, crockery and cutlery, and also refrigerated storage for perishables. Large establishments can have walk-in-coolers and refrigerators with varying degrees of temperature for storage of meat products, dairy products, and vegetables.

Preparation area: It should be designed between storage and cooking area where activities such as sorting, peeling, slicing, chopping, mincing, and kneading can be carried out. It should be equipped with double sink with draining board, worktops, peelers, and grinders.

Cooking area: It should be designed between preparation area and service area. It should house cooking ranges, bulk cookers, baking ovens, *chappati* puffers, and frying equipment.

Service area: Here prepared food is received and assembled into food trays. Refrigerators, table tops, and cupboards for storing trays, prepared food, cutlery, etc. should be available here.

Food distribution area: Depending on the place, required provisions should be made. For instance in a hospital, a cart or a trolley should be provided for storage, loading, distribution, and for receiving of the food trolley.

Dishwashing area: It should have provision for automated dishwashing machines, storage racks, stainless steel sinks, and drainers as per workload.

28.5 CONSTRUCTION REQUIREMENTS
28.5.1 LOCATION AND SURROUNDINGS

Food establishment shall ideally be located away from environmental pollution and industrial activities that produce disagreeable or obnoxious odor, fumes, excessive soot, dust, smoke, chemical or biological emissions, and pollutants, and which pose a threat of contaminating food areas that are prone to infestations of pests or where wastes, either solid or liquid, cannot be removed effectively. In case there are hazards of other environment polluting industry located nearby, appropriate measures should be taken to protect the manufacturing area from any potential contamination. The manufacturing premise should not have direct access to residential area. No person shall manufacture, store or expose for sale or permit the sale of any article of food in any premises not effectively separated to the satisfaction of the licensing authority from any privy, urinal, sullage, drain, or place of storage of foul and waste matter.

28.5.2 STRUCTURE

The building must be of sound construction and of adequate size to accommodate the equipment, food and food-related products, and the various activities involved with operating a food service establishment. All exterior doors and windows must be tight fitting (preferably self-closing) and capable of restricting the entrance of insects and rodents.

28.5.3 KITCHEN HEIGHT

As per National Building Code, the height of a kitchen measured from the surface of the floor to the lowest point in the ceiling (bottom slab) shall not be less than 2.75 m, except for the portion to accommodate floor trap of the upper floor.

28.5.4 SIZE

The area of a kitchen, where separate dining area is provided, should be not less than 5.0 m^2 with a minimum width of 1.8 m. Where there is a separate store, the area of the kitchen may be reduced to 4.5 m^2. A kitchen, which is intended for use as a dining area also, shall have a floor area of not less than 7.5 m^2 with a minimum width of 2.1 m.

28.5.5 FLOORS

Floors and floor coverings of all food preparation areas, equipment and utensil washing area, walk-in-refrigeration units, and washrooms shall be constructed of smooth, durable material which is either seamless or with seams that are heat sealed or chemically bonded. All floors and floor coverings shall be maintained in a clean condition and in good repair. The floor coverings, equipment and utensil washing areas, and washrooms shall be coved up the wall to a height of not less than 10 cm (4 in.) and sealed. There shall be adequate drainage. It should be easy to clean and disinfect. Floors shall be sloped appropriately to facilitate drainage flow in a direction opposite to that of food preparation.

28.5.6 WALLS, PARTITIONS, AND CEILINGS

All walls, partitions, and ceilings of food preparation areas, equipment and utensil washing areas, walk-in refrigeration units, and washrooms must be primarily light colored, smooth, nonabsorbent finish, and easily cleanable. Acoustical material free of porous perforations, smooth and durable enough to be washed with a cloth or sponge may be used, provided ventilation is adequate to minimize soiling. Concrete blocks used for interior wall construction shall be finished and sealed to provide an easily cleanable surface. Studs, joists, and rafters in food preparation areas and walk-in refrigeration units must be covered and not exposed.

28.5.7 DOORS AND WINDOWS

Doors shall also be made of smooth and nonabsorbent surfaces so that they are easy to clean and wherever necessary, disinfect. Adequate control measures should be in place to prevent insects and rodents from entering the processing area from drains. Windows, doors, and all other openings to outside environment shall be well screened with wire-mesh or insect proof screen as applicable to protect the premise from fly and other insects/pests/animals and the doors be fitted with automatic closing springs. The mesh or the screen should be of such type which can be easily removed for cleaning.

28.5.8 FINISHING MATERIALS

In areas where food is prepared, packaged, stored, or received and where tableware, utensils, and equipment are kept or cleaned, the floors, walls, and ceilings shall be constructed of materials that are easily cleaned, durable, impervious, light in color (to reflect light and facilitate proper cleaning), smooth, nontoxic, and noncorrosive. The joints between the floors and walls in the areas identified previously shall be coved and sealed to facilitate cleaning.

28.6 EQUIPMENT DESIGN AND INSTALLATION

Equipment and containers that come in contact with food or used for food handling, storage, preparation, processing, packaging, and serving shall be corrosion free to avoid imparting any toxicity to the food material. These should be easy to clean and/ or disinfect. Equipment shall be kept at all times in good order and repair. Every utensil or container containing food/ingredient shall be properly fitted with cover/lid or with a clean gauze net or other material of texture sufficiently fine to protect the food completely from dust, dirt, flies, and other insects. Equipment and containers for waste, by-products, and inedible or dangerous substances shall be specifically identifiable and suitably constructed. If required, a wastewater disposal system/effluent treatment plant shall be put in place. Containers used to hold cleaning chemicals and other dangerous substances shall be identified and stored separately to prevent malicious or accidental contamination of food.

All kitchen equipment must be of commercial grade quality and preferably be ISI certified. It is recommended that heavy and/or large food equipment that is not readily moveable be mounted on wheels wherever possible. Equipment that is not mounted on wheels and not readily moveable, must be sealed to the floor or mounted on legs providing a minimum height of 15 cm (6 in.) from the floor and located such that access can be gained to all sides of the equipment for cleaning purposes. All table or countertop mounted equipment that is not readily moveable and is not sealed directly to the table or countertop, must be set on legs allowing a minimum of 10 cm (4 in.) of space between the piece of equipment and the table or countertop.

28.7 SINKS

28.7.1 HAND WASH BASINS

A minimum of one separate hand wash basin with hot and cold water is required to be conveniently located in the food preparation area, and depending on the size of the kitchen and location of food preparation areas within the kitchen, additional hand wash basins may be required. The hand wash basin must be equipped with a mixing valve or combination faucet. A liquid soap dispenser and a single-use towel dispenser are provided at the hand wash basin. A hand wash basin in a washroom cannot be considered as a designated hand wash basin for a given food preparation area.

28.7.2 DISHWASHING

A conveniently located, stainless steel, three-compartment sink must be provided. The sink compartments must be large enough to permit the accommodation of dishes. The third sink must be large enough for total submersion of the equipment or utensils to be sanitized. The dishwashing area must be designed such that there is sufficient space for: handling dirty and clean utensils; maintaining an adequate separation distance to prevent them from coming in contact with each other and ensuring the flow is from soiled dishes to clean dishes. Dish tables and/or drain boards or racks are required and they must be noncorrodible; self-draining; and of sufficient size and numbers for the handling of soiled and clean utensils as well as air-drying of clean utensils.

Where a mechanical dishwasher is to be used, it must be of commercial grade, certified by NSF International (or equivalent) and approved by a Public Health Inspector. Properly sized grease traps or interceptors, servicing utensil washing sinks, must meet the Manitoba Plumbing Code and located such that they are accessible for easy cleaning and maintenance. A separate janitorial sink is needed for filling cleaning pails and disposing of cleaning wastes to prevent contamination of food and dishwashing areas.

If hot water is used to sanitize tableware, utensils, and equipment, maximum registering or paper thermometers (thermo-labels or temperature stickers) shall be used to ensure the required temperatures as specified previously are being met.

28.7.3 MANUAL DISHWASHING

Tableware, utensils, and equipment shall be washed, rinsed, and sanitized in the following manner:

1. washed in the first compartment with an effective detergent at a wash temperature not less than 44°C (111°F);
2. rinsed in the second compartment in clean water at a temperature not less than 44°C (111°F);
3. sanitized in the third compartment using one of the following treatments:
   a. immersion for at least 1 min in clean hot water at a temperature of at least 82°C (180°F);
   b. immersion for at least 2 min in a warm 24–44°C (75–111°F) chlorine solution of not less than 100 parts per million (ppm) concentration;
   c. immersion for at least 2 min in a warm 24–44°C (75–111°F) quaternary ammonium solution having a concentration of 200 ppm; or
   d. immersion for at least 2 min in a warm 24–44°C (75–111°F) iodine solution of between 12.5 and 25 ppm concentration.

Where chemicals are used for sanitizing, testing equipment shall be available for checking the concentration of the sanitizers regularly. If sanitizer concentrations exceed the concentrations noted previously, an additional warm water rinse is required to remove the sanitizer residual. Manufacturer's instructions for detergents and sanitizers shall be followed. Tableware, utensils, and equipment shall be:

1. air dried after being washed, rinsed, and sanitized and
2. handled and stored in a sanitary manner.

28.8 STORAGE SPACE

Adequate storage space is needed for the following:

- Staff changing room (where >5 employees are on duty at any one time, two change rooms are required: one each for female and male employees).
- Food preparation and service.
- Separation of raw food preparation from cooked food preparation and other ready-to-eat food preparation areas.

- Washing and sanitizing operations for utensils and equipment.
- Separation of food storage and handling areas from areas for chemical storage, toilets, waste storage, office areas, and other areas used for activities that could contaminate food or food preparation areas.
- FOG management systems.

The dry storage space required depends upon the menu, number of meals, quantities purchased, and frequency of delivery.

Shelving may be constructed of suitably finished wood but preferably of noncorrosive metal or plastic. Items should be spaced from walls sufficiently and raised at least 6 in. above the floor to allow for adequate maintenance and inspection of the facility. Storage facilities should be provided to store cleaned and sanitized utensils and equipment at adequate heights above the floor protected from splash, dust, overhead plumbing, or other contamination on fixed shelves or in enclosed cabinets.

Poisonous and toxic materials should be stored in areas designated for such use. Bactericides and cleaning compounds should never be stored with insecticides, rodenticides, or other poisonous materials. Insecticides and rodenticides should be kept in their original containers.

Location of storage areas should be away from food locations to avoid possible contamination. They must have adequate ventilation and maintained at 50–70°F. Entry and harboring of pests must be prevented in this area and no drains should pass through this area.

28.9 GREASE TRAPS

Grease traps must be sized for each site and each unit that is to be fitted to and they must have odor tight lids/airtight to stop airborne bacteria contaminating preparation/cooking surfaces. Surface mounted grease interceptors should be positioned to allow access for maintenance (and cleaning). Units are fitted to grease producing equipment within the kitchen to stop FOG from entering drainage system. Interceptors should be positioned at least 50 mm from walls for cleaning. A 5 A fused spur is required for installation of the dosing system. Although it is commonplace to fit interceptors beneath a sink drainer, they can be installed in any position on the pipe run where space is available.

Recommended lid clearance of around 250 mm is required to allow access for maintenance.

28.10 PIPE WORK

Main pipe work within commercial kitchens should be 50 mm (2 in.). Short connections of appliances to main pipe work should be 40 mm (1.5 in.). Wash hand basin connections to pipes should be 32 mm (1.25 in.). Pipe work should be a minimum of 20 mm (0.75 in.) from wall to allow for clearing underneath. All pipe runs should be provided with adequate supports. The fitting of concealed pipes in block walls "by chasing method" must be authorized by a structural engineer to ensure wall integrity. Extreme care should be taken to ensure walls are not chased too deeply on each side to accept services. Connections from fittings should enter walls between 200 and 250 mm from finished floor level.

28.11　GRATINGS

These are especially suitable for direct discharge from equipments. The free drainage area is up to 90% of the surface area achieving a virtually antisplash installation. Smooth mesh is available in standard and heavy duty for use in general drainage and fork lift area. Nonslip mesh is used in kitchen and production areas.

28.12　WASTE DISPOSAL SYSTEMS

Disposal systems are the actions performed to remove waste in a commercial kitchen from the premises. The kitchen's waste disposal system should be developed to prevent the occurrence of injury resulting from manual handling tasks (e.g., lifting of waste containers). Appropriate measures need to be taken to dispose food (previously served, unsafe, unsuitable, to be recalled, out of date), grease, garbage, recyclables, etc.

28.12.1　FOOD/GENERAL WASTE DISPOSAL SYSTEM

Waste disposal, when used in the kitchen should only be piped into a holding tank and under no circumstances should it go into the drainage. The holding tank should be located in a separate room from the kitchen. This will enable the emptying of the waste food and rule out any cross-contamination within the kitchen area. Waste should then be relocated to a dedicated waste disposal area, which is usually outside the back door or in a separate room. When transporting the waste, it should never pass through the kitchen or restaurant area. It is common practice to take the bins or green biodegradable bags outside to an industrial bin where they are emptied. If the entire bin were taken outside it should preferably be on wheels for easy transport.

Industrial bins must be enclosed, sealed tightly with a lid, and opened only when filling. Exposed waste left outside will attract animals and pests. To avoid unpleasant smells from the decomposition of waste, a garbage contractor should collect it at adequate intervals.

28.12.2　GARBAGE CHUTES

Where garbage chutes are used for waste disposal, they should be made of stainless steel. If any part of the chute is inaccessible, it should include a built-in washing facility.

28.13 **VENTILATION**

A public health engineer should check adequate ventilation to prevent accumulation of heat, condensation, smoke, dust, grease/oils, odors, or other contaminants within the facility. The ventilation system, which includes hoods, canopies, filters, and similar devices, shall be designed, installed, and maintained so as to prevent contaminants from collecting on walls and ceilings and from dripping onto food or onto food contact surfaces.

28.13.1 **NATURAL VENTILATION**

An adequate supply of clean air must be provided and maintained within a commercial kitchen. The simplest form of ventilation is natural, involving the use of windows, vents, and skylights. These should allow sufficient airflow to maintain a healthy working environment. All vents should be screened to prevent flies and be rodent proof.

28.13.2 **EXHAUST SYSTEMS**

These are made of a number of interdependent units. Exhaust hoods, exhaust fans, make-up air units, and packaged rooftop HVAC units all need to operate within defined parameters to complement one another and to maintain peak performance. The exhaust canopy must cover all cooking appliances with an overhang of 150 mm to capture the cooking fumes. An effective exhaust system should get rid of heat, particulate matter, grease-laden steam, and cooking vapors.

Exhaust hoods should be placed above cooking equipment and have the ability to capture and contain the airborne waste matter produced by cooking equipment. Exhaust fans must be capable of removing the collected airborne wastes at a rate equivalent to their generation, and make-up air units need to be capable of replacing an equivalent volume of the extracted waste fumes. The replaced air can be heated, cooled, or dehumidified as necessary by the HVAC unit working in unison with the other units.

28.13.3 **AIRFLOW PLAN AND DESIGN**

The extraction of waste air and intake of clean air should form a stable airflow pattern inside the kitchen.

28.13.4 **MECHANICAL AIR REPLACEMENT**

For this, intake must be located as far from the exhaust outlet from the kitchen as possible and consideration of the surrounding buildings, for example, paint/car body shops, due to the odors these create. Mechanical air replacement (MAR) should have a very fine filter to prevent the smallest flying insects entering the kitchen. These can destroy prepared food. Contractors must record all maintenance activity in maintenance schedules.

28.14 **LIGHTING**

The design of a lighting system should take into account the available natural light, required luminance levels (lux), reflectance of surfaces, and emergency lighting requirements.

28.14.1 **NATURAL AND ARTIFICIAL LIGHTING**

Although artificial lighting will normally be the main source of light, it is desirable to include natural light sources. Ideally, windows in the kitchen should not be less than 10% of the total floor area, and should look out onto the sky or open space.

Windows and skylights can provide views and allow light into a space, improving the staff's working environment, however they can also be problematic as a source of glare. Careful consideration should be given to the positioning of windows and the interaction between natural and artificial light levels.

28.14.2 **RECOMMENDED LUMINANCE LEVEL FOR A COMMERCIAL KITCHEN**

General working area 160 lux.
Food preparation and cooking 240 lux.
Washing areas 240 lux.
Dessert presentations and cake decorating 400–800 lux.

28.14.3 **REFLECTANCE**

Interfering reflections and glare can reduce visibility and become a source of distraction and annoyance. Light will reflect off walls, ceilings, floors, and work surfaces. Therefore, the colour, material, and type of finish of these surfaces should be carefully considered. The reflectance from these surfaces contributes to the overall luminance level of the area. Considering these, result in good visibility for the kitchen.

Ceilings occupy a substantial amount of the field of view. For large areas where there is indirect light penetrating the space, it is advisable to render the ceiling white or near white. Regardless of the size of the rooms the ceiling should have a reflectance level as close to 70% as possible.

Wall reflectance is important even though its contribution to the distribution of light is small. The colors and finishes should be selected taking into account contrasts between surfaces. For example, there should be a difference between the bench and wall. Finishes should not provide glare yet provide enough reflective light for staff to safely and effectively carry out their tasks.

Light fitting considerations: Lights should be installed so as not to contribute to food contamination. They should be designed to facilitate ease of cleaning. These can be recessed or surface mounted in ceilings as a sealed unit to stop the dust falling from the lights, for example, Perspex cover.

Suspended fittings will collect dust and become a source of contamination to food. Properly designed diffusers should be installed to assist with even distribution of light and contain fragments in the event of a globe shattering. The luminance level should be made at least 100 lux higher than the recommended level, because of a light loss factor that occurs over time.

28.14.4 EMERGENCY LIGHTING

It is not only a must for optimum functioning of the unit but it also ensures that people can be safely evacuated from the premises, if required. It must turn on automatically from its own power supply whenever there is a power failure. It must be positioned at vital functioning points, exits, and at any point where there is a potential hazard (e.g., a change in floor level) and at regular intervals to maintain minimum lighting levels.

28.14.5 ENERGY EFFICIENCY MEASURES

Various measures can be taken for energy efficiency. For new lighting in kitchen and storage areas, T5 fluorescent lamps with electronic ballasts are preferred. T5 fluorescent lamps are also used in signage and menu boards. For front of house areas consider replacing incandescent lamps with low energy lamps (CFL/LED). Existing fluorescent fittings can be re-lamped with triphosphor lamps and undertake re-lamping and cleaning programs on a regular basis. Movement sensors can be installed in storerooms, etc. to ensure lights are turned off when not in use. Managers may contact local energy providers to switching to green power. An energy audit can be done to develop an energy reduction strategy.

28.15 INSECT CONTROL

Despite best efforts and good practice, insects may still enter the kitchen. Flying insects can be eliminated through the use of insect control devices, for example, electronic insect killers. It is recommended that such devices are located at entrances to eliminate the insects on entry. Ideally the insect control device contains the insects but if the bug drops, consideration must be given as to where the insects will land to avoid contaminate food. Ventilation intake must have a filter fitted to stop flies and other airborne insects.

Crawling insects can be controlled through the use of baits. These should not be placed away from food areas.

The following measures can be taken to minimize the risk of pests entering food premises:

1. Prune back trees that hang over the roof.
2. Maintain a minimum 1200 mm wide free draining paved surface around the premises.
3. Avoid storing waste bins against external walls.
4. Install rodent proof strips at all entrance doors.
5. Install self-closing devices on entrance door.
6. Cover external vents with wire mesh.

7. Seal service penetrations (electrical and plumbing services) in external walls.
8. Install profiled sealing strips at the junction of roofs and external walls.
9. Ensure windows are tightly fitting and install fly screens where they can be opened.

28.16 BASIC FIRE PRECAUTIONS

All areas should be kept clean from dust, dirt, and oil to limit fire damage. All cooking and high-risk appliances should be watched when in use. Emergency devices should be dispersed around the kitchen, not all in the same area. These should be serviced at least once a year. The most common cause of serious fires in commercial kitchens is the overheating of deep fryer oil. This should be watched for.

Fire extinguishers: The number of extinguishers depends on the size of the kitchen and should be located in a main thoroughfare, preferably along the exit route, which can be easily accessed. They should be 100 mm high off the floor. For electrical fires, the extinguisher must be in proximity of the appliance. For cooking oils and fats, the extinguishers must be placed between 2 and 20 m from the risk. In a commercial kitchen, 20 m is too far away, it is safer to keep them within 10 m. All staff needs to be trained in their use and mock drills are recommended.

Fire hose: It is generally not used in a kitchen, because water is volatile against fats, oils, and electrical fires. But if used compliance with NBC guidelines needs to be ensured.

Gas suppression systems: These systems generally have a pipe system that goes up the wall with nozzles in the range hood. Though expensive, these are good investment as these are located directly over the hazard and are usually heat activated.

28.17 FIRST-AID KITS

One basic first-aid kit must be kept on the premises for every 50 staff members who are working at one time. For 50–100 staff an occupational first-aid kit is required. These should be easily accessible, for example, wall mounted. All staff must have a first aid kit within 100 m of their regular work position. At least one kit must be available on each level of a multilevel kitchen. All staff members need to be trained in their use. Mock drills are recommended. For over 400 staff, a first aid room is necessary.

28.18 CONCLUSIONS

When we visit an eating establishment the inside of the kitchen and preparation rooms are totally hidden from us and we only appreciate the aesthetics of the dining room! However, the overall design, layout, lighting, equipment and facilities like water, electricity, etc and their adequate availability and use, remains the Achilles Heel as far as food safety is concerned. Conceiving such facilities well at the time of inception, following laid down regulations, executing them on ground, adequate utilisation and their constant maintenance are vital for lasting food safety.

FURTHER READING

Building Act. UK, http://www.legislation.gov.uk/ukpga/1984/55; 1984.

Food Safety Guidelines for Food Banks, BC Centre for Disease Control; 2006.

FSSAI. Food safety & Standards (Licensing and Registration) of Food Business Regulation. New Delhi. http://www.fssai.gov.in; 2012.

Guidelines for the safe design of a Commercial Kitchen. Runcorn, Cheshire: Aluline (Group) Limited.

National Building Code of India. Bureau of Indian Standards. New Delhi, India; 2005.

Public Eating Establishment Standards. Saskatchewan Ministry of Health; 2010.

The Food Safety & Standard Regulations. Food Safety & Standard Authority of India; 2011.

The Workplace (Health, Safety, and Welfare) Regulations, http://www.legislation.gov.uk/uksi/1992/3004/regulation/1/made; 1992.

Food safety
in small eating
establishments and
in special situations

Safe cooking practices and food safety in home kitchen and eating establishment

29

P. Dudeja*, A. Singh**

*Department of Community Medicine, Armed Forces Medical College, Pune, Maharashtra, India; **School of Public Health, Post Graduate Institute of Medical Education and Research, Chandigarh, India

29.1 SAFE COOKING PRACTICES AND FOOD SAFETY IN HOME KITCHEN AND EATING ESTABLISHEMNT (EE)

Cooking is the art, technology, and craft of preparing food for consumption with or without use of heat. On one hand cooking can prevent many Food Borne Illnesses that would otherwise occur if the food is eaten raw on the other hand incorrect cooking practices renders food unsafe. The earliest records of cooking were left by the ancient Egyptians on various wall paintings that date back to 4000 BC depicting roasting, frying, broiling, and boiling. The Bible contains several references to ancient methods of cooking, and early carvings from Assyria and Babylonia show the use of charcoal in pans. Ancient Greeks mainly roasted meat. In the Roman Empire, cooking began to be considered an art. An indication of the eating habits of Middle Easterners during the 8th century AD can be obtained from a section in the Arabian Nights in Baghdad. In the medieval world most of the food for a common man was bought from a cook shop in contrast to the great houses, where huge kitchens were equipped with all kinds of kettles, saucepans, skewers, and other utensils. With industrial revolution the art and science of cooking also progressed.

India is a vast country with varied food habits across different states. Before the era of globalization, a traditional Indian woman would spend majority of her time in the kitchen as she was involved in preparation of raw material sifting, washing, drying, grinding grains, cooking per se, serving along with cleaning of kitchen and washing of dishes. All members of the family would consume hot freshly cooked food. The practice of storing and reusing leftover food, use of convenience foods, processed foods wasn't there. Eating out culture did not exist. With changing role of women in society the concept and type of cooking practices have drastically changed. Since the working females have less time to work in kitchen they usually hire domestic part

Food Safety in the 21st Century. http://dx.doi.org/10.1016/B978-0-12-801773-9.00029-7

time help for cutting/chopping vegetables, serving food and cleaning of kitchen and for washing of utensils. Some of the double income working couples with a comfortable salary even hire full time housemaids for cooking. Simultaneously universal availability of refrigerator to store food (raw and left over) and microwave to thaw and reheat food has also contributed to comfort of the lady who has to handle both her office and kitchen. Many other appliances like food processors, *chapatti maker*, toaster, grill, oven, air fryers, etc. are also being used in the kitchen. However, these time saving methods and gadgets have an impact on the safety of food. The judicious and correct use of these kitchen equipments, temperature control, and other safe cooking practices need to be followed to ensure that food is safe. The present chapter deals with various cooking practices/mechanisms which will ensure food safety in kitchen.

There are recipes for different food items and even different recipes for one dish. However, preparation of any dish involves some common generic steps (Fig. 29.1). Though for different dishes the sequence of steps varies.

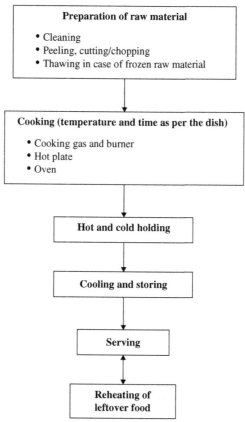

FIGURE 29.1 Generic Steps in Cooking

29.1.1 **PREPARATION OF RAW MATERIAL**

Food Safety precautions have to be observed at each step show in Fig. 29.1 to avoid occurrence of Food Borne Illnesses. Food safety precautions in each of the steps are given in succeeding paragraphs.

29.1.1.1 Cleaning

The practice of preparation of raw material is different in a home kitchen from that of a commercial or public EE. Generally for a home kitchen sorting of raw food is done mainly at the time of procurement. These are then stored in the refrigerator. When required they are washed, peeled, chopped, and cooked. Practices vary as per the prevailing custom in each home. Most of the people store fruits and vegetables in refrigerator without washing especially where women are working. While some discerning people may wash, dry and then store these items.

However, in an Eating Establishment the procedure is different. It even differs between a star rated and a small EE. Raw material reaches the kitchen following safe transportation practices. At the time of receipt, fruits and vegetables are sorted out in the reception area and cleaned of dust and dirt. This is done in three stages. The dirt is removed first, then it is rinsed with clean water and at third stage it is sanitized/washed with 50 ppm chlorine solution for 5 min or potassium permanganate solution. Special precautions are taken for the cleaning of green leafy vegetables. In case these are not washed properly these can be a source of infectious disease like neurocysticercosis in vegetarians (Varma and Gaur, 2002). Apart from this, physical impurities like mud, stones, grit, insects, etc. are removed from cereals and pulses by handpicking before cooking. Some inedible parts of vegetables and fruits are also removed. They are then stored at appropriate temperature (less than 5°C) in a refrigerator. However in small EEs (road side *dhabhas*, canteens, etc.) which don't have adequate capacity refrigerator and space for storage, procurement of raw material is done on a daily basis. Here the chopped fruits and vegetables are kept in open for a long time where these can be easily contaminated. Hence, if such a practice is followed it is mandatory that these cut vegetables are cooked, cooled and then stored in refrigerator.

29.1.1.2 Cutting and chopping

This process needs a chopping board and sharp knives. In some kitchens food processors may be used for this process. After having ensured the cleanliness of raw fruits and vegetables, the food safety issue in next step of cutting and chopping is related to the hygiene of cutting board and knives. Contaminated equipments can be a source of cross contamination and make raw material unsafe (Zhao et al., 1998). The cutting boards are available in different colors (red, green, blue, yellow, white) so that they can be distinguished for different foods.

Generally, green color is preferred for vegetables and fruits, red for mutton, blue for fish, yellow for poultry and, white for ready to eat precooked food. Each EE can develop its own system of identification. Similarly the knives can be identified for vegetarian and nonvegetarian foods by the color of handle or shape of the blade. This practice of keeping separate chopping boards and knives for vegetarian and nonvegetarian foods and/or cooked and raw foods prevents cross contamination and helps in maintaining the safety of food. Various precautions for using chopping boards to ensure food safety are as follows:

1. Separate board for vegetarian and nonvegetarian items.
2. Separate for cooked and raw foods.
3. Always use a clean cutting board for food preparation.
4. After each use and before next step while preparing food, clean the cutting boards thoroughly in hot, soapy water, then rinse with water and air dry or pat dry with clean paper towels.
5. After cutting raw meat, poultry or seafood on your cutting board, clean thoroughly with hot soapy water, then disinfect with chlorine bleach or other sanitizing solution and rinse with clean water. To disinfect cutting board, use chlorine solution or readymade sanitizers. Flood the surface with the solution and allow it to stand for several minutes. Rinse with water and air dry or pat dry with clean paper towels.
6. All cutting boards eventually wear out. Discard cutting boards that have become excessively worn or have hard-to-clean grooves. These grooves can hold harmful bacteria that even careful washing will not eliminate.

29.1.1.3 Thawing

There has been an increase in use of frozen precooked foods in the Indian middle class society. Moreover uncooked nonvegetarian items like chicken, mutton, fish, etc. are also available in frozen form. It is important to understand here that freezing does not destroy microorganisms. If a contaminated food is frozen, all the organisms are not destroyed during freezing. Rather, organisms do not grow in frozen food. But when such food is thawed they multiply rapidly. It is a common practice to leave frozen foods at room temperature for thawing (Jay et al., 1999; Sammarco et al., 1997).

Thawing frozen food correctly is important for keeping food safe to eat. The temperature of food should not exceed 5°C during the thawing process. Hence appropriate planning is required before cooking frozen foods. If during thawing food is allowed to enter the temperature danger zone of 5–60°C, bacteria will grow rapidly.

There are four ways for thawing food (Fig. 29.2).

The process of thawing frozen food needs to be monitored by checking the temperature of food using a thermometer. This is done at the end of refrigerator thawing. For thawing in running water, temperature of the food needs to be checked every 30 min. The cleanliness of the thermometer in use cannot be ignored in these steps. Once food is thawed it must never be refrozen. In case it is not used, cook the food item and then store rather than refreezing.

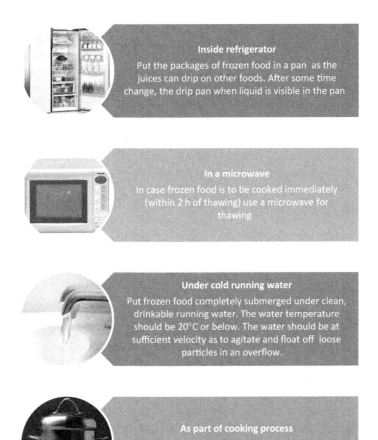

Inside refrigerator
Put the packages of frozen food in a pan as the juices can drip on other foods. After some time change, the drip pan when liquid is visible in the pan

In a microwave
In case frozen food is to be cooked immediately (within 2 h of thawing) use a microwave for thawing

Under cold running water
Put frozen food completely submerged under clean, drinkable running water. The water temperature should be 20°C or below. The water should be at sufficient velocity as to agitate and float off loose particles in an overflow.

As part of cooking process
For certain foods like frozen patties, nuggets, pizza, soup, and vegetables

FIGURE 29.2 Different Methods of Thawing Frozen Foods

29.1.2 COOKING

The most likely time when food becomes contaminated is during preparation and after cooking. Food can be contaminated from surfaces, utensils, clothing, sinks, chopping boards, hands, waste, or unclean equipment, by using contaminated foods as eggs with dirty or cracked shells and from pests such as cockroaches, flies or rats in food preparation areas. Another practice which can cause contamination is tasting cooked food with fingers. Instead, a clean tasting spoon can be used for same. In small EEs sometimes a ladle is used to put the food on the surface of hand for tasting. "Top up" containers should never be used for freshly cooked food instead a clean container for each new batch of food that is prepared or cooked must be used.

Always separate raw food from food that is ready to eat. All foods must reach a temperature of 75°C during cooking. The most important food safety aspect during cooking is the time and temperature control principle. Temperature control is the attainment of correct temperature to minimize the growth of bacteria. This means keeping chilled food at 5°C or below, and hot food at 60°C or above. So keep hot foods hot and cold ones cold. Most types of microorganisms that cause food poisoning grow in potentially hazardous foods at temperatures between 5 and 60°C called the *temperature danger zone* (Kim et al., 2013). To measure temperature of the food, a food grade thermometer is required (Takeuchi et al., 2005) (Fig. 29.3). The practice of using a thermometer is not common in our country.

So, in such places it is advisable to cook any food with liquid consistency till it boils. Boiling ensures sufficient rise in temperature to kill harmful bacteria. In case a thermometer is present it must be a food grade thermometer, which is accurate to +/–1°C. The thermometer must have a probe so that the internal temperature of food can be measured. The thermometer must be kept safely in the kitchen and must be cleaned and sanitized before every use. This is important to prevent contamination from one food to another.

Foods that are not potentially hazardous may become hazardous if they are altered in some way. For example, corn flour is not potentially hazardous because it is too dry for bacteria to grow. But when used to make soup by adding milk it can be contaminated. Most raw whole fruit and vegetables are not potentially hazardous because they do not allow any food poisoning bacteria to grow. But, when they have been cut, bacteria may be able to grow on the cut surface, and so cut fruit and vegetables should be stored chilled.

There are occasions when it is impractical to keep the food at 5°C or below, or 60°C or above. Some examples are buffets at parties/meetings/hostels messes, etc. The maximum time that potentially hazardous food can be in this temperature danger zone of 5–60°C is 4 h. After 4 h, any remaining food must be thrown away. The *4 h* must include any time that the food was between 5 and 60°C during handling, during

FIGURE 29.3 Food Grade Thermometer

preparation and processing, after processing, during transport and, in the case of buffets, the time setting up. If it happens so that the food is required to be held between 5 and 60°C and then refrigerated then it must not be in 5 and 60°C for longer than 2 h.

Another common practice in our households and EEs is that food handlers insert the finger into the dish to check the right temperature. It is unhygienic. This can transmit harmful bacteria from fingers to the food. In case the food is to be tasted or temperature is to be observed, a clean spoon needs to be used.

In case a microwave is being used for cooking following precautions must be ensured for safety of food:

- Stir food between cooking.
- Cover and cook to prevent loss of moisture.
- Let the food stand covered for 2 min after cooking so that temperature equilibrium is maintained.
- Heat the food to an additional 14°C above recommended temperature in conventional cooking.

29.1.3 **HOT AND COLD HOLDING**

Many a times food once cooked is not consumed immediately or later and needs to be displayed for hours together before consumption, for example, in buffets, hostel messes, parties, religious occasions, trade fairs, festivals, etc. In such situations, there is a time gap between the consumption of food by the first person and last one. During this time of display, the food needs to be held at a specified temperature for a certain duration to maintain the safety. Here also the principle of temperature danger zone (5–60°C) applies. The golden principle for holding cooked foods is to keep hot foods hot and cold foods cold (Garden-Robinson, 2009). Hot-holding equipment must be able to keep foods at a temperature of 65°C or higher and cold-holding equipment must be capable of keeping foods at a temperature of 5°C or colder. These food warmers should maintain the correct temperature rather than giving a false sense of security

When holding hot foods for service, following guidelines help to ensure safety of food:

- Stir at regular intervals, as it will help distribute heat evenly throughout the food.
- Cover as covering will help retain heat and eliminate potential contaminates from falling into the food.
- Measure internal temperature of food two hourly using food grade thermometer.
- Discard any hot food after 4 h if it has not been maintained at a temperature of 60°C or higher.

Another important precaution that must be observed is, not to use hot-holding equipment to reheat. In case food needs to be reheated it should first be heated to an internal temperature of 75°C and then transferred to the hot-holding equipment. Also, one should never mix/top up freshly prepared food with foods being held for service as this practice can result in contamination.

When holding cold foods for service, following guidelines help to ensure safety of food:

- Protect all foods from possible contamination by covering them.
- Use food thermometer to measure the food's internal temperature every 2 h.
- Never store food items directly on ice. All food items, with certain exceptions, should be placed in pans or on plates when displayed. Ice used on a display should have a proper drainage system. All pans and plates should be sanitized after each use.

In case there is a situation when one has to deal with questionable hot and cold-holding practice, the issue must be resolved in favor of food safety. It is better to discard foods than risk health or safety. The best one way to avoid discarding too much food is to prepare and cook only as much food as required.

29.1.4 COOLING AND STORING

In case the food is not to be consumed, immediately after cooking it must be cooled and then stored in the refrigerator. Food when hot should never be placed in the refrigerator directly. It must be cooled quickly and then stored for use later. Cooked food certainly should not be left longer than one to 2 h at room temperature before being placed in the refrigerator. This is so because bacteria start to develop within 2 h and then spread rapidly. It's a myth that keeping hot foods in refrigerator will warm other items or reduce the working efficiency of refrigerator. There are different ways of cooling food so that it can be put inside the refrigerator. The simplest is to divide food into smaller portions in shallow so that heat loss is facilitated. Another way is to place the cooked food in a sealed container and then run under cold water. However, in case the container is not closed properly food can be contaminated. If feasible cooked food can be kept over an ice bucket also. However if these cannot be followed still food must be placed in the refrigerator to avoid spoilage.

29.1.5 SERVING

Food can be contaminated during serving through the person doing service, environment (dust, flies, etc.) or through the utensils being used. The famous dictum of "There are many slips between the cup and the lip" applies here too. Before serving food the food handler or the service staff should practice personal hygiene as washing of hands with soap, ensuring cleanliness and paring of nails, etc. The serving dishes need to be checked for any dirt or remains of detergent or leftover food. It is a common practice in our country to wipe the cleaned and freshly washed utensils with a mop (cloth piece) to dry it. This practice can recontaminate the washed utensil. It is not uncommon to see this practice being done in home kitchens too! The serving dishes must be handled in a way to avoid touching of eating surface or mouth contact surface. This will prevent contamination from hands. For example, the plates must be handled from the bottom or edge, cups by handles or bottom, spoons by handles.

Do not serve food in chipped or cracked dishes as these areas can attract dust, food particle accumulation and latter growth of bacteria. Never pick up glasses or cups by inserting fingers inside to hold them.

Handling of food with bare hands should be avoided. Always use a service spoon, pair of tongs, or a fresh disposable glove to pick and serve. It is a common practice to distribute food with bare hands during religious occasions. This can also contaminate food in case the hands of person serving are not clean or not washed properly.

29.1.6 REHEATING OF LEFTOVER FOOD

Leftover or surplus foods must be stored safely in case they are required to be used in future. Those foods which are highly perishable and have been in danger zone of temperature for more than 2 h should be discarded. Most of other foods can be kept for a day if they have not been handled much, stored at correct temperature and reheated properly. Some foods contain spores which are not destroyed during normal cooking process. When these foods are kept in danger zone the spores germinate and bacteria began to multiply till food is refrigerated where they remain dormant. When such food is reheated and passes through temperature danger zone these bacteria multiply. Hence, it is important to reheat the food to 75°C and kept at a temperature of more than 65°C until it is served (Meggitt, 2003). It is a common practice in home kitchens to use microwaves to warm the leftover food taken out of refrigerator. Simply warming the food before serving will do harm as bacterial growth is favored in these temperatures. Another important food safety feature is that leftover food must never be mixed with fresh food. High-risk foods (discussed later in this chapter) should not be reheated more than once.

29.1.7 KEY POINTS FOR GOOD COOKING PRACTICES

- Do not prepare food long before serving time.
- Don't store perishable foods at room temperature for more than 4 h.
- Ensure appropriate cooling of food before storage.
- Provide sufficient temperature for reheating.
- Thaw frozen meat and poultry well before cooking.
- Prevent cross contamination between raw and cooked food during handling of various food items.

29.1.8 HIGH-RISK FOODS

Certain foods have risk of getting contaminated as microorganisms grow easily in them. These foods include:

- raw and cooked meat
- dairy products and dairy-based desserts such as fruit cream, custard
- cooked rice
- foods containing raw eggs like mayonnaise sandwiches, salad dressings

- salads and salad ingredients
- bread, toast, rolls, sandwiches, pizzas
- baked goods
- garnishes such as lemon wedges or pickles on plates
- fruit or vegetables for mixed drinks
- ice
- paneer as an ingredient
- seafood

For handling potentially hazardous food in the kitchen it is important not to use bare hands. Such foods at risk can be handled by using forks and spoons, napkins, spatulas, tongs, etc. If such foods at risk are stored then it is imperative to monitor the temperature of refrigerators and deep freezers or of the display units to ensure they keep food between 0 and 5°C. A record of such checks should be maintained. In case these have been left in the temperature danger zone (5–60°C) for up to *2 h* the food should be reheated/refrigerated or consumed. If they have been left in the temperature danger zone *for longer than 2 h, but less than 4 h,* they should be consumed immediately (Table 29.1). In case the time in danger zone exceeds 4 h the food must be thrown and discarded. Over and above special care must be observed while making following preparations of food.

29.1.8.1 Stuffed preparations

While making any stuffed preparations like *paranthas*, rolls, pattis, *dosas*, etc. the material for stuffing should be cooked first and then filled. This is so because the stuffing slows the process of heat penetration of these food items and the temperature in the center may not reach at desired levels of safety. Another food safety issue is that food items are exposed to bare hands while preparing them. The external temperature of the food gives a false sense of security, as it is hot from outside.

Often chopped slices of chicken, meat, etc. (cooked/raw) are kept in refrigerator to be used as stuffing when required. These are minced and mixed with other ingredients to make the stuffing. Mincing such potentially hazardous food items can transfer bacteria from surface to inside so that they get distributed throughout the entire mass of mince. Minced meat is also handled more and spoils much faster. Hence, minced meat must be thoroughly cooked at sufficiently high temperatures.

Table 29.1 Actions for Handling High-Risk Foods

Situation	Action
Left in temperature danger zone for up 2 h	Reheat and consume Refrigerate and consume Consume directly
Left in temperature danger zone for more than 2 h and less than 4 h	Consume immediately
Left in temperature danger zone for more than 4 h	Discard and throw

29.1.8.2 Coated preparations

Many food items are coated in batter. This batter can be made of *besan*, corn floor, bread crumbs, etc. At times food items are dipped in egg also before cooking. Examples are *pakoras*, cutlets, etc. These coatings act as insulators and reduce the transfer of heat to the food being cooked. The batter for coating is generally handled with bare hands for mixing. Often in commercial EEs the batter is left as such for hours. This way batter can get contaminated with microorganisms and spoil the entire food preparation.

29.1.8.3 Salads and sandwiches

When salad and sandwiches are prepared using highly perishable food like raw fruits and vegetables, salad dressings with raw eggs and poultry such food items must be stored in the refrigerator immediately after preparation. If bread is used it acts as an insulator and prevents the chilling of items inside or between bread.

29.1.8.4 Meat

Meat items are high-risk food and must be bought from a reputable butcher or retail shop. In case it is in frozen form the use by date must be conformed to. While storing in refrigerator, keep it away from ready to eat foods so that juices from meat do not cross contaminate these foods. These must be purchased at the end of shopping spree and kept first in the refrigerator. It is preferred to keep meat on the lowermost shelf of the refrigerator. The process of thawing can be done as described earlier in the chapter. Keep a separate chopping board and knife. During cooking of meat ensure that high enough temperature is reached and it is cooked not only from outside but also from inside. In case there are small pieces they must be moved so that each piece is cooked evenly. All meat should be checked visually to see if it is cooked thoroughly. To check for thorough cooking a fork is pierced through the thickest part of the meat and then the juices should run clear. Meat changes its color when it is cooked. There should be no pink meat left after it is cooked well.

29.1.8.5 Eggs

Extra care is required when preparing foods that contain raw egg, such as home-made mayonnaise. Bacteria, such as, *Salmonella* are present on eggshells (Doorduyn et al., 2006). Inside the egg these can contaminate food and cause food poisoning. It is advised to wash the eggs with clean water before storing these in refrigerator. Hands must be washed in case raw vegetables or ready to eat items are to be handled after handling egg. For example, if bread is to be toasted after making omelets then hands must be washed in between the two processes. Cracked eggs must be discarded.

29.1.8.6 Fish

Something is fishy! This proverb is used in a situation when one can sense something is wrong. It is well said as fish and fish products are spoilt easily because of decomposition. Fish deteriorates or loses its freshness because of autolysis which sets in after death

While procuring fish it is crucial that person should be able to identify for poisonous fish and fish parts. Sewage, bacteria and viruses (e.g., the virus of hepatitis type A) are concentrated in shellfish such as oysters, and fish may carry *Vibrio parahaemolyticus, Salmonella, Clostridium botulinum type E*, and other organisms. Consumption of certain fish may sometimes give rise to "fish poisoning." Fresh fish has following features:

- have a shiny, iridescent surface
- body is covered with a nearly transparent, uniformly spread, thin coating of slime
- eyes are bright and protruding with black pupil transparent cornea
- gills are bright free from slime
- odor is marine-like/seaweedy; fatty fish have a pleasant—margarine-like smell
- flesh is soft and flabby immediately but becomes firm after setting of rigor mortis

After freezing, good fresh fish have a delicate, pleasant odor and flavor when cooked. If the fish is a little older before freezing, it is insipid; a lack of odor or flavor is noticeable. Once procured, it is essential that the fish be unloaded as quickly as possible, minimizing bruising and rough handling, and conveyed to the initial processing area without undue delay. Fish can be cooked by grilling, frying, roasting, baking, poaching, steaming, and microwaving. Ensure adequate temperature is reached while cooking. The flesh should be opaque and separate easily with a fork when fully cooked.

Cooking is surely the first art that human beings ever attempted and it's still the most universal. Nevertheless, the art of safe cooking must be learnt and practiced in true spirit so that all food served is safe for consumption.

REFERENCES

Doorduyn, Y., Van Den Brandhof, W.E., Van Duynhoven, Y.T., Wannet, W.J., Van Pelt, W., 2006. Risk factors for *Salmonella* Enteritidis and Typhimurium (DT104 and non-DT104) infections in The Netherlands: predominant roles for raw eggs in Enteritidis and sandboxes in Typhimurium infections. Epidemiol. Infect. 134 (03), 617–626.

Garden-Robinson, J., 2009. Keep Hot Foods Hot and Cold Foods Cold: a Foodservice Guide to Thermometers and Safe Temperatures.

Jay, S.L., Comar, D., Govenlock, L.D., 1999. A national Australian food safety telephone survey. J. Food Prot. 62 (8), 921–928.

Kim, S.A., Yun, S.J., Lee, S.H., Hwang, I.G., Rhee, M.S., 2013. Temperature increase of foods in car trunk and the potential hazard for microbial growth. Food Control 29 (1), 66–70.

Meggitt, C., 2003. Food Hygiene and Safety: A Handbook for Care Practitioners. Heinemann, Oxford.

Sammarco, M.L., Ripabelli, G., Grasso, G.M., 1997. Consumer attitude and awareness towards food-related hygienic hazards. J. Food Safety 17 (4), 215–221.

Takeuchi, M.T., Hillers, V.N., Edwards, Z.M., Edlefsen, M., McCurdy, S.M., 2005. Gem No. 401. Food thermometer educational materials: "Now You're Cooking… Using a Food Thermometer!". J. Nutr. Educ. Behav. 37 (5), 271–272.

Varma, A., Gaur, K.J., 2002. The clinical spectrum of neurocysticercosis in the Uttaranchal region. J. Assoc. Physicians India 50, 1398–1400.

Zhao, P., Zhao, T., Doyle, M.P., Rubino, J.R., Meng, J., 1998. Development of a model for evaluation of microbial cross-contamination in the kitchen. J. Food Prot. 61 (8), 960–963.

Food safety in schools, canteens, hostel messes, mid-day meal scheme, ICDS

30

R.K. Gupta

Department of Community Medicine, Army College of Medical Sciences, New Delhi, India

We do sometimes hear in the news about food safety concerns at community eating establishments (CEE), such as a school/college canteen, food served through a mid-day meal or the *aanganwadi* in the Integrated Child Development Scheme (ICDS) program. But more often than not these instances are more of a sensational nature rather than having a great public health significance. For example, news items like "a lizard or a cockroach found in the school mid-day meal" is much more common than an instance of food poisoning cases. Given the innumerable CEE servings of food through government schemes (ICDS, mid-day meal, etc.) or private canteens; these handful of food safety breach instances seem miniscule. This notwithstanding, any breach in the food safety of a CEE calls for an investigation and remedial measures.

This chapter deals with the food safety issues in CEE, mainly concerning the school/college canteen, food served through mid-day meal or the *anganwadi* in the ICDS. We are dealing with two sets of CEE: (1) Government-run CEE (2) Privately run CEE. Both sets have their peculiarities.

The government run CEE [mid-day meal and the *anganwadi* (ICDS)] are meant to serve millions of children every day, across the country, free of cost, on a no-profit basis. Given this situation in a developing country like India, it would not always be correct to assume that the food supply chain of harvesting, selection of raw material, supply, transport, distribution, cooking, and serving would have been done in the most desirable and safe manner, all the time and at all places. This becomes even more understandable, since it is mostly the poor children who are being fed, who have minimum say in the matter. Moreover, in the prevailing circumstances, the authorities responsible for running these schemes are minimally accountable. Given the economic status of beneficiaries, the children/parents have no choice but to be happy, so long as they are getting "food." Therefore, representations about the quality/safety of food are seldom forthcoming, till there is a disaster.

The second group of CEE are generally privately owned contract-based system that caters to canteens (of schools/colleges), hostel messes, and the like. These are generally profit-making contractors, who run a predecided, fixed menu with rather

Food Safety in the 21st Century. http://dx.doi.org/10.1016/B978-0-12-801773-9.00030-3

standard protocols. Once these protocols with respect to acquiring raw materials, storage, menu, cooking, serving, etc. are well set the chances of deviation are minimal, thus ensuring the minimum level of "food safety." But in this situation too, so long as the prices are acceptable, food conforms to the taste of youngsters and there are no unforeseen situations (food poisoning or a lizard!), the business runs fine.

We see that in both the situations (government and private), the beneficiaries have accepted the minimum level of food safety that has been offered to them. This has primarily happened owing to the "constraints" of the clients. They are "poor" in the first instance, thus constrained to accept the given low-quality/low-safety food; and in the second instance, the competitive/low price makes them accept the mediocre food.

But the matters brought up in the previous discussion do not absolve the authorities or the owners from their responsibility of offering safe and healthy food to the clients. Let us see the general principles of food safety that must be followed in a CEE.

30.1 GENERAL PRINCIPLES OF FOOD SAFETY FOR COMMUNITY EATING ESTABLISHMENT (GUPTA, 2009)

30.1.1 PROCUREMENT AND STORAGE OF RAW MATERIAL

Food should be procured from a reliable supplier. The hygiene standards maintained by the supplier should be noted. Issues such as cold and chilled storage, separation and handling of raw and cooked foods, and cleanliness of premises and equipment, must be ensured. No more food should be purchased than the amount which can be stored in the available deep freeze cabinet or refrigerator (4°C).

Nonvegetarian foods: Raw and cooked meat and poultry, milk, cream, and fish should be bought in quantities sufficient for one day only. Meat and fish should not be refrigerated for more than 3 days.

Dry and canned foods: Dry goods and canned foods should be bought in reasonable quantities. Overstocking must be avoided; there is danger from vermin and deterioration where storage facilities are poor. These products should be stored in a dry, well-ventilated storeroom. Dented/leaking/puffed and damaged cans should be discarded.

Refrigerate perishable foods: Meat, poultry, dairy products, fish, and cooked rice must be refrigerated. Raw and cooked meat and poultry must be stored separately to prevent cross-contamination. The temperature of the *refrigerator* should be 1–4°C.

Storing nonperishable food: Nonperishable food must be stored in a storeroom. It should be kept cool, well ventilated, and dry. Adequate precautions must be taken for rodents, flies, and cockroaches. Surfaces of shelves, walls, and floors should be easy to clean.

Deep freeze unit: For longer storage deep freezer is used. A temperature of −18°C is desirable.

Vegetables: Vegetables should also be procured on a daily basis. They must be kept in a cool dry storage place away from direct sunlight. Most vegetables can be refrigerated.

30.1.2 FOOD HANDLERS' HYGIENE

Immaculate personal hygiene of the cooks and other food-handlers is of prime importance in maintaining food safety in CEE. Food handlers should take a thorough bath before starting daily work, keep their hair and nails clipped short and invariably scrub and wash their hands with brush, soap, and water after every visit to the toilet and before handling food or raw material. There should be access to toilets, hand-washing facilities, and wash rooms for the kitchen staff. Those suffering from open wounds, boils, diarrhea, typhoid, dysentery, etc. must not enter the kitchen. A regular medical examination and vaccination against enteric group of fevers is also recommended.

30.1.3 KITCHEN HYGIENE AND SANITATION

Besides personal hygiene, it is vital to have an immaculate kitchen hygiene as well.

Kitchen premises: The entire kitchen premises should be spacious, lighted, fly proof, rat proof, airy, and spotlessly clean at all times. The kitchen complex should have a separate cooking room (actual kitchen), a storeroom for fresh provisions, a preparation room, a scullery, and a room for the cooks' clothing.

Floors, walls, and ceiling: Floors and walls must have nonslip surfaces, should be impervious to moisture and easy to clean. Provision for exhaust/chimney-vents must be there.

Lighting: Lighting must be good, both natural and artificial, particularly over the work and preparation areas, sinks, and cooking equipment. Shadows must be avoided.

Ventilation: Natural and mechanical ventilation is necessary to prevent rise in temperature, smoke, and humidity.

Kitchen equipment: Ease of cleaning is an important factor in selecting all surfaces, equipment, and utensils. Keep surfaces, equipment, and utensils clean and in good repair. Slicing/mincing machines and can openers require frequent and thorough cleaning; they must be easy to dismantle and reassemble.

Preparation room: Preparation room is meant for the preliminaries of cooking such as peeling, cutting, and washing of food. The preparation of vegetables should always be done on a zinc-toped table or granite slabs fitted with a chopping board. A meat chopping block preferably of a special hardened plastic (high-density polypropylene) must be provisioned, that is, thoroughly washed and cleaned after use. It must be disinfected with a suitable agent (e.g. hypochlorite) and covered with a layer of powered salt and dried in the sun. Preparation room should be supplied with hot and cold water for which foot operation is preferable. A soap dispenser, kept in a hygienic condition is also a must. A nail brush with plastic or nylon back and bristle should be available. Hand drying should be done using individual methods such as paper towels. Common towels may cause cross infection.

Store room: A separate fly proof and airy storeroom for raw fresh food stuffs should be provided. Raw foodstuffs should be kept in baskets/crates ensuring free circulation of air and stacked on shelves. A cool room or refrigerator must be available where fresh fruits, vegetables milk, and curd can be stored. Meat, fish, and poultry should be kept refrigerated or frozen. Grains, pulses, flour, and other dry stuff should be kept in racks, away from the walls, either in neatly tied bags or in boxes in a separate well-ventilated store-room. Equipment and utensils should be stored separately. A room for the cooks clothing and other necessaries should be provided separately.

Scullery: The scullery should be dry, clean, and tidy. Sinks should be adequate, and draining boards should be sufficient and clean. All utensils after use should be thoroughly cleaned, washed, dried, and kept in clean places. Tables should be scrubbed with washing soda and water twice a day using a hard brush.

Washing arrangements: Efficient washing arrangements are necessary to clean and remove bacteria from all dining room and kitchen equipment. The essential provisions are a good layout of washing area, correct temperature of wash and rinse water and a good detergent suited to the type of water. Orderly methods of work in rinsing, stacking, racking, and storage are also necessary.

Waste disposal: Waste must be collected in pedal-operated bins which can be emptied regularly and washed out or in paper or plastic bags on pedal-operated stands. Bags can be sealed and put into dustbins, incinerated, or collected by the local refuse collector.

Vermin and fly control: Measures must be taken for rats, mice, flies, and cockroach control.

Conveniences: Toilets should not open directly into food-preparation rooms. Foot-operated flushes are desirable. Wash basins should be available in or adjacent to the toilet.

30.1.4 SERVING OF FOOD

Dining room: The dining room should be clean, fly proof, well lit, and ventilated. While serving food, it should not be exposed to flies or dust. It should be presented in a manner that will enhance the acceptability or appeal, and reduce wastage. An effort should be made to supply hot food.

Precautions to be taken during serving:

1. Avoid prolonged exposure of susceptible food to warm environment. This will encourage rapid bacterial growth and deterioration of food. Keep cold food cold, below 5°C.
2. Avoid warm storage of cooked food. Keep hot food hot, above 63°C or else below 4°C.

3. Do not reheat cold food to store in a warm holding apparatus (hot case/casserole). Place only fresh hot food in such equipment, that too for a short while before consumption.
4. Minimize handling of cooked foods with bare hands.

30.2 PRINCIPLES OF FOOD SAFETY SPECIFIC TO VARIOUS COMMUNITY EATING ESTABLISHMENTS/PROGRAMS

30.2.1 FOOD SAFETY AT MID-DAY MEAL PROGRAMS (GOVERNMENT OF INDIA, 2015a)

Mid-day meal is a government run school meal scheme, wherein one meal is provided free to primary students in government schools. The Government of India has issued elaborate guidelines on food safety for this scheme. Food contamination refers to the presence of harmful chemicals and microorganisms in food which lead to illness. Salient points of these are summarized here.

The grain stocks are issued by Food Corporation of India to departments of schools in each State. Three samples are to be drawn during supply of grains, that are retained for 3 months, to ascertain the veracity of complaints if any, at a later date. Only packed *dals*, salt, spices, condiments, and oil with "AGMARK quality symbol" should be purchased. Only "double fortified salt" should be used for cooking mid-day meals. Vegetables, fruits, and perishable food commodities should be procured fresh and storing for longer duration should be avoided. Raw materials should be purchased in quantities that correspond to storage/preservation capacity. All raw materials should be physically checked and thoroughly cleaned. Packaged raw material must be checked for "expiry date." Food grains such as wheat and rice should not be stored for more than a quarter year; and must be stored in airtight bins or stacked neatly in gunny bags in area free of rodents and insects. Salt, condiments, oils, soya bean, pulses, etc. should be stored in airtight containers.

Food safety measures for kitchen during cooking have been elaborated. These include maintaining temperature, using refrigerator, segregating vegetarian and nonvegetarian food, use of apron, gloves and caps by food handlers, etc. The tasting of the food by a teacher, just before serving and maintaining a record is mandatory.

The floors of kitchen and the slabs should be cleaned every day before and after the food is cooked. Cleaning of utensils, equipments, and other materials, for example, cloths, mops, and brushes should also be ensured. Tables, benches, boxes, cupboards, glass cases, etc. shall be clean. Cooks and helpers should maintain a high degree of personal hygiene and cleanliness.

Although other pest control measures may be undertaken, generally no pesticides/insecticides should be used in cooking area. Adequate time and facilities will be made available for hand washing of cooks and children as well. Adequate drainage, waste disposal systems, and facilities should be provided.

Hygiene of kitchen-cum-store should be maintained as elaborated in an earlier section in this chapter. Continuous supply of potable water should be ensured in the premises. In case of intermittent water supply, adequate storage arrangement for water used in food or washing should be made. Water needs to be tested for chemical as well as microbiological contamination. The testing of water can be done with Public Health Engineering Department.

30.2.2 FOOD SAFETY AT *ANGANWADI* CENTERS UNDER ICDS (INTEGRATED CHILD DEVELOPMENT SCHEME)

In ICDS there is provision of supplementary nutrition for children, 6 months to 6 years of age and for pregnant and lactating women. Freshly cooked hot food and a morning snack are provided to children of 3–6 years who attend the *anganwadi* center (AWC) daily for 300 days annually. Pregnant women, infants, and young children are especially vulnerable to infection hence utmost care should be taken at all stages for managing supplementary nutrition. The food supplied from ICDS involves both local preparation and processing, hence it is imperative that precautionary measures are undertaken at different levels.

The Government of India has issued guidelines for *anganwadi* centers under the ICDS regarding safety and quality of supplementary nutrition. It recommends that the supplementary nutrition should conform to Food Safety Standards Act 2006, and Regulations 2011.

There are several constraints in ICDS like no uniformity in design of *anganwadi* center, lack of space, high rent for AWC's in urban areas, lack of water, and sanitation facilities, etc., however basic food safety principles need to be followed. The kitchen is usually a side room where preparation of supplementary nutrition takes place. Usually a tube well is installed in the premises/nearby (Government of India, 2013).

30.2.3 GENERAL TIPS FOR MAINTAINING FOOD SAFETY AT AWCS

- All equipment and utensils must be cleaned as frequently as necessary.
- Adequate precautions must be taken to prevent the food item from being contaminated during cleaning or disinfecting of rooms, equipment or utensils.
- Floors and drains must not be cleaned while the food is being prepared.
- At the end of day's work, floors, drains, and walls must be thoroughly cleaned.
- Toilets must be kept clean and tidy at all times.
- Provision of safe disposal of stool and wastes must be made.
- Nearby area/surroundings in the immediate vicinity of the premises must also be kept clean.
- Before pesticides are applied, care must be taken to protect people, food, equipment, and utensils from contamination. Pesticides should always be kept in original containers, clearly marked and be stored in a locked storage separate from production.
- Fingernails of the *anganwadi* worker/helper must be trimmed. Nail polish or artificial nails should not be worn.

- No watches, rings, jewelry, and bangles should be worn during cooking, serving, and distribution.
- Glass in any form should not be allowed in the cooking area.

30.2.4 FOOD HANDLING AND SAFETY MEASURES FOR HOT COOKED MEAL

- Good-quality ingredients should be procured with certifications like AGMARK, ISI/BIS.
- More than 1 month's quantity of food should not be stored.
- Hands should always be washed before handling food.
- Provision of soap should be there, children should wash hands before eating.
- All utensils, equipment's should be cleaned before and after use.
- Disposal of leftover food should be done properly.
- In case of emergency or food poisoning AWWs should take quick action (as described in the guidelines) and take support of health system without delay.
- Water and food should be tested at regular interval.

30.2.5 MAINTAINING QUALITY ASSURANCE

It is the responsibility of local management committees of the respective CEE to ensure and maintain quality control of the unit; may it be a canteen, mess, school mid-day meal or an ICDS-related *aanganwadi* center CEE. A set of guidelines issued by the government of India meant for school CEE can be taken as a template for quality assurance for all such CEE (Government of India, 2015b). A committee could be set up to oversee each CEE. This would comprise of schoolteachers, parents, students, canteen operators, village head, *aanganwadi* worker, hostel warden, catering representative, and other stakeholders, as the case may be. This committee will coordinate, implement and monitor the CEE policy to ensure quality and availability of nutritious and hygienic food.

30.2.6 CHECKLIST FOR MONITORING AND CONTROLS

The following aspects with respect to the CEE must be monitored:

1. *Training*: Proper training is delivered, evaluated, and recorded for all concerned with respect to food safety and hygiene.
2. *Raw material procurement*: The committee will have a stake/authority in tendering and procurement process so as to ensure food safety and quality of raw materials before purchase, delivery, and usage.
3. *Inspection:* Regular inspection and assessment is done by the committee for hygiene controls for the food handlers and in the kitchen/related areas.
4. *Periodical checks*: Checks for cooked food and water quality are conducted through an external FSSAI approved laboratory, whenever required.

5. *Record keeping:* Records are maintained well, improvement areas are identified and pending points are tracked.
6. *Certification*: Certification from the FSSAI/other agency is facilitated by the committee.

REFERENCES

Government of India, 2013. Operational Guidelines on Food Safety and Hygiene in ICDS. Ministry of Human Resource Development, New Delhi.

Government of India, 2015a. Guidelines on Food Safety and Hygiene for School Level Kitchens Under Mid-Day Meal (MDM) Scheme. Ministry of Human Resource Development, New Delhi.

Government of India, 2015b. FSSAI. Draft Guidelines for Making Ivailable Wholesome and Nutritious Food to School Children. Ministry of Health & Family Welfare, New Delhi.

Gupta, R.K., 2009. Public health aspects of food hygiene and sanitary regulation of eating establishments. Bhalear, R., Vaidya, R., Gupta, R. (Eds.), Textbook of Public Health & Community MedicineArmed Forces Medical College & WHO, Pune, pp. 772–779.

Food safety issues related to street vendors

31

S. Malhotra

*Department of Dietetics, Postgraduate Institute of Medical
Education and Research, Chandigarh, India*

Street food vending is essentially an urban phenomenon. This is a very important segment of the unorganised sector of food industry. As a result of rapidly growing economy and better employment opportunities in urban areas, cities have become a center of work place, trade, and commerce. A large volume of floating population also visits cities daily. Most of them commute to cities a long distance from their hometowns often from early morning until late evening; they are away from their home. This trend is further expected to accentuate exponentially with the increasing urbanization and population growth, especially in developing countries. Due to this transition, food habits of people are seriously affected. In most cases lunch is away from home. Though some people bring tiffin from home, rest are daily customers of street food vendors. Street food has become a staple food for the commuters, workers, students, migrants and tourists. For example, in Connaught Place, New Delhi you see this trend during lunch hours.

Street food vending is prevalent all around the world; however there are variations within regions and cultures regarding scale and pattern in street food vending. In India in recent years there is an increasing trend in the sale and consumption of foods on the roadside. Indian street foods vary from region to region. The *chaat*-fare in North India consists of many tangy and spicy delicacies. In the eastern parts, a typical street food is "*chop*" which is like potato patties dipped in flour batter and deep fried. *Jhalmuri*, a delicacy made from puffed rice is a famous Kolkata street food. *Vada pav* and *pav bhaji* are the combinations famous from Maharashtra, the Western part of India.

Nowadays street food is very popular and is in demand because it saves time and energy. In fact, street foods have become the USP of many cultures. It reflects the traditional local culture. It is one of the best ways to experience the real culture of any community. Apart from this, street food is appreciated for the taste and flavor it offers at low, affordable price to the general population (Bhowmick, 2005). The Street vended food is of many types:

- Where no cooking is required, for example, ready to eat foods. Salads which are made primarily of raw vegetables and fruits.

Food Safety in the 21st Century. http://dx.doi.org/10.1016/B978-0-12-801773-9.00031-5

- Precooked food brought and assembled/tossed/mixed on the site (e.g., *tikki, paani-puri, idli*, etc.).
- Food cooked on the spot and served (e.g., *paranthas*, noodles, *chaat, dosa*, etc.).

Street vended foods include meat, fish, fruits, vegetables, grains, cereals, frozen produce, and beverages. They may be found in clusters around public places, such as the market or fair, place of work, schools, colleges, railway stations, hospital, and bus terminals and may be vended from roadside makeshift stalls, carts, or small establishments by hawkers or vendors. Street foods are served with the minimum amount of fuss in individual portion dished into take away containers. These containers can be disposable plastic, paper or Styrofoam plates, bowls, and cups.

According to FAO globally 2.5 million people eat street food daily. Street vendors are roughly estimated to be 10 million. They constitute approximately 2% of the metro populations.

31.1 DEFINITION OF STREET FOOD

FAO defines street vended foods as "Ready–to–eat foods and beverages prepared and/or sold by vendors and hawkers especially in street and other similar public place" (FAO, 1988). A street vendor is broadly defined as "A person who offers goods for sale to the public without having a permanent built up structure but with a temporary static structure or mobile stall (or head load)." Street vendors may be stationary by occupying space on the pavements or other public/private areas or may be mobile in the sense that they move from place to place carrying their wares on push carts or in cycles or baskets on their heads, or may sell their wares in moving trains, bus, etc. (Government of India, 2004).

31.2 INCREASING TREND OF STREET FOODS

The concept of traditional street food has acquired new dimensions in developed countries with food streets/food centers emerging as new tourist attractions. Though street food culture pervades almost all countries of developing regions, the Asian street food is considered as the best in the world.

The Time magazine survey of 2004 and the CNN report of March 2012 find Asia's 10 greatest street food cities and tell us how the street food make these cities economically bustling and socially charming. Malaysia's Penang, Taiwan's Taipei, Thailand's Bangkok, Japan's Fuloka, Vietnam's Hanoi, Korea's Seoul, Singapore, China's Xian, Philippines' Manila, and Cambodia's Phnom Penh find their places in this list of 10 top Asian street food cities. Nothing represents the rich tapestry of India's multicultural fabric better than street food. The cuisine of a place speaks volumes about the weather; cultural lifestyle and habit of the people. The convenience and low price make street food the most favored choice. Many itinerant workers who do not have proper housing and cooking facilities have no option but to depend upon street foods.

The rising popularity of street food vending, besides its social support system for the under privileged urban population is it's easily accessibility; variety in taste, low cost fresh and often nutritious attributes. People spend almost 50% of their income on food; among low-income groups this figure may go up to 70% (Bhandari et al., 1989). A study conducted in India in the city of Varanasi, observed that about 42% of working men and women in the age groups of 25–45 and 61% of the students in the age group of 14–21 consumed food from street vendors rather than carrying foods from home to the workplace. About 82% of people of all age groups prefer to go to street vendors against 18% only who prefer to go to the restaurants in the evening (Mishra, 2007).

Street food is convenient because it is easily available, cheap, and nutritious, easy to afford as well as tasty, authentic, and culturally enriching, thus it plays a very important role, particularly for millions in the middle and lower income groups. The FAO and the WHO have carried out several studies on street foods in different countries of Asia, Africa, and Latin America, and found that besides their convenience and employment potential, the greatest factor in their favor is the fact that they can provide per capita calorie and protein requirements within a cost of around $1. Thus it also plays an important socioeconomic role in serving the food and nutritional requirements of consumers at affordable prices. It requires a low capital investment, offers a chance for self-employment and provides business opportunities for developing entrepreneurs. It also contributes to local and national economic growth by supporting local agricultural producers and food processors. Thus the socioeconomic significance of street foods is immense.

Street foods are an attractive experience of varied foods for tourists. In developing countries, making and vending street food provides a regular source of income for vast number of men and particularly women, who lack education or skills. Also the street food vendors are valuable, because it is a large significant area supporting the livelihood of millions of urban poor.

31.3 FOOD SAFETY HAZARDS

In spite of numerous advantages offered by street foods there are also several health hazards associated with this sector. Food may look, smell, and taste good and appear wholesome. However, microorganisms, chemicals, and hard foreign subjects may be found in and on raw foods. Food safety hazards can be found throughout the food supply chain. The physical environment in which street vendors work typically lack proper infrastructure, such as clean running water, toilets, and hygienic kitchens. Small organisms have the potential to infest food at any time during the harvesting, processing, storing, and preparation processes. These can be the biological, physical, or chemical agents in food which may cause illness or injury. Food safety hazards are of three types:

Physical hazards are the foreign elements in food which can be due to accidental contamination and/or poor handling practices. These can be metals, glass, wood, hair, plastic, nails, etc. or residues of pesticides in vegetables and fruits.

Biological hazards are caused by microorganisms like bacteria, viruses, or parasites which are present in air, food, water, soil, animals, and humans. Microbiological contamination due to improper food handling practices, poor personal hygiene, or use of substandard water.

Chemical hazards are the residues of cleaning and sanitizing agents, lubricants, paints, coatings, fertilizers, etc. These can occur at any point of time during growing, harvesting, storage, or service. Toxic metals like copper, zinc, brass, etc. can be present in the coatings of galvanized containers used for storing purposes or use of aluminum and ironware for cooking and heating purposes or excess of intentionally added chemicals which help to maintain a food's freshness or to enhance its flavor in foods, or use of artificial colors in *jalebis, sherbets, laddus*, etc.

Table 31.1 reports current hazards and critical control points observed at the main steps along the street food production chain.

Table 31.1 Hazards and Critical Control Points Identified at Different Steps of Street Food Production Chain

Step	Hazards (Biological)	Critical Control Points
Primary production: raw food stuff	Initial contamination	Raw food stuffs from illegal sources and bad quality.
		Contamination of raw food stuffs by different elements
Storage:	Growth of bacteria from initial contamination	Storage in inadequate containers or stores without protection from rodents and flies.
		Contamination by excrement and other food wastes
Fragmentation and conditioning	Growth of bacteria	Addition of illegal ingredients.
		Further contamination via hands
Processing: precooking/ preparation	Cross contamination and survival of bacteria.	Bad separation of raw products with cooked foods.
	Initial contaminants occur in unlock foods	Contamination via hands or in other ways.
		Inadequate washing of raw food before preparation
Vending: cooking/ cooling/reheating	Survival of pathogens and spores.	Inefficient cooking or reheating temperature.
	Production of toxins	Cross contamination.
		Inefficient holding temp.
Exposure vending: serving/consumption	Contamination and growth of bacteria and spores.	Bad protection from flies and dust.
	Production of toxins	Bad handling.

Chirag et al. (2013)

31.4 **FOOD SAFETY PRACTICES PERTAINING TO STREET FOODS AND THEIR HEALTH IMPLICATIONS**

Safety of street foods is vital for food safety and nutrition security. With the increasing pace of globalization and tourism, the safety of street food has become one of the major concerns of public health. Despite the nutritional, social, and economic benefits of street food they may pose serious concerns for health of urban population. Major concerns are related to biologic agents and chemical substances in food products. There are other concerns like poor hygiene, inadequate access to potable water, and waste disposal means. Further, unsanitary conditions (like proximity to sewers or garbage dumps) and pollution from traffic adds to the public health risks associated with street foods. Ensuring food safety is significant for the welfare of an individual, a community, and a nation.

Foodborne illness associated with the consumption of street vended foods have often been reported in India. Foods exposed for sale on the roadsides may become contaminated either by spoilage or pathogenic microorganisms (Bryan et al., 1992; Ashenafi, 1995; WHO, 1984). According to WHO (1989a,b), food-handling personnel play an important role in ensuring food safety throughout the chain of food production and storage. Disregard to hygiene measures on the part of the food vendors is quite common. Various studies indicate poor personal hygiene, health and food handling practices (vendors not wearing clean clothes, are dirty, chewing tobacco during food handling, and not wiping hands after every service of food). Contamination is often due to dirty clothing, lack of water supply, unhygienic handling and serving practices, contaminated hands, and lack of knowledge of hygienic practices. Bhasker et al. (2004) reported that defective personal hygiene can facilitate the transmission of pathogenic bacteria found in environment and on people's hands via food to humans. Handling with bare hands may result in cross contamination, introducing microbes on otherwise safe food (FAO, 1997). Paulson (1994) also reported that outbreaks are generally caused by foods due to poor personal hygiene of the vendors, during preparation or storage of food. Unhygienic surrounding like sewerage, improper waste disposal system, and inadequate water supply attract flies and houseflies which further increase food contamination as reported by Chumber et al. (2007).

Moreover, many foods are highly perishable. They are easily contaminated when produced in an unhealthy and unclean environment. More often than not, the microbiological quality of street foods, especially that of prepared dishes and drinks is below standard. This indicates inappropriate sanitary and hygienic practices during preparation and handling. In tropical countries most street foods may not be protected from flies. This facilitates spread of foodborne pathogens. *Salmonella* species in particular have been labeled as a post modern food contaminating bacteria, causing a high number of outbreaks of foodborne illnesses worldwide.

In India, traditional methods of processing and packaging, improper holding temperature, poor personal hygiene of food handlers are still observed during food marketing and sales. Consumption of raw or inadequately processed animal foods can have a significant public health risk as such foods are frequently contaminated with pathogens and occasionally with toxic chemicals. Similarly vegetables, fruits, and grains may carry hazardous contaminants. Additional hazards may also be in the form of use of improper food additives (often unauthorized coloring agents), mycotoxins, heavy metals and other contaminants (pesticide residues) in street foods.

31.5 FOOD SAFETY REGULATORY REQUIREMENTS

With the booming street food industry in the developing world, there is an urgent need to ensure that food vendors adhere to hygienic practices to protect public health. Since this is an unorganized sector, there is felt need of its development. There is difficulty in controlling the large numbers of street food vending operations because of their diversity, mobility, and temporary nature.

Lack of knowledge on hygiene and basic infrastructure like potable water, storage facilities, and unsuitable environment for food operations (such as proximity to sewers and garbage dumps) can contribute to poor microbial quality of foods. Appropriate location and condition of vending stalls, observation of personal hygiene by vendors, employing washed and clean utensils, and proper drainage and waste disposal are some steps to be taken which can lead to hygiene and safe food.

The advances in food science and technology, innovations in storage and distribution systems, and the growing demand can prompt the industry and food control authorities to further intensify their efforts to better protect the consumer.

The standards of street food safety can be upgraded by the vendors through implementation of some basic good practices with respect to hygiene and food handling. Standards for food handler requirements such as hand-washing, working attire, personal hygiene, and personal behavior should be maintained. Standards for food-vending equipment; which must be food grade, easy to clean and sanitize, standards for food serving such as safe food cover and packaging, regular hygiene and sanitation inspection, and monitoring of good food handling practices should always be maintained. Kinton and Ceserani (1996) recommended that food stuffs of all kinds should be kept covered as much as possible to prevent contamination from dust and flies.

Microbial hazards and their solution, critical points, practical control processing measures, and monitoring procedures as well as principles of food processing microbiology and food safety need to be incorporated for safe street food preparation.

The FSSAI has started engaging with the National Association of Street vendors of India in systematizing and professionalizing street foods across cities in the wider interests of ensuring public health and protecting livelihood of street food vendors. FSSAI is also in the process of developing guidelines and regulations to ensure food safety and safe street foods for state governments to implement. The FSSAI has prepared checklist, guidelines, and prerequisites for registration of medium to small food vending establishment to ensure food safety and upgrading of existing conditions of eating establishments. Several agencies including the state governments, departments of public health, commerce, consumer affairs and food processing, local municipalities, and police administration have identified roles and responsibilities in these guidelines and regulations that are expected to be executed to ensure safe street foods.

To instil a professional face to street food operators, the street food safety management needs a Hazard Analysis Critical Control Point (HACCP) and the prerequisite system as good manufacturing practices (GMPs) and good hygiene practices (GHPs).

FAO and WHO recognizing this spiraling Asian trend, has promoted documents on "Food safety requirements for street vended foods" and also on "Training aspects" of safe food for small operator and inspectors. Presently countries like Singapore, Thailand, and Malaysia have put together a structured national program to promote food safety in street vended food.

31.6 POLICY ISSUES FOR PROVISION OF SAFE FOOD

WHO has developed few measures for street food vendors based on the principles of five keys to safe food. These can be incorporated and taught in Indian scenario. These are:

Key 1: Keep clean
Key 2: Raw and cooked food should be kept separated
Key 3: Destroy hazards when possible
Key 4: Keep microorganisms in food from growing
Key 5: Use safe water and raw material

The Ministry of Food processing industries has proposed schemes for "Safe Food Towns" and "Safe Food Streets." These schemes aim to upgrade the quality of street food by promoting Indian cuisines at affordable rates in locations of tourist importance through upgrading and creation of common standards.

31.7 CONCLUSIONS

Keeping in view the scale of operation of such enterprises, it is quite a challenge to bring about improvements in the street food scenario, especially in a country where having adequate food daily is a dream for many, and no wonder, food safety is the second priority. In addition to the packed food manufactured in the organized sector, which is comparatively easier to test and monitor, there are issues with unorganized sector as to how to monitor the safety of food served in large number of roadside *dhabas* and stalls. Policing action has to be coupled with training. Local bodies have also to provide facilities for safe water supply, garbage disposal etc. The HACCP approach can be applied to keep down hazards to consumers (Malik, 1981).

REFERENCES

Ashenafi, M., 1995. Bacteriological profile and holding temperatures of ready-to serve food items in an open market in Awassa, Ethiopia. Trop. Geogr. Med. 47, 244–247.

Bhandari, N., Bhan, M.K., Sazawal, S., Clemens, J.D., Bhatnagar, S., Khoshoo, V., 1989. Association of antecedent malnutrition with persistent diarrhea: a case control study. Br. Med. J. 298, 1284–1287.

Bhasker, J., Usman, M., Smitha, S., Bhat, G.K., 2004. Bacteriological profile of street foods in Mangalore. Indian J. Med. Microbial 22, 197–1197.

Bhowmick, S.K., 2005. Street Vendors in Asia: A review. Economic and Political Weekly: 2256–2265.

Bryan, F.L., Teufel, P., Riaz, S., Qadar, F., Malike, J., 1992. Hazards and critical control points of vending operations at a railway station and a bus station in Pakistan. J. Food Proc. 55, 534–541.

Chirag, G., Bala, K.L., Kumar, A., 2013. Study of Hygienic practices of street food vendors in Allahabad city, India and Determination of Critical control points for safe street food. The Allahabad Farmer. LXV III, 2.

Chumber, S.K., Kaushik, K., Savy, S., 2007. Bacteriological analysis of street foods in Pune. Indian J. Public Health 51 (2), 114–116.

FAO, 1997. Street foods. Food and Agriculture Organization Report, Rome. 1–4.

Food and Agriculture Organization. 1997. Street Food: Small Entrepreneurs, Big Business. Available from: http://www.fao.org/english/newsroom/highlights/1997/970408-e.htm

Government of India, 2004. National Policy on Urban Street Vendors, Department of Urban employment & Poverty Alleviation, Ministry of Urban Development & Poverty Alleviation.

Kinton, G., Ceserani, P., 1996. Children and street foods. Food, Nutrition and Agriculture. 17/8. FAO, Rome. 28.

Malik, R.K., 1981. Food—A priority for consumer protection in Asia and the Pacific Region. Food Nutr. 7 (2), 18–23.

Mishra, S., 2007. Safety aspects of street foods: a case study of Varanasi. Indian J. Prev. Soc. Med. 38 (1–2), 1–4.

Paulson, D.S., 1994. A comparative evaluation of different hand cleaners. Dairy Food Environ. Sanit. 14, 524–528.

WHO, 1989. Health surveillance and management procedures for food-handling personnel: WHO Technical Report, 785: 5–47.

WHO, 1989. Health surveillance and management procedures for food handling personnel. WHO technical report series, 785. Geneva. 52.

Food safety during travel

32

S. Kathirvel

Department of Community Medicine, School of Public Health, Post Graduate Institute of Medical Education and Research, Chandigarh, India

32.1 INTRODUCTION

Travel is the movement of people from one geographical area to another. Travel is part and parcel of human life. "Travel" the etymology came from French word "travail" which means "struggle" or mentions three stakes, that is, "to trouble, torture and torment" (Online Etymology Dictionary, n.d.) It refers to the extreme difficulty of travel in ancient times. It was through travel that many discoveries were made!

Now, globalization and technology has made traveling easy and fast. As per a WHO report, at any point of time there are 500,000 people on board in airways every day (World Health Organization, 2010). One of the necessities for travel is the availability of safe food. The recent news of growing vegetables (Veggie garden) in space by NASA substantiates this (Meggs, 2010).

Historically, people used to pack cooked food and carry it along. Some carried raw food materials and cook wherever supporting environment is available. Now the availability of commercial food outlets has changed this. Packing of food for festivals, temples, carnivals, visiting relatives, and friends, etc. still prevails in some parts of world like India.

Exploring and enjoying the local cuisine is always an essential part of any travel. An otherwise enjoyable trip could be marred if one develops an upset stomach due to unhygienic food. Such instances also affect the tourism and travel industry.

The propensity of humans to experiment with food could also lead to certain foodborne diseases, from the local food (e.g., wild and domestic meat), which the traveler was not exposed to at his native place (Cohen and Avieli, 2004).

Prudent food practices from "farm to fork" can prevent most foodborne diseases. Many developing countries lack the technical expertise and financial resources to implement food safety policies. In principle, the government, the travel agency, and the host are responsible for food safety. But the main responsibility lies with the travelers in choosing what to eat and drink. It requires planning, preparation, self-discipline, and vigilance.

32.2 FOOD BORNE DISEASES

32.2.1 TRAVELERS' DIARRHOEA

Diarrhea referred to as "travelers' diarrhea (TD)," "travelers' tummy," "Delhi belly," "Karachi couch" is the most common illness reported in travelers anywhere in the world (Fodor's Travel Talk Forums, 2016). It is reported in 20–50% of travelers (Arduino and Dupont, 1993). Around 10 million people per year are affected with TD. It is defined as passage of three or more unformed stools per day with one or more of these symptoms: nausea, vomiting, abdominal cramps, fever, tenesmus, or blood in stools. Mostly TD is self-limited lasting for 3–4 days. It is benign in nature but it may cause great inconvenience. In 8–15% cases, TD may last for about 7 days and in 2% of cases for up to months. Usually, TD occurs in the first week of travel, but it may occur even 7–10 days after returning home (Steffen, 2005). Bacterial entero-pathogens are the most common (~80%) cause of TD (Arduino and Dupont, 1993; DuPont, 2009). Other causes are by viruses, protozoan, and parasites. The severity and complications are determined by source, amount and duration of exposure, aetiol-ogy, seasonality, and various host factors. Infants, children, elderly, pregnant women, and immune-compromised people with cancer, HIV/AIDS, diabetes are more highly prone to foodborne diseases (New Zealand Food Safety Authority, 2006; Schlichter, n.d.; Steffen, 2005). They should take special precautions during travel.

32.2.1.1 Causes of travelers' diarrhea (DuPont, 2009)

1. Bacteria
    a. Enterotoxigenic *Escherichia coli* (ETEC)
    b. Enteroaggregative *E. coli* (EAEC)
    c. *Salmonella* species
    d. *Shigella* species
    e. *Staphylococcus* species
    f. *Clostridium* species
    g. *Bacillus* species
    h. Enterohemorrhagic *E. coli* (EHEC)
    i. *Campylobacter* species
2. Viruses
    a. Rotavirus
    b. Norwalk virus
3. Protozao
    a. Giardia
    b. *Cyclospora*
    c. *Entamoeba*
4. Others
    a. *Cryptosporidium*

32.2.2 OTHER DISEASES

Botulism, typhoid, cholera, hepatitis A and E, avian influenza (H5N1), Guillain-Barré syndrome, Montezuma's revenge and giardiasis are other common conditions.

Food cuisine operators often enjoy the advantage of little liability for serving unsafe food since the chances of complaints to official authority are very less. This is due to travelers' short duration of visit and poor awareness of official complaint system. Some people do raise complaints but with the refund of money for their food they might compromise. Even against official complaints, the action may be masterly inactivity by the authorities. It may not always be possible to attribute the illness conclusively to a particular food/meal.

32.3 **DETERMINANTS OF FOOD SAFETY DURING TRAVEL**

Exposure to unsafe food and occurrence of foodborne diseases depends on number of factors apart from various host factors such as

1. type of travel
2. destination
3. duration of stay
4. season of travel
5. place of stay
6. knowledge on local foods and
7. modes of travel

For discussion purposes, food safety during travel is divided according to mode of travel.

32.3.1 **TYPE OF TRAVEL AND DESTINATION**

Travel can be divided into following four major types:

1. *Local*: Intrastate/provincial travel
2. *Regional*: Interstate/provincial/regional travel
3. *National*: Interregional but within a country travel
4. *International* travel

The common type of travel is short-term local travel like traveling from one district to another. Regional travel is also similar in kind, since the individual has the knowledge of food facilities. So the chances of foodborne diseases are minimal but not completely absent. Mostly, the travelers are accustomed to the local and regional foods. In South India, *idli, dosa, vada, pongal* are common breakfast menu. For example, if a person belongs to Kanyakumari (Tamil Nadu—Southern end of India) who travels to Hyderabad (Telangana) or Bengaluru (Karnataka), he finds little difficulty in assessing a good restaurant and so there is minimal risk.

In interregional or long-distance national travel; like person from northern part of India traveling to southern part or vice versa, there are high chances of exposure to poor quality of food. The main staple food is wheat in North India and rice in case of South India. Language is another barrier. The methods of food preparation, serving, and eating are different.

32.3.2 **DURATION OF STAY**

Increase in duration of stay at any new place usually increases the chances of getting foodborne diseases, as is confirmed by a systematic review on global etiology of TD (DuPont, 2009). Developing immunity may reduce incidence of disease if the duration is further prolonged.

32.3.3 **SEASON OF THE TRAVEL**

The environmental conditions such as temperature, humidity, rain, etc. are important factors for precipitating disease. During rainy season the contamination of water and food is quite possible. At tourist places, the peak season and off season poses greater risk of illness. In peak season, owing to high demand, there are numerous street vendors and the fixed restaurants also compromise their quality and food safety. During off-season time, the choice of food facilities may be reduced like in the hills, people tend to be indoors during peak winters and majority of the food facilities may be closed. Tropical climate poses high risk for TD because of the growth of the microorganisms and vectors (von Sonnenburg et al., 2000).

32.3.4 **PLACE OF STAY**

A standard hotel accommodation may be good for getting reasonable quality of foods and not staying in hotel itself increases the chances of TD. There are opportunities worldwide to stay as paying guest (PG) in houses. These various places of stay determine the exposure to microorganisms and hence occurrence of foodborne diseases (Steffen, 2005). Detailed brochures are sometimes circulated among the travelers regarding various dos and don'ts including food as one of the topics, which might be helpful (Schlichter, n.d.).

32.3.5 **KNOWLEDGE ON LOCAL FOODS AND LANGUAGE**

Poor knowledge of local language and food items limits food availability to the traveler. Especially in non-English speaking countries, the food names could confuse the travelers. In that case, the traveler can use a phrase book for reference (Cohen and Avieli, 2004).

32.3.6 **MODE OF TRAVEL**

Food safety is also determined by the mode of travel. The most stringent implementation of quality control and food safety is in the airways followed by seaways, railways, and finally roadways. But the incidents of foodborne diseases also occur in airways and seaways. These are better reported since the surveillance system is better as compared to railways and roadways.

32.4 COMMON TIPS FOR ENSURING FOOD SAFETY IN ALL MODES OF TRAVEL

Woo (2011); Fodor's Travel Talk Forums (2016); New Zealand Food Safety Authority (2006); Schlichter (n.d.); Wikihow (n.d.); NY Times Health (n.d.); Cohen and Avieli (2004)

1. *Learn about the area to be visited*
    a. Search and read about the area to be visited in advance. Ask people who visited the place and ask them to share their experiences. Tell other people who are going to visit with you. Especially teach children, pregnant women, and elderly.
    b. Find out the restaurants and hotels that normally cater travelers.
    c. Ask help from concierge or crews for finding a safe place for eating.
    d. Beware of food adulteration.
    e. Pack adequate food from a good restaurant, while on move.
    f. *Website searches*: Website of the tourist place, hotel, and restaurants can help with regard to food safety. There are websites where people share their experiences on best or worst food.
    g. Official ratings or grading system on food safety may also help.
2. *General hygiene measures*
    a. Swimming pool
        - Water may look clean. Avoid water entering into mouth and nose
        - Do not rinse your mouth while showering and in pool
    b. Hand hygiene
        - Wash your hands with soap and water wherever possible. If not possible carry an alcohol-based hand sanitizer (check for allergy before using it) or disposable wipers.
        - Do not touch places while eating which may look clean but may be contaminated. If so, wash your hands or use hand sanitizer.
    c. Finding a restaurant (Food and water safety, n.d.; von Sonnenburg et al., 2000)
        - Extensive handling of food by the operators poses high risk for developing illnesses.
        - Generally a clean and busy restaurant is considered to be good for taking food because the chance of serving fresh food is high. But just a busy restaurant may not always ensure safe food!
        - Look for food, food handlers, utensils used, and other hygiene measures to assess the cleanliness. You may ask people who are eating.
        - Avoid food at street vendors.

    d. Package of food
  - Avoid food packed in newspapers and other types of used papers.
  - Take only reputed brand and sealed food items.
  - Check the physical, chemical, and nutritional content of the food and package materials. Renal system of thousands of infants was affected in China who took melamine (non protein chemical) adulterated milk (Yang et al., 2009). Bisphenol is an industrial chemical used in the production of baby bottles and food containers can cause development and reproductive toxicity at high doses (FDA, n.d.).
  - Pack food in an insulted lunch bag or cooler and keep it below 40°C.
  - Transport food in a cooler (packed with ice or ice packs).
  - Pack easy to transport, self-stable foods, and single serve boxes of cereals.

    e. Temperature maintenance (DuPont, 2008; National Sanitation Foundation, n.d.)
  - Do not eat food kept for more than 2 h at room temperature.
  - Avoid food served at room temperature. Hot food should be hot and cold food should be cold.
  - Beware of anything cold especially meat.
  - Do not take the foods from street vendors which are not cooked for at least 5 min.

    f. Eating and sharing food items (Fodor's Travel Talk Forums, 2016)
  - In India, sharing food with a cotraveler is a common culture. It should be avoided not only for general safety and security reasons but also for food safety reasons. The shared food may contain undesirable/contaminated agents.
  - Eat any prepared food in one sitting.
  - Do not save the left-over.

    g. Medical kit
  - It should contain oral rehydration salt (ORS), bismuth subsalicylate, antibiotics, antipyretics, and antispasmodics, etc.

    h. Immunization
  - Take the necessary immunization such as Hepatitis A, cholera, and typhoid vaccines.
  - Mental stress also reduces the immunity, so avoid being stressed.

3. *Water treatment* measures (Woo, 2011; Food and water safety, n.d.; DuPont, 2008)

    a. Use of boiled water for drinking purposes (rolling boil for 1 min) is ideal. In higher altitude, water boils at a lower temperature and so germs are less likely to be killed.

    b. Use bottled water in case it is not possible to boil water. Check the seal before use. In hotels ensure that the seal of the water opened in front of you. Use bottled water of a reputed brand. Do not leave the opened bottle without cap.

   c. Purifiers can be used because modern purifiers are transportable and very effective. But it may not filter viruses grossly contaminated water.

   d. *Chemically disinfected water (iodine/chlorine based drinking water tablets):* It does not have any effect on viruses. Purchase it form the place of origin because it may not be available at destination. Be cautious in using iodine-disinfected water in children, pregnant women, persons with sensitivity and thyroid disorders.

   e. Use only pretested water for even brushing and cleaning of mouth.

   f. Sealed carbonated water is considered safe because of little chance of refill of the used bottles. The acidic pH in carbonated water reduces the exposure to harmful bacteria.

   g. Before opening the bottle or can, wipe the container opening and drinking edge with a clean tissue paper.

   h. Do not take water or any other beverages or desserts with ice of unknown water source. Ensure that the ice is prepared from safe water. The organisms inside the ice survive and even alcohol may have little effect on the germs present in the ice.

   i. Do not drink fountain drinks

4. *Hot and cold beverages*

   a. Automatic hot or cold beverage vending machines are considered to be safe as there are lesser chances of handling, but the source of water and the cleanliness of the ingredient should be verified.

   b. Do not take ice creams with unknown source of water for ice preparation. You can take reputed/branded and packed ice creams instead.

   c. Do not take raw milk or other milk products. Use only pasteurized milk and juices.

5. *Vegetarian food*

   a. Wash fruits and vegetables with clean water thoroughly. These may be grown near the soil in which undesirable manure might have been used.

   b. Do not consume uncooked or undercooked food items. Eat only freshly cooked ones.

   c. Do not take salads and unpeelable fruits and vegetables. Do not consume cheese or fruits and vegetables that are peeled by someone else.

   d. Avoid vegetables and fruits with damaged skin which may accommodate worms and germs.

6. *Nonvegetarian food*

   a. Nearly all animal foods can be classified as high-risk foods such as eggs, poultry, red meat, and shellfish.

   b. Take only hard-boiled eggs (firm to touch). Do not take raw or runny eggs in any form, as salmonella may be present both in outer covering and inside egg. Salmonella free eggs are also available in the international markets.

   c. Do not eat uncooked or undercooked poultry meat products. There is a chance of transmission of avian influenza (H5N1) (Human infection with avian influenza A (H5N1) virus: advice for travelers, n.d.). Adequately and thoroughly cooked poultry products are safe.

d. Do not try bush meats such as monkey, bats, or other wild games (von Sonnenburg et al., 2000). It is good not to try new animal products when you are not used.

e. Eat small fish because the possibility of accumulation of chemical toxin is less (Schlichter, n.d.; Kresser, 2010). Especially avoid large reet fish. Avoid eating fish organs such as liver, intestines, eggs, and head which also have high level of toxins, compared to other parts of the fish.

f. Eating spoiled and fermented fish (certain species such as tuna, mackerel, *mahimahi*, etc.) can cause histamine or scombrotoxin fish poisoning which is a self-limiting condition within 24 h (Lawley, 2013).

g. Avoid seafood when there is high exposure to radio nucleotides like in case of Fukushima in Japan (Kresser, 2013).

32.5 MODE OF TRAVEL

32.5.1 ROADWAYS

Roads are most common route of travel in any country. The travel may last for hours to days. People may go on foot, bicycle, or any automobile. In case of a bus travel, we find food shops at bus stations and on highways. Local food vendors may board the vehicle. The safety and quality of the food also depends on the type of bus you are traveling in. Even bottled and sealed beverages can be found contaminated because of refill and use of expired items. The highway hotel's quality and safety may be better compared to local sellers. But long standing hours of buffets and poor hygienic status of the workers could compromise food. On board pantry facility is available on board some buses (eg. "Airawat Bliss" air-conditioned bus at Bangalore, India.

Tips for ensuring food safety on road:

1. Taking your food from home is the best option.
2. If not possible, use clean and busy restaurant.
3. Do not eat from street vendors especially colored and decorated food items.

32.5.2 RAILWAYS

India has the one of the longest railway route in the world. The passenger service is gradually improving. A discussion among foreign tourists goes like this: A traveler from Singapore asks "I have a pretty weak stomach (India Travel Forum, 2010). Can we eat stuff purchased off the train pantry or we need to stock up of food before we board? What is the usual practice of foreigners taking long haul trains?" The reply was given by a traveler from Belgium. "I don't really know if the food is safe or not but I always eat anything I can purchase on the trains. I just like all these different things. Yummy!!. Also the food you can get on the platforms seems ok to me. I've never been sick" he replied on the same day.

I have only taken a couple of long haul trains, always found food safe to eat" a Canadian traveler replied. "Stay away from the milk based sweets (*barfi* etc.) they often have a much higher than normal bacteria count" another traveler replied.

Other interesting answers and feedbacks:

"It depends, if the pantry car is on the train or food is picked up from stations. I got very sick once from a barely warm *samosa*...If you know your food is coming fresh off the cooker in the pantry car, then no problem.

"Look through the pantry car window and make up your mind. One look is enough".

"if there is a pantry car, are all foods guaranteed to be prepared there?..."

"No all foods are not guaranteed to be prepared there..."

I went inside pantry car a couple of times and saw a microwave where they are heating the snacks which they picked up from the station before"

"...catering on the *Rajdhani* and *Shatabdi* expresses are of a very high standard..."

"well, if this makes you feel comfortable-I have not heard anyone die from eating food in trains" (a traveler from USA).

"Be wary of food served on newspapers, sometimes the newspapers are collected from the ground and sold to the food sellers".

"I've also found that Indian families on the trains will insist on sharing food with you (a Canadian passenger)"

These discussions never end...

Nowadays food and beverages can be ordered online for train delivery. The station of traveler's choice (bid stations) can be selected and those companies assure the quality and safe food from a reputed restaurant (e.g., Travel Khana).

Food is available to train travelers at the following:

a. in-house pantry car where all food is cooked;
b. in-house pantry car only for storage and temperature maintenance;
c. food facilities at railway stations.

32.5.2.1 Tips to train travelers

1. Avoid window side seats while eating and, if unavoidable, take a seat opposite to the flow of wind.
2. Do not share any food or beverage with anybody.
3. Do not use water from train for rinsing mouth and brushing.
4. Do not touch any place of seat or others during eating.
5. Drink always reputed brand bottled water
6. Do not take colored food items.
7. Do not eat ice creams in train unless it is of reputed brand/from pantry car with proper storage.

32.5.3 SEAWAYS

Ships may accommodate thousands of travelers. The travel may last for days to months. Travelers from different parts of the world enjoy the special foods prepared on board. On a confined ship there may be no choice of food and no other commercial outlets present. Numbers of foodborne disease outbreaks were reported in the past from top cruises of the world due to poor sanitation condition. The very recent (January, 2014) gastrointestinal illness affecting 577 passengers and 49 crews of Royal Caribbean cruise "The Explorer of the Seas' Voyage" made to return ashore prematurely (Reuters, 2014). In 2000, the famous"Disney Magic" had an outbreak of gastro enteritis among 260 travelers due to contaminated shrimp served on lunch buffet (CDC, n.d.; Vessel Sanitation Program—Disney magic outbreak investigation, n.d.). Variety of bacterial and parasitic pathogens was isolated from stools of the affected travelers.

A traveler from a cruise complained about the limited menu and just tepid food. Another traveler from Pacific Dawn says that"...most part of the food had that reheated taste...lack of choice noticeable....texture of the vegetables like a wet sponge..." (P&O pacific dawn review by oldmouse3, 2015). Somebody in a discussion forum mentioned ships as a giant Petri dish (Hunter, 2014). Each year 19–21 million illnesses and 56,000–71,000 hospitalizations happen because of outbreaks on cruises (CDC, n.d.).

32.5.3.1 Food safety tips for sea travelers

1. Do not take buffet foods since they are left for hours because thousands of passengers are to be served.
2. Refer "Green Sheet" from Centre for Disease Control (CDC) website for the compliance on "Vessel sanitation program-Operational manual" to see the score of each cruise ships for travel decisions (Vessel Sanitation Program Operations Manual, 2005).
3. Spitting on the pools and drinking alcohol and eating are common inside the pools.
4. Check the number of alcoholic drinks because the alcohol content may vary from country to country.

The food facilities can be divided into

1. official restaurant of the ship
2. other commercial eateries
3. alcoholic and non alcoholic beverage (juice) centers
4. swimming pools, pool decks, and spa

32.5.4 AIRWAYS

There are incidents of foodborne diseases in well-reputed airways also (Mangili and Gendreau, 2005). Exposure to unsafe food may occur with

1. Self-packed food from home or other outside air port eateries
2. Food from inside airport lounge
3. Food on board

Food items brought from outside airport check in area are not allowed inside the airport lounge and flights in India. But it is allowed in most other countries. Even water more than 200 mL is not allowed on Indian domestic flights. In case of international flights the rules are different. On-board sky food shops have menu with limited items. There are high chances that the temperature maintenance system may be at fault or not kept at prescribed temperature and poor or uneven heating of food items.

All flights should follow the International Flight Service Association (IFSA) guidelines for ensuring food safety and other sanitary and hygiene measures (World Food Safety Guidelines, n.d.). An editorial of IFSA says that food safety is the top priority (Fowler, n.d.). Food and Drug Administration (FDA), USA found cockroaches, Gnats (too numerous to count), unrefrigerated food, utensil on dirty racks in the company Gate Gourmet which serves many top airlines (Morran, 2010; AOL Travel UK, 2012; The Examiner, 2012). Nearly the same or worse condition was observed in LSG Sky Chefs (Clark, 2010). The executive lounge of British Airways at Terminal 5 was rated poor in food hygiene and safety in 2013. But fortunately there was no reported illness (Dalton, 2013; Cohen, 2013).

32.6 CONCLUSION

Though neglected, food safety during travel will always remain as a contemporary issue. Food safety during travel is mostly taken as a chosen decision by the traveller. But the equal and shared responsibility and accountability of the government and food facility operators on ensuring the safety of food cannot be neglected. Stringent and universal quality control and regulatory measures, strong surveillance, and authentic information system will ensure safe food especially during travel. The need for region, country, and subcountry specific food guides and its dissemination to travellers is a need of the time.

REFERENCES

AOL Travel UK, 2012. Mice and cockroaches found by airline food inspectors. Available from: http://travel.aol.co.uk/2012/11/18/mice-and-cockroaches-found-by-airline-food-inspectors/

Arduino, R.C., Dupont, H.L., 1993. Travellers' diarrhoea. Baillieres. Clin. Gastroenterol. 7, 365–385.

Centers for Disease Control and Prevention. Vessel sanitation program—cruise ship outbreak updates. Available from: http://www.cdc.gov/nceh/vsp/surv/gilist.htm

Centers for Disease Control and Prevention. Food and water safety. Available from: http://wwwnc.cdc.gov/travel/page/food-water-safety

Centers for Disease Control and Prevention. Human infection with avian influenza A (H5N1) virus: advice for travelers. Available from: http://wwwnc.cdc.gov/travel/page/human-infection-avian-flu-h5n1-advice-for-travelers-current-situation

Centers for Disease Control and Prevention. Vessel Sanitation Program—Disney magic outbreak investigation. Available from: https://www.cdc.gov/nceh/vsp/surv/gilist.htm

Clark, T., 2010. Airline food catering companies exposed for 'poor hygiene'. Available from: http://www.dailymail.co.uk/travel/article-1290521/Gate-Gourmet-LSG-Sky-Chefs-Flying-Food-Group-slammed-poor-hygiene.html

Cohen, B., 2013. Poor food hygiene and safety rating for British Airways lounges in terminal 5 of Heathrow airport. Available from: http://www.flyertalk.com/the-gate/blog/25160-poor-food-hygiene-and-safety-rating-for-british-airways-lounges-in-terminal-5-of-heathrow-airport.html

Cohen, E., Avieli, N., 2004. Food in tourism. Ann. Tour. Res. 31, 755–778.

Dalton, A., 2013. British Airways Heathrow lounge service "poor". Available from: http://www.scotsman.com/news/transport/british-airways-heathrow-lounge-service-poor-1-3106334

DuPont, H.L., 2008. Systematic review: prevention of travellers' diarrhoea. Aliment. Pharmacol. Ther. 27, 741–751.

DuPont, H.L., 2009. Systematic review: the epidemiology and clinical features of travellers' diarrhoea. Aliment. Pharmacol. Ther. 30, 187–196.

Fodor's Travel Talk Forums, 2016. First time to India. Concerned about food safety. Available from: http://www.fodors.com/community/asia/first-time-to-india-concerned-about-food-safety.cfm

Food and Drug Administration. Bisphenol A (BPA): use in food contact application. Available from: http://www.fda.gov/NewsEvents/PublicHealthFocus/ucm064437.htm

Fowler, J. International Flight Services Association Statement on airline food safety. Available from: http://www.ifsanet.com/?page=AirlineFoodSafety

Hunter, M., 2014. Are cruise ships floating petri dishes? Available from: http://edition.cnn.com/2014/01/29/travel/cruises-sanitation/

India Travel Forum, 2010. Food on train safe? Available from: http://www.indiamike.com/india/indian-railways-f10/food-on-train-safe-t114734/

International Flight Services Association. World Food Safety Guidelines. Available from: http://www.ifsanet.com/?page=World_Guidelines

Kresser, C., 2010. Is eating fish safe? A lot safer than not eating fish! Available from: http://chriskresser.com/is-eating-fish-safe-a-lot-safer-than-not-eating-fish/

Kresser, C., 2013. Fukushima radiation: is it still safe to eat fish? Available from: http://chriskresser.com/fukushima-seafood/

Lawley, R., 2013. Scombrotoxin (histamine). Available from: http://www.foodsafetywatch.org/factsheets/scombrotoxin-histamine/

Mangili, A., Gendreau, M.A., 2005. Transmission of infectious diseases during commercial air travel. Lancet 365, 989–996.

Meggs, L., 2010. Growing plants and vegetables in a space garden. Available from: http://www.nasa.gov/mission_pages/station/research/10-074.html

Morran, C., 2010. Airline food might not only taste bad, it might make you sick. Available from: http://consumerist.com/2010/06/28/airline-food-might-not-only-taste-bad-it-might-make-you-sick/

National Sanitation Foundation. Protect against food poisoning while traveling. Available from: http://www.nsf.org/consumer-resources/health-and-safety-tips/food-safety-away-from-home-tips/protect-against-food-poisoning-while-traveling/

New Zealand Food Safety Authority, 2006. Food safety when you have low immunity. Available from: http://www.foodsafety.govt.nz/elibrary/industry/Safe_Eating-Zealand_Food.htm

NY Times Health. Travel abroad—in-depth report. Available from: http://www.nytimes.com/health/guides/specialtopic/travelers-guide-to-avoiding-infectious-diseases/print.html

Online Etymology Dictionary. Available from: http://www.etymonline.com/index.php?term=travail&allowed_in_frame=0

P&O pacific dawn review by oldmouse3, 2015. Available from: http://www.productreview.com.au/r/p-o-pacific-dawn/289721.html

Reuters, 2014. Over 600 fall ill on Royal Caribbean cruise. Available from: http://www.torontosun.com/2014/01/26/more-than-300-fall-ill-on-royal-caribbean-cruise-ship

Schlichter, S. Food safety: how to avoid getting sick while traveling. Available from: http://www.independenttraveler.com/travel-tips/safety-and-health/food-safety-how-to-avoid-getting-sick-while-traveling

Steffen, R., 2005. Epidemiology of traveler's diarrhea. Clin. Infect. Dis. 41 (Suppl. 8), S536–S540.

The Examiner, 2012. Roaches and other creepy crawlies found in facilities that make airline food. Available from: http://www.examiner.com/article/roaches-and-other-creepy-crawlies-found-facilities-that-make-airline-food

US Public Health Service, Centers for Disease Control and Prevention, National Center for Environmental Health, 2005. Vessel Sanitation Program Operations Manual. Available from: http://www.cdc.gov/nceh/vsp/operationsmanual/opsmanual2005.pdf

von Sonnenburg, F., Tornieporth, N., Waiyaki, P., Lowe, B., Peruski, L.F., DuPont, H.L., Mathewson, J.J., Steffen, R., 2000. Risk and aetiology of diarrhoea at various tourist destinations. Lancet 356, 133–134.

Wikihow. How to avoid having diarrhea during travel (with pictures). Available from: http://www.wikihow.com/Avoid-Having-Diarrhea-During-Travel

Woo, R., 2011. Avoid foodborne illness when traveling abroad. Available from: https://www.foodsafety.gov/blog/international_travel.html

World Health Organization, 2010. International travel and health: situation as on 1 January 2010. World Health Organization.

Yang, R., Huang, W., Zhang, L., Thomas, M., Pei, X., 2009. Milk adulteration with melamine in China: crisis and response. Qual. Assur. Saf. Crop. Foods 1, 111–116.

Food safety during fairs and festivals

33

G. Ghose, I. Saha

Department of Community Medicine, IQ City Medical College, Durgapur, West Bengal, India

33.1 INTRODUCTION

Fairs and festivals are a global phenomenon and are celebrated in every country or region of the world. They are celebrations, which propagate the cultural and/or religious heritage of communities and nations. They are a means to keep alive age-old traditions, which are passed on through generations. They also provide a break from the routine life and an opportunity to celebrate, bring happiness, and strengthen community feelings. Fairs and festivals are always associated with peace, harmony, and happiness, and therefore are an inescapable function of societies.

In India, the flow of fairs and festivals continues throughout the year. A large number of religious and national festivals occur which involve the whole nation. Many festivals are regional in nature, celebrated with great fervor in specific areas. Numerous fairs, called "*Mela*" take place throughout the year. There are religious fairs, historical fairs, animal worship fairs, cattle fairs, monsoon fairs, changing season fair, and many more. These celebrations comprise of rituals of prayers, ritual holy baths, exchanging goodwill, decorating houses, wearing new clothes and jewelry, music, sports activities, artistic performances, song and dance, and lastly, and very importantly, eating and drinking.

Community gatherings where food is provided have existed ever since early human communities celebrated important events—much before Government regulations on food safety were formalized (Schmidt and Rodrick, 2005). Safeguarding public health during these times is of utmost importance, and lapses in this respect may result in disastrous outcomes. A case in point is the epidemic, which occurred during *Haridwar Mela* in 1867 and reportedly spread to Persia and Russia (Mukherjee, 1987).

During these fairs and festivals, mass gatherings take place, comprising of people of different age groups arriving from various places to attend the event. Food safety during these festive times always poses a challenge to public health authorities entrusted with the responsibility to ensure a healthy environment for the population attending these functions.

33.2 FOOD SAFETY HAZARDS DURING FAIRS AND FESTIVALS

Despite continued progress in improving the quality and safety of foods produced all over the world, food-related illness remains a serious public health problem, particularly during transient large gatherings of people during fairs and festivals. Even in developed countries, such as the Unites States, during Jul. 2007, over 600 visitors to the annual food festival called "Taste of Chicago" fell ill with *Salmonella* linked to a hummus dish sold by a Chicago restaurant, "Pars Cove" (International Food Security Network, 2014). In another outbreak of salmonella food poisoning, symptoms of abdominal cramps, diarrhea, nausea, and vomiting were reported after food served at a church fundraiser in North Carolina on Sep. 7, 2013 was consumed (North Carolina State University, 2013).

In India, where generally people are minimally concerned about food hygiene and safety, any contamination of food during mass gatherings may lead to outbreak of food borne illness with consequential morbidity and mortality. In Feb. 2013, during celebration of *Saraswati Puja* in a school in Kamrup district of Assam, 400 students from adjoining schools got ill within 2–5 h of consumption of *khichri* and *prasad* and presented with fever, vomiting, diarrhea, and abdominal pain. Causative agent was salmonella and its preformed toxins (Sharma et al., 2014). In Oct. of the same year, 70 people presented with symptoms of fever, diarrhea, and vomiting after consumption of *prasad* prepared from sprouted seeds during *Ayudha Puja* in Chowdeshwari temple, Lakkashandra near Bangalore (Deccan Herald, 2013).

Certain festivals occur periodically every year and are associated with certain specific types of food items, which are consumed. For example, *jalebi* in *Diwali*, *thandai* drinks during *Holi*, or *paneer*- and *khoya*-based sweets during *Vijaya Dashami*. Due to sudden spurt in demand, large quantities of these items are manufactured and sold, often under unsafe and unhygienic conditions. Make shift roadside food stalls mushroom all over the towns and cities, where food is left exposed to contamination with dust, houseflies, or unhygienic handling. During the celebration of *Navratras* food is prepared from *kuttu ka atta* (buckwheat) and *singhare ka atta* (from water chest nut). Cases of food borne illness have been reported due to consumption of fungus infested *kuttu ka atta*. In Apr. 2011, some 300 people were hospitalized in Delhi after consumption of preparations made from contaminated buckwheat flour during the *Navratri* period (TNN, 2011). One of the patients, a 50-year-old Municipal Corporation of Delhi staff died due to the illness.

It is a particularly common practice during religious festivals in most parts of India, that community feeding is organized for which food is cooked on a mass scale and served to public. Preparation and storage of food in such situations is often unhygienic leading to local outbreaks of food borne infections. Many such incidents continue to occur throughout the year. Some recent occurrences are listed Table 33.1.

There are some mega fairs, held cyclically, which lead to the agglomeration of a large number of people from various parts of the country. Examples are the *Kumbh Mela(s)* and *Hariharchetra Mela* at Sonepur, Bihar. Here too, there is a great demand

Table 33.1 Some Recent Events of Food-Related Outbreaks During Religious Festivals in India

Date	Place	Event
May 2009	Mandava village, Jhunujhunu district, Rajasthan	150 People got ill mainly with complaints of stomachache and vomiting after eating in religious *Sawamani* festival (The Hindu, 2009)
Oct. 2009	Tulsibani village, Mayurbhanj district, Orissa	50 People admitted to hospital after consumption of rice in a religious festival in a village (Thaindian News, 2009)
Jul. 2010	East Midnapore district, West Bengal	300 People got ill after a mass meal in a religious event (Thaindian News, 2010)
Apr. 2011	Nabrangpur, Orissa	200 People fell sick after consumption of *prasada* (*bhog*) during *Satyanarayan Puja* (Orissa Diary, 2011)
Nov. 2011	Rajkot, Gujarat	100 people suffered from diarrhea and vomiting after consuming *basundi* (milk sweet) and ice cream in a religious gathering (PTI, 2009)
Dec. 2011	Ahmedabad, Gujarat	200 People suffered from food poisoning after consumption of milk shakes prepared from contaminated milk during Muharram (Times of India, 2011)
Mar. 2012	Karindalam Village, Kasaragod district, Kerala	50 People suffered from food poisoning after consumption of ice cream from make-shift stall outside temple premise during festival (The Hindu, 2012)
Jul. 2012	Kolkata, West Bengal	400 People were admitted to hospital and 2 children died due to gastroenteritis after consuming food in an Iftar party (News Track India, 2012)
Mar. 2013	East Godavari district, Andhra Pradesh	200 People fell ill after consuming contaminated curd during *Kalyanam* festival of Lord Venkateswara (The Hindu, 2013)
Apr. 2014	Nalgonda district, Telengana	350 People were admitted to health facilities after consuming *panakam* the special drink prepared with jaggery, served at the Sri Kodandarama Swamy temple during *Sri Rama Navami* festival (The Hindu, 2014)
May 2014	Solapur, Maharashtra	400 People suffered with vomiting, stomachache, and 2 died after consuming food in a temple after worship (Video Samachar, 2014)

for food, which is prepared in large quantity, usually with scant attention to hygienic practices. There are also various trade fairs and entertainment fairs organized all over the country, where again food stalls are an important component. People visit these fairs to enjoy and are prone to try out rich or new foods with gay abandon. Characteristics of these fairs are:

1. Held at particular/specified times of the year at designated places.
2. Organized for short durations of time.
3. Gives rise to concentration of a large number of people.
4. People traveling from distant places often unaware of local administrative or health infrastructure.
5. Establishment of temporary arrangements for stay and food.

As these fairs are held in temporary sites, food preparation, storage and transport can be problematic. Food sellers at most of these fairs try to cut costs to increase their profit. They add diluents, such as, water to products to increase volumes. Cheap cooking oil is mixed with expensive oil, tea waste is mixed with new tea, and anything from urea to blotting paper is added to thicken gravies, sauces, or milk products (Williams and Banerji, 2012).

Due to the previously mentioned factors maintenance of food safety is at risk. The usual safety controls that a permanent kitchen provides, such as, regulated and temperature controlled cooking, proper storage/refrigeration, and good washing facilities, may not be available when cooking and dining at these events. Hence the chances of the following adverse, though preventable consequences are increased:

1. Increased incidence of food borne diseases with resultant increase in morbidity and mortality.
2. There may be occurrence of common exposure point source epidemic.
3. Continued propagated epidemic till the depletion of the susceptible hosts.
4. Increased burden on health care delivery system—may be manifested as increase in outpatient and inpatient load in the clinics and hospitals.
5. Increase in financial burden of the community.

33.3 CAUSES OF FOOD BORNE ILLNESS DURING FAIRS AND FESTIVALS

Reasons for the increased risk of food borne illness during fairs and festivals are as follows:

1. Procurement of raw foods may be compromised. Due to the huge demand, raw food procured is often of poor quality, may not be fresh and may be contaminated. Also, due to inadequate arrangement of water supply, proper washing of raw food may not be maintained.

2. Inadequate cooking of foods—less time and improper temperature.
3. Improper storage of foods/inadequate refrigeration or maintenance of adequate heat.
4. Cross-contamination of foods due to improper storage, often consequent to limitation of space or to lack of knowledge/training.
5. Use of inadequately trained hired/casual staff to cook and serve food.
6. Washing facilities for utensils/kitchen and for consumers not available or inadequate.
7. Contaminated/unsafe water supply for preparation of food and drinking purpose.
8. Limitation of space leading to overcrowding, which hampers maintenance of cleanliness.
9. Poor environmental sanitation with housefly/rat/other pest nuisance.
10. Climatic conditions conducive for compromised food safety measures. Incidence is higher during summer/monsoon.

The United States Center for Disease Control and Prevention lists the following six circumstances as the ones most likely to lead to illnesses (Centers for Disease Control and Prevention, 2013).

1. *Inadequate cooling and cold holding.* More than half of all food borne illnesses is due to keeping foods out at room temperature for more than 2–4 h.
2. *Inadequate hot holding.* Cooked foods not held above 60°C until served can be a significant source of food borne illness.
3. *Inadequate reheating.* When previously cooked foods are not reheated to above 74°C, illness often results.
4. *Preparing food too far ahead of service.* Food prepared 12 or more hours before service increases the risk of temperature abuse.
5. *Poor personal hygiene and infected personnel.* Poor hand washing habits, and food handlers working while ill are implicated in one out of every four food borne illnesses.
6. *Contaminated raw foods and ingredients.* Serving inadequately cooked chicken/meat products, or raw milk and dairy products that is contaminated, or using contaminated raw eggs in sauces and dressings, has often led to outbreaks of food borne disease.

In addition to threat of contamination of food with various microorganisms, there is also a risk of adulterants and contaminants, which may be added to food by vendors, particularly during festivals when large quantities of food have to be prepared. Wide usage of nonpermitted colors, such as, rhodamine, orange II, and auramine have been detected during analysis of samples of sweets and confectionery collected during festivals (Sudershan et al., 2009). Even permitted colors have been found to be used in excess of permitted limits, or found in foods in which they are not permitted (Sudershan et al., 2009).

33.4 MEASURES TO ENSURE FOOD SAFETY DURING FAIRS AND FESTIVALS

Food borne illness is transmitted by the *5 F's*—Foods, Fluids, Fingers, Flies, and Fomites. Keeping these factors in mind, necessary measures need to be taken. The basic principles of food safety apply equally in festival settings, namely (International Food Security Network, 2014):

- Acquiring food from safe sources.
- Cooking and storing food at proper temperatures.
- Controlling cross-contamination.
- Practicing good personal hygiene.

Food safety during festivals is the responsibility not only of the festival planners and vendors, but also of the individual consumers. Guidelines for actions required at various levels to ensure food safety are discussed.

Actions by Mela Authority.

1. *Licensing and registration of food business.* As per Section 31 of Food Safety and Standards Act, 2006, (Ministry Of Law and Justice, 2006) "no person shall commence or carry on any food business except under a license. However, petty manufacturer who himself manufactures or sells any article of food or a petty retailer, hawker, itinerant vendor or a temporary stall holder relating to food business or tiny food business operator need not apply for a license but they shall register themselves with such authority and in such manner as may be specified by regulations, without prejudice to the availability of safe and wholesome food for human consumption or affecting the interests of the consumers." The appropriate authority should grant license or registration only after ensuring adherence to food safety measures by the vendors. Implementation of these measures, however, is a difficult task, as the regulatory authorities are grossly understaffed at present. Flouting of regulatory norms is rampant but only a handful of defaulters are prosecuted (Williams and Banerji, 2012).

2. *Requirements of site.* The event organizer the designated fair authority should be responsible for the site, the water supply, and the waste disposal. The organizer must ensure that the site and layout for the event are properly planned. The area selected should be dry and well drained, without dust or muddy conditions. Washroom facilities, potable water availability and waste disposal should be catered for adequately, taking into consideration the footfall expected. Pest control measures should be arranged as applicable.

Responsibilities of food vendors. All participating vendors should ensure they have updated license/registration from the appropriate Government authority. The following need to be implemented by all vendors:

1. *Site.* All food outlets/stalls should be clean and tidy. Surfaces where food is served, displayed, or consumed should be spotlessly clean and continuously maintained in that state.

2. *Food handlers*. All food handlers should practice good personal hygiene and hand-washing, wear clean outer garments, use head gear that confines the hair and must not use tobacco within the food premises. Persons with cuts or burns should not be employed for food handling. If that is unavoidable, then they must wear disposable gloves that are changed often (Thunder Bay District Health Unit, 2011). Above all, all food handlers should be trained adequately on measures for food safety.

3. *Food holding*. All food held/displayed for sale should be protected with covers, lids or suitable barriers to avoid contamination. Perishable/hazardous foods stored or displayed should be maintained at proper temperature as under:
   a. Cold holding: 4°C or less
   b. Hot holding: At least 60°C

4. *Cooling equipment*. Refrigerators and other cold storage equipment/cooling chambers should be maintained clean. Thermometers should preferably be available to readily check the holding temperature.

5. *Utensils*. Utensils used for cooking/serving should be scrupulously cleaned with adequate fresh water/soap/sanitizers. Washing area should be earmarked and away from the actual serving/eating area. Disposable plates, bowls, cups, spoons/forks, and knives should be catered for customer use.

6. *Garbage disposal*. Containers with lids should be provided at convenient locations for disposal of used plates/leftover food etc., and these should be cleared at suitable periodicity and disposed off at central garbage disposal areas.

Steps to be taken by individual consumers. Individuals visiting fairs and/or participating in festivals who partake in consumption of food/drinks need to be aware of the hazards of food related illness in such circumstances, and take the necessary simple measures are discussed for prevention of the same.

1. *Practice good personal hygiene*. Hand washing with soap and clean running water for at least 20 s, prior to and after eating. Use hand sanitizers or disposable wipes if no place is found for washing hands.

2. *Dispose used trays/plates and left over food at proper place*. Be aware of food safety, its importance and basic principles. A quick checklist is given in Table 33.2.

3. *Monitoring and surveillance*. Fair and festival authorities should ensure constant monitoring of the *food vending establishments*. Food samples should be collected periodically from different vendors and analyzed for safety. Occurrence of cases with symptoms of food related illness like abdominal pain, nausea, vomiting and/or loose motions, must be reported to the appropriate authority without any delay, so that the local health department can initiate immediate measures to locate the possible source of infection and implement necessary control measures. *Food Safety and Standards Authority of India* (FSSAI) supports development of training manuals and communication campaign to individuals and agencies involved in food safety related work.

Table 33.2 Food Safety Checklist for Consumers

No.	Point to Check
1.	Is the vendor licensed/registered? Has the vendor been inspected?
2.	Does the vendor have a clean or tidy workstation, such as, clean floors, walls, ceilings, etc.?
3.	Does the vendor have adequate food preparation and serving areas?
4.	Whether hand-washing facilities are present?
5.	Does the vendor have a basin or sink for food handlers to wash hands and utensils?
6.	Does the vendor have an adequate supply of potable drinking water?
7.	Do the employees wear cap, gown or apron, and gloves while handling food?
8.	Does the vendor have refrigeration on site?
9.	Does the vendor have adequate hot food-storage equipment?
10.	Whether foods are being served freshly prepared or hot?

33.5 CONCLUSIONS

Food is an integral part of many social events. Temporary food service will therefore always be a part of such events. These events expose populations to risk of food borne infections, as the standard safety measures available in permanent locations are difficult to maintain. The universal principles of food safety, that is, Clean–Separate–Cook–Chill are applicable for food preparation and distribution in festivals and fairs too. It is a challenge for public health authorities to ensure that these principles are followed so that food borne illness is avoided and festivals remain a happy occasion for the community.

REFERENCES

Centers for Disease Control and Prevention, 2013. Food safety at fairs and festivals. Atlanta, USA. Available from: http://www.cdc.gov

Deccan Herald, 2013. Food poisoning lands 70 devotees in hospital. Available from: http://www.deccanherald.com/content/363053/039food-poisoning039-lands-70-devotees.html

International Food Security Network, 2014. Food Safety Info Sheet. Available from: foodsafetyinfosheets.org

Ministry of Law and Justice (Legislative Department), 2006. Government of India. Food Safety and Standards Act.

Mukherjee, P.N., 1987. Dr. BN Ghosh's Treatise on Preventive and Social Medicine, sixteenth ed. Scientific Publishers Co, Calcutta.

News Track India, 2012. Iftar party: 2 kids dead, 400 hospitalized with food poisoning. Available from: http://www.newstrackindia.com

North Carolina State University, 2013. Food Safety Info Sheet. Available from: foodsafetyinfosheets.org

Orissa Diary, 2011. India—more than 200 people ill due to food poisoning. Available from: http://orissadiary.com

PTI, 2009. Food poisoning leaves 22 ill in Rajkot district. Available from: http://www.dnaindia.com/india

Schmidt, R.H., Rodrick, G.E., 2005. Food Safety Hand Book. John Wiley and Sons, New York.

Sharma, J., Malakar, M., Gupta, S., Dhandar, A., 2014. Food poisoning: a cause for anxiety in Lakhimpur district of Assam. Ann. Biol. Res. 5, 46–49.

Sudershan, R.V., Rao, P., Polasa, K., 2009. Food safety research in India: a review. As. J. Food Ag-Ind. 2, 177–189.

Thaindian News, 2009. Over 50 people in Orissa suffer food poisoning. Available from: http://www.thaindian.com

Thaindian News, 2010. Food poisoning affects 300 persons in Midnapore. Available from: http://www.thaindian.com

The Hindu, 2009. 150 take ill after eating contaminated food (Jhunjhunu). Available from: http://regionalnews.safefoodinternational.org

The Hindu, 2012. Over 50 children ill; food poisoning suspected. Available from: http://www.thehindu.com/news

The Hindu, 2013. 200 taken ill due to food poisoning. Available from: http://thehindu.com/news

The Hindu, 2014. 350 persons taken ill following 'food poisoning'. Available from: http://www.thehindu.com/news

Thunder Bay District Health Unit, 2011. Special event guidelines: operating guidelines for event organizers and food vendors, 2011. Available from: http://www.tbdhu.com

Times of India, 2011. 200 suffer from food poisoning. Available from: http://timesofindia.indiatimes.com

Times News Network, 2011. Adulterated *kuttu* kills MCD staffer, 300 ill. Available from: http://timesofindia.indiatimes.com/city/delhi/Adulterated-kuttu-kills-MCD-staffer-300-ill/articleshow/7892256.cms?referral=PM

Video Samachar, 2014. 400 people suffer food poisoning, one dead after consuming temple food in Solapur. Available from: http://www.videosamachar.com

Williams, M., Banerji, A., 2012. India's sour food safety record. Available from: http://in.reuters.com

Food safety during disasters

34

R.K. Gupta

Department of Community Medicine, Army College of Medical Sciences, New Delhi, India

It will not be an exaggeration to state that "developing countries" are more prone to effects of disaster than the developed countries. It is agreed that geographically and climatically, the two might have equal probability of being struck by disaster. But if we see the very definition of disaster, that is, *"Disaster is a serious disruption of the functioning of a community involving widespread human, material, economic or environmental losses, which exceeds the ability of the affected community to cope using its own resources",* it is clear that developing countries are unable to cope up with the effects of disaster, in the same manner as the developed countries do.

Hazard vulnerability profile of India indicates that earthquakes account for 57% of all natural disasters, followed by droughts (16%), floods (12%), cyclones (8%), and landslides (3%) (National policy on Disaster Management, 2009). Of the 35 states and Union Territories, 27 are disaster prone (Ministry of Health, 2006). There is no doubt about the fact that disasters (both man made and natural) are here to stay, or may be increase in the times to come. Disasters of unprecedented magnitude are expected to be triggered as a result of climate change and haphazard development (Ministry of Environment & Forest, 2007). In many parts of the world there are widespread humanitarian disaster situations with large-scale ethnic conflicts, civil wars, refugee crises, terrorist attacks, and the like.

Food safety is an inevitable casualty of disaster episodes. Whether a disaster lasts for few days (say terrorist siege of a building or a hijacked aircraft) or for weeks/months (like earthquake, floods, or famine), the priority of water and food safety and availability is second to none. In fact the limited food availability in times of disaster severely compromises food safety as well.

Irrespective of the disaster, in practice, we face one of the two food safety situations. First, and inevitable in every disaster, is the acute situation of compromised food safety owing to sudden destruction of infrastructure, exhaustion of meager home food stores, disruption of electricity and food supplies, and inability of the community to reach out to safer food that might be available at a short distance. Second, is the situation when the acute initial episode is somehow over, food supplies (precooked/community kitchens) are made available, yet food safety concerns persist. Accordingly we need to have appropriate food safety strategies to prevent

Food Safety in the 21st Century. http://dx.doi.org/10.1016/B978-0-12-801773-9.00034-0

and manage such crises: (1) preparation for food safety before a disaster; (2) food safety in aftermath of the disaster.

34.1 FOOD SAFETY ADVICE

During or following natural disasters, food in affected areas may become contaminated with microbiological and other agents. Consequently, community is at risk for outbreaks of foodborne diseases, including diarrheal diseases, hepatitis A/E, typhoid fever, cholera, and dysentery.

As mentioned earlier, food safety risks are linked to poor availability of food, unsafe food storage, handling, and preparation. In many cases cooking may be impossible during natural disasters due to the lack of facilities or fuel. Poor sanitation, lack of safe water, and toilet facilities, can compound the risks. Some people may already be at risk through malnutrition, exposure, shock, and trauma; it becomes essential that the food they consume is safe.

Disasters can mean destruction, spoilage, prolonged nonavailability of food, and power outages. Knowing what to do before and after an event can help tide over the situation better. By following simple guidelines, food availability can be maintained and spoilage minimized.

34.2 PREPARATION FOR FOOD SAFETY BEFORE A DISASTER (WHO, 2015B)
34.2.1 STORE FOOD

While preparing for an emergency food storage, it is important to have a ready cupboard of foods which have long shelf life and require little or no cooking, water, or refrigeration, in case utilities/power supply gets disrupted. It is important to meet the needs of entire family including babies, elderly, and those who are on special diet (e.g., diabetics/hypertensives). It is desirable to cater for pets' needs as well. Alcohol, caffeinated drinks, very salty and spicy foods must be avoided as they increase the need for drinking water, which may be in short supply.

Since disaster strikes unannounced and disrupts food supply, it is wise to plan to have at least 3-days food supply at hand. Dehydrated foods (cornflakes/cereals/oats/*sattu*/fish, etc.), pickles, dry chutneys, rusks, dry fruits (dates, cashew, walnuts, resins, apricot, figs), roasted, peanuts, biscuits, chocolates, milk powder, jams, jellies, flour, dry mixes, and canned foods will remain palatable for months to years. Canned cheese, ghee, butter, jaggery, etc. should also be part of inventory, as these are concentrated sources of energy. Bread may be stored with a condition of using/replacing it frequently.

Certain storage conditions can enhance the shelf life of canned or dried foods. The ideal location is a cool, dry, dark place. The best temperature is less than 20°C. Keep foods away from stoves or hot places, as heat causes many foods to spoil quickly. It is advisable to protect from rodents and insects. Airtight containers also prolong the life of food. It is advisable to store food on higher shelves that will be safe during floods. It is a good practice to put the date on each food item and replace to use before their expiry date.

Whatever "ready-to-eat" kind of food is stored, it does need some preparation, which might not always be easy in a disaster situation owing to damage to home and loss of electricity, gas, and water. So certain food preparation items must also be catered for as disaster stores: cooking utensils, knives, spoons, paper plates, cups, and towels, gas stove/cylinder, or charcoal and its grill/*angithi*/camp stove. Charcoal must not be burnt indoors for its fumes are dangerous.

34.2.2 STORE WATER

We must not forget to store drinking water. It should be stored @ about 6 L per person per day. More water may have to be catered for hot climates, pregnant women, or sick persons. It is important to recycle water bottles every 6 months. Moreover unscented liquid household chlorine bleach or bleaching power must also be stored to disinfect tap/available water at the time of disaster in the eventuality of stored bottled water supply getting over.

Water must be stored preferably in factory sealed water bottles, or in clean water containers that can be tightly closed. Avoid using containers that might break, such as glass bottles, containers that might have ever been used to store any toxic solid or liquid chemical or those with persistent odor.

34.2.3 BRACE UP FOR ELECTRICITY FAILURE (USFDA, 2015)

In times of disaster power outages can occur and may take from few hours to several days to be restored. Without electricity or a cold source, food stored in refrigerators and freezers can become unsafe. Bacteria in food grow rapidly at higher temperatures. The following precautions must be taken with the refrigerator:

- Make sure the freezer and the refrigerator are at stipulated temperatures.
- Keep sufficient ice/freeze containers of water to help keep food cold in the freezer/refrigerator, after the power is out. A nonfrostfree freezer might be better than a frostfree one, in this respect.
- Freeze items such as leftovers, milk, and fresh meat/poultry immediately—this helps keep them at safe temperature for longer.
- Do not overload the fridge.
- Avoid keeping "room temperature stable" foods such as butter, ghee, chocolates, tomato sauce, oil, etc. in fridge, to avoid overload it.
- Never leave the fridge doors open/ajar. Always keep it closed. This helps to maintain cold temperature. Even when there is no electricity, full freezer will hold the temperature for approximately 48 h.
- Defrost fridge regularly as part of maintenance for efficient functioning.
- An appliance thermometer indicating temperature can help determine the safety of food.

- Discard refrigerated perishable food such as meat, poultry, fish, soft cheese, milk, leftovers, and delicate items after 4 h without power.
- Food may be safely refrozen if it still contains ice crystals.

34.3 PRINCIPLES OF FEEDING PROGRAMS DURING DISASTERS (UNHCR, 1999)

It is worthwhile to outline the principles involved in feeding and food distribution during disasters. The various feeding strategies are outlined as follows:

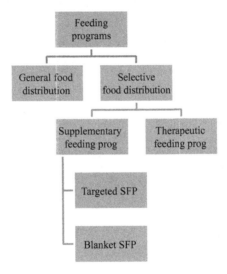

During disasters, the feeding programs may be of two types, namely general and selective food distribution.

General food distribution: Aim of general food distribution is to provide food to all for a basic level of survival. It is done as one of the three modes:

1. As "emergency" energy-dense-food distribution immediately in the aftermath of disaster;
2. As mass cooked food distribution through community kitchens; or
3. As "dry" ration distribution for self-cooking by the affected community, a little later when situation stabilizes.

Selective food distribution: It is meant for specific purposes, commonly with one of the following aims:

1. *Supplementary nutrition*: It caters for deficiency in basic diet, for example, for malnourished children, pregnant and lactating women.
2. Therapeutic feeding is meant for medical/therapeutic feeding for severely malnourished.

Taking the recent real-life example of Chennai floods in India, even though Chennai is a major Indian (metro) city, there was food shortage, for few days. Besides the National Disaster Relief Force, Armed Forces and local government agencies, many religious (temples/mosques/gurudwaras) and nongovernmental organizations helped in preparation and distribution of cooked food, water, snacks, baby foods, milk, etc. The distribution hubs were railway and bus stations, relief camps, hospitals, slum areas, and poor localities.

In remote areas, food aid is provided through road transport and airdrops. These mechanisms have their own limitations. So far as road transport is concerned, it may be disrupted later due to progressive disaster (e.g., roads destroyed due to aftershocks/effects of earthquakes, landslides, worsening rains, or floods). Major limitation of airdrop of foods is that the relief may not always reach the most needy and it may be "lost." Moreover, logistics and cost may limit indefinite number of such relief sorties. Even though, during the Uttarakhand floods and Nepal earthquake, Indian Air Force did hundreds of helicopter relief sorties, carrying food/relief material in the onward sortie and evacuating stranded people on return.

Although it might sound paradoxical, but many a times so much food relief pours in from various quarters—international, national, regional, governmental, and nongovernmental that it languishes in stores and godowns with no mechanism for its efficient distribution. There is a case, when trains full of grain came as "food aid" from Punjab to Rajasthan, to tide over the drought situation in year 2000, but there was no authority to take over the grain at Bikaner railway station. The food supplies were unloaded in a rail yard, where it rotted, while many suffered from hunger. Such waste of resources has to be managed through astute administrative coordination between various agencies, since lack of foodstuff in affected areas is an important determinant of food safety.

Another function of the technical/administrative authorities dealing with receiving, distributing, and cocoordinating food aid is to ensure the quality of food received as donations/aid. Substandard quality food distributed to needy people may create havoc in the form of bacteriological or chemical food poisoning. Hence it is vital to ensure this aspect of food distribution/monitoring as well.

34.4 FOOD SAFETY MEASURES IN THE AFTERMATH OF NATURAL DISASTERS (WHO, 2015A)

No matter what is the type of disaster and what are the feeding modalities adopted, food safety always remains a concern during such eventualities. It is owing to the fact that the availability of food is limited, utilities, supplies, water, transport, and electricity are disrupted, severely compromising hygiene and food safety. During and following natural disasters, particularly floods and tsunamis, food may become contaminated by surface water, silt, dust, gravel, etc. Surface water is the most notorious as it percolates to the last corner. It may itself have been contaminated by pathogenic bacteria from sewage, wastewaters, and dead animals or humans. Therefore, it is important to inspect the food and decide which is good to use.

34.4.1 INSPECTING AND SALVAGING FOOD

If feasible, all food stocks should be inspected and assessed for their safety. Food must be checked for physical hazards, such as glass, wood splinters, and stones that may have been introduced. Food is affected immediately by water. So in areas that have been flooded, whatever intact foods remain should be moved to a dry place, preferably away from the walls and off the floor. Discard canned foods with broken seams, serious dents, or leaks; and jars with cracks. Undamaged canned goods and commercial glass jars of food are likely to be safe. However, if possible, containers should be sanitized before opening them for use. Foods that are exposed to chemicals should be thrown away. Moldy food should not be consumed as it may contain toxic substances. The likelihood of mould growth on stored dried vegetables, fruits, and cereals is greater in a humid environment and where food has become wet.

Refrigerators must be inspected to determine if it has been affected by the lack of electricity or floodwaters. Where food has remained cold and otherwise unaffected, the food is probably safe to consume. Where power is not available, try to use refrigerated food before it is held in the danger zone (5–60°C) for more than 2 h, especially meat, fish, poultry, and milk. Food should definitely be discarded if it shows signs of spoilage (off odors, colors, or textures).

34.4.2 WATER HYGIENE

Water for drinking and food preparation should be considered as contaminated unless specifically confirmed as safe. Use uncontaminated bottled water for drinking, as far as possible. If bottled water is not available, all water should be boiled or otherwise made safe before it is consumed or used as an ingredient in food. If water cannot be boiled, and the available water is turbid/unsafe, filter it through clean cloth after allowing it to settle, and draw off the clear water. It should then be disinfected and made worthy of drinking by using household bleach/bleaching powder. Add 1/8 teaspoon (or 8 drops) of regular, unscented, liquid household bleach or one pinch of bleaching powder for 5 L of water, stir it well and let it stand for 30 min. Store it in clean containers with covers. Resultant mild chlorine odor is acceptable.

34.4.3 KITCHEN HYGIENE

In the aftermath of a disaster, when the kitchen is flooded by dirty water, thoroughly wash all utensils with soap and water. Rinse and then sanitize them by boiling in clean water or immersing them for 15 min in a solution of 1 tablespoon of unscented, liquid chlorine bleach/a pinch of bleaching powder in 5 L of clean/drinking water. Thoroughly wash kitchen platforms with soap and water, using hot water if available. Rinse and then sanitize them as described previously. Allow to air dry. Salvage undamaged factory sealed glass/plastic food containers, cans, bottles by thoroughly washing them with soap and water, brushing any silt, and rinsing them cans clean with water. These can be further sanitized by immersion in boiling water for 2 min, or using bleaching powder solution as described previously. Consumers should always follow basic safe food handling rules to protect themselves and ensure that the foods they eat are safe.

34.4.4 PROVISION OF FOOD AID AND TEMPORARY COOKING FACILITIES

After a natural disaster as soon as families have re-established their capacity to cook, any food they may be given is usually distributed in dry form for them to prepare and consume in their homes or temporary shelters. People may not always be familiar with all kinds of dry foods. When given, they should be shown how to prepare dry foods especially to use safe water if the food is not cooked. In addition to safe water for food preparation, safe water for washing hands and utensils is needed. Shortage of fuel for cooking may also be a major constraint and is essential for ensuring adequate cooking and reheating of cooked food. In some cases, as an alternative to mass feeding, it may be possible to help households by setting up temporary shared neighborhood kitchens where people can prepare food for their own families or in groups. Where basic infrastructure is lacking, shelf-stable rations that do not need cooking or hydration should be provided.

34.4.5 IDENTIFICATION AND RESPONSE TO FOODBORNE DISEASE OUTBREAKS

It is vital to detect foodborne disease outbreaks early. There are some indications of a foodborne disease outbreak like an increase in patients of diarrhea or a surge in demand for antidiarrheals and antiemetics. The patients must receive timely treatment. The suspected/contaminated food must be removed from circulation. Immediate epidemiological investigation must be undertaken for rapid identification of causative agent and the timely provision of information to the public and the actions they should take to minimize risks.

34.4.6 COMMUNITY EDUCATION AND INFORMATION

Following a disaster, community should be advised to take special care regarding food safety when procuring food and water, for example, about source of water. General information and advice about foodborne diseases should also be provided to the population at risk.

34.5 CONCLUSIONS

"In the rainy season villages situated near water should live away from the level of floods. And should keep a collection of wooden planks, bamboos and boats. They should rescue a (person) from floods using gourds, skin-bags, canoes, tree-stems and rope braids. For those who do not go to rescue, will be fined twelve coins..." (Gautam, 2013)

~Chanakya, 300 BC

What the great Indian philosopher, *Chanakya* had said 2500 years back could not be more true even today, ... that it is not the government's responsibility alone, to save community from disaster; it is for everyone to rise to the occasion, if they have to survive and for that we have to be prepared always!

REFERENCES

Gautam, P.K., 2013. Role of Indian Military in Disasters, IDSA Comment. Institute of Defence Studies And Analyses, Delhi.

HPSDMA, 2009. Disaster risks in India. National policy on Disaster Management 2009. Available from: http://www.hpsdma.nic.in

Ministry of Health, 2006. Disaster risk reduction: The Indian model. Ministry of Home Affairs (NDMA Division), Delhi, p. 3.

UNHCR/WFP, 1999. Guidelines for selective feeding programmes in emergency situations. United Nations High Commissioner for Refugees, Geneva.

United States Food and Drug Administration (USFDA), 2015. Protecting & Promoting your health. Available from: http://www.fda.gov/Food/ResourcesForYou/Consumers/ucm076881.htm

WHO, 2015a. Ensuring food safety in the aftermath of natural disasters. Available from: http://www.who.int/foodsafety/foodborne_disease/emergency/en/

WHO, 2015b. Environmental health in emergencies and disasters. Available from: http://www.who.int/water_sanitation_health/hygiene/emergencies/emergencies2002/en/

Domestic regulatory scenario of food safety and interface of food safety laws, standards, regulations, and policies at the international level

Relevant food safety regulations and policies

35

D.P. Attrey

Central Military Veterinary Laboratory, Meerut, Uttar Pradesh, India; High Altitude Research, Defence Research and Development Organisation, Leh, Jammu and Kashmir, India; Amity Institute of Pharmacy, Amity University, Noida, Uttar Pradesh, India; Innovation and Research Food Technology, Amity University, Noida, Uttar Pradesh, India; Amity Institute of Seabuckthorn Research, Amity University, Noida, Uttar Pradesh, India; Lala Lajpat Rai University of Veterinary and Animal Sciences, Hisar, Haryana, India

35.1 INTRODUCTION

For providing good health to all citizens and for maintaining it properly, countries need to ensure safety and quality of food consumed by their respective populations. Moreover, unscrupulous traders may also indulge in adulteration and fraud to make easy money, for example, synthetic milk in India, powdered infant formula with melamine in China, and adulteration of beef in Europe. These issues can be resolved to a great extent, through appropriate food laws and policies. A country should not only have appropriate statutory, regulatory legislations, policies in place for the supply of safe and good quality food at all times, but the same should also be implemented seriously. Such legislations and policies should also be in harmony with the statutory and regulatory requirements of international trade. Harmonization of food laws with WTO requirements is also essential. While the developed countries follow the laws strictly and seriously, the developing and underdeveloped countries lack enforcement mechanisms, administrative machinery, and even the political will to implement the food laws.

The ambit of Food Security Act may be to provide food for filling empty stomachs of people but not necessarily ensuring balanced diet to all citizens, common man in India wants *"Roti, Kapada, aur Makan"* (food, clothing, and shelter). These are fundamental necessities of life. Provision of safe food and water at affordable cost to all, is the fundamental duty of any elected government. Modern food laws, which are pillars of effective food control system, must be based on international principles and standards. In addition, minimum quality requirements should also be included in the food laws to ensure that foods produced are unadulterated. Food laws should cover entire food chain beginning with primary production, provisions for animal feed, on-farm controls, early processing, and distribution till the food reaches the consumer. In addition, countries should continue to update their food laws to be in harmony with international agreements such as WTO and Codex standards, which are the benchmark for food safety at the international level (FAO, 2016).

Although the voluntary standards, like ISO standards and good quality practices, etc., provide a systematic tool to achieve food safety and quality, yet dispute are resolved on the basis of official legislations. For producing safe food, industry generally follows "Good Manufacturing Practices" (GMP) and may voluntarily establish, and update their food safety management system (FSMS) based on any new scientific developments. Science-based solutions to food safety problems are needed for formulating evidence-based policies. But these sometimes create confusion in the minds of general public since there can be different scientific opinions due to scientific uncertainty, generally due to insufficient data and different justifications/interpretations, for example, some scientists in the United States maintain that growth promoters in beef production are perfectly safe while EU scientists argue that they are unsafe. Similar is the case with GM foods and irradiated foods. It becomes difficult for consumers to have confidence in scientific arguments (FAO/WHO, 2002).

35.2 EVOLUTION OF FOOD LAWS

Food legislations had started evolving for last sixty years to regulate food safety among nations in the world. Common man understands health protection as protection only from microbial, chemical or physical contaminants. But it is much more. Common man does not realize the ill effects of unbalanced food. Provision of a healthy (balanced) diet is more important for health protection. Overall objectives for developing national food laws are (FAO/WHO, 2002):

1. To provide a high level of health protection for consumers.
2. States have an obligation to ensure that only safe food and feed are placed on the market.
3. Where appropriate, animal health and welfare, protection of environment, and plant life should be taken care of.

The objectives should also include provision of balanced nutritious diet to all the citizens.

Traceability (identification of the source) of food, feed, ingredients, and food-producing animals should also be ensured by producers, which should be monitored by the state. Food laws need to be based on high-quality scientific inputs by conducting risk analysis through risk assessment, management and communication. Role of risk assessors and risk managers must be separated. In countries where risk assessors and risk managers belong to the same organization it was essential to ensure a functional, well-documented and transparent separation between their respective functions.

As per FAO/WHO (2002), many food scares are not associated with adverse health effects and may present little or no risk, but sensational and dramatic coverage by the media can precipitate a consumer reaction out of proportion to the risk. Risk management should be proportional to the risk to consumers' health, rather than to the intensity of media coverage. Perception is reality for consumers and unless the level of understanding of the issues is raised with the public through media, consumer confidence will continue to be damaged. Moreover, the level considered acceptable for a specific risk can vary between groups of consumers, groups of scientists, and in different countries.

To achieve realistic food safety, food risks must be reduced along the entire food chain, including primary production, as a policy. Risks, if any, must be communicated and end users must be assured as to how the risk shall be managed. Adequate awareness and education of all the stakeholders in the food chain is crucial for raising the level of competence of all in the food business. Educating consumers and those involved in food production and preparation is a major challenge to increase the level of understanding and raise standards. Food laws should have provision for mass awareness programmes in food safety. Making people aware is the starting point, increasing knowledge comes next, followed by changing attitudes, and finally most difficult of all, changing behavior. Food safety and nutrition education may be included in the curriculum of both primary and secondary schools to educate the consumers and future workers in the food industry. The involvement of Ministries of Education with Ministries of Health and Ministries of Agriculture in an integrated approach is essential (FAO/WHO, 2002).

35.3 INDIAN FOOD LAWS

In India, Food Safety and Standards Act, 2006 was enacted on Aug. 24, 2006 to consolidate the laws relating to food and to establish the Food Safety and Standards Authority of India (FSSAI) for laying down science-based rules, regulations, standards and other relevant instructions for articles of food and to regulate their manufacture, storage, distribution, sale, and import, to ensure availability of "safe and wholesome food for human consumption" (FSSA, 2006). FSSAI or the Food Authority of India was established on Sept. 5, 2008, which appointed various committees and the expert panels for drafting the rules and regulations. Food Safety and Standards Rules (FSS Rules) were finalized and notified on May 5, 2011. After enactment, the FSS Act, 2006 also came into force from Aug. 5, 2011, with the enactment of the Food Safety and Standards (Licensing and Registration of Food Businesses) Regulations, 2011 (FSS (L&R), 2011), on Aug. 5, 2011. All statutory regulations, rules, instructions, and their amendments, if any, are brought to the notice of general public through "Gazetted Notifications" by FSSAI on behalf of Govt. of India (FSSAI, 2016).

To discuss food safety regulations and policies it is essential to first understand the definition of Food Safety, Food Audit, and the FSMS. As per FSA, 2006, "Food Safety" means assurance that food is acceptable for human consumption according to its intended use. The food safety audit means a systematic and functionally independent examination of food safety measures adopted by manufacturing units to determine whether such measures and related results meet the objectives of food safety and the associated claims. The FSMS means the adoption of GMP, Good Hygienic Practices (GHP), Hazard Analysis and Critical Control Point (HACCP) and such other practices as may be specified by regulation, for the food businesses. Bureau of Indian Standards (BIS) follows the international standard IS/ISO 22000: 2005, which is designed to allow all types of organizations within the food chain to implement a FSMS ,which is little more comprehensive compared to HACCP system through guidelines for applicants and other documents for FSMS certification.

In fact, the food safety laws can only be implemented if sufficient and sufficiently trained manpower is available at all levels, including the Regulators (Designated Officers, Adjudicating Officers and Food Safety Officers), Food Auditors, Food Quality Analysts (Instrumentation/Microbiology/Chemical), Food Handlers (Manufacturer, Wholesaler, Retailers, etc. Even the Consumers (General Public, Children and Housewives) also require sufficient awareness. As per the food laws, FSSAI and the State Food Authorities have to maintain a system of controls, involving risk communication, food safety surveillance and other monitoring activities covering all stages of food business. The Act also empowers the Food Authority to recognize any agency to conduct food safety audits, based on FSMS. Food testing infrastructure requires a network of Testing/Analytical Laboratories with basic modern analytical facilities and technical manpower. Thus, the Act includes systematic education and capacity building among employees of the industry and the regulators, and also among general public (FSSA, 2006).

As per law, food safety is the responsibility of everyone in the food chain. The onus of safe production of primary food, its processing, distribution and sale, including imports, lies with the FBO.

35.4 FOOD SAFETY POLICIES

Due to a number of reasons the government is not able to sell all the food grain stock it holds. Government had decided to appoint a High-Level Committee (HLC) for Food Corporation of India (FCI, the agency that stocks food grains) restructuring. In the beginning of 2015, this HLC had observed that during the last five years, on an average, buffer stocks with FCI have been more than double the buffer stocking norms. HLC had further observed that the current system is ad-hoc, slow and expensive. A transparent liquidation policy is the need of the hour, which should automatically take care of the situation when FCI is faced with surplus stocks, which are more than the buffer stocking norms. FCI should get greater autonomy and flexibility and work professionally like a business organization, having freedom to operate in Open Market Sale Scheme and even to go for exports (Das, 2015).

In India, "Minimum Support Price" (MSP) of various commodities has remained unprofitable to the farmers as the policy on providing adequate MSPs to farmers remains vague and unprofitable to the farmers.

Certification and accreditation are the hallmarks of quality, which are voluntary in nature. ISO 9001 Quality Management System (QMS) certification is considered the mother of all quality standards, which can be adopted by any type of organization. The main QMS which facilitates Food Safety Certification is ISO 22000, which is the basic FSMS. It has two parts namely the HACCP and the Pre Requisite Programmes (PRPs). As per FSS Act 2006, FSMS should be adopted by all food business operators (FBOs) and should implement GMP, GHP, HACCP, and all other relevant statutory regulations. These are, however, applied by the FBOs engaged in processing and manufacturing of food products. Since food safety problems may start right from the stage of primary production, adequate quality control must be adopted from this stage onwards through all the relevant good quality practices like Good Agriculture Practices (like Good Harvest and Collection Practices and Good Veterinary Practices), Good Hygiene Practices, etc.

35.5 PROCEDURE FOR IMPLEMENTING FOOD LEGISLATIONS/ INSTRUCTIONS

As per FSSR (L&R) (2011) a "Petty Food Manufacturer" or "Petty FBO" is a person who manufactures and sells or distributes any article of food himself or a petty retailer, hawker, vendor or stall holder; except a caterer; or such other food businesses "*who has an annual turnover of less than Rs 12 lakhs*" and/or whose production capacity of food (other than milk and milk products and meat and meat products) does not exceed 100 kg/L per day or procurement or handling and collection of milk is up to 500 L of milk per day or slaughter capacity is 2 large animals or 10 small animals or 50 poultry birds per day or less.

FBOs having "*an annual turnover of more than Rs 12 lakhs*" and the following food businesses have to obtain license under Section 2.1.16 of FSSR (L&R) (2011), Schedule 1 and Section 2.1.2 (3), which fall under the jurisdiction of DO:

1. Dairy units including milk chilling units equipped to handle or process more than 50,000 L of liquid milk/day or 2500 MT of milk solids per annum.
2. Vegetable oil processing units and units producing vegetable oil by the process of solvent extraction and refineries including oil expeller unit having installed capacity more than 2 MT/day.
3. All slaughter houses equipped to slaughter more than 50 large animals or 150 or more small animals including sheep and goats or 1000 or more poultry birds per day.
4. Meat processing units equipped to handle or process more than 500 kg of meat per day or 150 MT/annum.

5. All food processing units other than mentioned under (I)–(IV) including relabellers and repackers having installed capacity more than 2 MT/day except grains, cereals and pulses milling units.
6. 100% Export Oriented Units.
7. All Importers importing food items including food ingredients and additives for commercial use.
8. All FBOs manufacturing any article of food containing ingredients or substances or using technologies or processes or combination thereof whose safety has not been established through these regulations or which do not have a history of safe use or food containing ingredients which are being introduced for the first time into the country.
9. Food Business Operator operating in two or more states.
10. Food catering services in establishments and units under Central Government Agencies like Railways, and Airports, Seaports, and Defence, etc.

As per clause 5 of the FSSR (L&R) (2011), "Registering Authority" means Designated Officer/Food Safety Officer or any official in Panchayat, Municipal Corporation or any other local body or Panchayat in an area, notified as such by the State Food Safety Commissioner, for the purpose of registration as specified in these Regulations. "State Licensing Authority" means Designated Officers appointed under Section 36(1) of the Act by the Food Safety Commissioner of a State or UT. As per clause 2 of the FSSR (L&R) (2011) all FBOs in the country will be registered or licensed in accordance with the procedures laid down further:

35.6 REGISTRATION OF PETTY FBOs

Besides other provisions, all petty FBOs shall register themselves with the Registering Authority by submitting an application for registration in Form A under Schedule 2 of these Regulations along with a fee as per Schedule 3. Also these petty FBOs shall follow the basic hygiene and safety requirements as per Part I of Schedule 4 of these Regulations and provide a self attested declaration of adherence to these requirements with the application in the format provided in Annexure-1 under Schedule 2. The concerned Registering Authority shall carry out food safety inspection of the registered establishments at least once in a year. A primary producer of milk, who is a registered member of a dairy Cooperative Society, and supplies or sells the entire milk to the Society, shall be exempted from getting registered with the Registering Authority.

35.7 LICENSING OF FBOs

As per FSS (L&R) (2011), no person shall commence any food business unless he possesses a valid license, provided that any person or FBO carrying on food business on the date of notification of these Regulations, under a license, registration or permission, as the case may be, under the Acts or Orders mentioned in the Second

Schedule of the Act shall get their existing license converted into the license/registration under these regulations by making an application to the Licensing/Registering Authority after complying with the safety requirements mentioned in Schedule 4 contained under different parts dependent on nature of business, within one year of notification of these Regulations.

All FBOs engaged in selling anything edible are supposed to obtain licenses/registrations.

BIS offers two Certification schemes to the food industry:

1. HACCP Stand-alone Certification against IS 15000:1998
2. HACCP-based Quality System Certification, which further provides for two Certifications: one audit Certification of Quality System against IS/ISO 9000 and the other Certification of HACCP against IS 15000:1998.

Looking at increasing demand for "Contract Farming" in India, many international retailers have now started entering Indian farming sector and markets. Food Safety today generally focuses on four areas (FAO/WHO, 2002):

1. Microbiologic safety due to *Salmonella*, *Escherichia coli*, *Listeria*, etc. Modern systems of farming, processing, distribution and retailing have inadequate safeguards. The "modern food chain" means that infection on a single farm can soon become very widely distributed.
2. Chemical safety of food with long-standing concerns about pesticides and heavy metal contamination, especially during primary production.
3. Genetically modified organisms, novel foods, and processes.
4. Nutritional quality of diet: Nutritional quality of the diet must be included in food safety legislations to provide a nutritious diet to all.

Although the public perceives correctly that nutrition is of major importance for their health, yet the immediate and rapidly emerging health risks like the "dioxin," "BSE," and "GMO" crisis, etc. take higher precedence, draw higher attention, and become immediate concerns. As such this situation continues and the main aim of providing safe and balanced food to all citizens of India, under the national legislations, especially the Food Security Act, remains a distant dream. In fact importance of nutrition and balanced diet has been highly diluted.

Since India is a huge country following recommendations based on Pan-European Conference, 2002 are relevant for adaptation by India.

1. Improvement/establishment of National and States level networks for collection, compilation and sharing of information and data on food quality and safety, food risks and contamination and foodborne diseases.
2. Collaboration between national- and state-level authorities to strengthen and harmonize integrated and transparent systems for surveillance, outbreak investigation, reporting systems and diagnostic methods on food safety and quality.
3. Enhanced cooperation between the health, agriculture, fisheries, and food production sectors for food safety surveillance and monitoring.

4. Food safety strategies should be risk based.
5. An integrated and multidisciplinary policy approach to food safety and quality should be applied with participation of all governmental and nongovernmental stakeholders in the whole food chain, including primary production.
6. Prevention-oriented regulation and control systems for reduction of foodborne disease, reduction of food- safety risks and protection of environment should be developed and coordinated.
7. In policy making, consideration should be given to ethical, religious other legitimate concerns in addition to risk assessment.
8. In case of scientific uncertainty or where risk assessment is not conclusive, provisional risk management measures may be adopted based on precautionary principle.
9. With particular regard to scientific advice, risk assessment and risk communication, an independent, transparent, and effective national food safety authority is recommended.
10. Effective and independent risk assessment is essential and which should be carried out in an independent, open, and transparent manner and should also address new or unforeseen risks.
11. In some cases, it might be advantageous to give responsibility to a single agency for official food monitoring and control along the whole food chain.
12. Improved education and training in food hygiene should be used to increase the competence of the workers and effectiveness of inspectors throughout the food chain. Education of consumers should begin at school.
13. Open consultation and public debate involving consumers and all other stakeholders is needed in order to increase the confidence of consumers in the safety of food and to develop a comprehensive, transparent, and integrated approach to food safety and nutrition policy.
14. Information on results of all official monitoring tests and on the outcome of official food inspections and other official food control activities should be made public as openness and transparency builds consumer confidence.
15. Primary production must come under the purview of food safety regulations, since it is the biggest source of unsafe food in India.
16. Top most priority should be given in Legislation as well as at policy level to keep people healthy, to maintain their health and prevention of health problems through balanced food, since food not providing adequate health is also unsafe.
17. Vigorous public awareness campaign should be launched and continued against adulteration and food frauds (FAO/WHO, 2002).

35.8 INTERNATIONAL SCENARIO ON FOOD SAFETY

The International Standard ISO 22003 defines FSMS as the set of interrelated or interacting elements to establish policy and objectives and to achieve those objectives, which are used to direct and control an organization with regard to food safety.

The key elements of the FSMS are:

Good practices/PRPs,
Hazard analysis/HACCP,
Management element/system,
Statutory and regulatory requirements, and
Communication.

Internationally there are many Food Safety Certifications which meet these requirements like HACCP, ISO 22000, FSSC 22000 and many more. These are voluntary certifications to strengthen the food safety system. Many countries are integrating the HACCP based system into their legislations and food inspection service programmes. However, for achieving still higher levels of safety and quality of food products, ISO 9000 system may be combined with the HACCP system. ISO 22000, (particularly ISO 22003) is a QMS for food and is popularly called as FSMS. It has two components namely ISO 9001 and ISO 22000. The ISO 22000 further has two components namely PRPs and HACCP. PRPs are good quality practices, which differ from process to process. Most food industries, the world over, have so far been practicing HACCP, which is an excellent tool for achieving food safety. However, it may not be possible to address all food safety issues through HACCP alone, for example, problems arising due to agriculture chemicals, pollutants, contaminants and natural toxins during the primary production. HACCP programmes are usually applied by industries (FSMS, 2006).

Besides HACCP, FSMS is now becoming more popular both at national as well as international levels since it is a systematic adoption of new scientific developments along with appropriate follow up of the statutory food laws, regulations, rules, instructions and other relevant orders on food safety through enforcement by the National Food Authority.

Main aim of farm and food certification is to enlighten the customers regarding methods of food production on farm during primary production. Through this certification, customers are assured of minimizing detrimental environmental impacts of farming operations during food production, by regulating use of chemicals and ensuring a responsible approach to workers' health and safety as well as animal welfare.

Although "Organic Food Certification" is also gaining popularity in many developed countries yet one has to be careful about intentions of producer to implement standards ethically and impartially. Processed food exports from India are declining due to poor food quality and safety. However, manipulation of certification regulations as a way to mislead or outright dupe public is a very real concern. There are examples where nonorganic inputs are allowed as exceptions without loss of certification, status and standards are interpreted in such a way that they meet the letter, but not the spirit/intention of particular rules.

As per FAO/WHO, 2005, many emergencies related to food and food productions have occurred in the past in EU countries. This has forced EU to strengthen its food safety systems. The food safety hazards mainly comprise of foodborne diseases, zoonoses, residues of unwanted substances in food and dangerous animal diseases. EU "White paper on food safety" establishes current food safety policy in EU and is based on risk analysis approach. In this framework, EU General Food Law of 2002 has led to establishment of European Food Safety Authority and several legal measures to be enforced by Member States.

Prohibition of adulterated and misbranded foods is the foundation for food safety laws. Although the goal of zero risk foods is not achievable, since it would be too expensive, yet governments must assure the public that everything reasonable is being done to minimize the risk. US FDA has issued guidance and regulatory information with links to Federal Register documents on food safety programs, manufacturing processes, industry systems, and import/export activities. The guidance documents represent FDA's current thinking on a topic. These are only for guidance and are not binding on any one, that is, FDA or the public. These regulations have binding obligations and have the force of law (FDA, 2015). Main regulations in USA are:

1. FDA Food Safety Modernization Act (FSMA)
2. Food Facility Registration
3. Current Good Manufacturing Practices (CGMPs)
4. Hazard Analysis & Critical Control Points (HACCP)
   a. Dairy Grade - A Voluntary HACCP
   b. Juice HACCP
   c. Retail & Food Service HACCP
   d. Seafood HACCP
5. Retail Food Protection
6. Federal/State Food Programs
7. Food Protection Plan 2007

Imports & Exports Regulations: In Europe, European Union's Council Directive (DIR/93/43/EEC), makes it obligatory for FBOs to ensure that adequate safety procedure are implemented, maintained and reviewed on the basis of HACCP. The hygiene directive 93/43/EEC has been in the meantime fully implemented into European Member States' national legislations. In Turkey also legislative provisions have been made to facilitate implementation of HACCP (FAO/WHO, 2002).

World Trade Organization (WTO) is successor to General Agreement on Tariffs and Trade (GATT) (which was established after Second World War). The Uruguay Round of Multilateral Trade Negotiations (1986–94) has led to creation of WTO, which has also taken up food safety issues at international level, especially the international trade. GATT and the WTO have helped to create a strong and prosperous trading system contributing to unprecedented growth (FAO/WHO, 2002).

Main mandate of the European Food Safety Authority (EFSA) is to perform risk assessment. Rather than developing new major legislative initiatives, the Commission believes that it is necessary to ensure that existing law is correctly applied. To this end, training of those who must verify the implementation of Community law is essential.

REFERENCES

Das, S., 2015. FCI still bogged down by excess buffer stocks of rice, wheat. Financial Express. Available from: http://www.financialexpress.com/article/markets/commodities/fci-still-bogged-down-by-excess-buffer-stocks-of-rice-wheat/82404/

FAO, 2016. Food regulations. Available from: http://www.fao.org/food/food-safety-quality/capacity-development/food-regulations/en/

FAO/WHO, 2002. FAO/WHO Pan-European Conference on Food Safety and Quality, Budapest, Hungary, February 25–28, 2002 Final Report. PEC/REP 1. Available from: http://www.fao.org/3/a-y3696e.pdf

FDA, 2015. Guidance & Regulation, Guidance Documents & Regulatory Information. U.S. Food and Drug Administration (FDA). Available from: http://www.fda.gov/Food/GuidanceRegulation/

FSMS, 2006. Manual of Food Safety Management System, FSS Act, 2006. Food Safety and Standards Authority of India (FSSAI). Available from: http://www.fssai.gov.in/Portals/0/Pdf/manual%20of%20food%20safety%20management%20system,%20fss%20act%202006.pdf

FSS (L&R), 2011. Food Safety and Standards (Licensing and Registration of Food Businesses) Regulations, 2011. Available from: http://www.fssai.gov.in/Portals/0/Pdf/Food%20safety%20and%20Standards%20(Licensing%20and%20Registration%20of%20Food%20businesses)%20regulation,%202011.pdf

FSSA, 2006. Food Safety and Standards Act, 2006. Available from: http://www.fssai.gov.in/portals/0/pdf/food-act.pdf

FSSAI, 2016. Gazetted Notifications. FSSAI, Ministry of Health and Family Welfare, Govt of -India. Available from: http://www.fssai.gov.in/GazettedNotifications.aspx

Food safety policies in agriculture and food security with traceability

36

D.P. Attrey

Central Military Veterinary Laboratory, Meerut, Uttar Pradesh, India; High Altitude Research, Defence Research and Development Organisation, Leh, Jammu and Kashmir, India; Amity Institute of Pharmacy, Amity University, Noida, Uttar Pradesh, India; Innovation and Research Food Technology, Amity University, Noida, Uttar Pradesh, India; Amity Institute of Seabuckthorn Research, Amity University, Noida, Uttar Pradesh, India; Lala Lajpat Rai University of Veterinary and Animal Sciences, Hisar, Haryana, India

36.1 INTRODUCTION

As per Yadav (2010), the National Agricultural Policy of India was announced on July 28, 2000. It was formulated as it was absolutely essential to build on the inherent strength of agriculture and allied sectors to address the constraints and to make optimal use of resources and opportunities emerging as a result of advancement in science and technology. This policy intended to explore the vast untapped potential of Indian agriculture by strengthening rural infrastructure to support faster agricultural development, promote value addition, accelerate the growth of agro-business, create employment in rural areas, improve standard of living of the farmers, agricultural workers and their families, discourage migration to urban areas and face the challenges arising out of economic liberalization and globalization in the World Trade Organization (WTO) regime.

The main objective of the policy was to achieve more than 4% growth in agriculture sector. The policy, however, did not succeed in achieving the growth of over 4% on sustainable basis.

"Agriculture is the science, art, or practice of cultivating the soil, producing crops, and raising livestock and in varying degrees the preparation and marketing of resulting products." As per International Labor Office (1999), "agriculture" is the cultivation of animals, plants, fungi, and other life forms for food, fiber, biofuel, medicinal, and other products used to sustain and enhance human life. Further, as per dictionary, a "policy" is a set of ideas or a plan of what to do in a particular situation that has been agreed to, officially by the concerned agency (or government) (Attrey, 2008).

A look at India's National Agricultural Policy and various 5-year plan documents reveals that the policy makers and India's agricultural experts have given priority to foods of plant origin and detailed planning was done for crop agriculture.

Food Safety in the 21st Century. http://dx.doi.org/10.1016/B978-0-12-801773-9.00036-4

36.2 HAS NATIONAL AGRICULTURAL POLICY TAKEN CARE OF "FOOD SAFETY"?

According to Research Reference and Training Division, Ministry of Information and Broadcasting (RRTD, 2012), main objectives of the National Agriculture Policy of India—2000 (NAP-2000), were to attain the following:

- A growth rate of more than 4% per annum in the agriculture sector.
- Growth that is based on efficient use of resources and conservation of soil, water, and bio-diversity, with equality, that is, growth which is widespread across regions and farmers, and which is demand driven and caters to domestic markets and maximizes benefits from exports of agricultural products in the face of the challenges arising from economic liberalization and globalization.
- Growth that is sustainable technologically, environmentally, and economically.

Region wise development of plantation crops, horticulture, floriculture, roots and tubers, aromatic and medicinal plants, bee keeping, and sericulture were proposed. Livestock breeding, dairying, and poultry were also proposed to be promoted through generation and dissemination of appropriate technologies. Government also proposed an adequate and timely supply of quality inputs such as seed, fertilizers, plant protection chemicals, biopesticides, agriculture machinery, and credit at reasonable rates to farmers.

As per Chand (2016), the policy also envisaged to evolve a "National Livestock Breeding Strategy" to meet the requirement of milk, meat, eggs, and livestock products and to enhance the role of draught animals as a source of energy for farming operations.

Though many issues have been addressed by the policy, food safety during primary production does not find a place.

36.3 FOOD SAFETY AND FOOD SECURITY AS PART OF NATIONAL AGRICULTURAL POLICY

Food safety starts from the beginning of food production at the farm and ends at consumer's plate, that is, from farm to plate. Food is considered unsafe when it contains microbiological and chemical contaminants, etc. Planners in the Ministry of Agriculture probably did not pay much attention to food safety since they considered it to be the exclusive domain of Ministry of Health. The Ministry of Health on the other hand, concentrated more on the end products of agriculture, that is, finished food products, which ignored food safety in production. Both these ministries have to prepare a joint policy to ensure adequate provision of safe and balanced food.

In view of the aforementioned, a document envisaging the issues highlighted may be prepared. The policy should consider safety of primary foods produced through the application of quality management system and use of prerequisite programs such as good agricultural, veterinary, hygiene, and other good-quality practices; development of cold chain, where required; policy on appropriate safe storage of produce as per good-quality practices; development of farmer friendly agri-markets having location and infrastructure for prevention of food borne hazards and risks, risk analysis, veterinary public health, and public health policy for monitoring food safety, etc. Methodology on enhancing productivity and safety of primary foods should be included in the policy.

Food safety must accompany food and nutrition security (Chan, 2014). Despite progress in reducing undernutrition, our planet's population is still affected by many food-related challenges, including vitamin and mineral deficiencies, obesity, and noncommunicable diseases. During Second International Conference on Nutrition (ICN-2) at Rome, all participating nations, including India, had committed to "eradicate hunger and prevent all forms of malnutrition worldwide, particularly undernourishment, stunting, wasting, underweight and overweight in children under five years of age and anemia in women and children, among other micronutrient deficiencies …. etc." (WHO, 2014).

Food safety is not only an important part of public health but it is regulatory requirement also, which works through inspection, education, and surveillance, whereas food security is a community-based program in which farmers constitute the base of the program. There are differences among public health professionals and community workers during implementation of these two programs, although both have the same aim, that is, "safe accessible food for all." Adequate coordination between food security and food safety authorities is essential to achieve this aim (Wanda, 2004).

The policy of encouraging production of only rice and wheat on available cultivable land may not be sufficient. This may lead to scarcity of affordable animal proteins. Farmers must use the cultivable land for producing "requirement-based agricultural commodities" to provide affordable balanced food to the entire population of India, which may be ensured by the government, through a modified policy. Politicoreligious approaches to meat eating have further compounded this problem by distorting the production and consumption patterns of meat in India. Milk and meat production are intricately linked to each other. Therefore, policy on development of milk and meat has to be prepared together. Safety of foods of animal origin is directly linked with their availability and affordability to common man. Poor people cannot afford to eat a balanced diet containing animal proteins. Hence there is essential requirement of having not only sufficient availability but also accessibility at the affordable rates. People below the poverty line are so poor that safe and reasonably good-quality foods of animal origin continue to remain unaffordable to them.

An agricultural policy should contain food safety as its integral part. Food safety has to start right from primary production at the farm level, where microbial and chemical risks and misuse of agrochemicals, pesticides, growth hormones, and veterinary drugs, etc. can be monitored and controlled. Contaminated water with industrial/farm wastes, when used for irrigation of crops or due to its spillage in cultivated land, could ultimately result in production of unsafe food. Prerequisite programs such as good agricultural and other good-quality practices can help in reducing the level of microbial and chemical hazards in food raw materials. The 12th 5-year plan estimated that food grains storage capacity may be expanded to 35 MT. But it should be monitored, controlled, and reviewed continuously from food safety point of view.

Genetically modified foods: Production and use of Genetically Modified (GM) crops and cloned animals for enhancing availability of food will also require monitoring from food safety point of view. To prevent food from becoming unsafe, appropriate policy measures and legislations must not only be enacted, but implemented also. Agricultural policy must have adequate provisions to ensure that such contaminants and adulterants do not enter the food chain and appropriate provisions for cold chain must be in place, where required. Adoption of HACCP and other GAPs, good hygiene practices, etc. must be integrated into the policy.

A debate is continuing on the safety of GM foods of plant origin. People are now having suspicion on foods of animal origin also, which are produced through genetic cloning and other techniques. This needs more clarification. Clones are genetic copies of an identified/selected animal. They are similar to identical twins, but born at different times. Cloning is an assisted reproductive technology now being used widely by livestock breeders. It is an extension of other technologies such as artificial insemination, embryo transfer, and in vitro fertilization. The mechanism of cloning involves transfer of nuclear material from somatic cells to the egg of the female, which is then developed into an embryo in the laboratory. The embryo is implanted in the uterus of a surrogate mother, which carries it to term and delivers it like her own offspring. Cloning allows farmers to upgrade the quality of their herds by providing more copies of their best animals—those with naturally occurring desirable traits, such as resistance to disease, high milk production, or quality meat production. These animal clones are then used for conventional breeding, and their sexually reproduced offsprings become the food-producing animals.

FDA has concluded that cloning poses no risk to people eating food from clones of animals traditionally consumed as food or their offsprings. As is mandatory in the case of GM foods, labels for food obtained from cloned animals do not have to state that food is from animal clones or their offspring, since main use of clones is to produce breeding stock and not the food directly (FDA, 2008).

36.4 TRACEABILITY

All stakeholders should be able to trace back source of their food to know its level of safety and quality and to take corrective action in case of nonconformities. As per FAO/WHO (2005), traceability/product tracing is an important concept. European Union's (EU's) General Food Law also makes traceability compulsory for all food and feed businesses. It requires that all food and feed operators implement special traceability systems. They must be able to identify where their products have come from and where they are going and to rapidly provide this information to the competent authorities.

For implementing a food safety management system, it is mandatory for an organization to establish and apply a traceability system that enables the identification of product lots and their relation to batches of raw materials, processing, and delivery records. As per ISO 22000:2005, traceability system helps in identifying incoming material from the immediate suppliers and initial distribution route of the end product. As per the standard, traceability records must be maintained for a defined period for system assessment to enable the handling of potentially unsafe products and in the event of product withdrawal. Records shall be in accordance with statutory and regulatory requirements and customer requirements and may be based on the end product lot identification. In a large country like India, where farms are of small size and raw materials are procured by processing units or exporters through huge number of small traders, traceability/product tracing to the farm is very difficult. However, it is not impossible. If planned properly it can be done (FSSAI, 2010). Traceability should, therefore, form part of agricultural policy.

Traceability system comprises of two primary capabilities, the ability to track movements and to trace custody of a food product in the food chain. In defining traceability, it is important to distinguish between the terms—tracking and tracing. "Tracing" is the ability to recreate the history of a product in the food chain and to identify the origin, movements, and relevant associated information of a particular unit and/or batch of product located within the supply chain by reference to records held upstream. "Tracking" is the ability to trace the destination of a product in a food chain and to follow a path of a specified unit and/or batch of product through the supply chain as it moves from organizations towards the final point-of-process, point-of-sale, point-of-service, or point of consumption. In other words, it is the movement of the product forward through the food chain to understand where it has gone, what it has gone into and what it has come into contact with (FSSAI, 2010). A system should be developed to trace back the pooled raw materials of liquids (such as milk) or commodities (such as grains, etc.), at the farm gate, at a collection center or a processing unit, through adequate coding. In such cases, even if it may not be possible to trace the origin of the samples immediately, a probable area of origin can be pin-pointed, which can be further investigated to know the exact source of problem/hazard.

REFERENCES

Attrey, D.P., 2008. A Mini Text for Beginners—ISO 9001: 2000, Quality Management System. Attrey's Mini Text Series on Quality Management Systems and Good Quality Practices; developed for training at appropriate levels in INMAS, under the project "Design, development and establishment of suitable Quality Management System for INMAS for subsequent certification in ISO 9001:2000 & NABL Accreditation by a certifying body"; Institute of Nuclear Medicine and Allied Sciences (INMAS), (DRDO) as Faculty, Institute of Defence Scientists and Technologists; CEFEES, Defence Research and Development Organization (DRDO), Ministry of Defense, Government of India, Delhi.

Chan, M., 2014. Food safety must accompany food and nutrition security. Lancet 384 (9958), 1910–1911, Available from: http://www.thelancet.com/journals/lancet/article/PIIS0140-6736(14)62037-7/abstract.

Chand, S., 2016. What are the main features of the National Agricultural Policy of India? Available from: http://www.yourarticlelibrary.com/economics/what-are-the-main-features-of-the-national-agricultural-policy-of-india/2768/

FAO/WHO, 2005. Food export control and certification. Proceedings of Second FAO/WHO Global Forum of Food Safety Regulators; Agenda Item 4.5; GF 02/8a, October 12–14, 2004, Bangkok, Thailand. Available from: http://www.fao.org/docrep/meeting/008/y5871e/y5871e00.htm#Contents

FDA, 2008. Animal Cloning and Food Safety. U.S. Department of Health and Human Services; U.S. Food and Drug Administration, USA. Available from: http://www.fda.gov/ForConsumers/ConsumerUpdates/ucm148768.htm.

FSSAI, 2010. Training Manual for Food Safety Regulators, vol. 5. Key Aspects to ensure food safety; 2010. The Training Manual for Food Safety Regulators who are involved in implementing Food Safety and Standards Act 2006 across the Country. Available from: http://www.fssai.gov.in/Portals/0/Training_Manual/Volume%20V-%20Key%20Aspects%20to%20ensure%20Food%20Safety.pdf

International Labour Office, 1999. Safety and health in agriculture. International Labour Organization, USA. p. 77. Available from: http://books.google.com/books?id(GtBa6XIW_aQC&pg(PA77; http://www.capitalco.com.au/Portals/0/Docs/Fertiliser_Chemical/Agriculture.pdf

RRTD, 2012. New Agriculture Policy. Research Reference and Training Division, Ministry of Information and Broadcasting, Government of India. Available from: http://rrtd.nic.in/agriculture.html

Wanda, M.R.N., 2004. Exploring food safety & food security tensions. Does food safety conflicts with Food security. The safe consumption of food. Working paper 04-01. Food Industry Centre, University of Minnesota, USA; CPHERI, India; School of Nursing, University of Victoria, British Columbia, Canada. Available from: http://www.uvic.ca/research/groups/cphfri/assets/docs/Food%20safety%20food%20security%20Wanda.pdf

WHO, 2014. WHO. FAO/WHO Second International Conference on Nutrition (ICN2). Available from: http://www.who.int/nutrition/topics/WHO_FAO_announce_ICN2/en/ and http://www.who.int/foodsafety/Lancetfoodsafety_nov2014.pdf; http://www.fao.org/about/meetings/icn2/en/

Yadav, K., 2010. National Agricultural Policy. Agropedia. Available from: http://agropedia.iitk.ac.in/content/national-agricultural-policy

Food safety in international food trade—imports and exports

37

D.P. Attrey

Central Military Veterinary Laboratory, Meerut, Uttar Pradesh, India; High Altitude Research, Defence Research and Development Organisation, Leh, Jammu and Kashmir, India; Amity Institute of Pharmacy, Amity University, Noida, Uttar Pradesh, India; Innovation and Research Food Technology, Amity University, Noida, Uttar Pradesh, India; Amity Institute of Seabuckthorn Research, Amity University, Noida, Uttar Pradesh, India; Lala Lajpat Rai University of Veterinary and Animal Sciences, Hisar, Haryana, India

37.1 INTRODUCTION

The world has become a smaller place due to faster means of communication and modern technologies. The international food trade (import and export) has increased manifolds in the past two decades, due to the adoption of each other's tastes and food habits and new scientific developments in food science and technology, meeting the requirements of people globally. At the same time, increased international trade is also becoming a potential source of increased health risks through potential safety threats in the traded food. However, increased consumer awareness about food safety and quality is a good sign, which helps to reduce the chances of development of health risks (FAO/WHO, 2005).

The World Trade Organization (WTO) provides adequate opportunities to all countries through fair trading practices in the world market. WTO agreements, namely, Technical Barriers to Trade and Sanitary and Phytosanitary Measures (SPS) expect that all member countries implement at least the minimum rules and maintain minimum trading discipline in food safety to protect the health of consumers by achieving the appropriate level of protection (ALOP). At the same time, these agreements also try to create a balance between consumer safety and trade economy by ensuring that the official rules, regulations, and standards of the member countries do not create unnecessary barriers to trade, while permitting them to impose safety standards in their respective countries to protect their populations from potential foodborne hazards. Both agreements encourage member countries to recognize each other's "conformity assessment" systems (namely, inspection, certification, and testing), based on international standards, so that products certified in one country are accepted without the need for further inspection/testing by the other country through

"Equivalence" or "Mutual Recognition Agreements" (FAO/WHO, 2005). Before discussing the food-safety issues in imports and exports, it is also essential to understand the basic definitions of related terms, which have been given in Appendix A.

As WTO agreements encourage member countries to recognize each other's conformity-assessment systems, based on international standards, it is essential to understand "conformity assessment" (CAC/GL 20, 1995). According to MacCurtain (2015), conformity assessment involves a set of processes that show the service or system meets the requirements of a standard. It helps the regulators to ensure that health, safety, or environmental conditions are met. Conformity assessment involves testing, certification, and inspection (see Appendix A).

Conformity assessment cannot be understood without having a good idea of the Quality Management System (QMS). ISO 9001 is considered as the base of all QMSs, ISO has produced standards to help make conformity-assessment activities as uniform as possible, across industries and across the world. Among other things, this reduces the need for duplication of testing when importing or exporting, thus facilitating global trade. A Mutual Recognition Agreement (MRA) may increase confidence in conformity assessment between countries as it formally recognizes the results of each other's testing, inspection, certification, or accreditation, reducing duplication of conformity-assessment activities. The international standard "ISO 9001:2015" meets specific requirements for a QMS and helps in assuring *conformity* to the customer and applicable statutory and regulatory requirements. All requirements of ISO 9001:2015 are generic and are intended to be applicable to all organizations, regardless of type, size, and the product. An ISO Standard is "not a product standard" but a QMS standard. Most of the industries produce products, which are restricted to the fulfillment of "specifications." However, technical specifications do not give guarantee that the expectations of the customer are consistently and continuously fulfilled. This necessitates the shift from "quality of end product" to "quality of system." All QMSs are based on the fundamental concepts of ISO 9000 family of standards. Before establishing conformity assessment, one should understand these concepts. Although these standards are voluntary in nature, but in case these are adopted, one has to follow these simple concepts (Attrey, 2008) as follows:

- "write what you do";
- "do what you write";
- "check what you have done"/"prove what you have done" through audit;
- "remove nonconformances" through corrective action; and
- "improve the system continuously" by taking preventive actions.

These fundamental concepts of a QMS remain the same for all standards. All the QMSs are voluntary. However, if one chooses to adopt a QMS, all the requirements must be met. Unfortunately, in India most are not aware of the QMS. While implementing a Food Safety Management System (ISO 2000), these concepts must be followed by all the concerned to achieve the objectives of food safety and quality in a food-control organization at each level.

1. Identify all important processes/activities you perform in your business or at the farm, factory, or during distribution and marketing chain. Then "write what you do"

at all these places in the form of written operating procedures. Finally, ask yourself whether your operating procedures conform to the statutory and nonstatutory requirements as per the related standard and related good-quality practices (such as, GMP/GHP/HACCP, etc. and try to incorporate them in the operating procedures).

2. Then "do what you have written" in the operating procedures, that is, implement the operating procedures.
3. Prove "what you have done" through audit, that is, check what has been done is "as per what you have written."
4. Remove nonconformances, if any (by corrective actions).
5. Improve the system (by corrective and preventive actions and amending the operating procedures as per the audit results).
6. Reaudits bring continual improvement, which should again become the starting point of your operating procedures.

Amendments in the quality manual and operating procedures as a result of continuous auditing bring continual improvement in such a way that these are in line with what the organization is actually doing or wants to do, their method of working, and work culture so that understanding, adoption, and implementation of the QMS becomes possible, easy, and simple, requiring minimum change. The quality people believe that "if it is not written it is not done."

Specific QMSs, such as, Food Safety Management System, ISO 2000 (in a food control organization), intend that all identified activities and operating procedures conform to the system requirements given in the quality manual as per related standard and good-quality practices so as to achieve the planned objectives. However, these operating procedures must be designed in such a way that a balance is maintained between the achievement of planned objectives as per the requirements of the food-control system and implementation of fair practices to conduct the international food trade smoothly.

Food import- and export-control systems tend to be reciprocators. An exported product becomes an imported product for the importer.

Food safety in a country is officially controlled by its food control organization through various rules, regulations, standards, and practices. Inspections and certifications are an integral part of the Food Control System through which the procedures/processes are implemented as required. But the domestic enforcement of food laws, however, many a times, becomes an obstacle in the smooth conduct of international food trade. Use of basic principles and guidelines of food inspection and certification systems help to ensure that foods and their production systems meet the requirements of food safety besides achieving fair trade practices. Codex principles and guidelines provide a very good basis for drafting regulations and procedures for the Food Control Organization to facilitate the conduct of imports and exports smoothly (CAC/GL 20, 1995).

Codex Alimentarius Commission (CAC) has prepared a framework for the development and operation of an import-control system to protect consumers, facilitate fair practices in international food trade, and reducing the chances of introducing unjustified technical barriers to trade (CAC/GL 47, 2003). Principles of international trading in food and food products developed by Codex, help in maintaining a "balance" through adequate implementation of domestic food laws and conduct of international trade smoothly. Although it is a tricky situation, yet Codex has made a very good beginning by designing the basic principles and guidelines for imports and exports.

37.2 PRINCIPLES AND GUIDELINES OF INSPECTION AND CERTIFICATION SYSTEMS FOR IMPORTS AND EXPORTS

Official inspection and certification systems are important means of food control worldwide. The confidence of consumers in the quality and safety of their food supply depends to a large extent on their perception as to the effectiveness of food-control measures, especially the inspections and the certifications. The worldwide trade in foods of animal origin generally depends on the use of inspection and certification systems. But it has been seen many times that these requirements, if not based on international principles and guidelines, may create barriers to international food trade. Use of design and application of these systems as per international principles and guidelines, will ensure adequate consumer protection in addition to the facilitation of food trade (CAC/GL 26, 1997).

37.3 PRINCIPLES OF FOOD IMPORT AND EXPORT INSPECTION AND CERTIFICATION (CAC/GL 20, 1995)

These include:

- assessing objectively the risks appropriate to the circumstances;
- nondiscriminatory approach in the level of risk;
- sufficient means to perform tasks efficiently;
- use of international standards and guidelines, such as, Codex for harmonization;
- signing of equivalence agreements, where possible (obligation for equivalence rests more with the exporting country); and
- respect for the legitimate concerns of the other country while remaining transparent as far as possible.

37.4 GUIDELINES

Guidelines on Design, Operation, Assessment, and Accreditation of Food Import and Export Inspection and Certification Systems have been provided in CAC/GL 26 (1997). According to these guidelines, the controls program/procedures should not compromise the quality or safety of foods, particularly in the case of perishable products.

37.5 GUIDELINES ON FOOD IMPORT-CONTROL SYSTEM

In India, the Food Control System is operated by the Food Control Organization, which is the Food Authority (Food Safety and Standards Authority of India, FSSAI), established under the Food Safety and Standards Act, 2006. According to this Act, no person shall import any unsafe or misbranded or substandard food or food containing extraneous matter, or any article of food into India, for the import of which a license is required, except in accordance with the conditions of the license (FSA, 2006). As per the Food Safety and Standards (Food Import) Regulation, 2016, no person shall import any food without an import license and no food article shall bebe cleared from the customs unless it has a 60% shelf life at the time of its clearance from the customs. In addition to the food-business operators (FBO) License for the import of food, the Food Importer shall also register himself with the Directorate General of Foreign Trade and possess valid a Import–Export Code (FSSAI, 2016). General steps involved in the Food Import Clearance System (FICS) have been provided in the form of a "flow diagram" by the FSSAI (FICS, 2013).

As the exporting FBOs in India were having problems in producing the documentary evidence for the consent of the Indian Food Authorities, (as it was required by many regulatory bodies of importing countries), FSSAI has created a separate category, that is, "Exporting FBOs" for the issue of license under the Food Safety and Standards (Licensing and Registration of Food Businesses) Regulations, 2011. It provides criteria for the licensing of FBOs involved in food export, where manufacturers of food items with a 100% export are not required to submit a certificate of conformity but have to submit an undertaking that they will provide the Certificate of Compliance of the food product as per regulation. In case a new food product is intended to be introduced for domestic sale, the FBO shall first obtain a Food Product Approval from FSSAI (FSSAI, 2015).

The Food Authority, adopts a risk-based approach and a risk-based inspection process for the clearance of imported food articles by profiling the Importer, Custom House Agents, imported product, manufacturer of the imported product, country of origin, source country of the consignment, port of entry, compliance history, and any other parameters deemed fit for profiling the risk associated with the commodity. The Food Authority may also review the documents before arrival of the consignment to reduce the time of clearance from the customs by restricting/prohibiting the import of food articles based on risk perception or outbreaks of disease, etc.

According to the Indian Manual on FICS, various sections, such as, 22, 23, 25, 26, 43, 47, and 67, of the FSS Act 2006, cover related provisions on food imports, while some other sections provide for penalties for violations (FICS, 2013).

At the international level, (CAC/GL 47, 2003), provides a framework for the development and operation of an import-control system to protect consumers and facilitate fair practices in food trade while ensuring unjustified technical barriers to trade are not introduced. According to this guideline, physical checks during imports should normally be avoided, except in justified cases, such as, products associated with a high level of risk; a suspicion of nonconformity for a particular product; or a history of nonconformity for the product, processor, importer, or country.

Where an imported product does not conform to specifications, necessary action should be taken as per criteria explained in guidelines of (CAC/GL 26, 1997) to ensure that the action is proportionate to the degree of public health risk, potential fraud, or deception of consumers.

General characteristics of food import-control system (CAC/GL 47, 2003) are:

- requirements for imported and domestic foods should be the same;
- responsibilities of competent authorities should be clearly defined;
- legislation and operating procedures should be clearly defined and transparent;
- there should preferably be a precedence to protect consumers; and
- the importing country should have a legal provision for the recognition of the food-control system of the exporting country.

Food import-control systems should include provisions for the recognition of food-control systems of an exporting country by facilitating entry of goods, including the use of memoranda of understanding, MRAs, equivalence agreements, and unilateral recognition. Such recognition should include controls applied during the production, manufacture, importation, processing, storage, transportation of the food products, and verification of the export food-control system applied (CAC/GL 47, 2003).

Most developed countries base their import-inspection programs on internationally recognized standards and principles to ensure that imports of food products meet the desired food-safety requirements. Products not meeting regulatory requirements attract a penal action. Importers are responsible for the safety of foods that they import. Highly perishable foods of animal origin, for example, meat, fish, etc., have the potential to present higher levels of risk due to microbes or veterinary drugs.

Import controls generally include:

- assessment of inspection systems in exporting countries;
- assessment that certified products meet importing country's or equivalent regulatory provisions, performance-based assessment of documentation or shipments to verify that certification provisions are met; and
- statistics-based product sampling and analysis to confirm that residues, if present do not exceed regulatory standards.

37.6 FOOD EXPORT-CONTROL SYSTEM

The exporting country should provide detailed information to prove that its own safety-control system achieves the importing country's objectives and/or level of protection.

- Equivalence agreements for food-safety (sanitary) control measures are entered into after an importing country determines that an exporting country's control measures, even if different from those of the importing country, achieve the importing country's appropriate level of health protection.
- Equivalence agreements for other relevant requirements for food are entered into after an importing country determines that the exporting country's control measures, even if they different than those of the importing country, meet the importing country's objectives.

The development of equivalence agreements is facilitated by the use of Codex standards, recommendations, and guidelines by both parties. The importing and exporting countries should facilitate consultative process by exchanging information on:

1. legislative framework, including the texts of all relevant legislation, which provides the legal basis for the uniform and consistent application of the food-control system;
2. control programs and operations, including the texts of all the exporting country's operative measures relating to control programs and operations;
3. decision criteria and action;
4. facilities, equipment, transportation, and communications, as well as, basic sanitation and water quality;
5. laboratories, including information on the evaluation and/or accreditation of laboratories and evidence that they apply to internationally accepted quality-assurance techniques;
6. details of the exporting country's systems for assuring competent and qualified inspection personnel through appropriate training, certification, and authorization, including number and distribution of inspectors;
7. details of the exporting country's procedures for audit of national systems, including assurance of the integrity and lack of conflict-of-interest of inspection personnel; and
8. details of the structure and operation of any rapid alert systems in the exporting country.

Export-certification system is an integral part of food export-control systems, which helps to ensure that the exported food is safe and meets the sanitary requirements of the importing country as per CAC/GL 26 (1997). At the same time, an effective certification system depends on the existence of an effective inspection system. It has been observed that export-control systems of most countries are generally weak and are very minimal, providing only sanitary, phytosanitary, or health certificates, that too when insisted on by the importing country (FAO/WHO, 2005).

For strong import and export-control systems, importing countries should co-operate and recognize the genuine export-control certification systems of exporting countries without much fuss, as these issues tend to become barriers to trade due to the reluctance of importing countries to recognize export-control certificates of their trading partners through equivalence agreements. SPS Agreement expects member countries to establish formal systems of import control as per Codex standards along with adequate legislative framework (FAO/WHO, 2005).

37.7 EQUIVALENCE

Importing and exporting countries may sometimes operate different food inspection and certification systems, incorporating different technical requirements. These requirements may relate to control of production and processing systems, conformity-assessment systems, language(s) used to label products, and mechanisms for the prevention of fraud. Reasons for these differences include differences in prevalence of particular food-safety hazards and risks, methods of risk management, and differences in the historic development of food-control systems. In such cases to facilitate trade, there is a need to determine the effectiveness of sanitary measures of the exporting country in achieving the appropriate level of sanitary protection of the importing country. This has led to the recognition of the principle of equivalence as provided by WTO on the Application of SPS Measures, which states the following:

> *Members shall accept the SPS Measures of other members as equivalent, even if these measures differ from their own or those used by other members trading in same product, if the exporting member objectively demonstrates to importing member that its measures achieve the importing member's appropriate level of sanitary or phyto-sanitary protection. Members shall upon request, enter into consultation with the aim of achieving bilateral or multilateral agreements or recognition of equivalence of specified SPS measures.*

"Equivalence" (of sanitary measures), as per the SPS Agreement, is thus defined as the "the state wherein technical requirements applied in an exporting country, though different from requirements applied in an importing country, achieve the country's stated objective for that technical requirement." As per CAC/GL 26 (1997) "equivalence," is defined as "the capability of different inspection and certification systems to meet the same objectives."

An importing country should recognize equivalence of inspection and certification of the exporting country in accordance with these guidelines. For determination of equivalence, governments should recognize that:

- inspection and certification systems should be organized for the risk involved, considering that the same food commodities produced in different countries may present different hazards and,
- control methodologies can be different but achieve equivalent results (e.g., environmental sampling and the strict application of good agricultural practices, with limited end-product testing for verification purposes, may produce a result equivalent to extensive end-product testing for the control of agriculture chemical residues in raw products).

Controls on imported food and domestically produced foods should be designed to achieve the same level of protection. The importing country should avoid unnecessary repetition of controls where these have already been validated by the exporting country. The exporting country should provide access to enable the inspection and certification systems to be examined and evaluated, on request of the food control authority of the importing country.

Countries should have the legislative framework, controls, procedures, facilities, equipment, laboratories, transportation, communications, personnel, and training in place to support the objectives of the inspection and certification program.

Where different authorities in the same country have jurisdiction over different parts of the food chain, conflicting requirements must be avoided to prevent legal and commercial problems and obstacles to the trade later.

Food-exporting countries should have a well-developed food quality-control system for producing safe and quality food products for exports to reduce impediments at the importing end; which are as follows:

- to minimize and even eliminate rejection or noncompliance at the point of import;
- to avoid duplication of inspection, sampling, and tests at the exporting and importing ends, which results in better usage of the collective resources more efficiently and effectively;
- to reduce financial burdens due to recalls and cost of testing/destruction of consignments at the importing end;
- to reduce variations in quality due to production by small farmers, fishermen, or enterprises; and
- to improve the credibility of the country by ensuring that inferior quality products are not exported by unscrupulous traders. Such problems can be minimized with mandatory export certification, for example, export certification has become mandatory in the Indian dairy sector and products produced only by approved units (implementing food-safety management systems) are permitted for exports.

37.8 INSPECTIONS

Inspection is the examination of food or systems for control of food, raw materials, processing and distribution, including inprocess and finished product testing, in order to verify that they conform to requirements. The food authority should take all necessary steps to ensure integrity, impartiality, and independence of official inspection systems and to ensure that the inspection program contained in the national legislation is delivered to a prescribed standard.

Inspection services should utilize laboratories that are evaluated and/or accredited under officially recognized programs to ensure that adequate quality controls are in place to provide reliable test results. Validated analytical methods should be used wherever available.

Inspection systems' laboratories should apply the principles of internationally accepted quality-assurance techniques to ensure the reliability of analytical results.

37.9 CERTIFICATION SYSTEMS

Certification is the procedure by which official certification bodies and officially recognized bodies provide written or equivalent assurance that food or food-control systems conform to requirements. Certification of food may be, as appropriate, based on a range of inspection activities, which may include continuous online inspection, auditing of quality-assurance systems, and examination of finished products.

Certification should provide assurance of conformity of a product or that a food-inspection system conforms to specified requirements, and will be based on:

- regular checks by the inspection service;
- analytical results;
- evaluation of quality-assurance procedures linked to compliance with specified requirements; and
- any inspections specifically required for the issuance of a certificate.

These inspection and certification bodies/services should carry out self-evaluation or have their effectiveness evaluated by third parties periodically at various levels, using internationally recognized assessment and verification procedures. Assessment and verification of systems of an exporting country should be conducted as per (CAC/GL 26, 1997).

Apart from above, the tendency of domestic traders to give misleading impression of better quality or safety of exported or imported products in comparison with domestic products, should also be curbed by the food-control authorities by ensuring that no differences are observed in the quality of products meant for exports/imported products when compared with products meant for domestic consumption.

According to FAO/WHO (2005) all exporting countries should develop a clear export policy taking into account:

- the specific products;
- type of parameters (health, safety, and quality);
- voluntary requirements, if any, along with mandatory ones;
- the authority;
- the systems of inspection and certification to be followed; and
- signing of MRAs or equivalence agreements for recognition of its export certification with the aim to ensure that a safe and good-quality product is exported.

All specifications, inspection, testing methods, and operating procedures need to be documented. In fact, the entire inspection and certification system for exports needs to be documented. Each inspecting official should have a copy of the operative part of the legislation so that they are available at the point of use. All specifications, inspection, and testing methods should also be transparent and accessible to interested parties overseas.

Self-certification by industries needs to be encouraged. Only those exporting units should be approved that follow QMS and total quality-management approach, that is, implementing GMP/GHP/HACCP, as well as, conformance to the requirement of international specifications and those of the importing country. The certification should be such that rejections at the importing end are reduced, as rejection is the main concern of the exporters. There should be a single legislation for food-quality control to include both export and import and this should contain clear defined roles of various authorities with a view to avoid overlap. Inspection systems should be aligned with ISO 17020/Guide 65 and CAC/GL 26 (1997).

Importers should enter into equivalence agreements with exporters for the recognition of each other's food export inspection and certification systems according to the SPS Agreement. Such agreements would facilitate exports and reduce inspections and rejections of products in overseas markets.

A system of accreditation needs to be developed in all countries for both inspection (ISO 17020) and certification bodies (ISO Guides 62 and 65), as well as laboratories (ISO 17025), and all organizations performing such activities. This would provide credibility to the inspection and certification activities.

All concerned personnel should be trained regularly and systematically to create awareness about inspection and the latest testing techniques, methods of sampling, risk analysis, HACCP, document and record control, auditing techniques, etc. All organizations need to be networked so that information can be coordinated and accessed by all those who are concerned.

Inhouse laboratories with basic test facilities for implementing ISO 9000/14000 and HACCP must be developed.

Appropriate guidelines must be developed for exporters/importers as per international standards. As per FAO/WHO (2005) clear areas must be identified for developing domestic, import, and export inspection and certification systems to meet international requirements to include:

- strengthening of laboratories in terms of equipment and training;
- implementation of ISO/IEC 17020 for inspection activity, ISO/IEC Guides 62 and 65 for certification, and ISO/IEC 17025 for testing;
- empowerment of human resource, especially inspectors in area of inspection and audits; and
- usage of a systems' approach in export certification.

As per Article 9 of the SPS Agreement, importing developed countries may provide technical assistance in these areas. In case of rejections, an export-control body should enter into dialogue with importing authorities to resolve problems that may arise due to the rejection of a consignment in the importing country. In general, rejections occur due to differing standards/conformity-assessment operating procedures, lack of transparency, etc. Export certification provides data and the entire background of the exporter.

37.10 **CONCLUSIONS**

Effective food-control measures through inspections and certifications create confidence in consumers regarding safety and quality of their food supply. But these may, sometimes, become obstacles in the international food trade as greater stress is given on the implementation of rules and regulations. Importing and exporting countries often operate different food inspection and certification systems due to different food-safety hazards/risks and their management procedures. If exporting and importing countries work together to make sanitary measures of an exporting country more effective to achieve ALOP of an importing country, they can achieve ALOPs as per WTO. Significance of export-certification system must be recognized by countries and export-control systems should be established as per Codex guidelines.

APPENDIX A. RELATED DEFINITIONS AS PER CAC/GL 20 (1995), CAC/GL 47 (2003), CAC (2006), FSSAI (2016)

1. ALOP is the level of protection deemed appropriate by the country establishing a sanitary measure to protect human life or health within its territory. (This concept may otherwise be referred to as the "acceptable level of risk.")
2. Audit is a systematic and functionally independent examination to determine whether activities and related results comply with planned objectives.
3. Certification is the procedure by which official certification bodies and officially recognized bodies provide written or equivalent assurance that food or food-control systems conform to requirements. Certification of food may be, as appropriate, based on a range of inspection activities, which may include continuous online inspection, auditing of quality-assurance systems, and examination of finished products.
4. Inspection is the examination of food or systems for control of food, raw materials, processing, and distribution, including inprocess and finished product testing, to verify that they conform to the requirements.
5. Act refers to the Food Safety and Standards Act, 2006 (34 of 2006).
6. Authorized Officer means a person appointed as such by the Chief Executive Officer of the Food Authority by an order for the purpose of performing functions under Section 25 of the Act.
7. Balance shelf life means the period between the date of import (Import General Manifest) to "best before" and "expiry date" as the case may be.
8. Bill of entry means the bill of entry filed by the importer under the provisions of Section 46 of the Customs Act, 1962 (52 of 1962).
9. Customs airport means any airport appointed under clause (a) of Section 7 of the Customs Act, 1962 (52 of 1962) to be a customs airport.
10. Customs area means the area of a customs station and includes any area in which imported goods or export goods are ordinarily kept before clearance by the customs authorities.
11. Custom house agent means a person defined under the Custom House Agent Regulation, 2004.
12. Customs port means any port appointed under clause (a) of Section 7 to be a customs port and includes a place appointed under clause (aa) of that section to be an inland container depot.
13. Fees means the charges prescribed by food authority for clearance of imported food consignments.
14. Food Importer means a FBO importing or desirous of importing food into the Indian Territory, who is duly licensed as an Importer under the Food Safety and Standards (Licensing and Registration of Food Businesses) Regulations, 2011 notified under the Act.
15. Food Analyst means an analyst appointed under Section 45 of the Food Safety and Standards Act, 2006.
16. Import means bringing into India any article of food by land, sea, or air.
17. Nonconformance report means a certificate or report issued to the customs authorities and the food importer by the authorized officer or any other officer specifically authorized for this purpose by the food authority on account of nonconformance to/with the Act and the rules and regulations made thereunder of the consignment of the food importer.

18. No objection certificate means a certificate or report issued to the customs authorities and the food importer by the authorized officer or any other officer specifically. Authorized for this purpose by the food authority on account of conformance to/with the Act and the rules and regulations made thereunder of the consignment of the food importer.
19. Packing list means the itemized list of food giving the description, quantity, and weight of each imported article of food.
20. Prearrival document scrutiny means and refers to the scrutiny of documents submitted by the food importer to the authorized officer in advance, before the actual arrival of the article of food in order to facilitate faster clearance of food imported at the customs port, the food importer may be required to furnish documents as notified by the food authority from time-to-time.
21. Prohibited Food means that article(s)/category(ies) of food as declared by the food authority time-to-time and published on the website of the food authority.
22. Prohibited sources means the particular locations for which conditional or absolute restrictions are mentioned by the food authority on its website for import of food.
23. Review application fee means the fee levied by FSSAI toward disposal of a review application submitted by the food importer against the order of an authorized officer pertaining to the clearance of food imports.
24. Review officer means the CEO or an officer authorized by the CEO of the food authority for review of the orders of the authorized officer. Such a review officer shall examine any order of the authorized officer passed with respect of any import at the Customs Port(s), if the food importer approaches the review officer with respect of any such reviewable order.
25. Shelf life means the period between the date of manufacture and the "best before" or "expiry date" whichever is earlier as printed on the label.
26. Stuffing list means a list of f items and its actual physical arrangement inside the container, cartons, pallets, or skids.
27. Transit country list means the list of countries through which the imported food transits before it reaches the Indian Territory.
28. Unclaimed food means an imported food consignment not having a claimant or bill of entry or both.
29. Uncleared food means an imported food consignment of which the delivery is not taken by the importer within the period specified in the no objection certificate issued by the food authority.
30. Visual inspection means the process of inspection by the authorized officer or an officer deputed by him for the purpose by which the physical condition of the food consignment, scrutiny of documents, and compliance of Packaging and Labeling regulations are ascertained for the food-safety compliances prior to drawl of samples.
31. Official accreditation is the procedure by which a government agency having jurisdiction formally recognizes the competence of an inspection and/or certification body to provide inspection and certification services.
32. Official inspection systems and official certification systems are systems administered by a government agency having jurisdiction empowered to perform a regulatory or enforcement function or both.
33. Requirements are the criteria set down by the competent authorities relating to trade in foodstuffs covering the protection of public health, the protection of consumers, and conditions of fair-trading.

34. Officially recognized inspection systems and officially recognized certification systems are systems, which have been formally approved or recognized by a government agency having jurisdiction.

REFERENCES

Attrey, D.P., 2008. A mini text for the beginners. ISO 9001: 2000, quality management system. Attrey's Mini Text Series on Quality Management Systems and Good Quality Practices. Ministry of Defense, Government of India, Timarpur, Delhi.

CAC/GL 20-1995, 1995. The principles for food import and export inspection and certification. Adopted by the Codex Alimentarius Commission at its 21st Session, 1995. Available from: http://www.fao.org/docrep/009/y6396e/Y6396E01.htm

CAC/GL 26-1997, 1997. Guidelines for the design, operation, assessment and accreditation of food import and export inspection and certification systems. Available from: http://www.fao.org/docrep/009/y6396e/Y6396E03.htm

CAC/GL 47-2003, 2003. Guidelines for food import control systems. Available from: http://www.fao.org/docrep/009/y6396e/Y6396E03.htm

CAC, 2006. Guidelines for Food Import Control Systems; CAC/GL 47-2003. Available from: http://www.fao.org/docrep/009/y6396e/Y6396E02.htm; DL on 28-02-16

FAO/WHO, 2005. Food export control and certification. Proceedings of the Second FAO/WHO Global Forum of Food Safety Regulators (Agenda Item 4.5; GF 02/8a), Bangkok, Thailand, October 12–14, 2004. Available from: http://www.fao.org/docrep/meeting/008/y5871e/y5871e00.htm#Contents

FICS, 2013. Manual on Food Import Clearance System (FICS) implemented by the Food Safety and Standard Authority of India (FSSAI). Ministry of Health and Family Welfare, Government of India. Available from: http://www.fssai.gov.in/Portals/0/Pdf/Import_Manual%20(17.10.13).pdf

FSA, 2006. Food Safety and Standards Act, 2006; No. 34 of 2006; passed by the Indian Parliament and received assent of the President of India on August 23, 2006. Available from: http://www.fssai.gov.in/portals/0/pdf/food-act.pdf

FSSAI, 2015. New government order dated January 23, 2015, regarding grant of license to exporting FBOs. Available from: http://www.fssai.gov.in/Portals/0/Pdf/Order_Grant_License_FBOs.pdf

FSSAI, 2016. Food Safety and Standards (Food Import) Regulations. Available from: http://www.fssai.gov.in/Portals/0/Pdf/Draft_Import_Regulation.pdf

MacCurtain, S., 2015. What is conformity assessment? CASCO Secretary; ISO Central Secretariat and Head of Conformity Assessment at the International Organization for Standardization (ISO). Available from: http://www.iso.org/iso/home/about/conformity-assessment.htm; http://www.iaf.nu/upFiles/IAFRepsLiaisonsContacts20Nov2015.pdf; http://www.iso.org/iso/Casco

Regulation of advertisement for food products in India— advertisement for food products

38

S. Bajaj

Department of Community Medicine, Armed Forces Medical College, Pune, Maharashtra, India

38.1 INTRODUCTION

One day last month I came home feeling good after taking a lecture on benefits of healthy eating for a group of housewives and see my children happily munching on a certain brand of chocolate. Still under the influence of the talk I had just delivered, I told them that they should munch on something healthy like fruits, if they are hungry. Pat came the reply from the younger one, "It's full of goodness mom, the aunty in the advertisement said so!"

We all have faced this scenario in our homes. Advertising has a deep impact on the viewers especially the impressionable minds of young children. "Catch them young" has been a pet advertising mantra. Today kids have more access to answers to their queries from media due to unlimited access to it (Nair, 2006). Under the influence of media advertisement blitzkrieg they use this "pester power" to force their parents for buying a certain product. Most of the time they gain access to them through television and internet. Boys in south India often imitate Rajnikant, the popular film actor, who had a particular style of flipping a cigarette to his lips. They innocently start trying to flip the cigarette in the same manner. Many took to smoking through this playful, imitative initiation. Advertisements take them to a fairy world.

38.1.1 DEFINITION

Advertising is a means of communication with the users of a product or service. Advertisements are messages paid for by those who send them and are intended to inform or influence people who receive them, as defined by the Advertising Association of UK.

Food Safety in the 21st Century. http://dx.doi.org/10.1016/B978-0-12-801773-9.00038-8

38.1.2 **SURROGATE ADVERTISING**

Even when the government ban advertisements of tobacco and alcohol, many companies come up with other products in the same name such as sodas, mineral water, music CDs, etc. which are then advertised.

38.1.3 **PRODUCT PLACEMENTS**

Most of the children who watched the Hindi movie "Krrish" were influenced by the health drink which was endorsed by the protagonist of the film to gain herculean physical strength. This is a marketing strategy of the companies to advertise their products. The director of the film apparently showed these products not because of the benefits to the viewers, but because he was paid hefty amounts for this.

Advertising, though originally used to market products, now, unfortunately, seems to market feelings, sensations, and styles of life—an astounding revolution in manners and morals. Advertising creates and sustains an ideology of consumption. It is a strong social force affecting Indian homes today. Advertising is a mass-marketing technique. No company can survive in today's world without advertising. Its foremost social responsibility in free market society is to sell products and sell them efficiently. It stimulates competition and leads to product improvement (Lomeli et al., 2004).

However, advertisers must be aware of the social consequences of the advertisements. For example, advertisements of the cold drink, *Mountain Dew* show daring stunt actions where the statutory warning is either not written or written in very small illegible fonts. Any kid getting motivated by seeing this advertisement may rehearse the same action with dire consequences.

Misleading, false, and offensive advertisements can lead to resentment among the consumers and have a negative impact on the society. To regulate the advertising agencies an association called Advertising Agencies Association of India (AAAI) has been formed. It has a governing board called Advertising Standard Council of India (ASCI, 1985). It is a commitment to honest advertising and fair competition in the market place.

38.2 **ADVERTISEMENT POLICIES FOR FOOD SAFETY**
38.2.1 **FSSAI ON ADVERTISEMENT POLICIES FOR FOOD SAFETY**

The Food Safety & Standards Act, 2006 seeks to regulate the law relating to advertising and unfair trade practices in the food sector. Section 3(1) (b) of the act defines "advertising" as "any audio or visual publicity, representation or pronouncement made by means of any light, sound, smoke, gas, print, electronic media, internet or website and includes through any notice, circular, label, wrapper, invoice, or other documents." Section 24 of the act places restrictions of advertisement and prohibits unfair trade practices. It lays down the following general principles:

a. No advertisement can be made of any food which is misleading or deceiving or contravenes the provisions of the Act, the rules and regulations made thereunder;

b. No person can engage himself in any unfair trade practice for purpose of promoting the sale, supply, use and consumption of articles of food or adopt any unfair or deceptive practice including the practice of making any statement, whether orally or in writing or by visible representation which:

　i. falsely represents that the foods are of a particular standard, quality, quantity, or grade composition;

　ii. makes a false or misleading representation concerning the need for, or the usefulness;

　iii. gives to the public any guarantee of the efficacy that is not based on an adequate or scientific justification thereof;

c. In any case where a defence is raised to the effect that such guarantee as described above is based on adequate or scientific justification, the burden of proof of such defence shall lie on the person who raises defence.

Sections 52 and 53 of the act prescribe the punishment for selling misbranded food and also for misleading advertisements.

Section 52 lays down that any person who whether by himself or by any other person on his behalf manufactures for sale or stores or sells or distributes or imports any article of food for human consumption which is misbranded, shall be liable to a penalty which may extend to 3 lakh rupees.

Section 53 prescribes that any person who publishes, or is a party to the publication of an advertisement, which falsely describes any food or is likely to mislead as to the nature or substance or quality of any food or gives false guarantee shall be liable to a penalty which may extend to 10 lakh rupees.

Various jurisdictions around the world have specific guidelines/codes laying down minimum standards for food advertisements. Although most of these codes are self-regulatory in nature, they can act as an effective deterrent to prevent misleading advertisement from being disseminated to the general public.

38.2.2 **THE ASCI CODE**

ASCI has drafted and implemented a Code for Self-Regulation in Advertising (*ASCI Code*) in India. Its purpose is to control the content of advertisements and not to hamper the sale of products which may be found offensive. It has been drawn up after wide industry consultation. It has been accepted by individuals, corporate bodies, and associations engaged in or otherwise concerned with the practice of advertising, as basic guidelines with a view to achieve the acceptance of fair advertising practices in the best interest of the ultimate consumer. This applies to advertisers, advertising agencies, and media. The code makes it clear that it does not seek to usurp or replace the other legal provisions that might affect advertising. Its purpose is only to complement such laws (ASCI, 1985).

38.2.3 INTERNATIONAL ADVERTISEMENT POLICIES

Most countries around the world have sought to monitor and/or regulate the content of food and beverage advertisements to ensure that commercial communication is conducted responsibly, physical well being of its citizens is maintained and not affected by wrong or misleading claims and also to enable consumers to make informed decisions.

38.3 FOOD SAFETY GUIDELINES

The following guidelines are proposed by FSSAI to be adopted to promote high standards in food and beverage communications and advertising:

1. *FSSAI*: A statutory regulatory authority of Government of India (2006) (Ministry of Health & Family Welfare) was set up under the FSSA, 2006 for laying down science based standards for articles of food and to regulate their manufacture, storage, distribution, sale, and import to ensure availability of safe and wholesome food for human consumption and for matters connected therewith or incidental thereto.
2. Communications and advertisements related to food and beverages can have a significant impact on the lives of the public in general and their physical and material well-being in particular.
3. It is imperative that food and beverage communications and advertisements fulfill their intended roles and FBOs adopt strict principles of self-regulation and not mislead the general public in any manner detrimental to their well-being.
4. The issue of labeling would be governed by explicit provisions under the act and the rules on labeling being developed and recognizing the need to
   a. promote high standards of business ethics to ensure that commercial communications to consumers are responsible;
   b. provide honest and truthful information about food and beverage products.

38.3.1 DEFINITION OF ADVERTISING FOR THE PURPOSE OF THE GUIDELINES

Advertising means any commercial communication by a Food Business Operator (FBO) to the public other than through a label which is published or broadcast or disseminated using any medium in India for payment or other valuable consideration to promote directly or indirectly the sale and intake of food and beverages in any manner.

38.3.2 GENERAL PRINCIPLES

1. Advertising and communication for food and beverages should not be misleading or deceptive. This means that claims about particular ingredients in a food and

beverage product or the underlying health benefits thereto should have a sound, authentic, scientific basis, and supported by evidence whenever required.

2. Advertising and/or marketing communications for food and/or food and beverage products that include what an average consumer, acting reasonably, might interpret as health or nutrition claims shall be supportable by appropriate scientific evidence and should meet the requirements of the basic food standards laid down under the FSSA, 2006 wherever applicable.

3. Advertisements should not disparage good dietary practice or the selection of options, such as fresh fruit and vegetables that accepted dietary opinion recommends should form part of the average diet.

4. Advertisements should not encourage excessive consumption or inappropriately large portions of any particular food. They should not undermine the importance of healthy lifestyles. Advertisements should rather try to promote moderation in consumption and the need to consume in suggested portion sizes.

5. Care should be taken to ensure advertisements do not mislead as to the nutritive value of any food. Foods high in sugar, fat, TFA, and/or salt should not be portrayed in any way that suggests they are beneficial to health.

6. The nature of the audience should be taken into account particularly when selling products in rural areas, to urban poor or to children. Advertisements and communications should not exploit their lack of experience or knowledge and always provide truthful information. In such cases, nutritional or health-related comparisons should be based on an objectively supportable and clearly understandable basis.

7. Communications for *Food* and/or *Beverage Products* including claims relating to material characteristics such as taste, size, suggested portions of use, content, nutrition, and health benefits shall be specific to the promoted product/s and accurate in all such representation.

8. Advertisements should not mislead consumers especially children to believe that consumption of product advertised will result directly in personal changes in intelligence, physical ability, or exceptional recognition unless supported with adequate scientific evidence.

9. Advertisements containing nutrient, nutrition or health claims and advertisements directed at children should observe a high standard of social responsibility.

10. Communications for food and/or beverage products not intended or suitable as substitutes for meals shall not portray them as such.

11. Claims in an advertisement should not be inconsistent with information on the label or packaging of the food.

12. Advertisements for food and beverages should not claim or imply endorsement by any government agency, professional body, independent agency, or individual in particular profession in India unless there is prior consent, the claim is current and the endorsement verifiable and the agency or body named.

13. Celebrities or prominent people who promote food should recognize their responsibility toward society and not promote food in such a way so as to undermine a healthy diet.

14. Advertisements should not undermine the role of parental care and guidance in ensuring proper food choices are made by children.

15. Advertisers and communicators must recognize their social and professional responsibility toward promoting a healthy lifestyle and strive to achieve high standards of public health. All advertisements and communications should be thus truthful, legal, decent and honest reflecting their social and professional responsibility (FSSA, 2006).

38.4 FOOD SAFETY AND STANDARD REGULATIONS (FSSR), 2011

1. These were notified by Competent Authority of Central Government. The same has been enforced with effect from August 5, 2011. The provisions contained in earlier legislation on the subject are now repealed.

2. The various false claims made by the FBO about food articles and consequent violation, if any, are punishable under the provisions of FSSA, 2006.

3. Violations related to food items, seriously jeopardize public health as well lead to unfair gains to Food Business.

4. Misleading advertisement related to food items are imputed with malafide intent on the part of person making the claim and is normally made to misguide a consumer to purchase food item without disclosing the complete details on the advertisement. Companies (Corporate bodies including firm or other association, individual) are also covered u/s 66, FSSA, 2006.

5. The burden of proof lies on the person wilfully making false claims or engaged in misleading advertisement.

6. An advertisement is defined u/s 3 of FSSA, 2006 as: any audio or visual publicity, representation or procurement made by means of light, sound smoke, gas, print, electronic media, internet, and website and included through any notice, circular, label, wrapper, invoice to other documents;

   As per Section 24, Restrictions of advertisement and prohibition as to unfair trade practices of FSSA, 2006:

   1. No advertisement shall be made of any food which is misleading or deceiving or contravenes the provisions of this Act, the rules and regulations made thereunder.

   2. No persons shall engage himself in any unfair trade practice for purpose of promoting the sale, supply, use and consumption of articles of food or adopt any unfair or deceptive practice including the practice of making any statement, whether orally or in writing or by visible representation which

      a. Falsely represents that the foods are of a particular standard, quality, quantity, or grade composition;

    b. Makes a false or misleading representation concerning the need for, or the usefulness;

    c. Give to the public any guarantee of the efficacy that is not based on an adequate or scientific justification thereof.

7. Provided that where a defence is raised to the effect that such guarantee is based on adequate or scientific justification, the burden of proof of such defence shall lie on the person raising such defence. Further, any person who publishes, or is a party to the publication of an advertisement, which

    a. falsely describes any food; or

    b. is likely to mislead as to the nature or substance or quality of any food or gives false guarantee, shall be liable to a penalty which may extend to ten lakh rupees.

8. All FBO as well as any person dealing with food articles are advised to be careful as well as alert and must strictly follow provisions contained in FSSA, 2006 and Regulations thereof eschewing misleading claims which is not established by scientific evidence and validated by science as proof beyond reasonable doubts. Food items under Section 22 of the Act including Nutraceuticals, health supplements, functional food which have not taken product approval or operating only on the basis of NOC pending approval of sale cannot make any claim in their advertisement with any Health Claim, Nutraceuticals Claims, or Risk Reduction Claim (FSSA, 2006, 2011).

38.5 JAGO GRAHAK JAGO

Jago Grahak Jago is a consumer awareness program from Ministry of Consumer Affairs, Government of India. The slogan *Jago Grahak Jago* has now become a household name as a result of the publicity campaign undertaken in the last few years. Through the increased thrust on consumer awareness in the 11th 5-year plan, the government has endeavored to inform the common man of his rights as a consumer. As part of the consumer awareness scheme, the rural and remote areas have been given the top priority. The government has used multiple channels to create awareness about it.

38.6 WHO CAN FILE A COMPLAINT?

1. Complaint in relation to any goods or services may be filled by a consumer
2. Any voluntary consumer association registered under the Companies Act, 1956 or under any other law for the time being in force
3. The central government or any state government
4. One or more consumers, where there are numerous consumers having the same interest
5. In case of death of a consumer, his legal heir or representative
6. A power of attorney holder cannot file a complaint under the act

38.7 HOW TO FILE A COMPLAINT?

A complaint can be filed on a plain paper. It should contain the following:

- the name description and address of the complainant and the opposite party
- the facts related to complaint and when and where it arose
- documents in support of allegations made in the complaint
- the relief which the complainants is seeking
- the complaint should be signed by the complainants or his authorized agent
- no lawyer required for filing the complaint
- nominal court fee

38.8 WHERE TO FILE A COMPLAINT?

It depends upon the cost of the goods or services or the compensation asked:

- *District forum*: If it is less than Rs 20 lakhs
- *State commission*: If more than Rs 20 lakhs but less then Rs 1 crore
- *National commission*: If more than Rs 1 crore

38.9 THE HORLICKS CONTROVERSY

The trademark Horlicks filed a civil suit in the Calcutta High Court in August, 2004 alleging the disparagement of their product by an advertisement of the licensed users of the trade mark Complan. The advertisement had depicted the two cups including one cup with the alphabet "H." In the said advertisement, Complan cup was shown as growing in height as compared to the cup with the alphabet "H." The appellants (Horlicks) succeeded in getting injunction orders against the respondent (Complan) restraining the respondent to continue with the said advertisement or any other advertisement which reflected adversely on the appellants product Horlicks. The second set of litigation was instituted in the same year in the Madras High Court by the appellants alleging that a series of advertisements had been issued throughout the country in August, 2004 disparaging the products "Horlicks" and "Boost" with false and misleading comparison with the product "Complan." The advertisement showed two cups on either side bearing alphabets "X" and "Y" with white color liquid and chocolate color liquid which was suggested to be indicative of "Horlicks" and "Boost." The children consuming Complan were shown to grow taller.

In December, 2008, the respondent introduced an advertisement in the print media which according to the appellants sought to give an impression to the readers that Horlicks was a cheap and ineffective product which did not give balanced complete planned nourishment to the child. The lower price of Horlicks is sought to be attributed to use of cheaper and inferior quality ingredients and the question posed to a mother of a child is whether the cheaper price or a child's complete growth was important while choosing a health drink. The appellants thus contended that the advertisement sought to convey that though Horlicks was cheaper in price it also compromised on a child's growth. Such a comparison was sought to be made more apparent by putting a choice to the mother as to whether she knew the difference between what is good or what is cheap.

38.10 **THE WAY FORWARD**

The cornerstone of the practices followed by other countries is self-regulation guided by some handholding in terms of prescription of basic criterion by the governments/legislatures/consultative bodies respectively. This is especially true for food products, owing to the vast nature, array of products, and the distinct requirements for each of such products, requiring use of large number of ingredients, additives and chemicals, effect of which are varied, far reaching, and dependent on number of criteria.

The FSSAI could adopt the suggested framework as a guideline and allow an independent body such as ASCI to formulate the industry code as applicable to food and beverage communications/advertisements and treat the same as benchmark for compliance with the provisions of the act. Such an approach would also allow the industry and the regulations to adapt themselves to ever-changing standards and operational/business processes.

38.11 **CONCLUSIONS**

Advertising can cause positive as well as negative impact on the society. Products which are heavily advertised are expensive due to the cost spent on advertisement. In today's materialistic world advertising has become a necessary social evil. Advertisements influence the consumers and can lead to contentment or discontentment depending on what the advertisement proclaimed and what the final product turned out to be. The FSSA, 2006 and ASCI Code seeks to regulate the law relating to advertising and unfair trade practices in the food sector.

REFERENCES

Advertising Standard Council of India (ASCI), 1985. Available from: http://www.ascionline.org/index.php/ascicodes.html

Food Safety and Standards (FSS) Act, 2006. Rules & Regulations, 2011.F. No. 6/FSSAI/Dir (A)/Office Order/2011-12.

Government of India, 2006. The Food Safety & Standards Act, 2006.

Lomeli, J.L., Goddard, E.W., Lerohl, M.L., 2004. Effects of advertising, food safety and health concerns on meat demand in Canada. J. Food Distrib. Res. 35 (1).

Nair, T., 2006. Children loosing childhood very fast. Statesman.

Food safety concerns in context of newer developments in agriculture/ food science/food processing

Nutritional labeling

39

P. Dudeja*, R.K. Gupta**

**Department of Community Medicine, Armed Forces Medical College,
Pune, Maharashtra, India; **Department of Community Medicine, Army
College of Medical Sciences, New Delhi, India*

39.1 INTRODUCTION

Globalization and urbanization have caused a shift in the eating patterns of Indian society. There has been a marked increase in consumption of packaged food items especially by the Indian middle class where working women find them convenient to use and store too. This has also paralleled with increase in knowledge about food-borne illnesses (FBI) and noncommunicable diseases (NCDs) through mass media. Also, there is growing awareness among people about attaining wellness through diet and nutritious food. In order to ensure that the food is safe, nutritious, and in accordance with the dietary choices of the consumers, the food package label holds special importance.

A food label may literally be a label—a piece (or pieces) of printed paper attached to a food package—or it may comprise all or part of the printed or lithographed exterior surface of the package. From the point of view of manufacturer or food business operator (FBO), it as an instrument of marketing and product promotion among the discerning literate consumers. It is a way by which any manufacturer can introduce and promote his product among the target consumers. The food label also gives credibility to the manufacturer as well as the food product in the market. It is a medium to reduce the information gap between producers and consumers.

From public health point of view, a label is a tool to promote health by providing accurate nutritional information so that consumers can make informed dietary choices. The WHO global strategy on diet, physical activity, and health endorsed in May 2004 by the World Health Assembly, states that providing accurate, standardized, and comprehensible information on the content of food items is conducive to consumers making healthy choices. Reading of nutrition labels is an important tool in hands of public health specialists for health promoting lifestyle profile (HPLP) in the community. With gradual increase in literacy level in India, nutritional labeling has emerged as an important aspect of consumers' food purchase decisions. Overall, countries and areas can be characterized as having one of four types of regulatory environment as given in Table 39.1.

Table 39.1 Categorization of Countries as per Labelling Requirements

Category Type	Regulation
1	Mandatory nutrition labeling on all prepackaged food products
2	Voluntary nutrition labeling, which becomes mandatory on foods where a nutrition claim is made (most countries also mandate labeling on foods with special dietary uses)
3	Voluntary nutrition labeling, which becomes mandatory on foods with special dietary uses
4	No regulations on nutrition labeling

39.2 NUTRITION LABELING AS PER FOOD SAFETY AND STANDARDS ACT (FSSA) 2006

With the advent of government notification on labeling requirements under the FSSA 2006 for food processors, food labels are now seen on most of the packed food items. As per FSSA 2006 every prepackaged food should carry a label. The label should not be false, misleading, or deceptive, or create an erroneous impression regarding its character in any respect. It shall be applied in such a manner that it should not become separated from the container. Contents on the label shall be clear, prominent, indelible, and readily legible by the consumer under normal condition of purchase and use. In case the container is covered by a wrapper the wrapper shall carry the necessary information or the label on the container shall be readily legible through the outer wrapper or not obscured by it. As per Food Safety and Standards Regulations (FSSR) 2011, following details are required to be furnished on the nutrition label of a packaged food.

1. Product name and category of food
2. Ingredients list in descending order of weight
3. Logo for vegetarian/nonvegetarian food (figure)
4. Nutrition facts panel or information which includes energy, protein, carbohydrate (sugars), and fat
5. *The shelf life (use by or best before date)*: The shelf life of a product is the time period within which it is to be consumed. It can be described as either "use by" date or "best before" date.
   a. *Use by*: The "use by" code is used for foods that are microbiologically highly perishable and deteriorate and become dangerous to human health after a short time (e.g., chilled foods, cooked chilled meals which should be refrigerated).
   b. *Best before*: This type of code is used for products where "use by" date is not applicable or required. The best before date must be expressed as a day, month, and year in that order. Some products are not required to be date

marked, for example, wines and spirits which have long shelf life, fresh fruits, and vegetables, etc.

6. *Storage conditions*: For example, "store in a cool dark place/store in a refrigerator once opened"

7. Name and address of the manufacturer, packer, and/or seller

8. The country of origin (in case of imported foods)

9. The weight

The nutrition facts table is a mandatory requirement for the package of food material. It has an ingredient list which is mandatory and nutrition claims which are optional.

39.3 **NUTRITION FACTS PANEL**

The Nutrition Facts panel is in the form of a table which gives information about the amount of 13 core nutrients and calories in an amount of food. It helps the consumer to choose products more easily, compare two products to make better food choices, learn about the nutrition information of the food, management of special diets, Increasing or decreasing intake of any nutrient and avoid certain items in particular disease conditions. Prototype table is given (Table 39.2). It looks the same on most foods. This makes it easy to find and easy to read. There are certain foods which do not require having a nutrition table. For example, fresh vegetables and fruits, raw meat, and poultry (except when it is ground); raw fish and seafood; foods prepared or processed at the store (bakery items, salads, etc.); sweets sold at *halwai* shop; foods that contain very few nutrients such as coffee, tea, herbs, and spices; foods served for immediate consumption as in cafeterias, airplanes, food service vendors, and vending machines; medical foods such as those used to address the needs of patients with certain diseases and alcoholic beverages.

Table 39.2 Pattern of Nutritional Information on the Label

Nutritional Elements	Amount
Energy	# kcal
Protein	# g
Carbohydrates sugars	# g
Fat	# g
Saturated fatty acids	# g
Polyunsaturated fatty acids	# g
Monounsaturated fatty acids	# g
Trans fatty acids	# g
Cholesterol	# mg

# denotes the quantity of the respective element.

39.4 INGREDIENT LIST

It is a list of all the ingredients in a food. The ingredients are listed in order of weight, from most to least. This means that the food contains *more* of the ingredients at *the beginning* of the list and *less* of the ingredients at the *end* of the list. Food companies have to put the ingredient list on packaged foods. For example, bran cereal—ingredients are whole wheat, wheat bran, sugar/glucose–fructose, salt, malt (corn flour, malted barley), vitamins (thiamine hydrochloride, pyridoxine hydrochloride, folic acid, d-calcium pantothenate), minerals (iron, zinc oxide) (Table 39.3).

The ingredient list may be found anywhere on the food label. However, most of the time, it is close to the Nutrition Facts table. The ingredient list can help the consumer to get to know if a food product has a specific ingredient and avoid certain ingredients in case of a food allergy or intolerance.

Table 39.3 Common Nutrient Claims

Claim	Interpretation (Serving of Product Contains)
Calorie free	Less than 5 cal
Sugar free	Less than 0.5 g of sugar
Fat free	Less than 0.5 g of fat
Low fat	3 g of fat or less
Reduced fat or less fat	At least 25% less fat than the regular product
Low in saturated fat	1 gram of saturated fat or less, with not more than 15% of the calories coming from saturated fat
Lean	Less than 10 g of fat, 4.5 g of saturated fat and 95 mg of cholesterol
Extra lean	Less than 5 g of fat, 2 g of saturated fat and 95 mg of cholesterol
Light (lite)	At least one-third fewer calories or no more than half the fat of the regular product, or no more than half the sodium of the regular product
Cholesterol free	Less than 2 mg of cholesterol and 2 g (or less) of saturated fat
Low cholesterol	20 or fewer milligrams of cholesterol and 2 g or less of saturated fat
Reduced cholesterol	At least 25% less cholesterol than the regular product and 2 g or less of saturated fat
Sodium free or no sodium	Less than 5 mg of sodium and no sodium chloride in ingredients
Very low sodium	35 mg or less of sodium
Low sodium	140 mg or less of sodium
Reduced or less sodium	At least 25% less sodium than the regular product
High fiber	5 g or more of fiber
Good source of fiber	2.5–4.9 g of fiber

39.5 NUTRITION CLAIMS VERSUS HEALTH CLAIMS

Nutrition claims on foods are of two types, nutrient content claims and health claims. A *nutrient content claim* can help consumer to choose foods that contain a particular nutrient which is required in more. For example:

- *Source*: *"source of fiber"*
- *High or good source*: *"high in vitamin A"* or *"good source of iron"*
- *Very high or excellent source*: *"excellent source of calcium"*

Similarly the nutrient content claim can also be used to choose foods that contain a nutrient which is required *less of*

- *Free*: *"sodium free"* or *"trans fat free"*
- *Low*: *"low fat"*
- *Reduced*: *"reduced in calories"*

However, the consumer needs to be educated about the fact that nutrient claims highlight one nutrient, nutrition facts table needs to be referred, to make better food choices.

Health claim helps the consumer choose those foods that can be included as part of healthy diet to reduce risk of chronic diseases. For example, we come across certain brands of oil with the health claim "good for your heart, good for health." Health claims are optional and only highlight a few key nutrients. Hence, it is worthwhile to refer to the nutrition facts table to make better food choices

39.6 OTHER CLAIMS

Other claims, often referred to as general health claims, have appeared in recent years on front-of-package labeling. They include broad "healthy for you" or "healthy choice" claims as well as symbols, logos, and specific words. These claims are not developed by the government. Instead, they are developed by third parties or corporations. It is required that the information provided should be truthful and not misleading however, consumers should not rely only on general health claims to make informed food choices. Some of the packed food items may not have all the aforementioned information on their food label as the label requirement depends on the type of food and packing size, the details of which are available in various government notifications.

Apart from the mandatory nutrition declarations some food labels display contents of cholesterol, mono unsaturated fatty acids (MUFA), poly unsaturated fatty acids (PUFA), trans-fatty acids, vitamins, minerals, dietary fibres, and sugar alcohols. These nutrients are included as the food label either has a health claim or a nutritional claim. Another reason for this inclusion is that internationally, the food labels are as per Codex Alimentarius guidelines. The requirements for food label at the international level are vast and include many of the additional nutrients which are mentioned previously. The Indian food regulations so far have been restricted only to total carbohydrates, sugar, fat, protein, and energy facts. However, considering the global requirements the Indian regulations may get broad based with time. The producers are therefore proactive and prefer displaying additional information. This one-time investment on their product label helps them to benchmark their products with international brands.

Food labeling differs from one country to other, but most of prepackaged foods have the following labels:

- name or description of food;
- net quantity (weight, volume, or number of items in packet);
- ingredients (arranged in descending order of weight);
- date mark (indicating by when it should be consumed);
- instructions for use;
- name and address of supplier; and
- place of origin.

39.7 PERCENTAGE (%) DAILY VALUE

Some food labels display % daily value (DV). It is not mandatory as per the Indian guidelines. It is provided for each nutrient except sugar and protein. It represents the percentage of nutrient in a food product compared with total amount of that nutrient a person should have in 1 day, based on 2000 calorie diet. For example, a label may show that a serving of the food provides 30% of the daily recommended amount of dietary fiber. This means one still needs another 70% to meet the recommended goal. It is recommended that total fat, saturated fat, and cholesterol in diet to be limited. Foods with a lower % DV for these nutrients should be preferred. On the other hand foods with a higher % DV for vitamins, minerals, and fibers should be included in the diet. Some of commonly used specific nutrient claims are given in Table 39.2. Example of a nutrition label is given in Fig. 39.1. General guidelines for the nutrient content are as follows:

- "Free" means a food has the least possible amount of the specified nutrient.
- "Very low" and "low" means the food has a little more than foods labeled "free."
- "Reduced" or "less" mean the food has 25% less of a specific nutrient than the regular version of the food.

FIGURE 39.1 Symbol for Nonvegetarian and Vegetarian Food Items

It appears that nutritional information affects purchasing behavior because it influences valuations and perceptions of the product. In that context, several surveys have studied the effect that claims create on personal evaluations. Health claims in the front of the package have been found to create favorable judgments about a product. For example, when a product features a health or nutrient content claim, consumers tend to view the product as healthier and are then more likely to purchase it, independent of their information search behavior. Other studies, however, have found that health claims have a weak effect on disease risk perceptions. Most importantly, one has to keep in mind that in the food choice process, there will always be a taste-nutrition trade off. Consumers may prefer the immediate gratification offered by a tasty product rather than the long run benefits of a nutritious product. Dietary indulgence is often cited as consumer's self-control problem (Fig. 39.2).

On the other hand, many of our consumers are not able to read and understand English. These labels are of no use to them. In order to achieve the full potential of these labels it would be a step forward to have these labels in the regional languages of the country for benefit of all.

FIGURE 39.2 Example of a Nutrition Label

39.8 NUTRITIONAL INFORMATION AND DIETARY CHANGES

Some researchers have argued that provision of health related information does not always lead to healthier consumption. Most empirical research, however, suggests that provision and use of information can significantly change dietary patterns. Several studies have found that nutritional label use contributes to a better dietary intake or to reduced consumption of "unhealthy" foods. Nutritional label use is also associated with diets high in vitamin C, low in cholesterol, and lower percentage of calories from fat. Other studies have found nutritional label use to increase dietary quality of consumers, with higher improvements detected when health claim information was used.

39.9 NUTRITION LABELS IN PACKAGED FOOD: "EAT" IN BETWEEN THE LINES

If a consumer reads a label on a burger stating that it is "fat-free" he would assume that the contents of the labeled container are free of fat. According to FDA, if there is less than 0.5 g of fat in a serving, the food is labeled as "fat-free." However, there is misbranding of food articles in the name of 97% fat-free food when the food actually contains more fat for example a food product X is marketed to be 97% fat free. The product is labeled as: serving size—56 g; servings/containers—4; total weight—224 g. The nutritional information on the label is calories/serving—70; calories from fat—15; total fat per serving—1.5 g. The aforementioned information clearly indicates that the percentage of calories from fat is 21.4% (15/70*100) which is quiet high fat. This is far from the claim of food companies of the product being "97% fat free." However, the companies get away with this very smartly by highlighting the percentage of no fat by weight in the food item, that is, it is computed by considering total fat by weight in the package as 6 g (1.5 g*4) and calculating its amount (percentage) with respect to the total weight of the product (6/224*100) which equals to 2.67% of fat by weight which the companies project as "97% fat-free." The consumers are misled as they are given the percentage of the "weight" that is fat-free rather than the percentage of "calories." Thus, instead of the going for the product by the big print in the front, always look for the calories per serving and what the companies refer to as serving size. This technique is used to sell 2% fat milk as 98% fat-free.

39.10 CONCLUSIONS

The famous government-led awareness campaign "*Jago Grahak Jago*" in India has awakened the masses through print as well as electronic media, highlighting issues of maximum retail price (MRP), labeling, and standardizing of various products including food items. However, a gap still exists in using food labeling as a reference among consumers for making their dietary choices. Still many a times food products are purchased without referring to the "best before date" or expiry date or loose food products such as oil, etc. are brought which results in consumers falling prey to FBI. Moreover consumers are befooled by the big prints of "fat-free" or "97% fat-free" on the front of the food product. Examining food labels is an easy way to detect breach in food safety as no specific lab tests are required and is a self documentary evidence for identifying the infringe in the food supply chain. Thus, consumers need to become self conscious and incorporate food labeling in their daily purchases so as to keeping themselves as well as their families healthy.

Nutrition labeling can be an effective means of helping consumers to make healthful food choices. However, existing evidence concerning the effect of health claims on diet and public health is insufficient. Regulations can play a crucial role in enhancing the potential for nutrition labeling and health claims to promote health. The effectiveness of nutrition labeling and health claims in improving national dietary patterns relies largely on a motivated and educated public to make healthful choices. Yet this approach has its limitations. If there is to be significant change, action on nutrition labels and health claims need to be part of an integrated approach. It should tackle the increasing rates of diet-related noncommunicable diseases at a population level. Simultaneously proper nutrition labeling also educates individuals.

FURTHER READING

FAO/WHO, 1993. Codex Guidelines on Nutrition Labelling (CAC/GL 2_1985) (Rev.1_1993). Food and Agriculture Organisation of the United Nations/World Health Organisation, Rome.

Training manual for food safety regulators, 2010. Foods Safety and Standards Authority of India. Ministry of Health & Family Welfare, Delhi.

WHO, 2003. Diet, nutrition and the prevention of chronic diseases. Report of a Joint WHO/FAO Expert Consultation. World Health Organisation, Geneva. WHO Technical Report Series, no. 916. Available from: http://www.who.int/hpr/NPH/docs/who_fao_expert_report.pdf

Nutraceuticals

40

P. Dudeja*, R.K. Gupta**

*Department of Community Medicine, Armed Forces Medical College, Pune, Maharashtra, India; **Department of Community Medicine, Army College of Medical Sciences, New Delhi, India

40.1 INTRODUCTION

The three basic requirements for humans to survive are food, clothing, and shelter. Among these food tops the list. Humans can eat almost anything. The early man lived by gathering, hunting, and fishing and later cultivation. Various discoveries transformed them from food gatherers to producers and allowed them to grow food and lead a settled life. They ate mainly natural food. Cereals were used by all civilizations in the past, for example, rice and millet were the principal foods in the far east and wheat, oat, and rye in Europe. Beginning about 5000 years ago, a far more complex way of life began to appear in some parts of the world. With advent of industrialization man started living in towns and cities, where industry and commerce flourished.

Grains are the most abundantly grown crops on earth and the English word "meal" which we often use for food, literally means "grain." With a change in time the way we use these grains has also changed. Initially, the grains were used in the unpolished form or crushed to make further consumable forms. In the modern society we have started processing the grains. It removes the nutrient-rich outer layers to yield a fiber-free grain. Along with this, overall consumption of cereal has declined. Consumption of meat, refined sugar, and processed foods has increased. This deviation in diet from natural foods has led to deficiency of essential nutrients in people.

But of late, people have realized that along with taste, convenience, and value, health is an important concern. Nutritive value of food items has become a growing concern. The demand now is that any food product should be "as natural as possible," which parallels the common man's knowledge and perception of what is healthy. People in developed world are aware of the fact that processed food is nutritionally deficient, so they would take vitamin supplements to compensate for the loss of nutrients. Now the trend has changed. Rather than going for pharmaceuticals, people prefer natural substitutes such as herbals. The pharmaceutical companies have

Food Safety in the 21st Century. http://dx.doi.org/10.1016/B978-0-12-801773-9.00040-6

exploited this preference of the people and have promoted benefits of natural products in the form of capsules and tablets (nutraceuticals). This can be aptly called as *"medicalization in nutrition,"* where the desired nutrients are acquired in the form of tablets, capsules, or drinks rather than natural food items. There is growing awareness about health in our country especially in the growing Indian middle class. This class has also seen rise in purchasing power of the people and have been the target for sale of nutraceuticals.

40.2 NUTRACEUTICAL

Hippocrates said *"let food be your medicine and medicine be your food,"* to predict the relationship between food for health and their specific therapeutic benefits. In Ayurveda, the ancient Indian science of medicine, a lot of emphasis is given to role of nutrition in health and disease. The term "nutraceutical," derived from the terms nutrition and pharmaceutical was coined in 1989 by Dr. Stephen De Felice, (Chairman of the Foundation for Innovation in Medicine). The term is intended for a nutritional supplement that is sold with the intent to treat or prevent disease and does not have any regulatory definition. Hence, a "nutraceutical" is any substance that may be considered a food or part of a food which provides medical or health benefits, encompassing, prevention and treatment of diseases. Isolated nutrients, dietary supplements, and diets to genetically engineered "designer" foods, herbal products, and processed foods such as cereals, soups, and beverages may be included under the umbrella of the term nutraceuticals.

There seems to be a thin line distinguishing terms such as nutraceuticals, pharmaceuticals, dietary supplements, functional foods, etc. These terms have often been used interchangeably. Items such as "yogurt," "probiotics," etc. feature on the weekly shopping lists of many households but few are acquainted with the term "functional foods" or "nutraceuticals." The term has no legal or regulatory status.

Nutraceuticals are sometimes also referred as *"functional foods,"* but in this case the difference between food and medicine is not very clear.

When food is being cooked or prepared using "scientific intelligence" with or without the knowledge of how or why it is being used, then the food is called as "functional food." Thus, functional food provides vitamins, fats, proteins, carbohydrates necessary for health of an individual. When functional food aids in the prevention and/or treatment of disease(s)/disorder(s) other than deficiency conditions like anemia, it is called a "nutraceutical."

Just like functional foods, another term often synonymous with nutraceuticals is "dietary supplements." According to the Dietary Supplements Health and Education Act (DSHEA), a dietary supplement is a product (other than tobacco) that is intended to supplement the diet that bears or contains one or more of the following dietary ingredients: a vitamin, a mineral, a herb or other botanical, an amino acid,

Table 40.1 Difference Between Food, Pharmaceuticals, and Nutraceuticals

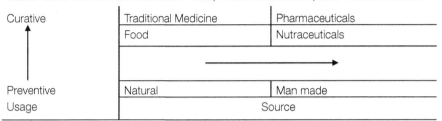

Curative ↑	Traditional Medicine	Pharmaceuticals
	Food	Nutraceuticals
	⟶	
Preventive	Natural	Man made
Usage	Source	

a dietary substance for use by man to supplement the diet by increasing the total daily intake, or a concentrate, metabolite, constituent, extract, or combinations of these ingredients. The product should be intended for ingestion through a pill, capsule, tablet, or in liquid form, and is not represented for use as a conventional food or as the sole item of a meal or diet and is labeled as a "dietary supplement." It includes products such as an approved new drug, certified antibiotic, or licensed biologic that was marketed as a dietary supplement or food before approval, certification, or license.

A very thin difference exists between nutraceuticals and dietary supplements—nutraceuticals should not only supplement diet *but also prevent and/or treat disease/disorder.* Also they are represented for use as a conventional food or sole item of meal or diet. Table 40.1 describes the difference between food, pharmaceutical, and nutraceutical.

Though it may take some time for such foods to be available in the departmental stores but certainly with time the interface between these terminologies will be more transparent. *"Pharmaceuticals"* may be considered as drugs used mainly to treat diseases, while "nutraceuticals" are those that are intended to prevent disease. Pharmaceuticals have (or have had) patent protection as a result of expensive testing to conform to the specifications of respective governments. However, many nutrients may never receive government approval since no one could justify the expense of testing requirements for substances that cannot be protected by patent laws.

40.3 CLASSIFICATION OF NUTRACEUTICALS

Nutraceuticals can be classified on the basis of either their chemical nature, the type of food items or whether they are traditional or not. Table 40.2 gives the classification of nutraceuticals with examples.

1. *On the basis of the chemical constituents*: phenols, alkaloids, fibers, fatty acid, minerals, terpenes, etc.
2. *Traditional/nontraditional*: fruits, vegetables, fortified juices.
3. *Type of food items*: vitamins, minerals, dairy products.

Table 40.2 Classification of Nutraceuticals

S. no.	Category	Details	Example
1.	**Based on food items**		
a.	Nutrients	Have established nutrient functions	Vitamins, minerals, amino acids
b.	Herbals	Herbs or botanical products as extracts or concentrates	Aloe vera, ginger, garlic
c.	Dietary supplements	Which are added to the food for a specific purpose and administered orally	Minimally refined grains, phytoestrogens (soya), dairy foods
2.	**Traditional/nontraditional**		
a.	Traditional	When the food is taken as whole food or without any change in it	Fruits, vegetables, dairy products
b.	Nontraditional	Outcome of agricultural breeding or addition of specific ingredients	Folic acid added to flour, calcium added to juices

40.4 MODE OF ACTION OF NUTRACEUTICALS

It has been stated that they function by increasing the supply of important building blocks to the body. The supply of these essential building blocks can be done by two ways: (1) by reducing signs of the disease as buffering agents for relief and (2) by directly providing benefits for health of the individuals.

40.5 BENEFITS OF NUTRACEUTICALS

There has always been apprehension about the benefits of nutraceuticals and the side effects that come along with them. Research has shown that nutraceuticals have proven to be immunity boosters (flavonoids, green tea, quercetin in onion) and aid in recovering from degenerative diseases (vitamin E, creatine, turmerine). Apart from the aforementioned benefits they also prevent or are helpful in chronic diseases such as cardiac ailments. To name a few useful compounds: anti oxidants, fiber, minerals, vitamins, flavonoids, etc. are beneficial. They have benefits against obesity—soy foods, flavonoids, green tea extracts. Some of them such as green tea, dietary fibers, antioxidants have proven to be antidiabetic also. Some are anticancer agents such as lycopene, soy foods, saponins from spinach, tomato, and potato.

40.6 **REGULATION GOVERNING NUTRACEUTICALS**

As already mentioned that nutraceuticals are not a distinct category of foods and the term has no regulatory meaning, as a result the FDA regulates it as all other food products. In 2005, National Academy Institute of Medicine and National Research Council formed a Blue Ribbon association that gave a framework for the Federal Food and Drug Administration to evaluate dietary supplements, without distinguishing nutraceuticals as separate items. Both pharmaceuticals and nutrients can cure and prevent disease(s) but only pharmaceuticals have governmental sanction.

The primary set of rules governing the human nutraceutical market is the DSHEA, 1994. This act does not permit FDA to consider a new product a "drug" or "food additive" if it falls under the definition of a "dietary supplement," which includes among other substances any possible component of the diet as well as concentrates, constituents, extracts, or metabolites of these components.

The other major component of this act shifts the burden of safety. The FDA now has to prove a substance is unsafe rather than the manufacturer proving the substance safe.

The DSHEA rules do not apply to nutraceuticals intended for animals. In a nutshell, the federal government has cited differences in metabolism of substances between humans and animals and potential safety issues with nutraceuticals used in food-producing animals as reasons to exclude animals from provisions of the DSHEA. Therefore, expressed or implied claims relating use of a product with the treatment or prevention of disease, or with an effect on the structure or function of the body in a manner distinct from what would be normally ascribed to "food."

In India, nutraceuticals have been defined under Clause 22 of the Food Safety and Standards Act (FSSA), 2006. However at the regulatory end many issues are unresolved. One of these issues is labeling standards and health claims. Lack of regulatory framework also makes substandard companies enter the market and compromise the quality of products.

40.7 **CONCLUSIONS**

A change in the dietary pattern in our society has led to emergence of chronic diseases such as diabetes, cardiovascular diseases, etc. Pharmaceutical companies exploit this situation by promoting nutraceuticals with claims of providing benefits for prevention of noncommunicable diseases. These markets are nearing maturity, with exceedingly high per capita expenditure on nutraceutical products. This has forced nutraceutical manufacturers to target developing countries like India and China, as key growth regions because of their sheer population.

Nutraceuticals have both advantages as well as few side effects. On one hand, advances in science and technology have led to unhealthy lifestyle with poor eating habits, on the other hand there is raised awareness about the importance of nutrition and health to counter such ill effects. India has very few pure nutraceutical-oriented industries. So, it is a fertile ground for the growth of nutraceutical industry. Moreover India has the advantages of a qualified human resource, world class R&D and abundant raw material. India aims to be a part of 0.9% of the 117 billion US$ industry of the world.

Companies such as *Dabur* have already established themselves with products such as *Chavanprash* and almond oil. Newer groups are joining the race as India's nutraceutical market is expected to grow manifold in the coming years. At this stage when new companies are entering with a plethora of products into the market, India is reeling under the burden of nutritional deficiencies, where a large part of the population does not have enough purchasing power to fulfill the daily calorie needs, leave aside the nutrients. On the other hand, there are people who consume large amount of calories but not adequate nutrients. To maintain a balance in the calorie intake of the people and reduce the disparity, the agenda should be equity in nutritional interventions. Sooner or later because of market forces nutraceuticals will be used for such interventions. But 70% of the Indian population still lives in rural areas hence, interventions on the nutraceuticals will cover only urban population unless it becomes a part of the nation's nutrition agenda and is not treated as a luxury for city dwellers.

FURTHER READING

Brower, V., 2005. A neutraceutical a day may keep the doctor away. EMBO Rep. 6, 708–711.

Crandell, K., Duren, S., 2007. Nutraceuticals: what are they and do they work? J. Biotechnol. 34(3), 29–36.

FICCI and Ernst and Young, 2009. Nutraceuticals: critical supplements for building a healthy India. FICCI–Ernst and Young.

Gupta, S., Chauhan, D., Mehla, K., Sood, P., Nair, A., 2010. An overview of nutraceuticals: current scenario. J. Basic Clin. Pharm. 1, 55–62.

Kalra, E.K., 2003. Nutraceuticals—definition and introduction. AAPS Pharm. Sci. 5, 1–2.

Rajasekaran, A., Sivagnanam, G., Xavier, R., 2008. Nutraceuticals as therapeutic agents: a review. Res. J. Pharm. Technol. 1, 328–340.

Reardon, T., Minten, B., 2011. The quiet revolution in India's Food supply chain 2011. IFPRI Discussion paper 01115.

Sarin, R., Sharma, M., Singh, R., Kumar, S., 2012. Nutraceuticals: a review. Intl. Res. J. Pharm. 3, 95–99.

Regulatory requirements for labeling, health, and nutritional claim

41

D.P. Attrey

Central Military Veterinary Laboratory, Meerut, Uttar Pradesh, India; High Altitude Research, Defence Research and Development Organisation, Leh, Jammu and Kashmir, India; Amity Institute of Pharmacy, Amity University, Noida, Uttar Pradesh, India; Innovation and Research Food Technology, Amity University, Noida, Uttar Pradesh, India; Amity Institute of Seabuckthorn Research, Amity University, Noida, Uttar Pradesh, India; Lala Lajpat Rai University of Veterinary and Animal Sciences, Hisar, Haryana, India

41.1 INTRODUCTION

Food trade is regulated through "Labels" and "Claims on the Label". Labeling regulations are very important for ensuring food safety and quality. As per law, all prepackaged foods must have a label and no food product can be sold without proper labeling [FSSR (L&R), 2011]. Regulatory requirements for labeling of food products, claiming health and nutritional benefits must be understood both by the industry as well as the scientists, for development of new food products. The Food Business Operator (FBO), intending to use that new product has to obtain regulatory approval, before bringing the same in the market.

41.2 GENERAL REQUIREMENTS FOR MANDATORY LABELING

Every prepackaged food sold in any manner (including direct selling) shall carry a label containing required information under the law, which shall not be false or misleading. It will be ensured by the FBO that label does not get separated from the container under normal conditions of handling and distribution. Contents on the label shall be clear, legible, indelible, unambiguous, prominent, and conspicuous to the consumer under normal conditions of purchase and use. The declarations should be printed or inscribed on the package in such style or type of lettering so as to be boldly, clearly, and conspicuously present in distinct contrast to the background of the label. No person is permitted to adopt any unfair or deceptive trade practices. The burden of proof of the correctness of the claim lies on the owner of the product. Also, no advertisement is permitted, which is misleading or deceiving or contravenes the provisions of (FSSA, 2006).

Food Safety in the 21st Century. http://dx.doi.org/10.1016/B978-0-12-801773-9.00041-8

An FBO can display any information or matter on the label provided that it is not in conflict with the requirements of the FSS Act/Regulations. The information required under these regulations shall be given on the principal display panel of the package or container. The labels should not contain reference to act or rules or regulations contradictory to required particulars/words.

Further, no advertisement shall be made of any food which is misleading or deceiving or contravenes the provisions of law. No person shall promote the sale, supply, use, and consumption of articles of food or adopt any unfair or deceptive practice including the practice of making any statement, whether orally or in writing or by visible representation in advertisement, which

1. falsely represents that the foods are of a particular standard, quality, quantity or grade-composition;
2. makes a false or misleading representation concerning the need for, or the usefulness; and
3. gives to the public any guarantee of the efficacy that is not based on an adequate or scientific justification.

Any food for which a nutrition or health claim is made should be labeled with a nutrient declaration in accordance with Section 3 of the Codex Guidelines on Nutrition Labeling (CAC, 1997).

41.3 CLAIMS

All substances intended for human consumption are either a food or a drug. According to Food Safety and Standards (FSSA, 2006), "Food" is any substance, which is intended for human consumption. Food products, including those claiming "health benefits on the structure and function of the body", come under the purview of FSSAI. Substances "intended to be used for or in the diagnosis, treatment, mitigation or prevention of any disease" are called "Drugs" (D&C Act, 1940). These come under the purview of Drug Controller General of India (DCGI).

Nutritional and *health claims* are defined as follows (CAC, 1997):

1. *Nutrition claim* means any representation which states, suggests, or implies that a food has particular nutritional properties including energy value and content of protein, fat, carbohydrate, vitamins, minerals, and other permitted listed nutrients. These claims can be of two types:
   a. *Nutrient content claim* is a nutrition claim that directly or indirectly describes the level of nutrient contained in food (e.g., "contains/source of ..."; "high in...", "rich in...", "low in... etc.)
   and
   b. *Nutrient comparative claim* is a claim that compares the nutrient levels and/ or energy value of two or more foods (e.g., "reduced"; "less than"; "fewer"; "increased"; "more than", etc.).

The following do not constitute nutrition claims:

1. the mention of substances in the list of ingredients;
2. the mention of nutrients as a mandatory part of nutrition labeling; and
3. quantitative or qualitative declaration of certain nutrients or ingredients on the label if required by national legislation.

Health claim means any representation that states, suggests, or implies that a relationship exists between a food or a constituent of that food and health. A health benefit claim for diagnosis, treatment, mitigation or prevention of any disease is called a "*disease claim*". Products with disease claim come under the purview of the DCGI and not the FSSAI. Health claims can be of three types:

1. nutrient function claims;
2. other function claims; and
3. disease risk reduction (DRR) claim.
   a. *A nutrient function claim*: A claim which describes the physiological role of the nutrient in growth, development, and normal functions of the body. For example nutrient "A" (naming a physiological role of nutrient "A" in the body in maintenance of health and promotion of normal growth and development) or food X is a rich source of nutrient "A", "Nutrient A is good for health" (particular function).
   b. *Other function claims*: A claim that describes the specific beneficial effects of the consumption of food(s) or their constituents, in the context of the total diet or normal functions or biological activities of the body. Such claims relate to a positive contribution to health or to the improvement of a function or to modify or preserving the health. For example "Substance A (naming the effect of substance A on improving or modifying physiological function or biological activity associated with health), food Y contains X grams of substance A."
   c. *Reduction of disease risk (DRR) claim*: Claims that state, suggest or imply that consumption of such food(s) or food constituents in the context of the total diet, reduce the risk factor of developing disease or health related conditions. DRR means significantly altering a major risk factor(s) for a disease or health related condition. Diseases have multiple risk factors and altering one of these risk factors may or may not have beneficial effects. The presentation of risk reduction claim must ensure, for example, by use of appropriate language and relevance to risk factors, that consumers do not interpret them as prevention claims (CAC, 1997).

41.3.1 DISEASE CLAIM

A claim for a product to diagnose, treat, prevent, cure, or mitigate a diseases is considered as a disease claim. Menopause, aging, and pregnancy are not themselves diseases but certain conditions associated with them could be diseases, if they are recognizable to consumers or health professionals as abnormal. Hot flashes, common symptoms associated with the menstrual cycle, ordinary morning sickness associated with pregnancy, mild memory problems associated with aging, hair loss associated with aging, and noncystic acne may be health problems but not disease. Claims to mitigate such health problems shall fall in the nondisease-related effect on the structure or function of the body. Such claimed effects are derived from a food attribute, such as nutritive value, and such claim is considered as a food claim, for which prior authorization is not required.

FDA has defined various types of claims which can be made on the label of a pre packaged food product viz "health claims," "nutrient content claims," and "structure/function claims." "Health claims" describe a relationship between a food substance (a food, food component, or dietary supplement ingredient), and reduced risk of a disease or health-related condition. A "health claim" by definition has two essential components: (1) a substance (whether a food, food component, or dietary ingredient) and (2) a disease or health-related condition (FDA, 2016a).

"Nutrient Content Claims" describe the level of a nutrient in the product, using terms such as free, high, and low, or they compare the level of a nutrient in a food to that of another food, using terms such as more, reduced, and lite. Most nutrient content claim regulations apply only to those nutrients that have an established daily value. "Healthy" is an implied nutrient content claim that characterizes a food as having "healthy" levels of total fat, saturated fat, cholesterol and sodium, as defined in the regulation. Percentage claims for dietary supplements are another category of nutrient content claims.

Structure/function claims and related dietary supplement claims describe the role of a nutrient or dietary ingredient intended to affect the normal structure or function of the human body, for example, "calcium builds strong bones." In addition, they may characterize the means by which a nutrient or dietary ingredient acts to maintain such structure or function, for example, "fiber maintains bowel regularity," or "antioxidants maintain cell integrity." Statements projecting a role of a specific substance in maintaining normal healthy structures or functions of the body are considered to be structure/function claims.

General principles for making claims on prepackaged foods [FSSR (L&C), 2012]:

1. Nutrition and health claims should be based on scientific documentation about nutritional requirements of the target groups/individuals in the country or the national nutrition policy.
2. Claims must be truthful, unambiguous, meaningful, and not misleading.
3. The person marketing the food should be able to justify the claims made. Claims that cannot be scientifically substantiated should not be made.
4. The claims should not encourage or condone excess consumption of a particular food.
5. The claims should be clear and meaningful and shall help consumers to comprehend the information provided, to understand the relative significance of such information and choose wholesome diets. Safety of nutrients used in the claims must be established or should have a history of safe use.
6. Food constituent forming basis of the claim must be validated with approved method.

7. All nutritional and health claims on products intended for children shall be accompanied by a declaration in bold "Regular balanced diet is important for children in their growing up years."
8. Claims shall not state, suggest or imply that a balanced and varied diet cannot provide appropriate quantities of nutrients as required by the body.
9. Claims containing adjectives such as "natural," "fresh," "pure," "organic," "original," "traditional," "premium," "finest," "best," "authentic," "genuine," "real," etc. when used, shall be as per law. Words or phrases like "home made", "home cooked" etc. shall not be used.
10. Claims related to religious or ritual practices like *Halal, Jhatka* can be made provided the food conforms to requirements of the appropriate religious or ritual authorities and is certified by concerned authorized veterinary agencies.
11. Above conditions are also applicable where claim is made on the finally prepared form of the food. Label of such a food must also contain information and appropriate instructions to ensure that the finally prepared product meets the claim made on it.
12. Food authority may ask the FBO any time to substantiate the claim scientifically and/or furnish details regarding the nutrients added or modified for their safety and efficacy.
13. All disclaimers related to a claim should appear in the same field of vision as that of the claim.

Competent national authorities should reevaluate health claims either periodically or following the emergence of significant new evidence that has the potential to alter previous conclusions about the relationship between the food or food constituent and the health effect.

The relevant data and rationale that the constituent for which the health claim is made is bioavailable, should be provided. If absorption is not necessary to produce the claimed effect (e.g., plant sterols, fibers, lactic acid bacteria), the relevant data and rationale that the constituent reaches the target site or mediates the effect is required (CAC, 1997). Health claims must be based on current relevant scientific substantiation. Level of proof must be sufficient to substantiate the type of claimed effect and the relationship to health as recognized by generally accepted scientific norms and review of data.

Health claim must consist of two parts viz information on physiological role of the nutrient or on an accepted diet-health relationship and information on composition of product, relevant to physiological role of the nutrient or the accepted diet-health relationship. Added nutrients in per day serve must remain below 50% of the scientifically established upper safe levels for each nutrient. A claimed benefit should be based on statistically significant results from appropriate scientific research studies or a well designed, randomized double blind clinical study. Such study should be conducted by or under guidance of established research institutions, in line with the principles of GCP and peer reviewed or published in a peer reviewed reputed scientific journal with an impact factor of not less than 1.0 at the time of submission. DRR claims should be made only from the list of approved claims (Annexure IV).

A new claim shall be made only after taking prior approval from the authority [FSSR (L&C), 2012].

According to US FDA, "Label Claims", come under three categories:

1. health claims,
2. nutrient content claims, and
3. structure/function claims (FDA, 2015).

A *health claim* is made on the label of a product containing a food or a dietary supplement, that expressly or by implication characterizes the relationship of any substance to a disease or health-related condition. Implied health claims include those statements, symbols or other forms of communication which suggest that a relationship exists between the food and a disease or health-related condition (FDA, 2015).

41.3.2 NUTRIENT CONTENT CLAIMS

The Nutrition Labeling and Education Act of 1990 permits the use of label claims that characterize the level of a nutrient in a food (i.e., nutrient content claims) if they have been authorized by FDA and are made in accordance with FDA's authorizing regulations. Nutrient content claims describe the level of a nutrient in the product, using terms such as free, high, and low, or they compare the level of a nutrient in a food to that of another food, using terms such as more, reduced, and lite. The term "Healthy" is an implied nutrient content claim that characterizes a food as having "healthy" levels of total fat, saturated fat, cholesterol, and sodium, as defined in the regulation authorizing use of the claim. "Percentage claims" for dietary supplements are another category of nutrient content claims. These claims are used to describe the percentage level of a dietary ingredient in a dietary supplement and may refer to dietary ingredients for which there is no established daily value, provided that the claim is accompanied by a statement of the amount of the dietary ingredient per serving. Examples include simple percentage statements such as "40% omega-3 fatty acids, 10 mg per capsule," and comparative percentage claims, for example, "twice the omega-3 fatty acids per capsule (80 mg) as in 100 mg of menhaden oil (40 mg)."

As per FDA (2016), structure/function claims are made on the labels of conventional foods and dietary supplements as well as drugs. Structure/function claims are the statements that describe the effect of a dietary supplement on the structure or function of the body. The Dietary Supplement Health and Education Act of 1994 established some special regulatory requirements and procedures for

- structure/function claims,
- general well-being claims, and
- nutrient deficiency disease claims.

Structure/function claims may describe the role of a nutrient or dietary ingredient intended to affect the normal structure or function of the human body, for example, "calcium builds strong bones." In addition, they may characterize the means by which a nutrient or dietary ingredient acts to maintain such structure or function, for example, "fiber maintains bowel regularity," or "antioxidants maintain cell integrity." "General Well-Being Claims" describe general well-being from consumption of a nutrient or dietary ingredient. "Nutrient Deficiency Disease Claims" describe a benefit related to a nutrient deficiency disease (like vitamin C and scurvy), but such claims are allowed only if they also say how widespread such a disease is in the country.

The above three types of claims are not preapproved by FDA, but the manufacturer must have substantiation that the claim is truthful and not misleading and must submit a notification with the text of the claim to FDA not later than 30 days after marketing the dietary supplement with the claim. If a dietary supplement label includes such a claim, it must state in a "disclaimer" that FDA has not evaluated the claim. The disclaimer must also state that the dietary supplement product is not intended to "diagnose, treat, cure or prevent any disease," because only a drug can legally make such a claim.

Structure/function claims for "conventional foods" focus on effects derived from nutritive value, while structure/function claims for "dietary supplements" may focus on nonnutritive as well as nutritive effects. FDA does not require conventional food manufacturers to notify FDA about their structure/function claims, and disclaimers are not required for claims on conventional foods. The regulation also provides criteria to help in determining when a statement about a dietary supplement is a disease claim, that is, a claim to diagnose, cure, mitigate, treat, or prevent a disease. Disease claims require prior approval by FDA and may be made only for products that are approved drug products or for foods under separate legal provisions that apply to claims called "health claims."

FSSAI follows almost similar approach as that of US FDA to ensure that foods, both domestically produced as well as imported, are safe, wholesome, and properly labeled. FSSAI also requires FBOs to apply for approval before marketing a new product. The labeling of food supplements shall comply with the packaging and labeling requirements as laid down by law under the Food Safety and Standards (Packaging and Labeling) Regulations, 2011. The label shall not attribute to food or health supplements the property of preventing, treating or curing a disease, or refer to such properties. The statements relating to structure or function or for the general well-being of the body are allowed as long as they are truthful and are also supported by generally accepted scientific data and in addition, the product shall bear a statement, *"This product is not intended to diagnose, treat, cure or prevent any disease(s)."*

Every package of food or health supplements shall carry the following information on the label.

1. the words "Food or Health Supplement;"
2. the common name of the Food or Health Supplement, or a description sufficient to indicate the true nature of the food supplement including the common names of the categories of nutrients or substances that characterize the product;

3. the amount of the nutrients or substances with a nutritional or physiological effect present in the product shall be declared on the label in numerical form in descending order;
4. the term "Not for Medicinal Use" shall be prominently written on the label;
5. the quantity of nutrients shall be expressed in terms of percentages of the relevant recommended dietary allowances as prescribed in India by the Indian Council of Medical Research and shall bear a warning "Not to exceed the recommended daily dose;"
6. a statement to the effect that the food or health supplement should not be used as a substitute for a varied diet;
7. a warning or any other precautions to be taken while consuming, known side effects if any, contraindications and product-drug interactions, as applicable; and
8. a statement to the effect that the products shall be stored out of the reach of children.

Any nutrient function or other function claim for promotion of normal growth and development or for maintaining good health, can be made only if substantiated by scientific data. Specific health claims indicating maintenance of normal health condition by use of specific nutrient shall be accompanied by the following indication-"For maintenance of optimum functions and not for therapeutic use. Claims on foods which are high in salt or sugar or fat, shall alongside the claim also mention the fact that the food is high in one or more of the these ingredients in the manner "This food is high in ...."(to be filled in by sugars and/or salt and/or fat) as the case may be [FSSR (L&C), 2012].

However, no DRR claims shall be made on products which are high in fat, salt or sugar. When a claimed benefit is attributed directly to the product or used on labels, advertisements or any other means as a mode of communication to the consumer, it shall be based on statistically significant results from appropriate scientific research studies or a well designed, randomized double blind (unless technically not feasible) clinical study, conducted by or under guidance of established research institutions, in line with the principles of GCP and peer reviewed or published in a peer reviewed reputed scientific journal with an impact factor of not less than 1.0 at the time of submission [FSSR (L&C), 2012].

FBO shall submit all the relevant substantiating documents along with the report of the peer review mentioning in detail the composition of the peer review group along with their designations or the copy of the published article as the case may be to the authority before launch of the product for scrutiny. Following the scrutiny, if the claim is found unsubstantiated, authority may order amendment or withdrawal of the claim. Scientific research on a particular nutrient shall cover its efficacy over the age group, geographical situation, social and economic criteria which the concerned product intends to target. DRR claims shall be made from the list of approved claims (Annexure IV). Any new DRR claim shall be made only after taking prior approval from the authority [FSSR (L&C), 2012].

Approved DRR claims are as follows [FSSR (L&C), 2012]:

- Calcium, vitamin D, and osteoporosis—food requirements: high in calcium, vit D, phosphorus content cannot exceed calcium content (w/w basis).
- Sodium and hypertension—Low sodium for "high blood pressure."
- Dietary fat and cancer—Low fat.
- Dietary saturated fat and cholesterol, and risk of coronary heart disease
- Coronary heart disease or heart disease and elevated blood total or LDL—cholesterol.
- Fiber-containing grain products, fruits, and vegetables and "cancer."
- Fruits, vegetables and grain products that contain fiber, particularly soluble fiber, and "risk of coronary heart disease."
- "Saturated fat" and "cholesterol" and "heart disease" or "coronary heart disease."
- Fruits and vegetables and "Cancer."
- "Total fat" or "Fat", "Some types of cancer."
- Folate and "Neural tube defects."
- Dietary noncariogenic carbohydrate sweeteners and dental caries.
- Soluble fiber and "reduced risk of heart disease."
- Soy protein and "Risk of coronary heart disease."
- Plant sterol/stanol esters and "Risk of coronary heart disease."
- Whole grain foods and "Risk of heart disease" and "Certain cancers." Diets rich in whole grain foods and other plant foods and low in total fat, saturated fat, and cholesterol may reduce the risk of heart disease and some cancers.
- Potassium and the risk of high blood pressure and stroke. "Diets containing foods that are a good source of potassium and that are low in sodium may reduce the risk of high blood pressure and stroke."

41.4 PROHIBITIONS

1. No claims shall be made which refer to the suitability of a food for use in the prevention, alleviation, treatment or cure of a disease, disorder, or particular physiological condition.
2. No product shall claim "added nutrient" if such nutrients have been added merely to compensate the nutrients lost while processing of the food.
3. Foods for infants shall not carry any claim unless specifically permitted under product standard or specific labeling requirement.
4. Claims which could give rise to doubt or suspicion about the safety of similar food or which could arouse fear in consumer shall not be made.
5. The health claims which suggest that health could be affected by not consuming the food shall not be allowed (CAC, 1997).

41.5 CONCLUSIONS

Regulatory requirements for labeling, health, and nutritional claims have been discussed. Anything packed in absence of consumer and having predetermined value (either weight/content or price), needs a label, since a label on the product is essential to develop consumer confidence. Claim is part of the label. Food trade is regulated through "Labels" and "Claims." Labeling regulations are essential for ensuring food safety and quality. Label information must be scientifically correct and not false/misleading. Claims on products intended for children must accompany a declaration in bold "Regular balanced diet is important for children in their growing years."

REFERENCES

CAC, 1997. Nutrition and Health Claims; Guidelines for use of Nutrition and Health Claims, (CAC/GL 23-1997). Revised in 2013. With kind permission from FAO. Available from: https://www.foedevarestyrelsen.dk/SiteCollectionDocuments/25_PDF_word_filer%20 til%20download/07kontor/Maerkning/Codex%20guidelines%20nutrition%20and%20 health%20claims.pdf; file:///C:/Users/lenovo/Downloads/CXG_023e%20(1).pdf

D&C Act, 1940. Drugs and Cosmetics Act, 1940, amended up to 1995, and D & C Rules 1945/ Apr 2003. Ministry of Health and Family Welfare, Government of India. Available from: http://www.indianmedicine.nic.in/writereaddata/mainlinkFile/File222.pdf

FDA, 2015. Label Claims, Types of Claims: Definitions, Guidance, Regulatory Information, and Permitted Claims. Available from: http://www.fda.gov/Food/IngredientsPackagingLabeling/LabelingNutrition/ucm2006881.htm; http://www.fda.gov/Food/IngredientsPackagingLabeling/LabelingNutrition/ucm2006873.htm

FDA, 2016. Structure/Function Claims. Dietary Supplements. Conventional Foods. Available from: http://www.fda.gov/Food/IngredientsPackagingLabeling/LabelingNutrition/ ucm2006881.htm

FDA, 2016a. Label Claims for Conventional Foods and Dietary Supplements. Available from: http://www.fda.gov/Food/GuidanceRegulation/GuidanceDocumentsRegulatoryInformation/LabelingNutrition/default.htm; http://www.fda.gov/Food/IngredientsPackagingLabeling/LabelingNutrition/ucm111447.htm

Food Safety and Standards (Packaging and labelling) Regulations, 2011. Ministry of Health and Family Welfare (Food Safety and Standards Authority of India); Notification No F.No. 2-15015/30/2010. Available from: http://www.fssai.gov.in/Portals/0/Pdf/Food%20Safety%20and%20standards%20(Packaging%20and%20Labelling)%20regulation,%202011. pdf; DL 15-11-15.

FSSA, 2006. Food Safety and Standards Act, 2006 (No. 34 of 2006). Ministry of Law and Justice (Legislative Department), Government of India; New Delhi. Available from: http:// www.fssai.gov.in/portals/0/pdf/food-act.pdf.

FSSR (L&C), 2012. FSSAI draft regulations on labeling (claims). Available from: http://gain. fas.usda.gov/Recent%20GAIN%20Publications/FSSAI%20Publishes%20Draft%20Regulations%20on%20Labeling_New%20Delhi_India_12-28-2012.pdf

FSSR (L&R), 2011. Food Safety and Standards (Licensing and Registration of Food Businesses) Regulations. Available from: http://www.fssai.gov.in/Portals/0/Pdf/Food%20safety%20and%20Standards%20(Licensing%20and%20Registration%20of%20Food%20 businesses)%20regulation,%202011.pdf

Genetically modified (GM) foods: the food security dilemma

42

J. Dutta

Human Resource Development Centre, Panjab University, Chandigarh, India

In any moment of decision, the best thing you can do is the right thing.
The worst thing you can do is nothing.
Theodore Roosevelt

Humans have been using the fields as laboratories when terms like "science, labs, experiments, genetics" did not even exist. Farmers, through hundreds of years of experience, experiments and insights, were able to evolve crops of diverse species and develop and perpetuate certain desirable traits also. As the global population increased the next challenge for agriculture shifted from diversifying to the achievement of high yield.

Scientific developments and a greater understanding of agriculture and genetics gave rise to the green revolution, from 1940s to the late 1960s which ensured that food could reach populations around the world. High yielding crop varieties, introduction of modern farming technologies, more effective irrigation systems and to a great extent excessive use of synthetic fertilizers and pesticides helped in unprecedented enhancement in crop yield. It is estimated that due to green revolution annual wheat production rose from 10 million tons in 1960s to 73 million tons in 2006 (BBC News, 2006) and the world grain production increased by over 250% (Kindall and Pimentel, 1994). Several countries and huge populations were saved from famine and starvation due to the dramatic increase in the crop yields during green revolution.

However, green revolution considered to be a miracle of science ultimately proved that it was not an "all rosy story." The side effects of the drastic methods and heavy and indiscriminate use of chemical pesticides and fertilizers had harmful impact on environment and health. Green revolution reduced agricultural biodiversity due to its heavy reliance on a few high-yield varieties. Many traditional varieties with unique genetic traits evolved over thousands of years were lost. Over-usage of chemicals to kill pests and noncompliance of safety measures in handling and spraying of pesticides led to their leaching into the soil and water, soil erosion and lowering of soil fertility. These chemicals ultimately found their way into food chain and entered the bodies of animals and humans resulting in pesticide poisoning and various kinds of cancers. These unforeseen, undesirable and fatal effects led the scientific community onto a fresh voyage of discovery of safer and better methods of reaching the goal of food for all. Genetic engineering technology presented itself as one such method with huge potential to fulfill the wishes of mankind. It seemed like a magic wand which could bestow upon humans the power to produce any kind of crop with all desirable traits and no disadvantages at all.

The traits of an organism are the manifestation of the proteins encoded in its DNA. If the DNA can be altered the traits can also be changed. Genetic engineering gave a biological technology through which the DNA could be manipulated. In order to introduce the desired characteristic in the genetic profile of an organism, a DNA segment of another organism coding for a specific trait is inserted into its DNA. Heritable material can also be removed from the original genetic makeup to delete an undesirable trait. The organism with this added or deleted DNA segment will thus be genetically modified and would behave differently from the organism with the original, nonmanipulated DNA. Though there are other methods also through which genes can be manipulated, such as traditional animal and plant breeding methods, in vitro fertilization, polyploidy, mutagenesis and cell fusion techniques but only those organisms qualify to be called genetically modified organisms where recombinant nucleic acid techniques are used.

The technique used for genetic engineering can be simplified into three steps that is cut, insert, and express. The first step in the process is to choose and isolate the gene of interest that will be inserted into the genetically modified organism. Recombinant nucleic acid (DNA or RNA) techniques are used for cutting the specific gene segment. Restriction enzymes cut DNA into fragments and gel electrophoresis separates them out according to length. Polymerase chain reaction can be used to amplify a gene segment. Well studied gene segments or the donor organism's genome may be present in a genetic library. If the DNA sequence is known, but no copies of the gene are available, it can be artificially synthesized.

The gene to be inserted into the (to-be-genetically-modified) organism is combined with other genetic elements and can be modified at this stage for better expression or effectiveness. The gene contains a promoter and a terminator region as well as a selectable marker gene. The promoter region initiates transcription of the gene and can be used to control the location and level of gene expression, while the terminator region ends transcription. The selectable marker, which in most cases confers antibiotic resistance to the organism it, determines which cells are transformed with the new gene. The constructs are made using recombinant DNA techniques, such as restriction digests, ligations and molecular cloning. The manipulation of the DNA generally occurs within a plasmid.

This is followed by the incorporation of that genetic material into the host organism either indirectly through a vector system or directly through microinjection, macroinjection and microencapsulation techniques and can be inserted within the host genome either randomly or at a specific location in the host genome or by generating mutations at desired genomic loci capable of knocking out genes. If genetic material from another species is added to the host, the resulting organism is called transgenic. If genetic material from the same species or a species that can naturally breed with the host is used the resulting organism is called cisgenic (Morgante et al., 2005). The gene then starts expressing itself manifesting the desired trait in the GM organism.

Genetic engineering techniques found application in numerous fields including research, industrial biotechnology and medicine and genetically engineered bacteria, fish and mice etc. have been created since 1974. It was not initially used for targeting improved crop yield or quality enhancement but scientists soon started realizing the vast potential and scope of the technique of direct manipulation of DNA of crop plants. Since then plants have been genetically modified for insect protection, herbicide resistance, virus resistance, enhanced nutrition, tolerance to environmental pressures and the production of edible vaccines (James and Anatole, 1996).

The history of the development of the idea of GM crops is quite interesting and gives an understanding of how these crops can be an answer to the perpetual problem of food security.

In 1901, bacteriologist Ishiwata Shigetane while investigating the cause of a disease outbreak in silkworms which were dying in large number found that it was due to an unidentified species of bacteria named *Bacillus thuringiensis* or Bt. In 1911, Ernst Berliner isolated the bacteria and identified it as a cause of a disease in flour moth caterpillars. Later researchers tried to find out the specific attribute of the insect-killing character of the bacteria and it was found out to be the crystals of proteinaceous insecticidal endotoxins called crystal proteins or cry proteins encoded by cry genes. The cry genes are located on a plasmid and are not chromosomal genes in most strains of the bacteria. The cry proteins act as toxins against insect species of various orders such as butterflies and moths (Lepidoptera), beetles (Coleoptera), wasps, bees, ants (Hymenoptera) flies and mosquitoes (Diptera) (Khened, 2003).

In 1981 scientists from University of Washington, discovered that the insecticidal proteins were found in a crystal like body produced by the bacteria. After ingestion by the insects the insoluble toxin crystals, get denatured by the action of the alkaline pH of the digestive tract, the cry toxins are liberated and gets inserted into the insect gut cell membrane, paralyzing the digestive tract. The insect then starves to death. The toxin is a natural biological insecticides and the bacteria can be considered as a reservoir for producing it. The Bt genes trigged production of their toxic protein only when the bacteria starts producing spores. The recombinant DNA techniques were used to isolate genes that encode for insecticidal proteins and by 1989 more than 40 Bt genes responsible for producing proteins toxic to specific groups of insects had been pinpointed and cloned in different labs. The game plan was to insert this bacterial gene having insecticidal properties, into plant genes so that the plants will start producing insect toxin and kill all the pests attacking the plant in the process.

In 1985 the genes from Bt were inserted into the genetic sequences of tobacco plants making them insect tolerant by expressing the cry genes. This scientific experiment was turned into a useful application by a Belgian Company Plant Genetic Systems and the first field trials of genetically engineered plants occurred in France and the United States in 1986. The People's Republic of China was the first country to commercialize transgenic plants, introducing a virus-resistant tobacco in 1992.

In 1994, the European Union approved tobacco engineered to be resistant to the herbicide Bromoxynil, making it the first genetically engineered crop commercialized in Europe. In 1995 the application was repeated on potato plants which could produce CRY 3A Bt toxin. This became the first GM pesticide producing crop and was approved by the Food and Drug Administration and the Environmental Protection Agency of the United States (Potato Pro Newsletter, 2010).

Gradually several Bt genes namely cry 1A 105, Cry 1Ab, m Cry 3A, and VIP were engineered and were approved for crops like cotton, corn, mustard, and rice which were labeled as GM foods. By 2009, 11 transgenic crops were grown commercially in 25 countries, including the United States, Brazil, Argentina, India, Canada, China, Paraguay, and South Africa.

42.1 APPLICATIONS OF GM CROPS

Several applications of GM crops were discovered, the important one are being discussed in subsequent sections.

42.1.1 ENHANCING YIELD

The world population was predicted to reach 10 billion inhabitants by 2050 and therefore increased yields to feed the ever increasing hungry populations to make food security an achievable target was considered to be the prime application of GM crops. GM technology could directly improve yield by accelerating growth rates, or increasing size. For example, GM technique has increased the yield of rice, the world's most important cereal crop by 35%.

42.1.2 ENHANCING/INCULCATING PLANTS RESISTANCE

GM technology can be used to inculcate in the plants, resistance to environmental threats (such as improving salt, cold or drought tolerance) enhancing their ability to grow in extreme conditions, to pathogens (such as insects, fungi or viruses) and/or to herbicides. Implications of this goal are: the insect and weed management of crops becomes easier, more crop yield with less labor and less cost input, reduced usage of pesticides, herbicides, fertilizers causing lesser chemical pollution and increased ability to grow crops in previously inhospitable environments. These modified crops would also reduce the use of chemicals, such as fertilizers, herbicides and pesticides, and therefore cause much lesser chemical pollution (US Department of Agriculture, 2014).

42.1.3 ENHANCING QUALITY

The quality of produce can be modified such as increasing the nutritional content, and quality, improved taste and improved storage such as delayed ripening or providing more industrially useful qualities or quantities. Implications of enhancing the nutritional quality of the crops can be improved processing characteristics leading to reduced waste and lower food costs to the consumers. Examples are the Amflora potato which produces a more industrially useful blend of starches, cows engineered to produce more protein in their milk to facilitate cheese production and GM soybeans and canola to produce more healthy oils. One of the goals of food security, that is, to provide large populations with adequate nutrients can be achieved through this. A strain of GM rice-IR 6,8144 has been modified for high levels of iron, zinc, and vitamin-A by transplanting 3 genes that allow it to produce kernels with beta-carotene, a compound converted into vitamin-A. Since rice is the staple diet of a vast population in the world, consumption of GM rice could fight anemia and blindness (Khened, 2003).

42.2 PHARMING

Through genetic engineering such materials can be produced that are normally not made by the organism. This is called pharming. Examples are, crops as bioreactors to produce vaccines, drug intermediates, or drugs; cows and goats engineered to express drugs and other proteins in their milk (US Department of Agriculture, 2014).

However, it soon became clear that GM crops were not a case of manna from the heavens since many negative impacts of GM crops were also noticed, experimented on and extrapolated. Obviously these impacts were unintended nevertheless these were serious enough to warrant further deliberations before full fledged release of GM crops into the environment. Consequentially, in the late 1980s and early 1990s, guidelines on assessing the safety of genetically engineered plants and food were framed by organizations including the FAO and WHO (NASB, 2010).

42.2.1 IMPACT ON ENVIRONMENT

Unintended environmental impacts include harming nontarget and/or beneficial species in the case of crops with engineered insecticidal properties, as well as the development of new strains of resistant pests. Additionally, there is concern that pollen from genetically engineered herbicide resistant crops could reach wild, weedy relatives of the crop and create so called superweeds. This is of particular concern in the United States with crops such as canola and squash. Pollen from GM crops can also contaminate non-GM plants.

42.2.2 IMPACT ON HEALTH

At present, there is no evidence to suggest that GM foods are unsafe. However, there is no absolute guarantee, either. Unintended health impacts from GMOs concern allergens, antibiotic resistance, decreased nutrients, and toxins.

42.2.3 ALLERGENS

Since protein sequences are changed with the addition of new genetic material, there is a concern that the engineered or modified organism could produce known or unknown allergens. Development of improved methods for identifying potential allergens, specifically focusing on new tests relevant to the human immune system and on more reliable animal models are therefore recommended.

42.2.4 ANTIBIOTIC RESISTANCE

Plant genetic engineers have frequently attached genes they are trying to insert to antibiotic resistance genes. This allows them to readily select the plants that acquire the new genes by treating them with the antibiotic. Sometimes these genes remain in the transgenic crop that has lead critics to charge that the antibiotic resistance genes could spread to pathogens in the body and render antibiotics less effective. However, several panels of antibiotic resistance experts have concluded that the risk is minimal.

42.2.5 DECREASED NUTRIENTS

Since the DNA of genetically engineered plants is altered, there is concern that some GMOs could have decreased levels of important nutrients, as DNA is the code for the production of nutrients. However, it must be noted that nutritional differences also have been documented with traditionally bred crops.

42.2.6 INTRODUCED TOXINS

Residual toxins resulting from introduced genes of the bacteria *Bacillus thuringiensis* in so called Bt crops are unlikely to harm humans. This is because the toxin produced by the bacteria is highly specific to certain types of insects. Prior to its inclusion in GE/GM crops, Bt has been used as a biological insecticide, causing no adverse effects in humans consuming treated crops (The New Scientist, 2013)

42.2.7 NATURALLY OCCURRING TOXINS

There is concern that genetic engineering could inadvertently increase naturally occurring plant toxins. However, traditional plant breeding also can result in higher levels of plant toxins.

42.2.8 IMPACT ON MARKETS

The bottom line of GM crops is the economic benefit pocketed by the seed producing agricultural companies. Market impacts include lower prices, higher costs and lost markets for farmers. Farmers suicides in India have been linked to farmers taking out higher debts on the promise that GM seeds will be a bonanza and then lose everything when the harvest fails. This has led to ban on GM imports or moratoriums on approving new GM varieties/hybrids, mandatory labeling practices and separate handing of GM and non-GM grains again impacting the farmers (WHO, 1999; Fernandez-Cornejo, 2012).

42.3 GM CROPS IN INDIA

In India the usage of GM technology started around the year 2002 and the government constituted the Genetic Engineering Appraisal Committee (GEAC) the regulatory body for GM crops in India under the Ministry of Environment and Forest with a mandate to go in all aspects of the GM technology before giving its approval for its acceptance. In March 2002, the GEAC gave its approval for the release into the environment of three transgenic varieties of Bt hybrid cotton with certain conditions. after receiving a favorable report both from the Department of Biotechnology and from Indian Council of Agricultural Research. Since 2002, Bt cotton has steadily prevailed over India's cotton fields and presently almost 90% of the cotton cultivation area is under Bt Cotton. There have been myriad responses on this decision (GMO Compass, 2013). While several NGOs have protested against the move several bodies have urged the government to go ahead with the approval for trials of other GM food crops. However, on this matter government is yet to take a final decision, whether to allow GM crops in the country or not (Bren, 2003).

The fact that trials of GM food crops are yet to begin does not mean that GM foods are currently not being consumed in India. They are already part of the food chain through nonfood GM crops such as Bt cotton. Cottonseed oil is widely used in homes and commercially by bakeries and other snack food manufacturers. It is also part of the feed for the cattle that give us milk. India thus far has no mechanism for labeling of GM food which is also a matter of concern. India imports large quantities of oil from countries like Brazil and Argentina, which grow GM soya and corn. The government is yet to take a final call on labeling. There is a complete lack of post-market surveillance, for example of millions of tons of Bt cotton seed oil has gone into the food chain during the last 10 years (Sharma, 2014).

The debate on GM crops is as unresolved and it seems to have divided the world into two extreme camps labeling GM either as a redeemer or blight. While one side is promoting GM foods as a godsend to achieve world food security, the other side is painting it black as a Frankenstein like mistake. Soon issues like, hidden agenda, political manipulations, developed versus developing countries, farmer suicides, economic clout, capitalism versus socialism were coloring the arguments and it has become very difficult to separate chaff from grain. GM food is a complex issue, and has further been complicated with several agencies having different agenda. Along with concerns for the environment the issue of food safety should be at the top of the agenda while taking any decisions regarding GM food as it can prove to be a boon for the world population bringing in high yields and abundance but if its ill effects get an upper hand it can also ruin the environment, put food security into jeopardy and cause irreparable damage to the nature and its organisms. We need to undertake further research bringing in hard core evidence. We need open forums to debate, discuss, argue and thrash out the issue. Above all, we need to have open minds to see and think clearly and accept the final verdict.

REFERENCES

BBC News, 2006. The end of India's green revolution? Available from: http://news.bbc.co.uk/2/hi/south_asia/4994590.stm

Bren, L.,2003. Genetic Engineering: The Future of Foods? FDA consumer magazine US Food and Drug Administration. Available from: http://permanent.access.gpo.gov/lps1609/www.fda.gov/fdac/features/2003/603_food.html

Fernandez-Cornejo J.,2012. Adoption of Genetically Engineered Crops in the US Recent Trends. USDA Economic Research Service. Available from: http://www.ers.usda.gov/data-products/adoption-of-genetically-engineered-crops-in-the-us/recent-trends-in-ge-adoption.aspx#.VCuYEldicwo

GMO Compass, 2013. USA: Cultivation of GM plants, maize, soybean, cotton: 88 percent genetically modified. Available from: http://www.gmocompass.org/eng/agri_biotechnology/gmo_planting/506.usa_cultivation_gm_plants_2013.html

James, C., Anatole, F.K., 1996. Global review of the field testing and commercialization of transgenic plants: 1986 to 1995—The first decade of crop biotechnology. The International Service for the Acquisition of Agri-biotech Applications, 1–44. Available from: http://www.isaaa.org/kc/Publications/pdfs/isaaabriefs/Briefs%201.pdf

Khened, S.M., 2003. Genetically modified (GM) products: saviour or scourge. Dream 2047, 26–30.

Kindall, H.W., Pimentel, D., May 1994. Constraints on the expansion of the global food supply. Ambio 23 (3), 198–205.

Morgante, M., Brunner, S., Pea, G., Fengler, K., Zuccolo, A., Rafalski, A., 2005. Gene duplication and exon shuffling by helitron-like transposons generate intraspecis diversity in maize. Nat. Genet. 37 (9), 997–1002.

NASB (National Agricultural Statistics Board) Annual Report, 2010. Acreage NASS. Available from: http://usda.mannlib.cornell.edu/usda/nass/Acre/2010/2010/Acre-06-30-2010.pdf

Potato Pro Newsletter, 2010. The history and future of GM Potatoes. Available from: http://www.potatopro.com/news/2010/history-and-future-gm-potatoes.

Sharma, V., 2014. GM crops trial by fire. The Tribune, 12.

The New Scientist, 2013. Grow your own living light. Available from: http://wwwnewscientist.com/article/mg21829156.500-one-per-cent-grow-your-own-living-lights.html#.VCuaQldicwo

US Department of Agriculture, 2014. Adoption of Genetically Engineered Crops in the US Economic Research Service. Available from: http://www.ers.usda.gov/data-products/adoption-of-genetically-engineered-crops-in-the-us.aspx#.VCuaoFdicwo

WHO (World Health Organization), 1999. Bt monograph. Available from: http://whqlibdoc.who.int/ehc/who_ehc_217.pdf

Organic farming: is it a solution to safe food?

43

M. Bansal

School of Public Health, Post Graduate Institute of Medical Education and Research, Chandigarh, India

"All truths are easy to understand once they are discovered. The point is to discover them."

Galileo Galilei

43.1 INTRODUCTION

With increasing population it has become a matter of concern to meet the demand. The Indian agriculture which is based on the traditional knowledge and practices could not produce enough to feed the entire population. The green revolution fulfilled our aspirations by changing India from a food importing to a food exporting nation. But this achievement was at the expense of ecology and environment and to the detriment of the well-being of people.

Adverse effects of modern agricultural practices on the farm, health and on the environment have been well documented. Application of technology, particularly in terms of the use of chemical fertilizers and pesticides all around us has persuaded people to think aloud. Their negative effects on the environment are manifested through soil erosion, water shortage, salination, soil contamination, genetic erosion as well declining crop productivity, damage to environment, chemical contamination, etc. The necessity of having an alternative agriculture method which can be friendly to eco-system while also sustaining, and increasing the crop productivity is realized now. Policies designed to improve the environmental sustainability of agriculture also include ban on pesticides, funding for research in efficiency improvement (e.g., fertilizers) or damage abatement technologies. Organic farming is recognized as the best alternative to the conventional agriculture. Organic farming is based on principles similar to our traditional agriculture but without causing any harm to the environment, and to human life.

43.2 DEFINITION OF ORGANIC AGRICULTURE

The term "organic" was first used in 1940 by Northbourne in the book *Look to the Land*. "The farm itself must have a biological completeness; it must be a living entity, it must be a unit which has within itself a balanced organic life." Clearly, Northbourne was not simply referring to organic inputs such as compost, but rather to the concept of managing a farm as an integrated, whole system (Scofield,1986).

The international food standards, Codex Alimentarius, states: "Organic agriculture is a holistic production management system which promotes, and enhances agro-ecosystem health, including biodiversity, biological cycles, and soil biological activity. It emphasizes the use of management practices in preference to the use of off-farm inputs, taking into account that regional conditions require locally adapted systems. This is accomplished by using, where possible, agronomic, biological, and mechanical methods, as opposed to using synthetic materials, to fulfill any specific function within the system" (Ramesh et al., 2005).

43.3 ADVERSE EFFECTS OF MODERN AGRICULTURAL PRACTICES: INDIAN CONTEXT

The role of agriculture in economic development in an agrarian country like India is a pre-dominant one. Modernization of Indian agriculture began during the 1960s which resulted in the green revolution making the country a food-grain surplus nation from a food deficit nation. Modern agricultural practice was based on the use of high yielding varieties of seeds, use of chemical fertilizers and pesticides, multiple cropping systems. These methodologies put severe pressure on natural resources like, land and water. Table 43.1 describes the consumption of pesticides and fertilizers in India.

43.4 ORIGIN OF ORGANIC FARMING

Before the advent of pesticides and fertilizers agriculture industries, the organic agricultural practices were the only option for farmers. They worked within biological and ecological systems available. For example, the only source of fertilizer to replace nutrients from cropped fields was human and animal manure and leguminous plants. The research on organic agriculture started in the 19th century when it was discovered that it was the mineral salts contained in humus and manure that plants absorbed and not the organic matter. The key founders of this theory were Sir Humphrey Davy and Justus von Liebig, who had published their ideas in *Elements of Agricultural Chemistry* and *Organic Chemistry in its Application to Agriculture and Physiology*. According to the theory inorganic mineral fertilizers could increase the production and efficiency of crops and can replace manure. The agricultural revolution began in the 1840s and with it came the first commercial

Table 43.1 Pesticide and Fertilizer Consumption in India

Sr. No.	Year	Pesticide Consumption (M. Tons)	Fertilizer Consumption (M. Tons)
1	1970–71	24.32	2.18
2	1980–81	45.00	5.52
3	1990–91	75.00	12.54
4	1991–92	72.13	12.73
5	1992–93	70.79	12.15
6	1993–94	63.65	12.24
7	1994–95	61.36	13.56
8	1995–96	61.26	13.88
9	1996–97	61.26	14.31
10	1997–98	56.11	16.19
11	1998–99	52.44	16.80
12	1999–2000	49.16	18.07
13	2000–01	46.20	16.71
14	2001–02	44.58	17.54

Indian Agriculture in Brief.

production of inorganic fertilizers. In 1924, Rudolph Steiner, the founder of the philosophy of "Anthroposophy" gave his agricultural lectures in 1924 which were the foundation of biodynamic agriculture. The first organic certification and labeling system, "Demeter, was created in 1924 because of Steiner's actions (Rundgren et al., 2002).

43.5 THE CONCEPT OF FOOD QUALITY AND FOOD SAFETY

The quality of food can be reviewed in many different ways. For example, the quality of fresh produce is often judged by visual characteristics such as size, shape, color, and freedom from blemishes which, it could be argued, are enhanced by pesticide, and fungicide applications. There is a wider definition of food quality within the organic movement depending upon embracing functional, biological, nutritional, sensual, ethical, and "authentic" considerations. Under this definition, quality factors vary widely, ranging from how food tastes to the working conditions of those producing it. The main principles for organic farming and food processing include:

- production of food of high quality in sufficient quantities
- operation within natural cycles and closed systems as far as possible, drawing upon local resources
- maintenance and long term improvement of the fertility and sustainability of soils

- creation of a harmonious balance between crop production and animal husbandry
- securing of high levels of animal welfare
- fostering of local and regional production and supply chains
- provision of support for the establishment of an entire production, processing, and distribution chain that is both socially and ecologically justifiable

43.6 FOOD SAFETY IN ORGANIC AGRICULTURE: THE EVIDENCE

Food safety, in context of organic agriculture is determined by the extent to which organic, and nonorganic foods contain undesirable components such as potentially harmful chemicals, drug residues, and pathogens. While talking about organic farming and its relation to food safety, the following could be discussed.

43.6.1 USE OF PESTICIDES IN AGRICULTURE SYSTEM: IS IT REALLY A SAFE OPTION?

While many believe that pesticides are necessary to produce and protect crops, it is universally agreed that consumer exposure to these toxins should be minimized on safety grounds. Due to the persistent nature of many pesticides, air, water, and soil are inevitably contaminated by them. Contrary to public perception, while the use of pesticides is severely restricted in organic farming, a small number can be used. Soil association standards allow organic farmers restricted use of seven nonsynthetic pesticides that have been approved on the basis of their origin, environmental impact, and potential to persist as residues. They are copper ammonium carbonate; copper sulfate; copper oxychloride; sulfur; pyrethrum; soft soap and derris (rotenone). Some plant oils such as *neem* and microbial agents such as *Bacillus thuringiensis* (Bt) are also permitted. These pesticides are simpler substances than those used in nonorganic agriculture, and tend to degrade quicker.

There are concerns about the safety of these compounds permitted in organic agriculture though, as would be expected given the prohibition of routine pesticide applications, organically grown food is usually found to have no residues.

43.6.2 NUTRITIONAL QUALITY OF ORGANIC AND NON ORGANICALLY GROWN FOOD

Nutritional quality of organically and non organically grown food can be accessed through:

a. Primary essential nutrients such as water, fiber, proteins, fats, carbohydrates, vitamins, dry matter, and minerals

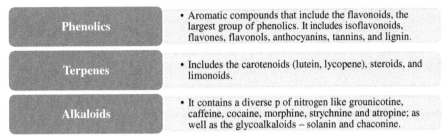

FIGURE 43.1 Secondary Metabolites or Phytonutrients

b. "Secondary metabolites" or "phytonutrients" in plants. There are some 5,000–10,000 secondary compounds in plants which are considered as health-promoting and protective and thus necessary for health. There are major four categories of phytonutrients phenolic, terpenes, alkaloids, and *sulfur containing compound* (Fig. 43.1).

Most studies comparing the primary nutritional components of organically and nonorganically grown crops, demonstrate significantly higher nutrient levels in organically grown crops than nonorganically grown crops (Brandt and Molgaard, 2001). However, some studies refute this finding. With the limited data, it is difficult to draw any final conclusion and more research is required to confirm this trend of higher nutrient levels in organically grown crops.

43.7 PUBLIC CONCERN ABOUT SAFE FOOD AND ITS QUALITY

Public concern about safe food and its quality has intensified in recent years. The controversy of genetically modified food, hazards of pesticides, and chemicals in food has pushed the demand for organically grown food. Consumers believe that organic food is qualitatively better: "it tastes better" and "it's better from the health point of view." But it is difficult to conclude if this perception is merely based on intuition or on conclusive evidence. People have favorable opinion about organic products for the following reasons as well, (from most to least important): good health (high nutritional value and minimal artificial chemical residues), safe environment, tastes better, animal welfare, minimal processing, and novelty. The consumption of organic products is much limited due to reasons like high price, limited availability in the market, scepticism about the credibility of product claims, poor appearance, nonawareness of organic foods, and contentment with existing products.

The potential **benefits** of the organic methods are immense.

43.7.1 **HEALTHY AND SAFE FOODS**

A study conducted in United States on the nutritional values of both organic and conventional foods found that the organic food, in general, had 20% less of the bad elements and about 100% more of the good elements.

43.7.2 **IMPROVEMENT IN SOIL QUALITY**

Soil quality is the foundation on which organic farming is based. Natural plant nutrients from green manures, farmyard manures, composts, and plant residues build organic content in the soil. Efforts are directed to build and maintain the soil fertility through the farming practices. For example, multicropping, crop rotations, organic manures, and minimum tillage are such methods. The soil under organic farming has higher water holding capacity, higher microbial biomass carbon and nitrogen, lower bulk density, and higher soil respiration activities compared to the conventional farms. This indicates that sufficiently higher amounts of nutrients are made available to the crops which further leads to higher nutritional level in organic produce.

43.7.3 **INCREASED CROP PRODUCTIVITY AND INCOME**

The results of field trials of organic cotton at Nagpur revealed that during the conversion period, in the first year the yield of cotton was low organically as compared to the conventional (using fertilizer and pesticides) and integrated crop management (using 50% each of organic and inorganic inputs) (Eyhorn et al., 2007. However, the yields of organic cotton started rising from third year. A study of 100 farmers in Himachal Pradesh during a period of 3 years found that the total cost of production of maize and wheat was lower under organic farming and the net income was 2 to 3 times higher. Another study of 100 farmers of organic and conventional methods in five districts of Karnataka indicated that the cost of organic farming was lower by 80% than that of the conventional.

43.7.4 **LOW INCIDENCE OF PESTS**

The study of the effectiveness of organic cotton cultivation on pests at the farm of Central Institute for Cotton Research, Nagpur revealed that the mean monthly counts of eggs, larva, and adults of American BoUworm were far lesser under organic farming than under the conventional method. Biocontrol methods like the *Neem* based pesticides to Ti-ichoderma are available in the country. Indigenous technological products such as *Panchagavya* (five products of cow origin) which was experimented at the University of Agricultural Sciences, Bangalore found to control effectively the wilt disease in tomato (Balasubramanian and Franco, 2009).

43.7.5 **INDIRECT BENEFITS**

Several indirect benefits from organic farming are available to both the farmers and consumers. While the consumers get healthy foods with better palatability and taste and nutritive values, the farmers are indirectly benefited from healthy soils and farm production environment.

43.8 **TECHNIQUE OF ORGANIC FARMING**

As a result of the restrictions on synthetic pesticides, organic farming systems have to manage the various weeds using a combination of indirect and direct methods of nonchemical weed regulation (Table 43.2). Indirect methods also called cultural methods, consist of all techniques that are aimed at improving crop performance, to diminish the distribution of weed seeds and to suppress the development of weeds in the standing crop. Direct methods also referred to as physical methods are designed to regulate weeds mechanically, manually, thermally or biologically. These techniques are used in conjunction with preventive and cultural methods for effective long-term weed management.

Table 43.2 Methods of Nonchemical Weed Management

Indirect Methods	Direct Methods
Crop rotation • competition • complementarity • allelopathy	Mechanical • hand weeding • various ploughs: chisel tines, discs, harrows • spring tines • rotary hoes • brush weeders • mulching
Farm hygiene • cleaning of seed supplies • cleaning machinery and tools	
Soil cultivation • tillage (turning/nonturning) • photobiology	Thermal • flame weeders • steam weeders • infrared weeders
Improvement of competitiveness • seed quality • morphology and vigor of cultivars • drilling design. density, row distance • sowing direction • strategic fertilization and irrigation	Biological • grazing with livestock • classical biocontrol • "bioherbicides," microorganisms as weed • pathogens

Table 43.3 IFOAM's Draft Revised Principles of Organic Agriculture

Principle of Health

Organic agriculture should sustain and enhance the health of soil, plant, animal, and human as one and indivisible.

Ecological Principle

Organic agriculture should be based on and work with living ecological systems and cycles, emulate them, and help sustain them.

Principle of Fairness

Organic agriculture should be built upon relationships that ensure fairness with regard to the common environment, and life opportunities.

Principle of Care

Organic agriculture should be managed in a precautionary and responsible manner to protect the health and well being of current and future generations and the environment.

43.9 BASIC IFOAM PRINCIPLES OF ORGANIC FARMING: COMPLEMENT TO FOOD SAFETY

The negative effects of modem chemical based farming system were first experienced by those countries, which introduced it initially. Organizations like International Federation of Organic Agriculture Movements (IFOAM) and Greenpeace have studied the problems of chemical farming methods and compared the benefits accruing from the organic farming with the former. Organic farming movements then adopted in European countries, America, and Australia have since spread to Asia and Africa too. The most widely recognized organic standards are those published by the IFOAM in 2002 (Table 43.3).

43.10 ORGANIC AGRICULTURE: INTERNATIONAL PERSPECTIVE

The organic food market in the world has grown rapidly in the past decade. International trade in organic foods showed an annual growth rate of about 20–22% during this period. Many retail chains and supermarkets are accorded with green status to sell organic foods. The important organic products traded in the international market are dried fruits and nuts, processed fruits and vegetables, cocoa, spices, herbs, oil crops and derived products, sweeteners, dried leguminous products, meat, dairy products, alcoholic beverages, processed foods and fruit preparations, cotton, cut flowers, etc. The FAO provides support to organic farming and attempts the harmonization of national organic standards, which is essential to increase international trade in organic products. FAO has, in association with the WHO, evolved the Codex Alimentarius for organic products.

43.11 ORGANIC FARMING AT NATIONAL LEVEL

Most of the countries lagging behind in the adoption of organic agriculture were from Asia and Africa. In terms of both in absolute and percentage area under organic farming, Tanzania, Zambia, Ghana, India, and South Africa are far behind many nations in organic agriculture. India also does not figure in the list of top 10 countries on the organic global map. Large sections of farmers in the developing countries are poor and their land holdings in many countries are small. Their access to external inputs is limited and their ability to improve production is low, given the economic and social disabilities to which they are exposed. Globalization and the opening up of the trade barriers have resulted in the decline of agricultural prices in the local markets to the detriment of the interests of small farmers. In India only nine states have clearly defined policies for organic farming, namely Karnataka, Kerala, Andhra Pradesh, Maharashtra, Madhya Pradesh, Himachal Pradesh, Uttarakhand, Sikkim, Nagaland, and Mizoram. Out of these Uttarakhand, Sikkim, Nagaland, and Mizoram have declared their intention to go 100% organic. Uttarakhand the third largest organic state with over 32,000 ha under organic cultivation has brought 47,000 farmers under this tag. With the help of Uttarakhand Organic Commodity Board, over 30 certified organic crop producer groups have come up here, producing wide range of organically safe food like amaranthus, basmati rice, maize, wheat, turmeric paddy, ginger, soyabean, rajma (kidney bean), finger millet, different type of pulses, medicines, and aromatic plants. Karnataka is the first state to announce an Organic Farming policy in 2004. State has set up the *Jaivik Krishik* Society that would facilitate farmers for certification and marketing. Here farmer can sell their produce directly to consumers. With the initiatives of Community Managed Sustainable Agriculture (CMSA), at Andhra Pradesh nonpesticide management movements, the state has freed 1.5 million ha and 1.5 million farmers from the tyranny of chemicals. By this movements 124 villages are declared pesticides-free and 26 villages deemed organic. The fundamental objective of CMSA is to provide healthy crops, healthy soil, healthy food, and healthy life to public by ensuring food security locally. In January 2016, Sikkim has become the first "fully organic" state. Use of chemical fertilizers and pesticides is banned and more than 75,000 ha of land is under organic cultivation.

43.12 PROMOTION OF ORGANIC FOODS: WHAT IS NEEDED?

In order to promote organic products, the organic industry needs to address various issues. Some of these are discussed in subsequent sections.

43.12.1 PRICING

Retail price premiums remain considerably higher than most consumers are willing to pay. Addressing this will remove a major barrier to organic food sales among less committed consumers.

43.12.2 **VISIBILITY**

Shopping habit is strongly influenced by the visibility. Thus, the appearance and visibility is the most important factor which needs to change if the habit of food purchasing is to change. Although organic foods certainly have increased their visibility, the appearance and layout of products needs more attention.

43.12.3 **LABELING**

Inconsistent and inadequate labeling reduces consumer confidence and trust in the integrity of organic claims. National and international harmonization or certified organic labels would be welcomed by most consumers if not by the certifying bodies who compete for farmers' business.

43.12.4 **AVAILABILITY**

The supply and quality of organic products must be consistent enough that buyers are not tempted to substitute them for conventional products. There may be circumstances when the organic industry is better served by a strategy of targeting a few key products rather than by attempting to provide a complete product range. Research has shown, for example, that few fresh fruit and vegetables account for most of the expenditure, suggesting that the organic industry could have the greatest impact on its overall sales by targeting the top-selling items.

43.13 **ORGANIC FOOD: CERTIFICATION**

It is widely recognized the world over that the certification of organic products should be based on the following principles:

- Organic production and processing standards should be clearly laid down.
- The conformation of production and processing to these stands must be verified.
- Organic labels should be permitted only to those produces, which are found conforming to the set standards.

India has more than 20 certifying agencies accredited by Agriculture and Processed Food Products Export Development Authority, also called third party certification. It is under the Ministry of Commerce, which set the norms for certification to help organic products find market abroad. These norms are based on the European Union Standards, known to be the toughest in the world. So, certificates in organic farming come at a huge cost and also require a elaborate documentation. In developing countries, it poses a serious challenge, most of whom are small landholder farmers and might be illiterate. There is a new alternative method developed to guarantee the organic integrity of products for small domestic producers, known as the participatory guarantee system. It is of low cost, involves minimal paper work, and makes farmers responsible for their success and integrity.

43.14 THE WAY FORWARD

There is a significant link between agricultural methods and food safety. The strategic investment in organic farming would have a major impact on public health, offering tangible benefits and an avoidance of the many potential and known risks posed by the continued use of artificial pesticides, fertilizers. There are valid and scientific evidence which indicates that organically grown foods are significantly different in terms of their safety, nutritional content, and nutritional value from those produced by nonorganic farming. Research into improving organic agriculture is also needed, as present day organic agricultural methods are far from fully developed and there is a lot of room for improvement, especially in the complex area of soil microbiological activity and promotion of symbiotic relationships between microorganisms and crops. In the biological and nutritional areas of food quality, the promotion of positive health is considered to be significantly dependant on the absence of residues of pesticides, toxins, and fertilizers in food. It should also be recognized by presence of primary and secondary nutrients in foods.

REFERENCES

Balasubramanian, A.V., Franco, F.M., 2009. Use of animal products in traditional agriculture. Centre for Indian Knowledge Systems, Chennai.

Brandt, K., Molgaard, J.P., 2001. Organic agriculture: does it enhance or reduce the nutritional value of plant foods? J. Sci. Food Agric. 81 (9), 924–931.

Eyhorn, F., Ramakrishnan, M., Mäder, P., 2007. The viability of cotton-based organic farming systems in India. Int. J. Agr. Sustain. 5 (1), 25–38.

Ramesh, P., Singh, M., Rao, A.S., 2005. Organic farming: its relevance to the Indian context. Curr. Sci. 88 (4), 561–568.

Rundgren, G., Vaupel, S., Crucefix, D., Blake, F., 2002. History of organic certification and regulation. In: Rundgren, G., Lockeretz, W. (Eds.), IFOAM Conference on Organic Guarantee Systems: International·Harmonisation and Equivalence in Organic Agriculture, 18-19 February 2002, Nuremburg. International Federation of Organic Agriculture Movements (I FOAM), Tholey,·Theley, pp. 5–7.

Scofield, A.M., 1986. Organic farming—the origin of the name. Biol. Agric. Hortic. 4 (1), 1–5.

Safety and quality of frozen foods

44

D.P. Attrey

Central Military Veterinary Laboratory, Meerut, Uttar Pradesh, India; High Altitude Research, Defence Research and Development Organisation, Leh, Jammu and Kashmir, India; Amity Institute of Pharmacy, Amity University, Noida, Uttar Pradesh, India; Innovation and Research Food Technology, Amity University, Noida, Uttar Pradesh, India; Amity Institute of Seabuckthorn Research, Amity University, Noida, Uttar Pradesh, India; Lala Lajpat Rai University of Veterinary and Animal Sciences, Hisar, Haryana, India

44.1 INTRODUCTION

A food that has been subjected to rapid freezing and is kept frozen until used is called "Frozen Food." Now, "TV dinner," a concept of complete meal right out of the freezer, has become popular as a convenience food in America and the frozen food market is likely to be worth $226.5 billion globally by 2015 (Foster, 2011).

As per FAO (2011), roughly one-third of food produced for human consumption (i.e., about 1.3 billion tons per year) is lost or wasted globally (mainly due to spoilage in under developed and developing countries). India is the largest producer and consumer of milk in the world. As per a news report (The Hindu, June 26, 2014), based on latest inputs from the National Dairy Development Board, milk production in India was 132.4 million tons in 2012–13, which rose to 140 million tons in 2013–14. As per APEDA, 2015, total meat production is estimated at 6.3 million tons in India, and India ranks 5th in the world in meat production, with about 3% of total world production of 220 million tons.

Total vegetable and fruit production in the world is 486 million and 392 million tons, respectively. India produces 32 and 71 million metric tons of fruits and vegetables, respectively. But the postharvest losses are almost 50% of the total fruits and vegetables production which badly affects their availability to the consumers. India loses almost Rs 13,300 crores due to spoilage of fruits and vegetables every year (Singh et al., 2014). According to the United States National Academy of Sciences about 10% of total production of cereals is also lost (generally due to spoilage) in developing countries (NAS, 1978).

Per capita food waste by consumers in Europe and North-America is 95–115 kg/year, while this figure in sub-Saharan Africa and Southeast Asia is only 6–11 kg/year (FAO, 2011). While the developed countries actually waste a huge quantity of consumable food, the causes of food losses and waste in low-income countries are mainly

Food Safety in the 21st Century. http://dx.doi.org/10.1016/B978-0-12-801773-9.00044-3

due to financial, managerial, and technical limitations in harvesting techniques, storage and cooling facilities in difficult climatic conditions, infrastructure, packaging and marketing systems. Reduction in food spoilage/losses, through good agricultural practices, adequate post harvest technology, processing and preservation, etc., could have an immediate and significant impact on their livelihoods (FAO, 2011).

44.2 FOOD SPOILAGE/DETERIORATION LEADING TO UNSAFE AND/OR POOR QUALITY FOOD

Primary fresh foods are highly perishable due to their high water content. Dried foods are generally considered as nonperishable due to low moisture content such as grains. Most preserved foods are semiperishable having medium moisture content. Fresh foods after freezing and storage at low temperatures fall in the category of semiperishable as long as the freezer temperatures are maintained properly. The stability and safety of the processed and stored food is directly influenced by its surrounding environment such as water activity, temperature, and/or the gaseous atmosphere. As per Dauthy (1995), food deterioration/spoilage usually occurs due to microbiological, biochemical (biological, chemical), physical, mechanical or external reasons. Microbiological activities in the food are considered as one of the most important reasons of food spoilage, which result in degradation of safety and quality of the food. Biochemical activities cause enzymic and nonenzymic deterioration of food. Physical processes responsible for food deterioration are physical changes such as dehydration and shrinkage/shriveling, crystallization, cold or freezing injury, exudation, etc. Mechanical reasons include mechanical injuries, physical pressure, bruising due to poor handling, etc. are also responsible for food deterioration. Many times, the external factors such as infestation/contamination by the insects, pests, vermins, rodents, etc. or their excreta, are also responsible for food spoilage and reduction in safety and quality of the food.

44.2.1 FOOD SPOILAGE DUE TO BIOCHEMICAL ACTIVITY

Prior to harvest, living animal or plant tissues generally have good defense mechanism against microbial attack. But after separation from living parent, the separated tissues become vulnerable to microbial attack. Microbiological changes in fresh, processed or stored foods are caused by microbes (bacteria and fungi). Bacteria can be classified according to their optimum temperature ranges viz mesophiles, thermophiles or cryophiles. Mesophiles grow best in moderate temperature, neither too hot nor too cold, typically between 10 and 40°C (Joanne et al., 2008). Most microorganisms fall in this range of temperatures. Organisms that prefer extreme environments are known as extremophiles viz thermophiles, between 41 and 122°C (Madigan and Martino, 2006) and psychrophiles or cryophiles, having optimum temperature from −20°C to +10°C (Feller and Gerday, 2003).

Biochemical activities such as enzymic and nonenzymic reactions continue in processed and stored foods even after harvesting. According to Ghaly et al. (2010), the cells start using their own stored energy to continue respiration and normal cell activity after harvesting, although limited respiration process and individual cell metabolic activities continue for sometime after harvesting the tissues.

Enzymic browning (as in apples) is caused due to oxidation of phenolic substances in plant tissues by endogenous enzymes such as phenolase. Enzymic lipid oxidation is also responsible for deterioration of stored foods since enzymatically generated compounds, derived from long-chain fatty acids in plant tissues, result in development of off flavors (Dauthy, 1995). Temperature, water activity, pH, and presence of added chemicals in the food, influence the enzymic activity.

Nonenzymic chemical changes in harvested plant tissues, which cause deterioration of sensory quality during processing and storage of foods, are "nonenzymic browning" (or maillard reaction) and "lipid oxidation," which may result in changes in color and flavor. Browning reaction occurs due to interaction between reducing sugars and amino acids, resulting in protein degradation, development of off flavors and darkening of products. Lipid oxidation is affected by light, local oxygen concentration, high temperature, the presence of catalysts (generally transition metals such as iron and copper), and water activity. Rancidity develops in the foods rich in unsaturated fatty acids in the presence of oxygen. Free radical development causes protein degradation, loss of vitamins and color changes in the food. Lipid oxidation occurs at higher rates at very low water activities (Dauthy, 1995; Attrey and Sharma, 1979).

Physical processes responsible for food deterioration are physical changes such as dehydration and shrinkage/shriveling, crystallization, cold or freezing injury, exudation, etc. Variation in temperature and moisture content may result in skin cracks in fruits and vegetables, or stress cracks in cereals.

Mechanical injuries such as bruising due to poor handling, careless manipulation, poor equipment design, excessive vibration or physical pressure from overlying material are also responsible for food deterioration. Automation in food handling must be carefully planned and designed to minimize mechanical injuries. External infestation/contamination by insects, pests, vermins, rodents, etc. are also the reasons for reduction in food safety and quality. Warm humid environments promote insect growth, although most insects will not breed if the temperature exceeds 35°C or falls below 10°C. Also many insects cannot reproduce satisfactorily unless the moisture content of their food is greater than about 11% (Dauthy, 1995).

44.3 METHODS OF FOOD PRESERVATION FOR ENHANCING SHELF LIFE OF FOOD

In general, foods are preserved by different methods as follows (Wagner, 2015):

1. Foods preserved by heat, that is, by retorted-pressure process and by acidification of foods.
2. Foods preserved by cold, that is, by refrigeration and freezing.
3. Foods preserved by reducing moisture, that is, by drying and by reducing water activity.
4. Other methods of preservations such as aseptic preservation and preservation through irradiation.

Physical treatments usually used in food preservation are curing, precooling, temperature treatments, cleaning, and waxing, whereas chemical treatments are disinfection, fumigation, and dipping.

Preservatives of various types have been used to preserve foods for many years. Smoking and salt were two of the earliest substances used, where smoking precedes salting. Preservatives can be divided into two categories according to function; (1) those which act to control or prevent growth of microbes, and (2) the ones used to control deteriorative chemical reactions such as rancidity (Wagner, 2015).

44.4 FREEZING AS A METHOD OF FOOD PRESERVATION

Freeze preservation maintains almost original quality of food products. The physical and chemical changes occurring during freezing and thawing affect the product quality and safety to a great extent. Water constitutes over 80–90% of the weight of most fruits and vegetables, which is held in rigid cell walls to provide a support structure and required texture to the tissues. Pure water freezes at 0°C. Soluble materials such as salts or sugars depress the freezing point by almost 2°C for every added mole (one molecular weight expressed in grams). Starch is one of the most commonly utilized ingredients for improving freeze/thaw stability. With products having starch are able to hold > 20 times their weight in water, starches provide cost-effective options for managing moisture while offering opportunities to enhance or modify texture (Foster, 2011).

Food products are complex, containing far more than a mole or two of solutes, such as salts, sugars, proteins, or fiber. These will affect the temperature at which a food system is devoid of free moisture—the eutectic point, which is not economically viable to achieve. But eutectic points of most foods are lower than those used commonly in commercial freezing processes, that is, about −50°C for meats, −55°C for ice cream, and −70°C for bread (Foster, 2011). Freezing of fresh foods basically involves freezing of the constituent water contained in its cells, except the monolayer water which cannot be frozen in normal circumstances (Attrey and Sharma, 1979); to maintain appropriate water activity for safe storage of the food.

When the water freezes, it expands and if the ice crystals are large they rupture the cell wall. Cold injury to cells starts appearing when ice crystals begin to form in the cells. Large ice crystals rupture the cell wall resulting in loss of turgor or firmness. The cellular fluids start escaping and release the degradative enzymes in the exudates during thawing, resulting in moisture loss, quality degradation, or flavor changes (Ghaly et al., 2010). This textural difference is especially noticeable in products which are usually consumed raw. Textural changes due to freezing are not as apparent in products which are cooked before eating because cooking also softens cell walls. The extent of cell wall rupture can be controlled by freezing the produce as quickly as possible. Rate of freezing is very important for controlling the formation of ice crystals. Rapid freezing results in the formation of smaller ice crystals with very less or no rupture in cell walls. That is why temperature of the freezer should be set at the coldest setting several hours before placing the foods in the freezer to keep the size of the ice crystals as small as possible.

Freezing and thawing processes involve heat transfer. From energy saving or improving quality and safety, new methods need to be developed. The novel freezing methods of high-pressure freezing and dehydrofreezing accelerate the freezing process, thus forming small and uniform ice crystals. Use of antifreeze and ice nucleation proteins improves freezing process directly by reducing the size of ice crystals. Thawing processes such as high-pressure thawing, microwave thawing, ohmic thawing and acoustic thawing can shorten thawing time, thus reducing drip loss and improving product quality and safety (Li and Sun, 2002).

Biochemical changes in tissues depend upon the temperature. The lower the temperature, the slower will be the rate of growth of spoilage microbes (bacteria and fungi) and slower would be the biochemical degradation. There may be limited bactericidal effect at very low temperatures. In fact spoilage rates would double for each 10°C rise, or shelf life would double with reduction of each 10°C in temperature of the product (Dauthy,1995).

As per Schafer (2014), the freezing process does not actually destroy the microorganisms which may be present on fruits and vegetables. While blanching destroys some surface microorganisms and there is actually a gradual decline in the number of these microorganisms during freezer storage. But sufficient populations of these microbes are still present to multiply in numbers and cause spoilage of the product when it thaws. It has been well established since long that microorganisms do not grow below −10°C and no preservative is required to be added for long-term frozen storage of foods, provided sufficiently low temperatures are maintained.

Fluctuating temperatures in the freezer cause migration of water vapor from the product to the surface of the container or build up of ice in the freezer compartment (may require frequent defrosting). Overloading the freezer with unfrozen products will result in a long, slow freeze and a poor quality product. Usually 2–3 pounds of vegetables per cubic foot of freezer space per 24 h is recommended. Moisture loss, or ice crystals evaporating from the surface area of a product, produces freezer burn-a grainy, brownish spot where the tissues become dry and tough. This surface freeze-dried area is very likely to develop off flavors. Packaging designed specifically for freezing foods will prevent freezer burn (Schafer, 2014).

Li and Sun (2002) have reported different methods of freezing as discussed in subsequent section.

44.4.1 **HIGH-PRESSURE FREEZING**

When water is frozen at atmospheric pressure, its volume increases. This increase in volume contributes to the volume of ice, which has a lower density than that of liquid water, resulting in a volume increase of about 9% on freezing at 0°C and about 13% at +20°C. This increase in volume causes tissue damage during freezing. However, with high pressure it is possible to produce ice having greater density than that of water, which will not expand in volume and cause tissue damage. In high pressure freezing water state would change to ice at 20°C, as pressure is reduced from 600 MPa to atmospheric pressure. With a good understanding of the water phase change, the use of high pressure can greatly help freezing process and improve product quality (Li and Sun, 2002).

The main advantage of high-pressure freezing is that the initial formation of ice is instantaneous and homogeneous throughout the product because of the high super-cooling reached on pressure release. Therefore, this technology can be highly useful to freeze foods with large dimensions in which a uniform ice crystal distribution is required and where thermal gradients are pronounced and damage of freeze-cracking would be possible when applying classical freezing methods, including cryogenic freezing. The use of high pressure facilitates supercooling, promotes uniform and rapid ice nucleation and growth, and produces smaller crystals (Li and Sun, 2002). One can achieve desired ice crystal formation by manipulating temperature or pressure as per the scientific principles. There is a water nonfreezing region (liquid state) below 20°C under high pressure. When pressure is released, a high super cooling is obtained and the ice-nucleation rate is greatly increased. Ice having a density of 1:31 may be formed at room temperature at a pressure of 900 MPa (Li and Sun, 2002).

44.4.2 CONCEPT OF MPA AND ATMOSPHERIC PRESSURE

1 Bar = 0.1 Megapascals; bar is the atmospheric pressure at the sea level. Megapascal is a metric pressure unit and equals to a force of 1,000,000 $N/m^2 r$. The abbreviation is MPa. Atmospheric air pressure is often given in millibars, where standard sea level pressure is 100 kPa or 1 bar. This should be distinguished from the atmosphere (atm), which is equal to 1.01325 bar. Common multiple units of the pascal are the hectopascal (1 hPa 100 Pa) which is equal to 1 mbar, the kilopascal (1 kPa 1000 Pa), the megapascal (1 MPa 1,000,000 Pa), and the gigapascal (1 GPa 1,000,000,000 Pa). On Earth, standard atmospheric pressure is defined as 101.325 kPa. Atmospheric pressure is the pressure in the surrounding air at or close to the surface of the earth. The atmospheric pressure varies with temperature and altitude above sea level. Standard atmospheric pressure (atm) is normally used as the reference when listing gas densities and volumes. The standard atmospheric pressure is defined at sea-level at 273°K (0°C) and is 1.01325 bar or 101325 Pa (absolute). The temperature of 293°K (20°C) is sometimes used. In imperial units the standard atmospheric pressure (atm) is 14.696 psi. 1 atm = 1.01325 bar = 101.3 kPa = 14.696 psi (lbf/in.^{2}) = 760 mm Hg = 10.33 m H_2O = 760 torr = 29.92 in.Hg = 1013 mbar = 1.0332 kgf/cm^2 = 33.90 ft. H_2O (Engineering Toolbox, 2015).

44.4.3 DEHYDROFREEZING

In this method of freezing, a food is first dehydrated partially to a desirable level of moisture and then frozen. Fresh fruits and vegetables contain more water than meat, and their cellular structure of cell wall is less elastic than cell membrane and is more susceptible to large ice crystal formation during freezing. Increasing freezing rate can reduce the formation of large ice crystals, but the tissue damage may still occur due to the presence of large amount of water. With dehydro-freezing one can reduce refrigeration load and costs of freezing. In addition, dehydro-frozen products could also lower the cost of packaging, distribution and storage, and maintain product quality comparable to conventional products. Some fruit samples have been reported to be successfully preserved with dehydrofreezing method by immersing in 68% (w/w) aqueous sucrose solution for 3 h (osmotic dehydration, better than air drying), then frozen in air-blast freezer with air velocity of 3 m/s at about −3°C. Intermediate moisture foods, however, can be stored without freezing (Dasgupta et al., 2011). It has also been reported that freezing starts at a lower temperature in the dehydrated product and the temperatures of dehydrated samples went down to −18°C in 19–20 min, 20–30% faster as compared with untreated samples which required the freezing time of 23–24 min. Brine solutions are used for osmotic dehydration of vegetables (Li and Sun, 2002).

44.4.4 ANTIFREEZE PROTEIN AND ICE NUCLEATION PROTEIN

Main worry of frozen food industry is how to control ice crystals in frozen foods. Antifreeze proteins can lower the freezing temperature and retard recrystallization on frozen storage, while ice-nucleation proteins raise the temperatures of ice nucleation and reduce the degree of supercooling. Antifreeze protein and ice-nucleation protein (INP), which are two functionally distinct and opposite classes of proteins, are directly added to the food to be frozen to control the size and structure of ice crystals within the food (Li and Sun, 2002).

44.4.5 ANTIFREEZE PROTEIN

Fish withstand the cold water in polar regions (freezing point of this water being close to −1.9°C). The plasma freezing point of fishes is about −1°C below the freezing point of surrounding water in Polar Regions. It has been reported that blood and tissues of the fish have AFPs, which prevent fish from freezing. Similarly many invertebrates, insects and certain higher plants, fungi, and bacteria also have antifreeze proteins. The function of antifreeze protein (AFPs) is to lower the freezing temperature and suppress the growth of ice nuclei, thus inhibiting ice formation. This property of antifreeze proteins is being utilized to improve or develop new food products. AFPs inhibit recrystallization of ice in dairy products such as ice cream and de-icing agents. The fine ice crystal in ice cream is very important to preserve its smooth and creamy texture. But recrystallization occurs inevitably when temperature fluctuates in storage or in transit, resulting in coarse texture of ice cream and damage in quality.

44.4.6 ICE NUCLEATION PROTEINS

Bacteria-induced ice nucleation has been recognized as a major contributing factor to frost injury in plants. Bacterial ice nucleation activators can reduce the degree of supercooling and catalyze ice formation in and on frost sensitive plants between −1 and −5°C, causing frost damage to many plants. Ice nucleation proteins (INPs) have great potential in the freezing of foods. They elevate the temperature of ice nucleation, shorten freezing time and change the texture of frozen foods, thus decreasing energy cost and improving the quality. Egg white samples without INPs froze below −15°C and formed small ice crystals. But the samples with the INPs froze at −3°C or slightly lower and formed large and long ice crystals.

44.5 METHODS OF THAWING

Thawing generally occurs more slowly than freezing. During thawing foods are subject to damage by chemical and physical changes, and by microorganism, making the food unsafe. The smaller the ice crystals during freezing, minimum shall be the rupture of the cell wall and the resultant drip loss during thawing, and minimum effect on quality and safety of frozen food for consumption. The quality and safety of the resultant thawed food shall be more acceptable. Now the emphasis is on development of newer chemical and physical methods/aids to freezing and thawing to save energy and/or to improve the quality and safety of frozen foods (Ghaly et al., 2010). For fruits and vegetables that are not seriously damaged by freezing, thawing at −1.1°C (30°F) seems to be most satisfactory. Fast thawing at higher temperatures results in more tissue damage. Since freezing renders products highly susceptible to bruising, they should be thawed with as little handling as possible. If the temperature drops several degrees below the freezing point of the food, it is said to have undergone "supercooling." In such situations, even slight vibrations/movements in the food matrix may cause immediate crystal formation, which may endanger textural integrity upon thawing. Tissues damaged by freezing generally lose rigidity, become mushy upon thawing, and appear water-soaked (Ghaly et al., 2010).

Consequently, the texture of the produce, when thawed, is much softer than when it was raw.

Quick thawing methods at low temperature are being developed as under to avoid rise in temperature and excessive dehydration of food to assure food quality.

44.5.1 HIGH-PRESSURE THAWING

High-pressure thawing of frozen meat requires only one-third of the time necessary at atmospheric pressure, maintaining all culinary qualities compared to conventionally thawed products. During high pressure thawing, the drip loss is minimal and no negative effects are observed on color or cooking loss. High cost, however, is the limiting factor for the high-pressure thawing.

44.5.2 MICROWAVE THAWING

The unique property of microwaves to penetrate and produce heat deep within food materials make them potential candidate for accelerating thawing process. Microwave thawing requires shorter thawing time and smaller space for processing, and reduces drip loss, microbial problems and chemical deterioration. However, localized overheating has limited the application of microwave thawing in food systems. The preferential absorption of microwaves by liquid water is also a major cause for run-away heating. In this case, food products take risk of excess water loss and thermally chemical deterioration. However, the improvement on the temperature uniformity during microwave thawing is necessary. The thawing rates of frozen samples in microwave thawing depend on material properties and dimensions and the magnitude and frequency of the electromagnetic radiation.

44.5.3 OHMIC THAWING

When electric current passes through conducting food with high electrical resistance, heat is generated instantly inside the food, thus increasing the temperature of the food item. This heating technology is termed as ohmic heating or electro heating. In the food industry, more attention has been paid to the application of ohmic heating on aseptic processing and pasteurization of particulate foods. In comparison with microwave heating, ohmic heating is more efficient because nearly all of the energy enters the food as heat and ohmic heating has no limitation of penetration depth. Ohmic heating also has advantages over conventional heating such as high heating rate, high energy conversion efficiency, volumetric heating, etc. Using ohmic heating to thaw frozen foods is an innovative method. In one method, negative electrons were introduced into a high voltage electrostatic field to thaw frozen foods. Using this method, frozen foodstuffs can be thawed rapidly in the temperature range $-3°C$ to $+3°C$. Frequency changes did not significantly affect thawing time and ohmically thawed samples showed reduced drip loss and improved water holding capacity when lower voltages were applied. Ohmic heating technology shows potential in supplying thawed foodstuffs of high quality.

44.5.4 ACOUSTIC THAWING

High-power ultrasound has also been reported to help in thawing the meat and fish. Acceptable ultrasonic thawing was achieved at frequencies around 500 kHz. Therefore, acoustic thawing requires investigation as another promising technology for thawing frozen foods and proper frequencies and acoustic power need to be established.

44.6 SAFETY AND QUALITY OF FROZEN FOODS

Quality should be optimal at the time of harvesting and aim should be to preserve the same quality and to keep the food safe during processing and storage. Rate of biochemical changes depends upon the temperature. Lower the temperature, slower would be the biochemical degradation and lower shall be the rate of growth of spoilage microbes (bacteria and fungi).

To maintain top quality, frozen fruits and vegetables should be stored at $-18°C$ (0°F) or lower. Major storage problem in meat is growth of spoilage microbes when one tries to avoid development of rancidity. Hence storing frozen foods at temperatures higher than $-18°C$ (0°F) increases the rate at which deteriorative reactions can take place and can shorten their shelf life.

44.7 MOISTURE LOSS/FREEZING INJURY

Moisture loss, or ice crystals evaporating from the surface area of a product, produces freezing injury or freezer burn, a grainy brownish spot where the tissues become dry and tough. This surface freeze-dried area is very likely to develop off flavors. Poor or inappropriate packaging can allow moisture to evaporate from the surface into the environment causing dehydration (Foster, 2011).

44.8 MICROBIAL GROWTH IN THE FREEZER

The freezing process does not actually destroy the microorganisms which may be present on fruits and vegetables. While blanching destroys some microorganisms and there is a gradual decline in the number of these microorganisms during freezer storage, sufficient populations still remain dormant, which multiply on thawing and cause spoilage of the product. For this reason it is necessary to carefully inspect any frozen products which have accidentally thawed by the fluctuating temperatures in the freezer.

44.9 EFFECT ON NUTRIENT VALUE OF FROZEN FOODS

44.9.1 VITAMIN C

Vitamin C is usually lost in a higher concentration than any other vitamin. Vitamin C may be lost (about 10%) during blanching, and some more during cooling and washing stages. The vitamin loss is not likely to occur due to freezing.

Vitamin B1 (thiamin): About 25% of vitamin B1 may be lost, since thiamin is easily soluble in water and is destroyed by heat during blanching and during moisture loss.

44.9.2 VITAMIN B2 (RIBOFLAVIN)

About 4–18% of vitamin B2 may be lost in green vegetables, during the preparation for freezing rather than the actual freezing process itself.

44.9.3 VITAMIN A (CAROTENE)

There is little loss of carotene during preparation for freezing and freezing of most vegetables. Most of the vitamin loss is incurred during the extended storage period.

44.9.4 **QUALITY**

One of the problems with freezing is the risk that deactivated pathogens may once again become active when the frozen food thaws. Water within flexible cellular structures throughout a food matrix expands as it freezes, and filling all the available spaces with ice. Upon thawing, the water is released and seen as syneresis (exudation). Rigid structures such as those of plant or vegetable tissues are more likely to simply break under the forces of the expanding ice crystals. In this case, visual and textural effects are observed. Incorporating stability and texture in frozen foods is the best way for enhancing frozen/convenience foods business (Foster, 2011).

Products can continue to undergo changes during frozen storage and transportation because they are below their eutectic points, that is, they are not completely frozen. Despite being frozen moisture is still moving. Temperature gradients that arise from fluctuations in temperature can drive water toward cooler temperatures typically found at the surface of a product. While temperature gradients may reverse, water that has found space to collect in product voids and packaging gaps will not return (Foster, 2011).

The major problem associated with enzymes in fruits is the development of brown colors and loss of vitamin C. Since fruits are usually served raw, they are not blanched such as vegetables. Instead, enzymes in frozen fruits are controlled by using chemical compounds which interfere with deteriorative chemical reactions. Ascorbic acid is used in its pure form or in commercial mixtures with sugars. Soaking the fruits in dilute vinegar solution or coating the fruits with sugar and lemon juice may also inactivate enzymes. But these methods do not prevent browning as effectively as the ascorbic acid. Landi et al. (2013) reported that polyphenol oxidase (PPO) and to a minor extent peroxidase are the key enzymes involved in enzymatic browning. Although ascorbic acid is frequently utilized as an antibrowning agent, its mechanism in the prevention of the browning phenomenon is not clearly understood. The concentration of endogenous ascorbic acid helps in reducing browning sensitivity. Addition of exogenous ascorbic acid (5 mmol/L) has been reported to have reduced PPO activity by about 90% by directly inhibiting the PPO activity.

Oxidative rancidity occurs in fatty foods through contact of frozen product with air. It can be controlled by using a wrapping material which does not permit air to pass into the product. Maximum possible air should be removed from the freezer bag or container to reduce the amount of air in contact with the product.

44.10 **EFFECT OF THAWING ON SAFETY AND QUALITY OF FOOD**

Thawing results in exudation, which is good medium for the growth of microbes, including pathogens. Growth of microbes on foods results in undesirable appearance and off-flavors of food. Safety of meat and fish may also be compromised if proper procedures are not followed during thawing. Although freezing puts the pathogens in suspended animation, it does not kill them. As soon as the food starts to become warmer at temperatures above 4°C (40°F), bacteria start multiplying. So, when thawing frozen food, it's important to keep it out of the "danger zone" of temperatures where bacteria thrive.

The following methods are recommended for thawing.

44.10.1 IN THE REFRIGERATOR

This is the easiest method but it takes a long time, so one must plan ahead. (A pound of meat can take an entire day to thaw.) Advantages to this method are that it's hands-off and the refrigerator keeps food at a safe temperature.

In cold water: Sealed packages of food may be thawed in cold water. Place the package under water in a bowl, pot, or sink and change the water every 30 min until the food is defrosted.

44.10.2 IN THE MICROWAVE

Microwave thawing may be uneven, leading to poor quality or even bacterial growth. It's best to use this method if food will be cooked immediately after thawing, or for frozen fruits and berries that shall be served immediately after thawing.

44.10.3 AT ROOM TEMPERATURE

Pastries, breads, and fruits may be thawed for 2–4 h at room temperature. Do not use this method for meat, fish, and vegetables.

Under running water: Meat may be thawed under cool running water.

44.10.4 COOKING WITHOUT THAWING

When there is not enough time to thaw frozen foods, or you're simply in a hurry, just remember: it is safe to cook foods from the frozen state. The cooking will take approximately 50% longer than the recommended time for fully thawed or fresh meat and poultry.

Never thaw frozen foods in a garage, basement, car, dishwasher or plastic garbage bag; on the kitchen counter, outdoors or on the porch due to hygiene considerations.

44.11 CONCLUSIONS

Huge quantity of fresh foods is spoiled every year across the globe which must be prevented and utilized to remove hunger and malnutrition. While a lot of consumable food is actually wasted in developed countries, the developing/underdeveloped countries do not have adequate resources and infrastructure to prevent the post harvest losses of animal and horticulture foods. Frozen foods are the most popular choice in developed world and these need to be propagated by developing adequate cold chain in poorer countries also by helping them with resources and infrastructure. During freezing and frozen storage, some physical, chemical, and nutritional changes may occur. As such, quality and safety of frozen foods must be kept in mind at all stages of freezing, frozen storage, and thawing. Freezing usually retains initial quality of products. To avoid loss of quality and to keep the frozen foods safe, basics of food preservation must be understood and applied during freezing and storage.

REFERENCES

Attrey, D.P., Sharma, T.R., 1979. Sorption isotherms and monolayer moisture content of raw freeze dried mutton. J. Food Sci. Technol. 16 (4), 155–158.

Dasgupta D.K., Attrey D.P., Vibhakara H.S., Bawa A.S. A Process for the Production of Shelf Stable Intermediate Moisture Apricots" (Indian Patent No. 245539 dated 25 January 2011).

Dauthy M.E.; 1995. Fruit and vegetable processing; FAO Agricultural Services Bulletin No.119; M-17; ISBN 92-5-103657-8 Food and Agriculture Organization of the United Nations; Rome, 1995; http://www.fao.org/docrep/V5030E/V5030E00.htm; Available from: http://www.vouranis.com/uploads/6/2/8/5/6285823/fao_fruit__veg_processing.pdf

Engineering Toolbox, 2015. Introduction to pressure—psi and Pa—online pressure units converter. Tools and Basic Information for Engineering and Design of Technical Applications. Available from: http://www.engineeringtoolbox.com/pressure-d_587.html

FAO, 2011. Global food losses and food waste—extent, causes and prevention. Rome. Available from: http://www.fao.org/docrep/014/mb060e/mb060e.pdf; DL on 18-12-15

Feller, G., Gerday, C., 2003. Psychrophilic enzymes: hot topics in cold adaptation. Nat. Rev. Microbol. 13, 200–208.

Foster, R.J. (Contributing Editor), 2011. Stabilizing frozen products. Natural Product Insider, October 17, 2011. Available from: http://www.naturalproductsinsider.com/articles/2011/10/stabilizing-frozen-products.aspx; DL 15-12-15

Ghaly, A.W., Dave, D., Budge, S., Brooks, M.S., 2010. Fish spoilage mechanisms and preservation techniques: review. Am. J. App. Sci. 7 (7), 859–877.

Joanne, W. M., Sherwood, L., Woolverton, C. J., Prescott, L.M., 2008. In: Prescott, Harley, and Klein's Microbiology. New York: McGraw-Hill Higher Education.

Landi, M., Degl'innocenti, E., Guglielminetti, L., Guidi, L., 2013. Role of ascorbic acid in the inhibition of polyphenol oxidase and the prevention of browning in different browning-sensitive Lactuca sativa var. capitata (L.) and Eruca sativa (Mill.) stored as fresh-cut produce. J. Sci. Food Agric. 93 (8), 1814–1819.

Li, B., Sun, D.W., 2002. Novel methods for rapid freezing and thawing of foods – a review. J. Food Eng. 54, 175–182.

Madigan, M.T.Martino, J.M., 2006. In Brock Biology of Microorganisms (eleventh ed.). Upper Saddle River, NJ : Pearson Prentice Hall. p. 136.

Schafer, W., 2014. The Science of Freezing Foods. Department of Food Science and Nutrition, University of Minnesota Extension, USA. Available from: http://www.extension.umn.edu/food/food-safety/preserving/freezing/the-science-of-freezing-foods/ DL on 21-12-15

Singh, V., Hedayetullah, M., Zaman, P., Meher, J., 2014. Postharvest technology of fruits and vegetables: an overview. J. Postharvest Technol. 02 (02), 124–135.

Wagner, A., 2015. Aggie Horticulture. TAMU 2134, Room 225, Horticulture Department, College Station, TX 77843-2134. Available from: http://aggie-horticulture.tamu.edu/food-technology/food-processing-entrepreneurs/getting-started/processing/

Ready to eat meals

45

R.K. Gupta*, P. Dudeja**

*Department of Community Medicine, Army College of Medical Sciences, New Delhi, India;
**Department of Community Medicine, Armed Forces Medical College, Pune,
Maharashtra, India

45.1 INTRODUCTION

There are times when one needs to have food quickly! It could be the whole meal, snacks, sweets, and savories or simply a cup of instant tea or coffee. In the present day's fast paced world such a situation arises more commonly, than what it used to be a decade or two earlier. In the typical Indian setting now, with the women-folk working outdoors, couples living separately, students/young professionals cooking on their own, the traditional elaborate kitchen is shrinking in favor of fast-foods, frozen-foods, tinned and canned foods, eating-out, pack and dine options, etc. The Ready to Eat Meals (REM) perfectly complement this changed lifestyle.

The ready to eat meals are prepared or cooked in advance needing no/minimal further cooking or processing before being eaten. So, these could be precooked, partially cooked, nonpreheated/frozen or preserved with a view to be eaten instantly.

45.2 HISTORY

The first modern day REM were manufactured for the Army. During the American Civil War, the military moved toward canned goods such as canned meat, pork, bread, coffee, sugar, and salt. During the First World War lightweight preserved meats were introduced. In 1963, the actual "meal, ready to eat," were developed by the United States Department of Defence (Mason, 1982). It was a ration that would rely on modern food preparation and packaging technology to create a lighter replacement for the canned food. A dehydrated meal stored in a waterproof canvas pouch was prepared, that could be quickly made "ready to eat." The REM have since been in the process of development, with dehydrated, semicooked, precooked, cooked, in cans, paper pouches, to plastic packaging. Some were supplied with additional heating equipment, some with chemicals producing enough heat to cook, on addition of water though exothermic reactions!

45.3 THE INDIAN SETTING

The Indian Defence Research Development Organization (DRDO) too worked on the Indian version of REM for soldiers. Technology of some of these developed products was transferred to private companies for commercial use. Today there are many food companies marketing REM such as Gits, MTR, Tasty Bite, Mother's recipes, Satnam Overseas, Godrej, Al-Kabeer, etc.

To suite the Indian cultural food habits, now REMs are available in prepared mixes of Indian foods such as *dosa, upma, and chutney mixes*. These are widely available in supermarkets. These are processed foods offering the convenience of eating off the shelf, eliminating the kitchen drudgery associated with making a meal at home. Some REM are considered better over other food products as they may not contain chemical preservatives and remain shelf-stable without refrigeration for one year or so. These changes are bringing a new revolution in processed food industry.

Families prefer these as it saves times, labor and tediousness of cooking. Another section of society which is dependent on these REM products are college students staying away from home or persons staying alone as the idea of cooking from scratch is impractical. REM could be useful for travel, tours, and excursion.

45.4 LIMITATIONS

Consumers have traditionally viewed the ready meals as less healthy than fresh foods. There have been concerns about the nutritional contents claimed and actually available in them. They may be high in saturated fats and salt content. With advances in food technology a wide variety of foods besides bread, jam, cheese, salted foods have been developed (e.g., candy, beverages, soft drinks, juices, processed meat and cheese, soups, pasta, potato chips, chapatis, sweets, vegetables, *pulao,* etc.), yet there are limitations. While these may be acceptable for one-off use, they may not be very desirable if they have to be consumed repeatedly, over time, as monotony might set in. The concept of a balanced diet may not really be a priority with such meals.

45.5 CLASSIFICATION

REM foods can be classified as given in Fig. 45.1.

45.6 SAFETY ASPECTS OF REM FOODS

REM products are available in packaged form. It is of utmost importance to be cautious of their safety as they are minimally processed thereafter. If these foods are contaminated at any stage from farm to fork they can be a cause of food borne illness. Following precautions must be taken with respect to food safety.

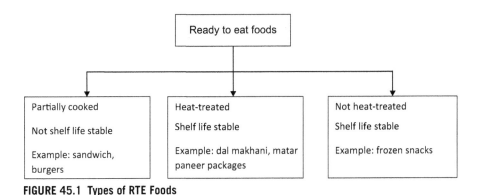

FIGURE 45.1 Types of RTE Foods

45.6.1 **RAW MATERIAL**

The microbial load of all the raw items to be used in preparation of RTE such as raw fruits and vegetables, liquid milk, meat, eggs, flour, cereal grains, etc. should be within acceptable limits. In case perishable items are to be used such as milk, meat then appropriate and adequate storage facilities (as outlined in Chapters 15 and 17 of this book) must be ensured.

45.6.2 **MANUFACTURING**

REM food-manufacturing units may be located in remote areas, from where food items are transported to various places. It is vital to follow safe transportation practices (refer Chapter 14 on safe transportation for details).

45.6.3 **EQUIPMENT**

For the safety of REM foods it is imperative that all food processing equipments viz. dough mixers, conveyors, rounders, dough dividers, racks, proofing equipments, oven, rollers, slicers, sifters, pasteurizer, homogenizer, retort, bottling unit, pulper, filtering screens, mixing vats, etc. should be clean, in good repair and free from evidence of rodent or insect activity. Time and temperature control of ovens, retort, heat exchangers and cooling area should be strictly adhered to, to ensure safety of food. Before using any equipment it should be cleaned. All vapor producing cooking equipment such as retort, ovens, grills, and fryers should be equipped with ventilation and an approved automatic extinguishing system to prevent unnecessary condensation in the working area. If this is not done microorganisms may continue to harbor within this equipment. Utensils such as spoons, beaters, pans, bowls, trays, spatulas, etc. should be sanitized after it is manufactured. For monitoring of food safety swabs of machine, working tables, utensils, food contact surfaces should be taken at regular intervals to ensure their microbial safety for food use. Antiseptic/disinfectant foot bath should be provided at the entrance of plant.

45.6.4 PACKAGING

Packaging material (pouches, films, laminates, cans, glass/PET bottles, closures, jars, cardboard boxes) should be kept and stored under hygienic conditions in a room intended for that purpose. This aspect is generally ignored and packaging material is kept on the floor. It is required that all packaging materials as bottles/closures should be sanitized before use. These closures should be labeled for the product inside. For the dispatch of all products first in first out system should be applied. For details refer to chapter on good storage practices.

45.6.5 DISTRIBUTION

Small EEs, tea shops, street vendors sell REM foods in open markets. It is imperative that these vendors are educated about the health hazard associated with flies, birds, rodents, and other vermin. The FBOs who sell such foods should observe basic hygienic measures to protect the consumer from environmental contamination and infections likely to be introduced during hawking. There should be a source of approved water available with them.

45.6.6 SMALL UNITS

Small bakeries making REM food such as sandwiches, burgers and hot dogs at homes/slums/streets may have poor hygiene and sanitation. These are generally unlabelled with no date of manufacture/expiry date. These could only be acceptable if their manufacturing process is hygienic and they are used fresh.

45.7 DETECTION OF THE FOODBORNE PATHOGENIC BACTERIA

The pathogens as follow should not be found in ready-to-eat food as they represent an unacceptable risk to health.

Campylobacter spp., *Escherichia coli* O157, *Salmonella* spp., *Shigella* spp.
Vibrio cholera, Bacillus cereus, Bacillus spp., Clostridium *perfringens*
Listeria monocytogenes, Staphylococcus, Vibrio parahaemolyticus

45.8 FOOD SAFETY CONCERN FOR SOME SPECIFIC REM
45.8.1 CHOCOLATES

Storage of cocoa beans, nuts, and coconuts should be checked for insects, rodents, and mycotoxins. Critical raw materials such as skim milk powder, milk, eggs, cocoa, etc. should be adequately heat-treated, pasteurized, or handled in such a way that bacterial contamination is eliminated or minimized. Many chocolate products are finished by hand-dipping so sanitation is of major concern.

45.8.2 **CUSTARD AND CREAM-FILLED FOODS**

Bacteria-sensitive materials, such as skim milk powder, milk and eggs, must have minimum bacteria levels, and must be stored, defrosted and handled well. Sanitation and good-quality raw materials are critical factors. The products are not subjected to a heat treatment after filling; the filling operations must, therefore, be conducted in the most sanitary manner possible. Equipment sanitation, clean-up procedures, and employee practices should be strictly as per the guidelines.

45.8.3 **FROZEN FOODS**

Raw materials should be subjected to a field examination. All those pieces (of the particular fruit, vegetable or other food), which are unsatisfactory because of mould, decomposition, insect and rodent, filth or foreign material should be sorted out. Report of unsatisfactory pieces as a percentage of the sample taken is made and the action is taken. Frozen conditions must be adequately maintained during transportation of the frozen foods.

45.8.4 **FRESH-CUT FRUITS, SHAKES, AND JUICES**

These are liable to contamination by fingers, flies, dust, water, vermin, and prompt decomposition. If they are on display, their immediate container, should not be in contact with the ground. Unsheltered displays should be high enough above the ground surface to prevent contamination. Dust and dirt on premises should be controlled. Bulk should be stored in a cold-store or room, or in an insulated container. Indications of spoilage as bad smell, unusual color and changed consistency should be used to discard the spoiled items.

45.8.5 **STATUTORY REGULATIONS**

The regulations under the FSS Law are as applicable to REM foods, as they are applicable to any other food product. All procedures of inspections, food adulteration regulations, sample taking, nutritional labeling, etc. must be utilized to make REM safe.

45.9 **CONCLUSIONS**

While ready to eat meals might be a boon to the busy people of today, in case enough attention is not paid to their hygiene, food borne infections will be inevitable. It is therefore imperative to provide better food manufacturing units and marketing infrastructure to serve safe processed and REM foods. RTE market in India, is worth Rs. 2900 crore, according to an analysis done by Tata Strategic Management Group. But the challenge of making it safe still remains to be met.

REFERENCE

Mason, V.C., Meyer, A.V., Klicka, M.V., 1982. Summary of Operational Rations. US Army Natick Research and Development Laboratory, Natick, MA. Technical Report TR-82/013.

Food packaging

46

R.K. Gupta*, P. Dudeja**

*Department of Community Medicine, Army College of Medical Sciences, New Delhi, India; **Department of Community Medicine, Armed Forces Medical College, Pune, Maharashtra, India

46.1 INTRODUCTION

Packaging which used to be a mere necessity few decades back has now turned into an art and a science. There are packaging courses and institutions, teaching packaging technology. While it primarily involves enclosing or protecting products for storage, distribution, sale, and use, today attractive packaging helps to draw customers towards the product. It also refers to the process of design, evaluation, and production of packages. While appropriate packaging is important to maintain the basic attributes of food (temperature, color, taste, texture, etc.), maintaining food safety is an important function of packaging. Here, it becomes equally important that the packing material itself is of food-grade and not a hazard to food safety.

46.2 HISTORY

Egypt seems to have pioneered food packaging. Mummies were packed with all articles of daily use including food. Egypt was one of the first countries to have used paper (from papyrus plant). There are references how paper was in use for packaging food as well (that included vegetables and spices). In the prehistoric times, however, only natural materials were available and used for packaging which included leaves, animal skin, bark, coconut shells, and dried vegetable skins. Subsequently, baskets of reeds, wooden boxes, wooden barrels, woven bags, etc. came into use. Pottery vases, water storage containers came later.

The use of tinplate for packaging dates back to the 18th century. The first corrugated box was produced commercially in 1817 in England. Gair discovered that by cutting and creasing he could make prefabricated paperboard boxes. Packaging advancements in the early 20th century included bakelite closures on bottles, transparent cellophane over wraps, that increased processing efficiency and improved food safety. Aluminum and plastics were also incorporated later.

Traditional packaging couldn't stand the rigors of World War II which led to Military Standard or tough military specifications in the 1940s. Polyvinyl chloride revolutionized food packaging. It formed a seal without clinging to itself, food or to the container. It has low permeability to oxygen, water vapors, and flavors, however there have been concerns about its toxicity. In 21st century food packaging has evolved as a specialized industry (Foodservice Packaging Institute, 2006).

46.3 WHY FOOD PACKAGING?

There are as many answers to this question as there are number of types of food packets. While a particular "mil-grade" type of tough food packaging developed during the World War II, when food got destroyed due to poor packaging; it was in the 1930s that single use cone cups, plates, etc. were widely used to feed workers on the remote dams, bridges, and roads of the works progress administration in United States, to cut cost and for ease of transport. It was Dr. Samuel J. Crumbine, a public health officer in Kansas, in 1908, who on a train witnessed one of his TB patients taking a drink of water from a common dipper and water bucket (a publicly shared way of drinking water) in the car. Right behind his patient was a young girl who drank from the same dipper and bucket. This inspired him to launch a crusade to ban publicly shared or common utensils in public places. Taking note of the trend Lawrence Luellen invented a disposable paper cup.

There are multiple needs, for the purpose of storing, carrying, transporting, preserving, maintaining temperature, economizing, glamorizing food, and so are diverse ways of packaging and carrying eatables.

46.3.1 ADVANTAGES OF FOOD PACKAGING

There are various advantages of food packaging. Good packaging protects against breakages, vibrations, temperature, heat, and humidity. Packaging acts as a barrier against water, dust, contaminants, direct touch, microorganisms, etc. All these attributes enhance the shelf life of the food product. Packaging gives a good look and glamorizes the packets for marketing, besides making it convenient for the user. The packets contain labels, which give specific information about the contents, dates of manufacture and expiry, nutrient values, details of manufacturer, etc. In addition there may be specific applications on some packets like the antitheft devices. Suitable packaging also helps in categorizing, grouping, and appropriate storage of articles.

46.4 PRINCIPLES OF PACKAGING

Packaging of food articles should be done in a way that chances of contamination, reaction with packed material, decomposition, etc. should be avoided. Packaging must be done in appropriate way or as per the norms so recommended by Food and Drug Authority India. It can be done in multiple layers as enumerated here.

Primary packaging: It envelops and holds the food product.

Secondary packaging: It is exterior to the primary packaging.

Tertiary packaging: It is the tough outermost covering that is used for bulk handling, warehouse storage and transport/shipping.

Advantages of food packaging: The major advantages of food packaging are enumerated in the following sections.

46.5 PACKAGING AND FOOD SAFETY

While food packaging is an integral component of food industry and helps to store food and beverages in hygienic manner, it can at times be a cause of concern for food safety. Some packaging materials such as certain types of plastic, polythenes, and styrofoam can release toxins when they are heated and can be dangerous to consumers. Packaging materials which are irradiated (along with food) can transfer unsafe nonfood substances into the food. Food packaging makes use of a variety of substances, including dyes for printing colorful labels, and glues and adhesives for keeping packaging closed. In order to protect consumers effectively, the relevant authority individually certifies each of these food packaging materials subjecting them to rigorous testing protocols.

46.6 PACKAGING MATERIAL

As mentioned above packaging has to be sturdy, attractive, economical, and yet non-toxic. It must act as a physical barrier to protect food from contamination and must also preserve the nutrients through avoiding interaction of food with oxygen, carbon dioxide, and humidity. Besides these the important properties of packaging material are their physical, chemical, biological and thermal stability, impermeability to liquids and special properties like X-ray resilience.

46.7 TYPES OF PACKAGING

The type of packing depends on various factors for example food item, the process of production, and quality of food, shelf life desired, transport considerations, etc. so it is important to consider the shape, size, color, stacking options, printing of labels, cost, environmental attributes (e.g., recyclability, carbon imprint), handling properties, etc.

46.8 PACKAGING MATERIALS

Various packaging materials are in use. Some of the important ones are discussed here.

46.8.1 GLASS

Glass is a popular packaging material, as it is nontoxic, nonleaching, easy to clean, nonreactive to food/chemicals, nonporous, and relatively cheap. It is environment friendly as it can be reused and also recycled. These attributes make it an ideal material, but for the fact that it is breakable, heavy, and brittle. It is used for production of bottles and jars and is widely used for liquids and sauces. Cold drinks, alcohol, pickles, jams, ketchups, and squashes are commonly stored in glass bottles.

46.8.2 ALUMINUM

Aluminum is the most abundant metal on earth. It is light weight, lustrous, reasonably strong, long lasting, and recyclable. It has good barrier properties. It is used for making cans, metallic trays, and forms for ready-to-cook food that are resistant for high and low temperatures, thus can be used for frozen and heated meals. Aluminum can be used as foils for direct food packaging and also for lamination of paper or plastic for better strength, heat stability and barrier against moisture, oils, air, and odors. These are commonly used for packaging soups, herbs, and spices.

46.8.3 PLASTIC

Plastic is a generic term for many related synthetic materials that are commonly used for food packaging. It is strong, long lasting, light weight, air tight, and recyclable. Plastic bags increase the shelf life and maintain the freshness of the product. Items that are extremely moisture free can be stored in plastic bags for long. It may be transparent, so we don't have to open it to find out what's inside. It can be used to make many types of packaging materials like bags, wraps, bottles, tubs, buckets, containers, resealable pouches, etc. Being airtight, bags can help to prevent food from getting soggy in humid areas. It has disadvantage of being nonbiodegradable. There are also concerns about leaching and diffusion of substances like bisphenol A and diethylhexyl adipate from plastics into food that may be carcinogenic.

46.8.4 PAPER

Paper is an age-old packaging item, prepared from cellulose based materials (e.g., wood). It is permeable to air, water vapor and gases (oxygen). It has low tear strength. A wide range of bags and boxes for different applications are prepared from paper. These are used for carrying dry food stuff such as sugar, salt, flour, bread, etc. Paper can also be used to make lightweight cartons that are used as a colorful outer cover for products packed in plastic or metal containers. Various types of cans may also be made out of cardboard to store snacks, spices, nuts, or even cups to drink liquids. Paper waste can be burned (with energy recovery), recycled, or biodegraded for composting.

46.8.5 **TETRA PACKS**

Classically, these were a tetrahedron shaped plastic-coated paper carton, with aseptic packaging technology made possible even a cold chain supply, substantially facilitating distribution and storage of dairy, beverages, cheese, ice creams, and prepared foods.

46.8.6 **POUCHES**

Various types of pouches prepared from high quality material that are durable and environmental friendly are available. These are spout pouches, zipper pouches and printed stand up pouches, reusable pouches, etc. Their attractive design enhances the appeal. In addition they also have the food labeling: manufacturing date, expiry date, nutrient content, logos, and messages.

46.8.7 **RETORT PACKAGING**

The retort pouch was invented by the United States Army Natick R&D Command, Reynolds Metals Company, and Continental Flexible Packaging. A retort pouch is made up of plastic and metal foil laminate pouch, with 3 or 4 wide seals usually created by aseptic processing, allowing for the sterile packaging of a wide variety of drinks, that can range from water to fully cooked, thermo-stabilized meals such as ready-to-eat meal that can be eaten cold, warmed by submersing in hot water, or through the use of a heater, lighter in weight and less expensive to ship. In this technique food which is first prepared (raw or cooked) is sealed into the retort pouch. The pouch is then heated to 240–250°F (116–121°C) for several minutes under high pressure, inside retort or autoclave machines. This process reliably kills all commonly occurring microorganisms (particularly *Clostridium botulinum*), preventing it from spoiling.

46.8.8 **ASEPTIC PACKAGING**

This is a technique in which the contents of a package and the packaging itself are sterilized separately.

46.9 **LABELING**

Chapter 39 is dedicated to product labeling in this book.

46.10 **TOXIC EFFECTS**

46.10.1 **LEACHING**

Some nonfood grade packaging materials exert certain toxic effects on food. A type of plastic, polyethylene tetraphathalate (PET), is commonly used as bottles for soft drinks, water, and juices. Chemical such as antimony trioxide may leach into drinks from PET bottles. Other toxins like di(2-ethylhexyl) adipate released may cause toxicity to liver and is suspected to cause cancer in humans. Polyvinyl chloride used in clear food packaging, has been described as a hazardous product, being an endocrine disruptor. Polystyrene used in egg cartons, disposable coffee containers, and packaging cheese and meat in supermarkets, leaches chemical styrene which can cause developmental and reproductive problems. Polycarbonates used in baby bottles, various food, and drink containers contains many chemicals including bisphenol-A that leach into food causing breast and prostate cancer, insulin resistance, and chromosomal damage (US Environmental Protection Agency, 2015).

46.10.2 METALLIC CONTACT

Certain metals (like aluminum, lead in olden days) may come in contact of food and cause toxicity, like metallic taste.

46.10.3 PHYSICAL AGENTS

Physical agents like packaging pins, metal pieces, wooden shrapnel, etc. can prove hazardous, if fall in food and eaten.

46.11 EMERGING TRENDS

There are many high technology driven innovative packaging trends that are emerging in today's industry. Oxygen scavenging agents (e.g., ferrous oxide, ascorbic acid, sulfites, catechol, etc.) that react with oxygen and reduce its concentration are being used. These would delay the oxidative degradation of food (e.g., fruits), thus increasing their shelf life. Carbon dioxide absorbers and emitters may be added to suppress microbial growth in fresh meat, poultry, cheese, and baked goods. Hygroscopic agents can help to control moisture and water activity, thus reducing microbial growth in products such as sweets and candy. Antimicrobials are also used to enhance quality and safety by reducing surface contamination of processed food. Self-heating packaging (calcium or magnesium oxide and water) generates an exothermic reaction. It has been used for plastic coffee cans, military rations, and on-the-go meal platters.

46.11.1 ACTIVE AND INTELLIGENT PACKAGING

Time temperature indicators, ripeness indicators, biosensors, and radio frequency identification are the examples of intelligent packaging components, that have a profound future.

Time temperature indicators integrate the time and temperature experienced by the indicator and adjacent foods. Some use chemical reactions that result in a color change while others use the migration of a dye through a filter media. To the degree that these physical changes in the indicator match the degradation rate of the food, the indicator can help indicate probable food degradation.

Radio frequency identification is applied to food packages for supply chain control and has shown a significant benefit in allowing food producers and retailers create full real time visibility of their supply chain, indicating for example the date of manufacture, time of loading, temperature fluctuations during transit, time of unloading, etc., through a bar-code like radio frequency sensor.

46.11.2 NANOTECHNOLOGY

Nanotechnology involves characterization, fabrication, and/or manipulation of structures, devices, or materials that are in 1–100 nm length range. This enhances polymer barrier properties, making the material stronger, more flame resistant, with better thermal properties and having favorable surface wettability and hydrophobicity. Nanotechnology innovation could produce remarkably new packaging concepts for barrier and mechanical properties, pathogen detection, and active and intelligent packaging.

46.12 CONCLUSIONS

Packaging is an integral part of food products, as it gives food the commercial shape, color, texture, transit opportunity and also shelf life. It helps to maintain the benefits of food processing after the process is complete, enabling food to travel to long distances. Labeling is integral to packaging rendering all the relevant information about the product. Concerns of cost and environmental degradation have to be always considered while selecting a particular food-package. Enormous technological advances are on the anvil that will make the food packaging almost as intelligent as the consumer.

REFERENCES

Foodservice Packaging Institute, 2006. A Brief History of Foodservice Packaging.
US Environmental Protection Agency, 2015. Reducing wasted food and packaging: a guide for food services and restaurants.

Information technology (IT) in food safety

47

R.K. Gupta

Department of Community Medicine, Army College of Medical Sciences, New Delhi, India

Imagine you point your smart phone at frozen chicken or a cauliflower in the food store and the phone tells you what farm it came from and the date it arrived in the store. It also tells you what temperatures were these exposed to during transit, and if it was safe to consume it!

Or suppose a consignment of dough from a famous pizza chain is contaminated with *Salmonella* and presume the dough was sent to 6 towns that further dispatched it to 20 pizza outlets which in turn would serve 4000 customers the next day. Any of these retail-outlets would be able to test the contamination independently by the click of a button. Further, on finding a contamination, it would be possible to alert other retail outlets ... and the surveillance chain cranks into action (Hardgrave, 2012).

Well these and many other similar, seemingly astonishing and unbelievable feats are close to reality with the advances in Information Technology (IT) involving various related technologies such as the Radio Frequency Identification (RFID), cloud computing and countless softwares which are innovations of IT.

47.1 WHERE ARE WE?

To put things in perspective, let us see where we are at the moment. Presently, a food safety plan is a manual management process that collects, collates, analyzes, and records data for better decision making. Like many other management systems (say, financial accounting, human resources, or salary systems), undoubtedly there is utility of IT in food safety as well. But so far IT has taken a back seat in food safety probably because the need of IT hasn't been fully realized or traditional software solutions have not addressed the real needs of the local food safety practitioner; and enterprise software solutions have proven too costly for most food businesses.

Food Safety in the 21st Century. http://dx.doi.org/10.1016/B978-0-12-801773-9.00047-9

47.2 **WHERE IT COMES IN?**

Many of the requirements of food safety management are repetitive in nature, involve work flows that are well established, and require routine actions. On the other hand, there are live systems that need to be revised on an ongoing basis as new information/ events come to light. Here we see information and data merging. All this lends itself well to the applications of IT (Howlett, 2012).

If we have standard/legal (global/national/regional) frameworks under which food safety standards can align; standard IT solutions could emerge. These solutions can be delivered on web or cloud based technologies. Web technologies make software available at cheaper rates and allow their wider use. Software can be quickly updated, and users can supervise and monitor work from anywhere in the world using a web-enabled device.

In this fast moving world there is an increasing need of doing the same job faster, with lesser resources, with no errors and more efficiently. IT seems to be the one step solution to these tall demands, more so for food safety scenarios, since it involves quick decision making as human health cannot be compromised.

47.3 **PREREQUSITES OF IT SYSTEM**

A well designed IT based food safety system should have these features. (Howlett, 2012):

- Should meet needs of the user and be easy to use.
- Fully integrated with requirement of retailers, consumers, inspectors and auditors.
- Capable of being quickly updated in line with changes in food standards/ regulations.
- Provide real-time data on lab results, inventory, outbreaks and changes in legislation.
- Allow users to remotely access and work on their food safety systems.
- Able to generate reports.

47.3.1 **PRINCIPLES OF IT UTILIZATION IN KEY APPLICATIONS (MOORE, 2013)**

The usefulness of an IT system, is based on certain principles. These would enable execute a task that would best help the stake-holders take timely decision. These are summarized below:

1. *Capability to operate from anywhere*: In today's mobile environment, it's imperative to deliver relevant information and notifications to users from anywhere, and current IT solutions should deliver precisely that. Smart phones, laptops and tablets enable this. This helps; being in the midst of work

environment, respond quickly to critical events, take quick decisions and minimizes risk while increasing efficiency. Access to real-time information empowers decision makers, for example, if a poor raw material is noticed or a failed lab sample is come across, immediate action can be taken thus preventing the issue from escalating.

2. *Trending*: The key to IT solutions is the ability to proactively recognize trends as they happen and take immediate corrective action as needed. Software can identify trends and indicate root-cause relationships, allowing quality improvements that will mitigate risks as they arise. If oven temperatures were not consistently being met for a product (increasing safety risk), operators can take immediate corrective action and adjust/compensate for the temperature drifts, mitigating risk, and ensuring safety.

3. *Problem predicting softwares*: Advanced software with built in integrative capacities of higher level analysis based on historical models, should be able to predict system failures and raise alarms. This protects quality and food safety. For instance, altered pH during a food process can compromise food quality. If the pH starts deviating the software detects it, alerts the operator, to adjust the process to prevent the critical damage.

4. *Use of standardized operating procedures*: It helps consistently adhere to recipes and comply with Hazard Analysis and Critical Control Points (HACCP). For example, workflow can help manage a HACCP plan by automatically triggering a software which can track data in real-time and automatically adjust work processes to meet specification thus maintaining food safety.

5. *Traceability*: Software should have capabilities that could trace a product through every step of a process. This would enable better control throughout the supply chain/process.

47.4 PRESENT USAGE OF IT IN FOOD SAFETY: DEVELOPING COUNTRIES

At least in the developing counties, at present the usage of IT in food safety is limited. It is essentially restricted to the larger management systems, of which food-safety might be an incidental beneficiary, rather than the primary stakeholder. For example, the use of IT in inventory management of food stores, IT equipment used for laboratory analysis of food samples and IT based systems being used for communicable disease surveillance, are all incidental to food safety and not specific to it.

Another area where IT use is expanding, is through the authorities exploiting the fetish of youth for apps and internet links to draw them towards food safety. A case in point is the *food safety helpline* that is introduced to have instant access to Food Safety and Standards Act (FSSA) updates in India. Food safety helpline has been established to help understand and implement the requirements of FSSAI. Over 4000 people read the newsletter every week. There are regular updates on the new and evolving laws. Free online training courses, webinars, and coaching sessions are also shared. It is a forum to interact with colleagues and expert to help implement food law (Food Safety Helpline, 2015).

There is a need to develop specific systems for food safety as is the case in developed countries. Let's have a look at how IT is being utilized in United States and Europe.

47.5 IT IN FOOD SAFETY: DEVELOPED COUNTRIES

In the developed countries specific IT based managerial solutions are in vogue particularly keeping food safety in mind. These could also be emulated in developing countries. Here are few examples discussed in subsequent sections.

47.5.1 SUPPLY CHAIN MANAGEMENT

IT based supply chain management systems provide stake-holders (manufacturers, suppliers, and consumers) with farm-to-fork visibility throughout the supply chain. It could be a cloud-based tool built upon a social network platform. It enables communicate quickly in real-time and verify data to assess risks. Any break in chain or a compromised food safety situation can be easily detected and corrective measures taken.

47.5.2 MEAT PACKAGING SOFTWARES

Specific softwares are available for animal products (fish, meat, and poultry). A record of temperature can be maintained throughout its transit over long distances. Products can be traced real time, spanning multiple locations, besides any specific requirement of the user. For example, product label can be changed or new label printed through a computer using a barcode identifier. This information enables to trace the carton over time, deciphering delays and temperature aberrations (if any) and the reasons thereof can be analyzed and rectified. This process helps maintain food safety.

47.5.3 FOOD ADULTERANT DETECTION SOFTWARE

A food adulterant detection software could be a boon for quick, easy and early detection of common food adulterants, thus immensely contributing to food safety even at a household level. Such a software rapidly screens for several adulterants using "near infrared" spectroscopy technology, for detection of adulterants. It exhibits simple color indicators (e.g., red and green) indicating if the food sample is safe to consume or not (adulterated or non adulterated). Such softwares are user-friendly and quick.

47.5.4 DETECTION OF CONTAMINATION: SMART-PHONE TECHNOLOGY

Now tests have been developed through IT that could identify multiple pathogens (e.g., *Listeria, E. coli* and *Salmonella*) in a single step which dramatically improves efficiency. The principle of the test is that the food sample is mixed with liquid crystals and antibodies. If the target pathogens are present, the antibodies will clump and distort liquid crystal matrix, and light shines through them and an optical reader detects it. Unlike the traditional methods, where it takes 24 to 48 h, in this technology results are displayed on a handheld device (e.g., a smart-phone) within 30 min (Giordone, 2012).

47.5.5 GENOME-SEQUENCING OF NEW PATHOGENS

Genome sequencing is a new technique which literally identifies a pathogen to its core. Softwares have now been developed to detect *E. coli* hybrid strains, *Listeria*, etc. This helps in epidemiological investigations and tracing the outbreak to its origin and thus ensuring effective and efficient control measures—a vital step in food safety of a region.

47.5.6 FLUORESCENT SENSOR FOR PATHOGEN DETECTION

There are softwares integrated for certain advanced fluorescent test systems. They utilize DNA sensors, for identifying pathogens such as *Salmonella* and *E. coli*, based on the principle of detecting metabolic DNA byproducts left behind by these bacteria. The probe for *E. coli*, for example, binds to a specific metabolic product from that bacteria and forces the DNAzyme to change its shape. Thus altered, the DNAzyme begins to fluoresce. The test is simple and time-efficient. It detects the traces of organism within hours to a day, as against traditional test that takes several days.

47.5.7 IT IN HACCP

HACCP plan is the foundation for most food safety programs. It provides for the food safety requirements that accomplish scientific risk assessment, and risk control management, for all activities related to food safety in an industry/plant/catering institution or even during transit.

For a HACCP plan, the food safety manager must identify and collect data on hazards. This data must then be used to assess risk. Data is validated for critical controls, and records maintained to demonstrate compliance with the plan. A plan to control potential hazards associated with sourcing, suppliers and raw materials is needed. Immense data might be needed such as HACCP plans, specifications, allergen data, questionnaires, inspection reports, etc. for the system to take off.

Now we have secure, cloud-based automation technologies available. Cloud solutions require no capital hardware/software investments and can be accessed anywhere, at any time, from mobile devices. This makes them both affordable and effective. The system is able to automate food safety plans for every product group and facility. It can access data from a single, central repository for use (Levin and Bernkopf, 2014).

Great details are available in the literature as to how HACCP automation works, what are its key benefits, etc., which is beyond the scope of this book.

47.5.8 **USE OF CLOUD COMPUTING TECHNOLOGY**

Cloud computing, or "on-demand computing," is an Internet-based computing, where shared resources, data, and information is provided to computers on-demand. It is a model for enabling ubiquitous, on-demand access to a shared pool of configurable computing resources (Hassan, 2011). Cloud computing and storage solutions provide users with various capabilities to store and process their data in third-party data centers.

Cloud-based applications are secure, user controllable, just like an online net-banking account, one can see own activity and balances, not those of everyone else who uses that same application, yet it is secure.

A comprehensive cloud-based system typically includes upstream, internal, and downstream functionality. Customized modules for supply chain management, food safety or quality management could be introduced. Cloud-based solutions may operate on mobile devices. These can be accessed from anywhere, at any time. They collect and analyze data at the point of origin: in the field, in the plant, or even on the road. Information can come in from suppliers, equipment and laboratories and can be processed to stakeholders.

Typically, a manual system is used for quality control of crop/plant/factory/laboratory. In such a systems, the manager collects, collates, analyzes and interprets data himself or through experts/assistants, investing considerable time, paper, and money. On the other hand, in the cloud based systems, observations are reordered on mobile device, if an aberration is detected during the real-time analysis, an auto-alert is sent to the food safety manager; data is analyzed and action insured immediately.

47.6 **CYBER SECURITY IN FOOD SAFETY (STRAKA, 2014)**

The food and beverage industry is as susceptible to cyber-security threats and attacks as any other industry. It is reported that, in the USA as many as 24% of all reported data breaches occur in food and beverage industry, second only to retail. There have been instances of loss of millions of credit/debit card numbers and stolen card data of thousands of customers from food retail outlets. This compromises confidence, credibility and of course causes huge financial loss.

Besides these, there is a risk of agro-terrorism with an intent of contamination of the food supply for terrorizing population and causing harm. If hackers gain access to a food supply network, they could introduce dangerous amounts of chemicals/radiation (normally used for disinfecting food produce) or poisons to food. Refrigeration systems can also be remotely shut down, compromising food safety.

Cyber-security is beyond the mere use of firewalls and antivirus softwares. It is a comprehensive plan that complies with safety objectives, requirements, and/or government regulations. A good cyber-security plan begins with a risk analysis to determine the current state of security and ends with securing, monitoring and constantly updating security systems.

47.7 LIMITATIONS OF IT

Having discussed the applications of IT in food sector especially. food safety, certain limitations will restrict IT's full potential. First, the developing countries will need some time and more importantly the correct mind-set to accept technology as a replacement to the existing manual systems. Secondly, availability of adequate infrastructure also poses a challenge; namely, internet connectivity, adequate band-width, un-interrupted power supply and IT-training. Thirdly, IT use may not be amenable for immediate use to small eating establishments, street vendors, hawkers, etc., till such time that IT's importance is appreciated by all concerned and the rules and legislations also support its unstinted and mandatory use. Therefore, particularly in the developing countries, its use may be restricted in rural areas, unorganized sector, small eating establishments and for the un-initiated, at least for the time being.

So far as developed countries are concerned, they are already in the process of utilizing basic IT systems. However, to used advanced IT systems, a limitation for their use is the cost-effectiveness. A company will adapt/subscribe to advanced (and costly) IT systems, only if it proves to be financially beneficial to it. Till then, they might languish in a "research" state only.

Cyber-security, is a limitation that will always pose a challenge for use of IT, to both developing and developed countries.

47.8 FUTURE OF IT IN FOOD SAFETY

Notwithstanding the limitations cited above, the future of IT in food safety is unfathomable. Given the advances in technology, occurring by leaps and bounds, utilization and implementation of IT will make the food more safe to eat through easing various hitherto complicated procedures.

For example take the future utility of radio frequency identification (RFID). RFID is a form of auto-identification, such as a bar code or a quick response code. An RFID tag contains a unique serial number, such as a license tag. It's presently used in automatic garage door openers, key fobs, and toll passes. It is also used by some retailers for inventory control. A handheld device can read the RFID tags on thousands of products and update the inventory in few minutes. These applications are just the tip of the iceberg. It is a matter of time when food will be tracked through the supply chain from producer to consumer, for its safety and quality, using RFID technology.

Such technology already exists. What is required is to marry it up with the need. And to install the required infrastructure in terms of electronic data recorders throughout the supply chain, its transmission to central data base, synchronization with relevant nodes, and interpretation as per requirement.

Lets return back to the example cited in the introduction. Suppose one had got sick from the pizza dough; using RIFD, it would be easy to trace back the source, through the supply chain. It would even be possible to locate the distribution center and the particular batch of the origin of pizza dough. It is even possible to determine where the flour used in that batch was processed and where the wheat was grown!

IT is a strong tool that integrates a safe, farm to fork chain. It functions independently; or as inter-sectoral link, vital to maintain food safety through the entire chain.

REFERENCES

Food Safety Helpline. http://foodsafetyhelpline.com

Giordone, G., 2012. Microbiological tests and kits: the latest advances. J. Food Qual. Saf.

Hardgrave, B., 2012. Radio frequency identification shows promise for food safety. J. Food Qual. Saf.

Hassan, Q., 2011. Demystifying cloud computing. J. Defense Software Eng., 16–21.

Howlett, G., 2012. IT solutions for food safety management. J. Food Qual. Saf.

Levin, B., Bernkopf, D. Automation Manages Robust HACCP (and HARPC) Programs, Food Qual. Saf. 2014.

Moore, K., 2013. Top five software capabilities in safety. J. Food Qual. Saf.

Straka, C., 2014. Cybersecurity in food. J. Food Qual. Saf.

Index

Printed in the United States
By Bookmasters